ERGEBNISSE DER MIKROBIOLOGIE IMMUNITÄTSFORSCHUNG UND EXPERIMENTELLEN THERAPIE

FORTSETZUNG DER
ERGEBNISSE DER HYGIENE, BAKTERIOLOGIE, IMMUNITÄTSFORSCHUNG
UND EXPERIMENTELLEN THERAPIE

HERAUSGEGEBEN VON

W. KIKUTH
DÜSSELDORF

K. F. MEYER
SAN FRANCISCO

E. G. NAUCK
HAMBURG

A. M. PAPPENHEIMER JR.
NEW YORK

J. TOMCSIK
BASEL

DREISSIGSTER BAND

MIT 63 ABBILDUNGEN

SPRINGER-VERLAG BERLIN HEIDELBERG GMBH 1957

ISBN 978-3-662-23733-5 ISBN 978-3-662-25832-3 (eBook)
DOI 10.1007/978-3-662-25832-3

Ursprünglich erschienen bei Springer-Verlag OHG. Berlin · Göttingen · Heidelberg 1957

Inhaltsverzeichnis

I. Multicellularity in Bacteria

By

K. A. Bisset

With 11 figures and 2 plates

Contents

1. Introduction

Advance in the study and understanding of the morphology and structure of bacteria has been remarkably rapid in recent years. At the same time, it has been, in a number of respects, exceedingly irregular. This irregularity has manifested itself in diverse ways, although it certainly cannot be held to betoken any lack of general and widespread interest. For example, the degree to which the more fully-accepted facts in the contemporary canon of knowledge on the subject are understood is very different indeed in different cultural spheres. Whereas, as might be anticipated, information of applied value, and especially that of medical interest, is much more equably distributed.

A different type of irregularity in the distribution of information upon the various aspects of bacterial cytology has an important bearing upon the subject of this review. This is the rather remarkable discrepancy between the speed at which information has spread and been assimilated, as between the different aspects of investigation upon the problems of bacterial cytology, and between the manner in which the information available upon different cellular components has been accepted. The enormous preponderance of work has been done upon the bacterial nucleus, and many, if by no means all the important advances which have been made in this field are reasonably well known. And although considerably less time and energy have been expended upon the flagella, much of what has been discovered has been fully recognised by the majority of those interested. The subject of what, for lack of a better term, may be called 'granular inclusions' in the bacterial cell, has also received a great deal of attention, but, as will appear in the later part of this introductory section, much of this work can only, howsoever reluctantly, be regarded as wasted; for a reason, not uncommon in science, that the investigators in question have sometimes failed to examine the premises of their arguments.

It may be that the attention which the foregoing problems have received is due, at least in part, to the fact that all of them have been the subject of controversies of a more or less violent nature. On the other hand, the one major field in bacteriology which has not proved at all controversial in itself, at least so far, has singularly failed to attract a similar degree of attention. However, if controversy is indeed the best form of advertisement in science, the problem of cellular structure in bacteria is not without hope of attaining notoriety. Not because any real disagreement seems likely to arise among those who have studied the matter, but because its proper understanding has a most important bearing upon most, if not all, of those controversies concerning the bacterial nucleus, flagella and granular inclusions, to which reference has already been made. This will appear later, from reference to Text-fig. 1.

It is a remarkable fact that although the existence of multicellular organisation in bacteria has been known for the greater part of that period of a hundred years which comprises the epoch of scientific bacteriology, it has never been understood, nor indeed so much as recognised by the majority of bacteriologists. It is now ten years since Robinow (*81*) published his remarkable testament of bacterial cytology, and by his action advanced general knowledge both of the nucleus and cellular structure of bacteria to a greater degree than has been achieved by any single publication before or since. But despite the very large number of papers dealing with problems in connection with the morphology, complexity, growth and composition of the cell envelopes of bacteria which has appeared since that date, many bacteriologists, even including a surprisingly high proportion of those actually engaged in the study of cytology, persist in treating bacteria as single cells, and in referring to them as such.

This does not apply only to those bacteria that, from lack of positive information to the contrary, may be regarded as single cells, either on the principle of choosing the simpler of two alternatives, or by analogy with known, unicellular genera, to which the bacteria under discussion may bear some real or fancied resemblance. It is also true of those genera, including most, if not all Gram-positives, the evidence for whose multicellular nature is beyond all dispute.

No better example than this could be found of the well-known principle that most observers, scientific or otherwise, see only what they expect to see. Or more explicitly, what they have been taught to expect to see. For if once familiarity with the concept has been gained, the division of, for example, a staphylococcus into two or four cells can easily be discerned in a Gram-stained preparation with a scholastic oil-immersion lens. Despite which, generations of bacteriologists have trained their successors to regard it as an undifferentiated sphere.

The amount of work which has been done upon the basis of this fallacious premise has automatically given rise to what can only be termed a vested interest in the traditional, unicellular concept. This is the sort of situation which occurs in all branches of science, and will doubtless continue to do so. In the present case, it is obvious that where the supporters of a particular school of thought have staked their professional reputations upon some interpretation founded upon the dogma that a bacterium is a single rod-shaped cell, they will tend to regard without enthusiasm the suggestion that many bacteria are, in fact, filaments of smaller cells, separated by cross-walls.

This is peculiarly true of those workers who rely too exclusively upon the electron microscope. This apparatus has provided some valuable information concerning the surface structures of bacteria, and especially concerning the flagella, but it is liable to provide a misleadingly convincing picture of unicellularity in bacteria which can clearly be seen, by the use of suitable cytological techniques with the classical light microscope, to be markedly multicellular. It must be admitted that, upon occasion, internal cell boundaries can be discerned in disrupted cells by this means [ANGELICO *et al.* (*1*), DAWSON and STERN (*38*), YOSHIDA, FUKUYA *et al.* (*106*)] but even in such cases it is a matter for question whether more than a small proportion of these structures can actually be rendered visible. Comparative studies by light and electron microscopy suggest that they cannot [BISSET (*23*)].

Too exclusive a reliance upon any single technique, not only upon the electron microscope, can lead to errors and misconceptions, especially if this narrow approach is to some degree guided by a preconceived view of the objective. It is axiomatic that a scientific hypothesis should be designed to explain observed facts, and that subsequent research should be designed to test the hypothesis. But it would be idle to pretend that these rules are invariably followed. It has happened repeatedly in bacterial cytology that workers employing a single technique, or a group of techniques directed to a single purpose, have promulgated some rash interpretation, usually by analogy with other types of cell, and have then expended much subsequent time and effort in the design of experiments intended, not so much to test their hypotheses as to protect them against inconvenient facts adduced from other sources by other workers. An excellent example of the spirited defence of an elegant theory against intrusive and incompatible evidence is given by HUNTER (*56*). This example is of peculiar interest in the present context, since, as in so many parallel cases, the original flaw in the theory of nuclear structure proposed by this worker and her collaborators was that the Gram-positive coccus under examination proved not to be a single cell, as was essential to the validity of the theory, but a complex of several cells, divided by septa [BISSET (*19*)]. Although this multicellularity in cocci has since been verified by a variety of methods [WEBB and CLARK (*101*), TOMCSIK and GRAC (*92*)], HUNTER has attempted to obviate this criticism of her work by demonstrating that it is possible so to stain the coccus in question that its cross-walls are not revealed. It is fair to comment that by exactly comparable means it has been proved repeatedly that bacteria are devoid of an organised nucleus! Failing to demonstrate the cytological structures of the bacterial cell presents no difficulties; the problem is to avoid such failure.

Even in the limited field of the cocci a number of further examples can be found of theories of nuclear behaviour which are destined to make shipwreck on the selfsame rock. One of these is that of KRIEG (*61*), who manifestly mistakes entire cells, stained with fluorescent dyes, for nuclei in a septate coccus which he assumes, in the absence of any evidence whatsoever, to be unicellular. Once more, too exclusive a reliance upon a single technique has proved dangerous. The parallel application of a cell wall stain, a relatively simple matter in the case of most cocci, would have supported or disproved KRIEG's ideas (Plate I, Fig. 8).

The examples just quoted illustrate how, in the absence of an understanding of the multicellular character which is typical of the majority of Gram-positive and some Gram-negative bacteria, errors may arise from failure to realise that all the components visible in a single coccus or bacillus may not be, indeed almost certainly are not, parts of a single cell, but usually belong to several different cells, and may comprise the major portions of small cells. This misconception has led especially to errors in interpretation of the nuclear behaviour of bacteria, but turns up also in the 'diagnostic' staining of bacteria [cf. Bisset (*11*)].

A second, parallel error arising from the same source, consists in the mistaking of elements which are actually derived from the cell envelopes for internal structures lying within the cell. This occurs in studies both of cocci and bacilli, but especially the latter. The septa and cross-walls which subdivide the multicellular bacteria may possess, or be associated with elements containing considerable components of nucleic acids. These are capable of combining with a variety of different types of basic dye, and frequently the most obvious structures demonstrable in this fashion are the granules at the junction of the cell wall or membrane and the developing cross-walls. These are clearly shown in the illustrations of many reliable cytologists, as for example Knaysi (*60*), Järvi and Levanto (*58*), and Robinow and Murray (*82*). The simple and straightforward interpretations of these authors, with which the present writer entirely concurs, are shown in Text-fig. 1 (*a*, *b*, *c*).

Also in Text-fig. 1 are shown some of the numerous alternative interpretations which have been made, by different authorities, to explain what are, beyond any reasonable doubt, the same structures. And it is quite obvious, not only that many of these are incompatible with the interpretation which has already been mentioned, and which is almost certainly the true one, but that they are also incompatible one with another. Indeed, in some cases, as will be observed, two or more interpretations proposed by the same author or authors may be in themselves mutually contradictory. The internal contradictions appear mainly to be due to a type of uncritical reliance upon a single technical method, supposed, often enough manifestly erroneously, to be specific for some particular element of the cell. Portions of the bacterial cell envelopes are capable of being stained by techniques stated by various authorities to be reliable methods of distinguishing, for example, nuclei, mitochondria, fat globules, and even reticulocytes [Bisset (*17*)].

The first three interpretations shown in Text-fig. 1, those of Knaysi, Järvi and Levanto, and Robinow and Murray, have already been referred to. That of DeLamater and Mudd (*40*) is shown only diagrammatically here (*d*). In the various figures published by these authors, the 'centrioles' which they claimed to demonstrate lie always in one or another of the sites illustrated in the figure; either on the mid-line at the poles of the bacillus, or in the comparable position on the complete cross-wall which represents the forthcoming fission of the bacillus in the centre; or more commonly at the junction of the cell wall and the cross-walls.

Stained by a different method, the material surrounding the cross-wall was equated with mitochondria (*e*) by Mudd and Wintersheid (*73*). On the other hand McGregor (*67*), while agreeing with the previous workers in their inter-

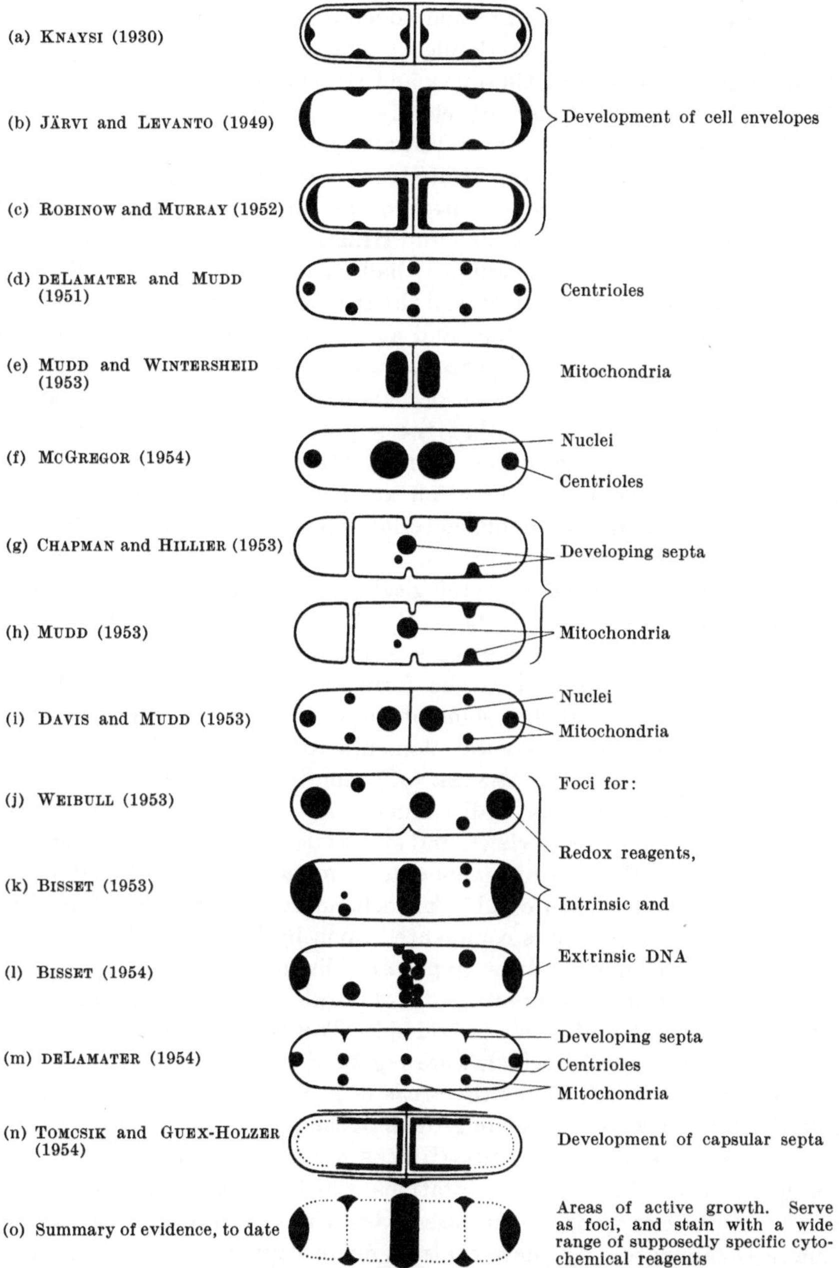

Text-fig. 1. Interpretations from bacterial cell envelopes. The figure shows diagrammatically the basophilic areas and granules which are commonly found at the points of cell division in septate bacteria, and the very diverse interpretations which have been based upon these appearances. Some of these, for example a, b, c, g, j, k, l, n, are sound and mutually corroborative. Others are, for the most part, fanciful and mutually exclusive. Discussion in text

pretation of the granules at the poles of the cell as centrioles, chose to regard those surrounding the cross-wall as nuclei rather than as mitochondria or further centrioles (*f*).

A great deal of valuable information concerning the bacterial cell envelopes has been obtained from the remarkable and beautiful electron micrographs of ultra-thin sections produced by CHAPMAN and HILLIER (*35*). These authors interpreted their results (*g*) in terms very closely comparable with those of KNAYSI, JÄRVI and LEVANTO, and ROBINOW and MURRAY, with certain exceptions which will be discussed in a later section. But MUDD (*71*), describing exactly the same material, (*h*), once more chose to interpret as mitochondria the elements of developing cross-walls which CHAPMAN and HILLIER, in accordance with the consensus of reliable opinion on the subject, had clearly described as cell-envelope material. DAVIS and MUDD (*37*) reserved the title of mitochondria for the smaller elements at the poles of the bacillus, and at the junction of cell wall and cross-walls, (*i*). The larger masses, disposed around the central cross-wall, they described as nuclei. These protagonists of bacterial mitochondria were attacked by WEIBULL (*102*), who showed that the supposedly specific reagents, which they used for their demonstrations, formed granular aggregations in exactly the same sites, (*j*). And this observation was paralleled by those of BISSET (*17*, *18*) who showed firstly that stainable materials with the reactions of DNA tended to become transferred from the nucleus, and possibly also from the cytoplasm of bacteria, under the influence of cytological techniques, and to form aggregates at the poles and cross-walls, (*k*). Secondly, he showed that DNA of extrinsic origin, derived in this case from herring-roe, when brought into physical contact with similar bacteria, appeared in the form of clusters of small granules at exactly the same places, and that some of these artifacts very closely simulated the types of controversial cytological elements with which this section is concerned (*l*). More recently DELAMATER (*39*) simultaneously described different portions of the polar and cross-wall elements as mitochondria and centrioles, in accordance with his previous views, and also as developing septa, in accordance with the criticisms of those of his opponents who had pointed out that this was, almost certainly, the true nature of his 'mitochondria' and 'centrioles', previously described, (*m*). This confusion is commented upon by MARSHAK (*66*), who points out the dubiety of claims to draw important distinctions on the ground that these tiny granules stain in different shades of purple. A further vindication of the views illustrated in *a*, *b* and *c* was provided by TOMCSIK and GUEX-HOLZER (*96*), who showed that those areas which were regarded by the authorities previously quoted and by BISSET (*17*, *21*) as the areas of growth of the cell envelopes, are certainly the areas in which the main growth of the capsule occurs, (*n*). And since the conclusions of TOMCSIK and GUEX-HOLZER were derived, not from stained and fixed preparations, with all the limitations which these conditions imply (as can be seen quite clearly from the discrepant views recorded in the diagram), but from the examination by phase-contrast microscopy of living bacteria treated with antibodies, the support which they provide is peculiarly valuable.

The final diagram (*o*) summarises what is really known about these areas in the developing, multicellular bacillus, and enables a distinction to be drawn between reasonable theories, supported by mutually corroborative evidence, and pure speculation, often mutually contradictory in a high degree, and supported only by analogy. These basophilic 'granules' are believed to correspond with the divisions between the cells of a multicellular rod, and mark the areas of growth and development of the cell envelopes. They serve as foci for the aggregation of reagents and of stainable materials in the cell; and partly for this reason, they stain by a very wide range of techniques erroneously considered to be specific for various types of cell component.

It is both interesting and also rather important to understand, in connection with the series of conflicting interpretations illustrated in Text-figure 1, that the straightforward views which we consider to be true were well known for many years before the more complex hypotheses were advanced. And yet the main argument which must be advanced against the acceptance of the latter is that they do not accord with the evidence available from other sources, but only with that derived from the observations on which they were based. And these, as has already been remarked, were themselves often enough based upon a single technical method, as applied to a single organism. In cytology, as in every other science, a hypothesis is tenable only when it is in accordance with all available evidence. The use of controls for experimental methods, although entirely ignored by many workers, is fully as appropriate in cytology as in other branches of investigation. The question of controls in bacterial cytology has been discussed by the present author in a recent review Bisset (*23*), and it is perhaps hardly relevant to consider it further at present. However, the problem is always one of importance, and it will be mentioned in the section on techniques, at a later point in this paper.

The examples just quoted could be expanded almost indefinitely, by reference both to contemporary publications and to those extending into the past for almost the entire period during which bacteria have been studied. These will serve, however, to illustrate the unfortunate state of confusion which has arisen in the study of bacterial cytology, because of this persistent and long-continued failure to appreciate the importance of multicellular structure in bacteria.

2. Organisation of the Bacterial Cell.

The present writer has already expressed the opinion [Bisset (*23*)] that, so far as can be determined by the evidence at present available, living organisms may be regarded as being divisible into three groups, according to the degree and type of cellular organisation which they exhibit. Of these three groups, by far the largest and (apparently) the most important, is distinguished by the possession of flagella, or at least, by evidence of having done so at some stage in their evolutionary history. The second group, widespread, and in its way successful, but showing far less evolutionary divergence than the first, comprises the blue-green algae. These have cells of a sort, and a nuclear apparatus, although very little is known about its workings, but they show no signs of ever having possessed flagella. The third group contains the viruses, which do not appear to have organised cells of any sort. They may not have a single origin, and indeed, may not be living organisms at all, as the term is usually understood.

The flagellate group includes the flagellate protista and the organisms of all types, plants, animals and fungi, which are believed to be descended from them. It also includes the bacteria, the most primitive of which have much in common with small, apochlorotic flagellates [Bisset (*14*, *21*, *22*)]. This point will be more fully discussed in the section upon morphology and systematics, at a later point in this paper.

It will be obvious that most of the members of the flagellate group, as thus defined, have long since lost their flagella; and this applies, of course, to many bacteria. However, it is reasonable to suggest that all the representatives of this vast group resemble one another more closely, in virtue of their presumed common phylogeny, than they resemble either the blue-green algae or the viruses. And despite the classical view that bacteria and blue-green algae form a natural group, because of their supposed lack of an organised nucleus, there is much more reason to expect that bacteria may resemble the other members of the flagellate protista, at least in the broad outline of the organisation of their cells. Although the suggestion has also been made that bacteria and viruses may possibly be related, either through the rickettsiae or, alternatively, through the organisms of the pleuropneumonia group, there is no evidence at present available which should lead us to take such a hypothesis seriously.

Although the bacteria may thus be expected to show many fundamental characters in common with plant and animal cells, it is unjustifiable to presume that they need necessarily be entirely alike in every respect. The dangers of thus deforming our knowledge to fit the Procrustean bed of preconceived notions of cell form is emphasised by MARSHAK in a recent review (*66*).

MARSHAK indicates that although the existence of a nuclear body of sorts, cell envelopes and flagella complete with blepharoplasts is well established in bacteria, recent claims further to describe such elaborations as mitotic centrioles and mitochondria are founded on little more than unsupported analogy, and are susceptible to some highly adverse criticism on both technical and theoretical grounds. VENDRÈLY (*100*) points out that one of the main arguments against such far-fetched speculations is the fact that many bacteria are known to be multicellular, and that this invalidates much of the theoretical interpretation upon which they are based [cf. also BISSET (*16*, *19*)].

Thus, although the cells of plants, fungi, animals and bacteria have many resemblances, each type has also its own peculiarities. It is out of place here to do more than enumerate some of the more obvious examples; such as cell division by constriction in animals, and by the formation of centrifugal and centripetal plates respectively in plants and bacteria; or the differences in the type and complexity of the centriole, which is asteroid in animals, simple in plants, migratory in fungi, and (probably) non-existent in bacteria.

Conversely, the more fundamental the character, the more probable it is that it will prove to be alike in all forms of cell. And although the subject of the bacterial nucleus will be considered in another review in this journal, it is necessary at this stage to refer to its form and behaviour in the bacterial cell, since this has an obvious bearing upon the significance of multicellularity. In any cytogenetic system the hereditary factors in a dividing nucleus must be so arranged that each factor is fully represented in both daughter cells. The manner in which this result is normally achieved in cells of the flagellate group is well known to all biologists. The hereditary factors (genes) are disposed in linear form along a chromosome which, when it splits along its length, serves to divide each gene occupying a position upon it. The process of mitosis serves to ensure that the chromosomes divide in harmony, so that each daughther cell contains its full complement of genes.

In bacteria, exactly the same principles are observed, but, possibly because of the relatively small size of the bacterial cell, the result is achieved in a less complex manner. The existence of a rod-like chromosome, dividing lengthwise, in the cells of bacteria, was first demonstrated convincingly by ROBINOW (*80*), (Plate I, Figs. 1, 5). Although these chromosomes frequently appear in pairs, it has been pointed out [BISSET (*11, 21, 23*)] that these are not true, diploid pairs. They do not divide equationally, with half of each member of the pair migrating to each daughter cell, according to the principles of mitosis in typical cells of the flagellate group. They divide reductionally, amitotically, and one member of the pair goes to each daughter cell. The pair consists, in fact, of recently separated sister nuclei. The cellular unit in bacteria is thus haploid but sometimes multinucleate. This system of reductional division of multiple, haploid nuclei has an important corollary when the question of multicellularity is again considered, because its effect is that mutant characters tend rapidly to be segregated, and that intermediate stages of the process, in multicellular bacteria, may be genetically heterogeneous. These conclusions, although founded upon purely cytological premises, have now been confirmed independently by investigators working upon purely genetical lines [LEDERBERG (*64*), WITKIN (*104*), RYAN and WAINWRIGHT (*83*)], and the single, reductionally-dividing chromosome may now be claimed to be a well-established cytogenetic entity.

The role of the blepharoplast in the bacterial cell serves also to separate it distinctly from other members of the flagellate group. The existence of this basal granule of the bacterial flagellum has now been conclusively proved [VAN ITERSON (*99*), BISSET (*21*), GRACE (*45*), PEASE (*75*)]. It is an approximately spherical, electron-dense body, varying from twice to four or five times the diameter of the flagellum itself. Unlike the typical flagellate blepharoplast, the bacterial blepharoplast appears to lie, not on or near the surface of the nucleus, but at the periphery of the cell, immediately below the cell wall, and on or in the semipermeable membrane. In this it probably resembles more closely the basal granule of an animal cilium.

In many flagellate protista the blepharoplast, which is, as has already been stated, closely applied to the nuclear membrane, plays an important role in the division of the cell. It is either closely associated with, or actually identical with the centriole of the dividing nucleus. But in bacteria this does not appear to be the case. There are several reasons for this state of affairs. Firstly, the division of the single bacterial chromosome takes place without the intervention of a complex mitotic apparatus. Secondly, the position of the blepharoplasts, near the surface of the cell and remote from the nucleus, makes any direct connection appear improbable. But thirdly, in this as in so many problems connected with the cytological behaviour of bacteria, the multicellular structure of so many bacteria very considerably affects the issue. As will become apparent in the next section, cell division and fission of the bacterium, although so frequently treated as synonymous, are by no means the same thing. In Text-fig. 2 can be seen two types of polar flagellate bacteria in process of fission. One, a *Pseudomonas* (A) normally has only one or two cells; the other, a *Spirillum* (B) is septate and multicellular. In the first, cell division and fission are approximately simultaneous (cf. Plate I, Fig. 2); in the second, cell division has greatly outpaced fission (Plate II, Fig. 16). However, irrespective of the degree of multicellularity, it is apparent that the new blepharoplasts, giving rise to the flagella of the daughter cell, either arise *de novo*, at the far side of one or more septa from the original blepharoplasts of the cell; or else, if strict continuity of these entities be assumed, must have migrated for considerable distances within the bacterium before settling down to produce a new flagellar outgrowth. In neither case is there anything to suggest a connection between the blepharoplasts and the division of the nucleus.

At this point, it might be constructive to consider the points of resemblance and difference between the more and less multicellular bacteria. A selection of both types is illustrated in Plates I and II, by means of photomicrographs. Figs. 1 and 2 show the nucleus and cell walls of a normally unicellular bacterium. This is the type of morphology which has been most closely studied, and is therefore best understood at present. The transverse chromosomes or nucleoids are disposed in the manner which has already been described in this section. Fig. 3 and 4 show the corresponding condition in a unicellular coccus. It should be emphasised that such cocci are relatively unusual, and that the multicellular type, illustrated in Fig. 8, must be regarded as normal among the commoner types of cocci. Figs. 5 and 6 show the nuclear structures and cell walls in a type of bacterium with a very pronounced degree of multicellularity, *Caryophanon latum*. The cells are little wider than disks, but each contains a pair of nucleoids. Figs. 7 and 8 show comparable structures in a septate coccus. In the preceding section, reference has already been made to the difficulties which may arise from failure to appreciate the complexity of structure which is to be found in Gram-positive cocci [cf. Bisset (*19*)]. Each of the two, four and more cells into which the coccus is divided (Fig. 8) contains, of course, its own nucleus (Fig. 7). But where this is not understood, the pattern of nuclear granules in each coccus may be subject to confusing misinterpretation. In most cases, such divisions at right angles are found only in spherical bacteria, *i.e.* those which are radially symmetrical. But they occur also, very exceptionally, in bilaterally symmetrical bacilli, as can be observed in Fig. 9. This is an unidentified pathogen, isolated from foot-rot in sheep, and is in some ways the most truly multicellular bacterium in the author's experience [Bisset and Thompson (*29*)].

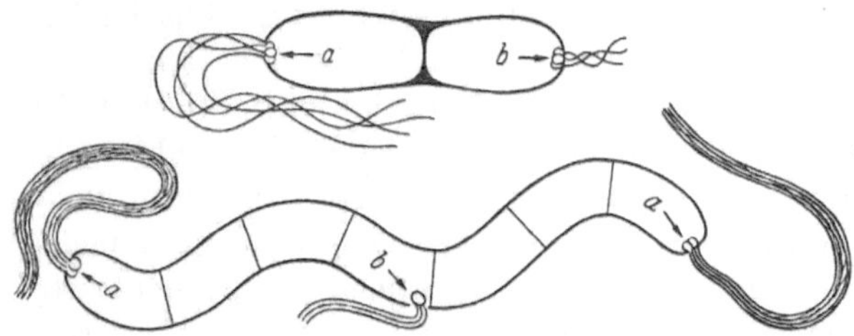

Text-fig. 2. Disposition of bacterial blepharoplasts. The blepharoplasts of dividing bacteria are indicated; in each case *a* represents the old, and *b* the newly arisen blepharoplast. It is apparent that they cannot be self reproducing, like those of protozoa, unless they are credited with the power of migrating across transverse septa, even in the case of the unicellular *Pseudomonas* (above). In the case of the multicellular *Spirillum* (below), the new flagella, at the point of incipient division, may be several cells distant from the original flagella at the poles

Although most aspects of the question of reproduction in bacteria can more conveniently be dealt with in the succeeding section, on cell division, there remains one aspect of the matter which can appropriately be discussed at present,

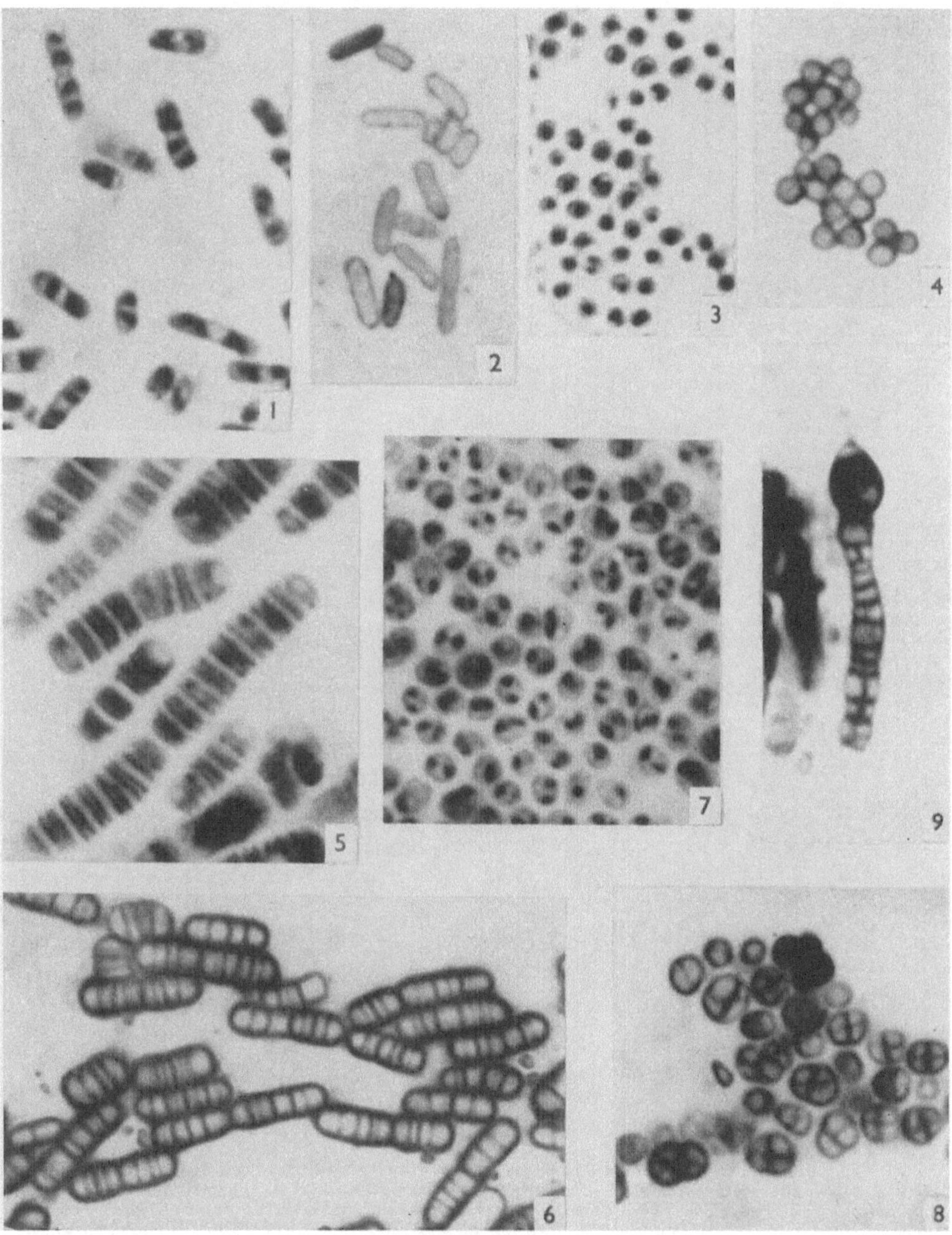

Plate I. Fig. 1. Nuclei of *Bacterium coli.* A typical unicellular bacterium, showing one or two pairs of transverse nuclear rods in each cell, dividing reductionally without true mitosis. HCl-Giemsa. ×3000. — Fig. 2. Cell walls of *Bacterium coli.* Each rod is a single cell. Tannic-acid-violet. ×3000. — Fig. 3. Nuclei of a unicellular coccus. Each coccus contains a spherical, central nucleus. HCl-Giemsa. ×3000. — Fig. 4. Cell walls of a unicellular coccus. Tannic-acid-violet. ×3000. — Fig. 5. Nuclei of *Caryophanon latum.* Each of the numerous disk-like cells contains a single paired nuclear body, fundamentally similar to those in Fig. 1. HCl-Giemsa. ×3000. — Fig. 6. Cell walls of *Caryophanon latum.* The numerous cross-walls are shown whereby these rods are divided into small cells. HALE'S method. ×2000. — Fig. 7. Nuclei of *Micrococcus cryophilus*, a typically multicellular coccus. Each cell contains a nuclear body. Similar configurations have been confused with mitotic figures by observers unaware of the multicellular structure. Trichloracetic acid and Giemsa. ×3000. — Fig. 8. Cell walls of a large coccus, similar to fig. 7. Each coccus contains two, four or more cells. HALE'S method. ×3000. — Fig. 9. Bacterium from foot-rot in sheep, showing complex cellular structure, with septa at right-angles. HALE'S method. ×3000

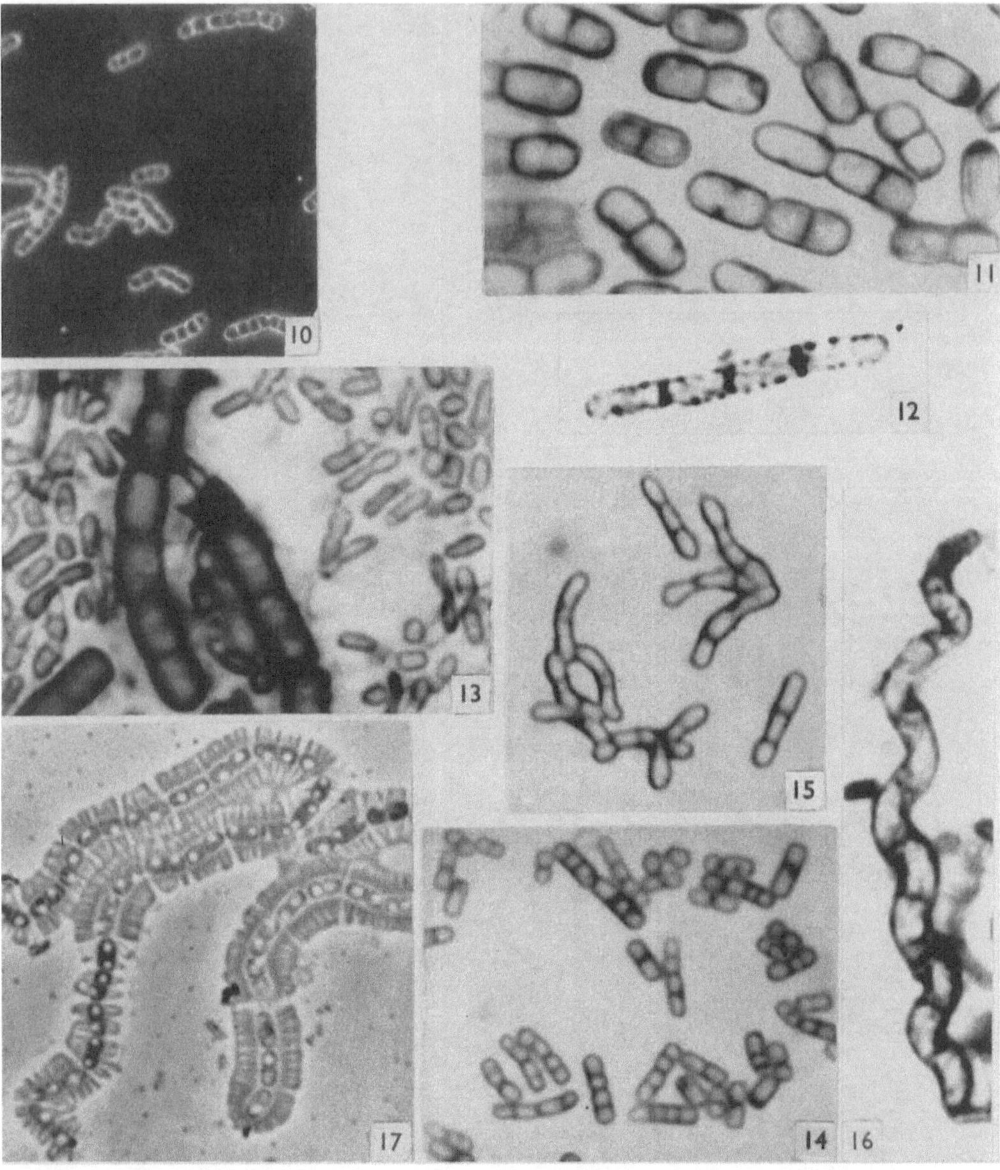

Plate II. Fig. 10. Cell membranes of *Bacillus megaterium*, by Newton's method, showing great cellular complexity. ×1500. (Reproduced from the J. Gen. Microbiol., by permission of Dr. B. A. Newton). — Fig. 11. Cell walls of *Azotobacter chroococcum*, showing mature cross-walls and incipient ingrowths. Hale's method. ×3000. (Reproduced from the J. Gen. Microbiol). — Fig. 12. *Bacillus megaterium* treated with DNA solution which has produced stainable aggregations on the walls, and especially the cross-walls. ×3000. — Fig. 13. Mother cells and swarmers of *Rhizobium* sp. Tannic-acid-violet. ×3000. — Fig. 14. *Lactobacillus brevis*, showing regular cross-walls. Hale's method. ×3000. (Reproduced from the J. Gen. Microbiol.). — Fig. 15. '*Lactobacillus*' *bifidus*, for comparison with fig. 14. showing irregular cross-walls and branching. Hale's method. ×3000. (Reproduced from the J. Gen. Microbiol.). — Fig. 16. *Spirillum* sp. Cell walls by modification of Hale's method. Showing cross-walls. — Fig. 17. Striated appearances in the structure of the capsule of *Bacillus* sp. by Tomcsik's method. Phase contrast. ×2500. (Reproduced from the J. Gen. Microbiol., by permission of Prof. J. Tomcsik)

in relation to cellular organisation, and that is the production of those small cells with a distributive function known as gonidia [Bisset (21)].

A great deal has been written concerning various types of reproductive or supposedly reproductive stages in bacteria, for example, the *G* and *L* colonies and the more or less complex cycles which have been described at different times in *Mycobacterium tuberculosis*. However, it is only in the nitrogen-fixing bacteria *Rhizobium* and *Azotobacter* [BISSET and HALE (*25, 26*)] and in some spirilla [PEASE (*76*)] that the small, motile cells with a truly distributive function have been fully described. In each case the gonidia are quite unexceptionable, small bacteria, with polar flagella, and are produced in large numbers within the bacterial cell. They are not, as has been suggested, merely the result of repeated cell division, but are formed in large numbers within the intact cell envelopes of the mother cell. The ripe mother cell of, for example, *Rhizobium*, is thus a multiple cell of considerable complexity (Plate II Fig. 13); and while by no means unusual in comparison with other members of the flagellate group, such an arrangement is very far removed from the sort of structureless cell, dividing by simple fission, that all bacteria are still supposed to be. The complexity is increased by the fact that, as can be seen in the same photomicrograph (Fig. 13), the gonidia of *Rhizobium* are produced from mother cells arranged in multicellular, septate filaments. So that the entire 'barred cell', as such filaments were previously described [THORNTON and GANGULEE (*90*)], is not in fact a cell, but a complex of several large cells containing several hundred small ones.

The implications of this type of phenomenon, in terms of cell organisation, are considerable. The mother cell must be capable of undergoing repeated cell divisions without alteration of external form, and probably succeeds in secreting large quantities of cell-surface structural elements (*i.e.* for the walls and flagella of the gonidia, within its lumen. And indeed it can clearly be seen that the transverse septa or cross-walls of such filaments are lined with thick aggregations of basophilic material [BISSET and HALE (*25*)]. These aggregations may reasonably by presumed to be concerned with the internal secretion of gonidial material; and it is their notable basophilia which gives the well-known 'barred' appearance to the filaments of mother cells. It was at one time believed that each 'bar' was capable of becoming transformed into a single 'flagellated coccus'(*i.e.* a gonidium) and of swimming off on its own. This notion was presumably the child of the dogma of bacterial reproduction by simple fission, but the true explanation is in much better accordance with what is found in other types of protista. And it is worthy of further emphasis that the type of gonidium which has been under discussion is entirely bacterial in behaviour and appearance, and not in any way different from what might reasonably have been expected from theoretical considerations.

3. The Composition of the Cell Envelopes

In any consideration of the multicellular structure of bacteria, the question of the nature of the cell envelopes and their physical and chemical structure is of primary importance.

As will appear when the bearing of these matters upon the problem of bacterial systematics is discussed, in a later section of this paper, the types of bacteria which are preponderantly unicellular and those which are preponderantly multicellular, have several important points of difference. In general, Gram-positive genera tend to belong to the latter, and Gram-negative genera to the former class [BISSET (*14*)]. This distinction is more exact than it appears to be on first examination, because certain bacteria which exhibit a multicellular, septate structure, although apparently Gram-negative, have proved either to have important taxonomic characters in common with the Gram-positives, or even to be slightly Gram-positive themselves at some stages of their growth cycle. Examples of this are *Caryophanon* (TUFFERY, personal communication), the 'barred cells' of *Rhizobium* [BISSET and HALE (*25*)], and also *Neisseria* [BISSET (*14*)].

The exact significance of the Gram stain is still a matter of dispute, although it is generally agreed that the substances responsible for the reaction lie in the cell membrane or surface layers of cytoplasm, and not the cell wall proper. But in addition to these, there are also differences between the composition of the cell walls themselves in the two types of bacteria.

To deal firstly with the Gram-complex itself, the theory of HENRY and STACEY (*50*), HENRY, STACEY and TEECE (*51*) still appears to the present writer to be reasonably well in accordance with the cytological facts. These authorities state that the difference between Gram-positive and Gram-negative bacteria lies in the proportions of ribose and deoxyribose nucleic acids present in the surface cytoplasm, the ribose form being predominant in the Gram-positives. Alternative theories are those of MITCHELL and MOYLE (*69*), who attribute Gram-positivity to the presence of a phosphoric ester, which they claim to be independent of the concentration of ribose nucleic acid, and of HOFFMAN (*53*) who suggests that the difference may be accounted for by a variation which he finds between the tyrosine content of the pentose nucleoproteins in the two types of bacteria. The problem is reviewed by VENDRÈLY (*100*), but his conclusions are indefinite.

According to BARTHOLOMEW and MITTWER (*3*), there is no structural difference between the cell membranes of Gram-positive and Gram-negative bacteria, apart from the types or concentrations of nucleic acids, but in their cell walls there are very decided differences. These have been described by SALTON and HORNE (*85*), SALTON (*84*), FISCHER and LAROSE (*42*) and ROBINOW and MURRAY (*82*). The consensus of this evidence is that Gram-negative genera have in their cell walls a higher lipid content and a much fuller range of amino-acids in the protein fraction. The cytochemical behaviour of their cell walls resembles that of intact wool proteins, whereas the cell walls of Gram-positive bacteria more closely resemble degraded wool proteins in this respect (FISCHER and LAROSE).

Since the lipid element is an important factor in the maintenance of the structural rigidity of the cell wall [HURST (*57*)], the difference in content of this class of substance suggests a possible explanation for the observed fact that Gram-negative bacteria normally produce elongated cells, and even non-septate filaments, which argues a very considerable structural rigidity. The effect of surface-tension forces is relatively enormous in the case of bodies whose smaller diameter may be much less than 1 μ, and if they lacked such a rigid structure they would rapidly be constrained into the form of an ovoid, or even a sphere. On the other hand, the Gram-positive bacteria, whose cell walls must be suspected, on the available evidence, of being much less rigid, do not form elongated cells at all. Their bacillary or filamentous forms are so much subdivided by cross-walls that the average cell is shorter than it is wide; sometimes many times shorter. It is appropriate to speculate, at this juncture, whether the entire question of multicellularity in bacteria, the subject of this review, may not be closely bound up with so apparently minor a point as the lipid content of the cell wall.

Yet another difference between Gram-positive and Gram-negative bacteria may be explicable in terms of the constitution of the proteins of the cell envelope, and this is the much more satisfactory behaviour of the latter when employed as antigens. This is almost certainly correlated with their chemical complexity [BISSET (*15*)]. The loss of antigenic specificity associated with the $S \rightarrow R$ variation in *Bacteriaceae* is accompanied by a change to a septate type of morphology and cell division, quite indistinguishable from that normally found in Gram-positive genera, and correspondingly unlike the unicellular from of the *S* variant [BISSET (*7, 8, 9*)]. Curiously enough, while the electron micrographs of sections of the Gram-positive *Bacillus cereus* CHAPMAN and HILLIER (*35*) clearly show the cell wall, and have given much useful evidence about it, the comparable sections of BIRCH-ANDERSEN (*5*) show almost no cell wall whatsoever in *Salmonella*; as is remarked by his collaborators LARK, MAALØE and ROSTOCK (*63*). However, this is not true of BIRCH-ANDERSEN, MAALØE and SJÖSTRAND's earlier sections of the very similar *Bacterium coli* (*6*), so that some hazard of technique may account for the apparent lack of a cell wall in that case.

The cell wall and cell membrane of both types of bacteria appear nevertheless to have a considerable lipid fraction, and stain with fat-soluble dyes [ROBINOW and MURRAY (*32*)]; an especially interesting demonstration of both cell wall and cell membrane can be obtained by the use of Victoria blue, which is at least partially lipid-soluble. However, the cell wall can also be stained by Alcian blue [TOMCSIK and GRACE (*92*)] and Janus green [HALE, personal communication; BISSET (*17*)], without any preliminary mordanting, and these dyes are 'specific' for polysaccharides and for mitochondria, respectively. After mordanting with phosphomolybdic acid, the cell wall stains very well with methyl green, which is 'specific' for nuclear structures [HALE (*47*), BISSET and HALE (*27*)]. These observations tend to show that too much reliance cannot be placed upon the type of information obtainable from such reactions. Whereas the bacterial cell wall almost certainly possesses chemical fractions of protein, polysaccharide and lipid, the preponderance of any particular fraction cannot readily be assessed from evidence of this nature. At the same time, it cannot be too often emphasised that elements derived from the cell wall are capable of disguising themselves as organelles of very many different kinds (cf. Text-fig. 1).

Owing to their small size, and to our continued lack of information concerning their cytophysical behaviour, it is difficult to assess the degree to which these various components may belong, not to the wall, but to the underlying cell membrane of the bacteria. There is no reason to doubt that this structure, which stains with lipid-soluble dyes [ROBINOW and MURRAY (*82*), BURDON (*33*)], may well have the same protein-lipid composition as the cell membranes of other members of the flagellate group [HARRIS (*48*), BOOIJ (*30*)]. Accordingly, the chemical fractionations and analyses which have been made of the cell wall, where these are not correlated with cytological observations, as indeed they very seldom are, are little more reliable than the aforementioned cytochemical tests as criteria of the true disposition of the complexes of different chemical substances which indubitably lie in the cell envelopes of bacteria. However, provided that this reservation is borne in mind, a considerable amount of information on the composition, and even upon the molecular structure of the cell wall, has recently been made available. Holdsworth (*54*) isolated a protein-carbohydrate complex from the cell wall of *Corynebacterium diphtheriae*. Holdsworth took some care to establish that the protein component differed in its amino-acid constitution from that of the intracellular proteins isolated from the same organism. He claimed the carbohydrate fraction to be an oligosaccharide. Others have described polysaccharides in the cell walls of *Corynebacterium diphtheriae* and other species of bacteria, both by chemical analysis and by cytochemical methods [Wong and Tung (*105*), Pennington (*77*), Järvi and Levanto (*58*)]. But the difference between a true polysaccharide and a close association of oligosaccharide units in a structure of this nature must be, for the most part, a matter of definition.

The work of Stacey on the nature and behaviour of the Gram-complex has also led him [Stacey (*88*)] to draw certain conclusions in respect of the disposition of the chemical components of the cell envelopes. He has pointed out that for the synthesis of the polysaccharide elements, the underlying, secretory nucleoproteins are probably required to carry with them a layer of polysaccharide or oligosaccharide which functions as a 'primer' and decides, or at least influences the chemical nature of the polysaccharide produced at the surface. This argument, as stated, applies mainly to capsular polysaccharides, but there seems to be no reason why it should not be equally true of any layer of envelope materials.

From the cytochemical works quoted above, it appears that the complete, mature cross-wall does not differ in its reactions from the remainder of the cell wall of which it is a part. Thus the septate bacteria are truly multicellular, with each cell enclosed in a true cell envelope. On the other hand, the structures which precede and secrete the cross-wall, and which will be described more fully in the following section, are apparently derived from the cell membrane. Although the two protoplasts may be entirely divided by a membranous septum of this kind, they will appear as a single cell in preparations designed to demonstrate the polysaccharide-containing, mature cell wall. Thus, even in recognisably multicellular bacteria, the true extent of this subdivision may fail to be recognised.

Of recent years, a surprising amount has been learned concerning the actual physical structure of the cell wall, in terms of its macromolecular organisation, by means of electron microscopy of disrupted fragment of cell envelopes. So far these observations have been made more frequently upon species of *Spirillum* than upon any other types of bacteria, but *Bacterium coli* [Yoshida *et al.* (*108*)] and one strain of an unidentified bacterium [Labaw and Mosley (*62*)] have also yielded information. In spirilla, a hexagonal pattern of macromolecules has been described by Houwink (*55*) and also by Salton and Williams (*86*), whereas Pease (*75*) and Yoshida *et al.* (*108*) suggested a striated structure in the same sense as the long axis of the organism, in *Spirillum* and *Bacterium coli* respectively. The cell walls of the bacterium described by Labaw and Mosley showed a pattern not unlike that illustrated by Houwink. However, the latter, in addition described an inner layer with a fine, striated appearance in the electron micrograph. Although they do not provide conclusive evidence in themselves, the appearance of these layers is consistent with the supposition that the outer, hexagonally-patterned layer consists of protein molecules, and the inner, striated layer, of polysaccharide; and this is borne out by YOSHIDA, FUKUYA *et al.* (*106*) who compared electron micrographs of *Bacillus subtilis* treated with phosphotungstic acid, before and after digestion with pepsin and trypsin. A rough structure appeared on the outer face of the undigested cell walls, but not if the phosphotungstic acid treatment took place after the enzyme treatment. These rather dissimilar studies thus confirm the hypothesis that the cell wall consists of a protein layer overlying a polysaccharide layer. The disposition of the lipid fraction, however, is not explained by these observations.

So far as the third element of the cell envelope, the capsule, is concerned, our knowledge of its structure and composition has been entirely revolutionised by the remarkable researches of TOMCSIK. And while reference is being made to TOMCSIK's methods, it is important to remember that, although so far applied mainly to problems of bacterial cytology, they have a universal application in all fields of biology. The original observations arose from his interest in the immunochemical reactions of the cell envelopes. Numerous efforts have been made, before and since, to examine microscopically the effects of antigen-antibody reactions upon the bacterial cell [PIJPER (*79*), MUDD and ANDERSON (*72*), YOSHIDA, FUKUYA *et al.* (*109*), YOSHIDA, TANAKA *et al.* (*111*)], but these have revealed little or nothing of the nature of the antigenically active structures in the cell, although a certain amount can be inferred [BISSET (*15*)]. TOMCSIK's results in this field, however, were achieved mainly by his skilful use of phase-contrast microscopy. And by this method he showed how the locality of action of antibodies prepared against polypeptide and polysaccharide fractions could be visualised by the changes in refractility which they produced in the bacterial cell wall and capsule [TOMCSIK (*91*), TOMCSIK and GUEX-HOLZER (*93*, *94*, *95*, *96*, *97*, *98*)]. The capsule of *Bacillus* species is thus revealed as consisting of laminae of polysaccharide and polypeptide elements, with reinforcements of polysaccharide at the poles and points of division of the bacillus (Plate II, Fig. 17; Text-fig. 4). In this way TOMCSIK and GUEX-HOLZER (*98*) have also confirmed that the cell wall itself consists largely of polysaccharide, and their observation tends to confirm once more that such protein element as exists in the cell wall probably overlies the polysaccharide element; otherwise it would, in all probability, have appeared as a distinct layer in TOMCSIK's photomicrographs; granting the assumption that, like the polysaccharides in question, it was represented antigenically in the antibody preparations employed. These workers subsequently demonstrated that the capsular materials were capable of entering into a salt-like combination with various proteins, at appropriate values of pH [TOMCSIK and GUEX-HOLZER (*97*)], and they employed a variety of enzymes and other agents to produce effects on the cell envelopes which will be discussed further in the two subsequent sections, in connection with the behaviour of the envelopes and especially the membrane, at cell-division. TOMCSIK's techniques will also be described in greater detail in the section concerned with technical methods.

4. Fission and Cell Division

Since the main subject of this review is the occurrence of multicellularity in bacteria, little further emphasis needs to be given to the fact cell division and fission, although habitually treated as synonymous, are very far from being so in fact. Plates I and II give illustrations of multicellular bacteria of various types. And in most cases it is apparent that cell division has proceeded rapidly, within the boundaries of a single bacillus or coccus which has not yet undergone fission. But it is also true that cells which have undergone complete fission may continue to be associated together in complexes of various types. Examples are shown in Text-fig. 3, *A* and *B*. The former is a group of cocci of the type usually known as a sarcina. Each coccus is multicellular, and composed of four cells in the plane of the diagram and possibly more in the third dimension (cf. also Plate I, Fig. 8, which shows a similar coccus, but not in sarcina arrangement). The multicellular cocci are themselves arranged in groups of four or eight, and the whole assemblage may be enclosed in a capsule. In an organisation of this type, the unit cells of the cocci have undergone cell division within the coccus, but not fission. It is the transformation of one four-celled coccus into two two-celled cocci which constitutes fission. In such a case, the two tend to keep more-or-less in step. Each coccus divides at a stage of development when its component cells are on the verge of division also, and the normal number of cells remains the same. At the same time, following these primary and secondary phases of reproduction, there must exist a tertiary phase, when the sarcina packet becomes inconveniently large and undergoes fission on its own account.

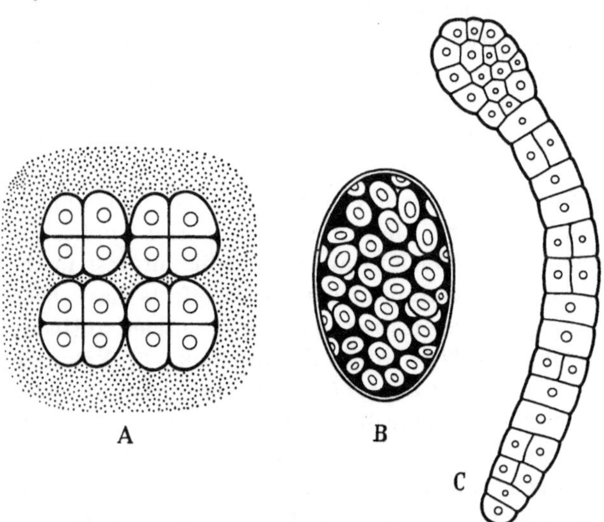

Text-fig. 3. Types of multicellularity in bacteria. *A* Sarcina type of coccus. Each coccus is itself multicellular, and these units are arranged in packets within a capsule. *B Azotobacter* mother-cell, packed with relatively tiny gonidia; each a perfect bacterial cell, complete with flagella. *C* Septate organisms from foot-rot in sheep with cross-walls in all senses, capable of fission both longitudinally and laterally

An entirely different type of aggregation of cells which have already undergone fission is illustrated in Text-fig. 3, *B*. This diagram shows the condition found in the swarmer mother-cells of *Azotobacter*. The larger mother-cell is packed with tiny swarmers, each one a complete, miniature bacterium. In Plate II, Fig. 13, is shown the same phenomenon in *Rhizobium*, which appears on cytological grounds to be closely related to *Azotobacter* [BISSET (*20*)], but in which the mother-cell is itself part of a multicellular bacillus, as has been described in a previous section.

The last diagram (*C*) in Text-fig. 3 shows the extreme type of multicellularity in rod-shaped bacteria (also Plate I, Fig. 9). In this unclassified organism from foot-rot in sheep [Bisset and Thompson (*29*)] the cells divide not only in the same sense as the long axis of the bacterium, but also at right-angles to it. There is no sign of any co-ordination between cell division and fission, and units of any number of cells may be formed, from pseudo-cocci with two or four, to the type of filament illustrated in the diagram and photomicrograph, which may have as many as a hundred tiny cells. Conclusions cannot, however, be too freely drawn from the behaviour of this organism, since very little is known about its systematic position and relationships.

The fission of the more typically multicellular bacteria, for example *Bacillus*, which takes place at the most longstanding (usually) of the septa subdividing the bacillus into four cells, may logically be regarded as the last stage in the process of cell division. But because, as in the sarcina already described, further cell division takes place so fast as to keep pace with or overtake the process of fission, the four-celled constitution is preserved (Text-fig. 4). In the same figure the arrangement of the cell envelopes in such a bacillus is illustrated as fully as is consonant with our present knowledge, and the section inset within the circle (*h*) represents one of the major areas of growth and development, as determined by the researches both of Tomcsik and Guex-Holzer on the capsule, and of Bisset on the cell wall and membrane.

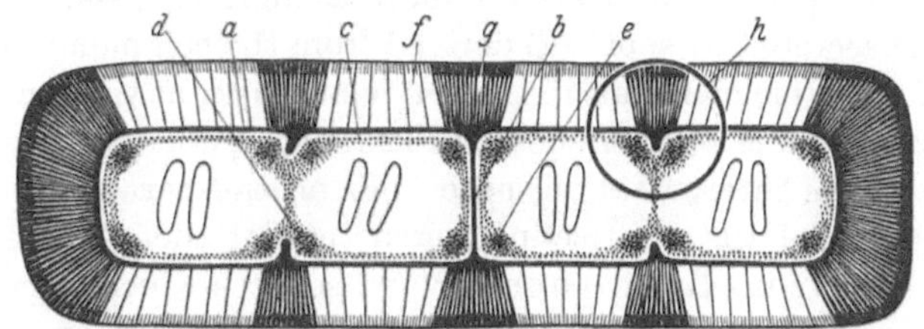

Text-fig. 4. Growth and development in a large *Bacillus* sp. *a* Cell wall; *b* cross-wall; *c* membrane; *d* septum from membrane; *e* growing-point at junction of cell wall and cross-wall (compare Text-fig. 1); *f* capsule; *g* polysaccharide reinforcement at growing-point of capsule; *h* main area of growth

The cell membrane (shown stippled in the diagram) is a most important element in this scheme, as will appear, and it is important to emphasise that the existence of this semi-permeable membrane as a positive structure in its own right, rather than simply as an interface or molecular layer on the surface of the cytoplasm, has been unequivocally demonstrated by a variety of different methods. Weibull (*103*) and Tomcsik and Guex-Holzer (*94*) have done so by lysozyme digestion. In the last stages of dissolution the protoplasts disappear and the empty membrane survives momentarily. Yoshida *et al.* (*110*) and others have shown the membrane quite clearly by the electron microscope. Newton (*74*) achieved a remarkable demonstration of the cell membrane of *Bacillus megaterium* in the undamaged cell, by means of a fluorescent derivative of the antibiotic polymyxin (Plate II, Fig. 10). That it is the cell membrane rather than the cell wall for which this substance has a selective affinity is shown by the number of cellular divisions which thus appear, six or eight per bacillus. Complete cell walls are seldom found to subdivide this type of rod into more than two or four cells. Compare, for example, Plate II, Fig. 11, which shows the rigid, polysaccharide-containing cell wall elements. In the latter case most bacilli have one complete cross-wall across the centre of the rod, and incomplete walls subdividing each half. By comparison of Plate II, Figs. 10 and 11, it will be apparent that the two or four completely divided cells shown must be subdivided by developing septa which do not appear in preparations stained for the cell walls. There is some divergence between the evidence upon this point provided by classical microscopy and by the otherwise very illuminating electron micrographs of very thin sections of bacteria illustrated by CHAPMAN and HILLIER (*35*). In the latter case the protein material which precedes the true cross-wall appears in the form of small granules. But it is the opinion of the present writer that this is due to coagulation of the membrane by the action of osmium tetroxide, to a strong solution of which the specimens had been exposed. This question is discussed in a recent paper on the cell envelopes of *Azotobacter* [BISSET (*20*)].

The process of division is illustrated in Text-fig. 4. The cell wall (*a*) forms cross-walls (*b*); these are produced as ingrowths (see inset *f*) and are both preceded and secreted by septa (*d*) derived from the cell membrane (*c*). The concentrations of stainable materials in the cell membrane, at the junctions of the cross-elements, are the same as those shown in Text-fig. 1, and have been responsible for numerous errors of interpretation, especially on the part of workers unaware of the significance of the multicellular structure of these bacteria.

There is evidence from several sources which suggests strongly that the main growth of the cell envelopes takes place in the area illustrated, at the point of origin of the developing cross-wall. Paradoxically, the clearest indications of this are obtainable from studies of those bacteria (*e.g.* the typical Gram-negatives) which seldom have more than two cells per bacillus, and usually only one (Plate I, Figs. 1 and 2). It has been shown [BISSET (*13*)] that in bacteria of this type there is a concentration of nucleoproteins at the poles of the cell, or often only at one pole, and at the point of division. And in peritrichously flagellate species it can be seen that the flagella borne by one half of a dividing cell are much more fully developed than those on the other half. From this and similar observations, it can be deduced that the cell envelopes grow from one tip of the cell, as well as from the point of division, and that effectively such bacteria do not reproduce by binary fission, but by the production of buds, each of which is comparable in size and appearance to the mother cell; except insofar as its flagella are less well developed initially. The role of the growing points, as the concentrations of nucleic acids already described are named by BISSET (*13*), is verified by the work of BERGERSEN (*4*) on the effects of chloramphenicol upon the growth of *Bacterium coli*. Under the influence of this antibiotic subsidiary growing-points developed in the cell envelopes at various points along the length of the rod, and grew outwards to form side branches. Growth continued so long as the growing-point persisted.

In multicellular bacilli, such as that illustrated in Text-fig. 4, the role of the growing-points in the tips of the rod is subsidiary to that of those at the points of division; that is, those (*e*) at the junction of the transverse septa. It is at these points (*h*) that the new cross-walls commence to grow as a centripetal ingrowth, and presumably the main development of the cell wall proper takes place in the same area. In fact, the ingrowth of the one and the outgrowth of the other may reasonably be regarded as reactions in opposite directions from a single secretory surface. And this entire concept is reinforced by the conclusions of TOMCSIK and GUEX-HOLZER (*96*) who, by the use of TOMCSIK's technique of phase-contrast examination of antigen-antibody reactions in bacteria, have come to the conclusion that it is precisely in this area that the main growth of the capsule also occurs.

In this case, as in others mentioned in this review, it can only be admitted that any one of these pieces of evidence is, by itself, slight and circumstantial. But where so many diverse observations, upon cell walls, flagella, capsules, and upon the effect of antibiotics on the growth of the cell, all lead to the same or parallel conclusions concerning the development of the cell envelopes, a considerable amount of confidence may reasonably be placed in them.

In Text-fig. 5 a reconstruction is given of the growth of the cell envelopes of three characteristic types of bacteria, two of which are markedly multicellular. The arrows show the direction of development from the growing-points. The first of these, A is in the form of a simple Gram-negative bacterium or pseudomonad, such as has recently been discussed. It is illustrated at a fairly advanced stage of cell division, and growth is thus taking place at one pole and at the point of division. Or, expressed in another way, at the right-hand pole both of the mother cell and the developing daughter cell or bud. The relative ages of the two main areas of cell wall, on the mother and daughter cells respectively, is indicated by the relative lengths of the flagella [*cf.* BISSET (*13*)]. At the free tip the direction of growth is outwards, so that the cell elongates progressively. At the point of division, this type of development, which contributes to the elongation of both cells, is combined with an inward growth of the envelopes to produce a constriction, whereby the two cells are eventually separated completely. In this instance cell division and fission are approximately synonymous.

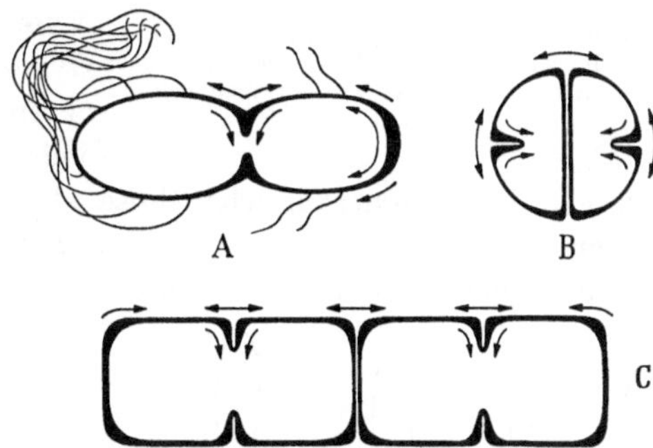

Text-fig. 5. Growth of cell envelopes in bacteria. *A* Unicellular bacterium; growing from tip of daughter cell and point of fission. *B*, *C* Septate coccus and bacillus; growing from points of junction of cell wall and cross-wall

The two remaining diagrams, B and C in Text-fig. 5, illustrate the analogous condition in the multicellular cocci and bacilli. In the bacillus the role of the free poles is much reduced, and in the cocci, of course, they do not exist. The main growth is at the points of junction of cell wall and cross-wall. This appears to be true, irrespective of whether the cross-wall is complete or is still in process of development; because the growth of these multicellular bacteria is markedly symmetrical, and does not proceed more rapidly in one area rather than another. In passing, however, it may be recalled that some types of multicellular bacteria, such as the mycobacteria and related genera, are much less symmetrical in this respect, and may show signs, for example, of cellular proliferation at the centre of the rod, and corresponding enlargement at the tips [BISSET (*10*)]. Some aspects of these differences will be discussed in Section 6, in relation to the systematics of bacteria, but so far as they affect the actual process of fission they may appropriately be described at this point.

The fission of a corynebacterium, as illustrated in Text-fig. 6, is fundamentally similar to that of a multicellular bacillus as shown in Text-figs. 4 and 5, but differs slightly in detail. *Corynebacterium* is among those bacteria the true morphology of which has been most completely obscured by the longstanding reliance which many bacteriologists have tended to place upon so-called diagnostic stains. Almost all the available information refers to *C. diphtheriae*, and much emphasis is laid upon the "metachromatic granules". The picture which is obtained by the use of, for example, NEISSER's stain, is utterly fallacious. Most strains of *C. diphtheriae* consist of filaments of from two or three to a dozen or more cells. The terminal cells are considerably enlarged by comparison with those at the centre of the filament, and it is the contents of these enlarged terminal cells which provide the stainable materials for the "metachromatic" granules. In connection with the latter phenomenon, it is not, perhaps, as well known as it should be that an excellent simulacrum of these "diagnostic" structures can be achieved by appropriate staining of sporing bacilli.

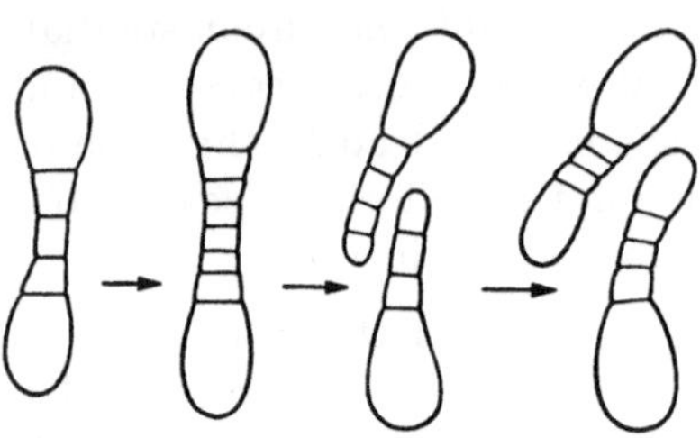

Text-fig. 6. Fission in *Corynebacterium*. Growth occurs by proliferation of small centre cells. After fission the new terminal cells enlarge

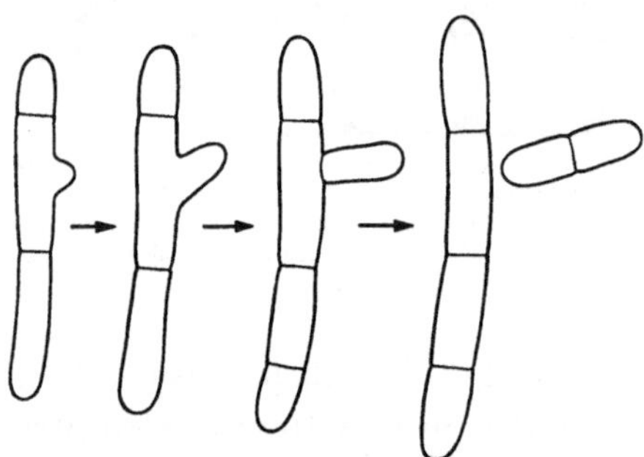

Text-fig. 7. Transient branching or budding. This condition is commonly found in the filamentous Gram-positive bacteria. A small, lateral branch forms, but is rapidly cut off by a cross-wall and eventually becomes completely separated from the main filament

The terminal cells do not appear to divide so freely as do the remaining cells but merely to increase in size. The small central cells proliferate, thus becoming progressively smaller, and eventually fission takes place at some point among them. The new terminal cells, at the recently-divided end of the filament, then proceed to enlarge, whereas the small cells at the centre continue to divide and proliferate. Alternatively, all the cells composing the filament may enlarge and become separated one from another; thus resembling a chain of cocci before they finally part company, and each cell commences again to divide and subdivide until a new filament is formed.

The multicellularity of bacteria of this type is well attested [BURDON (33), BISSET (*10*), HEWITT (*52*)], but a recent paper by DAVIS and MUDD (*37*) appears to credit such rod-like forms with the possession of only a single, central septum, and proceeds to describe a whole series of granular structures along its length, each cytologically and cytochemically different from the last!

Some bacteria exhibit true, if transitory branching. Occasionally the branch remains in connection with the parent filament for so short a time that the process is more closely analogous to budding. This occurs in certain Actinomycetales [BISSET and MOORE (*28*)], in *Corynebacterium diphtheriae* [BISSET (*11*)] and in *"Lactobacillus" bifidus* [HAYWARD, HALE and BISSET (*49*)]. Permanent branching is found, of course, in *Streptomyces*, whereas in *Actinomyces bovis* it appears that the primary mycelium shows temporary, and the secondary mycelium, permanent branching [MORRIS (*70*)]. Permanent branching, since by definition it is not correlated with cell division, does not concern us at the moment. Temporary branching is, however, a type of cell division, and is often followed by fission, as has been remarked upon. The process is, so far as is known, exactly comparable to normal growth and fission of a septate bacterium, except insofar as the polarity has become altered. It is illustrated diagrammatically in Text-fig. 7.

A further, and highly interesting observation, bearing upon the problem of cell-division in multicellular bacteria, has been made by MCQUILLEN (*68*), who states that the naked protoplasts of the unit cells derived from septate bacilli, when the cell wall is removed by lysozyme, are capable of dividing into two by budding and constriction.

5. Techniques

The demonstration of multicellular structure in bacteria is not simply a question of staining the cell wall. It has already been emphasised in the preceding sections that a full appreciation of the complexity of cellular structure which is capable of existing in a single coccus or bacillus can only be obtained by a study of all the components of the cell envelopes, indeed of all the components of the entire cell. The importance of the cell membrane is illustrated by the photomicrographs of NEWTON (*74*) (Plate II, Fig. 10).

It is a commonplace of bacterial cytology that the cell walls do not stain easily by the techniques most commonly employed for diagnostic purposes. The best evidence for this state of affairs is the almost total ignorance of the structure and behaviour of the cell walls which has been exhibited, even by experienced bacteriologists, for so many years. The measurements and dimensions embodied in the officially recognised descriptions of bacterial species refer almost inevitably to the naked protoplast, shrunken by drying and heat-fixation, and not to the overall dimensions of the cell wall. The majority of such recorded measurements are thus considerably too small.

In preparations stained by simple solutions of many basic dyes, the cross-walls of multicellular bacteria appear as unstained gaps, whereas the developing cytoplasmic septa appear either in their natural arrangement or as more deeply stained areas at the poles of the individual protoplasts. That is to say, in the second case, they appear to line the gaps in the bacillus which represent unstained mature cross-walls. A comparison of this appearance with a true demonstration of the cell walls is given diagrammatically in Text-fig. 8, and the corresponding appearances in a coccus of the septate type in Text-fig. 9. (Compare Plate I, Figs. 7, 8).

The use of so-called diagnostic staining methods can lead, in the case of multicellular bacteria, to the production of appearances which are misleading in the extreme and which have given rise to various misconceptions. In Text-fig. 10 the appearance of a *Corynebacterium* stained by a cell wall method (*a*) is compared with the appearances produced by, for example, Neisser's stain for metachromatic granules (*b*) or by Gram's stain (*c*). It is obvious that none of the hypotheses which have been advanced, equating these granules with nuclei, mitochondria, reserve food granules, etc., etc., can be valid, since each of them comprises the major portion of an entire cell [Bisset (*16*)]. Some of these problems, as applied to *Corynebacterium* and *Mycobacterium*, are discussed at length elsewhere [Bisset (*10*)].

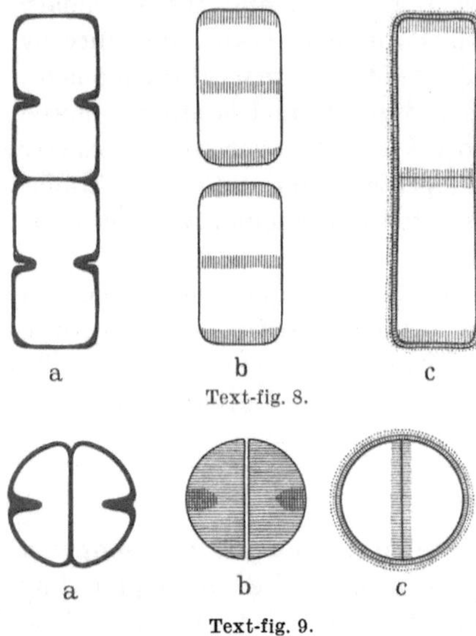

Text-fig. 8 and 9. Stained appearances in a septate bacillus and coccus. *a* Cell wall stain, *b* simple aqueous dye, *c* false cell wall staining by aggregation. (See text)

A fairly accurate demonstration of the bacterial cell wall is capable of being achieved rather more readily in Gram-positive than in Gram-negative bacteria. And since the former group includes most of the multicellular bacteria, this eases somewhat the problem of investigating their structure. There is not, so far as is known, any difference between the fundamental requirements for the staining of the cell walls of unicellular and multicellular bacteria, but in the following discussion it may be assumed, unless otherwise stated, that the multicellular forms stain more easily.

There exist simple dyes whereby the cell wall of bacteria can be demonstrated without any previous mordanting process. Examples of these are Janus green (Hale, personal communication) and Alcian blue [Tomcsik and Grace (*92*)]. In this laboratory we have found that, for example, septate cocci and *Azotobacter* are more readily stained by such simple methods than are the majority of common species.

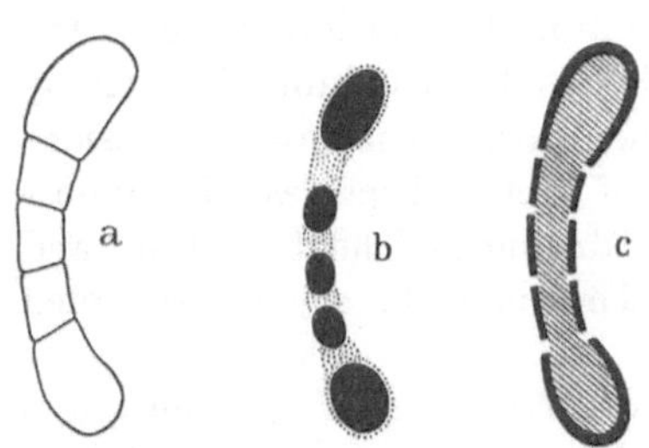

Text-fig. 10. Stained appearances in a *Corynebacterium*. *a* Cell wall stain, *b* 'diagnostic' appearance, showing granules, *c* Gram's stain

The best-known methods of staining the cell wall are the tannic-acid crystal-violet method of Eisenberg (*41*), Gutstein (*46*) and Robinow (*81*) and the phosphomolybdic-acid methyl-green method of Hale (*47*). The second is by far the better of the two, but both appear to depend upon similar principles. The mordant, whether tannic or phosphomolybdic acid, has the dual effect of denaturing and at least partially destroying the cytoplasmic proteins which, because of their much greater affinity for stain, may tend to obscure the complexities of the cell wall, and also of forming with the cell wall itself a stainable complex of some sort, (i.e. mordanting, as usually understood).

The formation of such stainable complexes can be, upon occasion, a source of error in the staining of cell walls, if their demonstration depends exclusively upon such a reaction. This is claimed by GIRBARDT and TAUBENECK (*44*) to be true of the congo-red cell wall stain of CHANCE (*34*), and it may apply to that of YOSHIDA, TANAKA *et al.* (*110*) which makes use of the same reagent. In these cases, as the present writer has already pointed out [BISSET (*17*)], and as GIRBARDT and TAUBENECK have now proved more conclusively, the effect is not truly to stain the cell wall, but to accrete upon it "flocculation membranes" which render it visible in outline (Text-fig. 8, 9), and in such cases, usually only the most mature septa are demonstrated.

Electron microscopy of the intact cell has proved singularly disappointing as a method of demonstrating the cell wall structure of multicellular bacilli and cocci. It has, in fact, proved positively misleading in those cases where it has been employed without the controls provided by adequate parallel preparations made by other methods. The reason for this is, of course, the very low penetrative powers of the electron beam. Much more information has been obtained by the electron microscopy of disrupted cell walls [ANGELICO *et al.* (*1*), DAWSON and STERN (*38*)], but even in this case it is apparent that only the most mature cross-walls are in fact rendered visible, and the existence of immature cell wall structure or membranous septa, too soft to stand up to the effects of desiccation *in vacuo*, which are inseparable from electron microscopy, may pass unsuspected. However, such softer structures do upon occasion appear in electron micrographs of disrupted cell walls [SCHULMAN *et al.* (*87*), YOSHIDA *et al.* (*107*)].

More advanced techniques for overcoming the low penetration of the electron beam, such as "adhesion partitioning" [BACKUS (*2*)] and sectioning [CHAPMAN and HILLIER (*35*), BIRCH-ANDERSEN, MAALØE and SJÖSTRAND (*6*), BIRCH-ANDERSEN (*5*), BRIEGER and GLAUERT (*32*), BRADFIELD (*31*)], have proved curiously erratic in their results. Of the papers upon electron microscopy of sectioned bacteria just quoted, only the first has really produced useful information about the structure of multicellular bacteria. One of the main problems in this connection is the necessity for increasing the contrast in these very thin sections by the use of fixatives, most of which are protein precipitants. BISSET (*21*) discussed this problem, and compared the appearance of the nuclei and cross-walls in septate bacteria which had been subjected to the action of osmium tetroxide solutions with the same structures under conditions of minimal treatment. MAALØE and BIRCH-ANDERSEN (*65*) state that buffered osmium tetroxide solutions are generally accepted as giving excellent preservation of ultrastructures, but give it as their own opinion that equally good contrast is obtainable by the use of formaldehyde as a fixative. These authors, who are probably among the most experienced in this field, thus imply, perhaps unconsciously, that they do not entirely concur with the general acceptance which they quote. It in the present writer's opinion [BISSET (*20*, *21*)] that such protein precipitants as osmium tetroxide and formaldehyde must of necessity be liable to produce distortion in the exceedingly tiny structures which are described by workers using this type of technique; and evidence of such distortion has been presented in the works quoted. There is no doubt whatever that sectioning techniques, as applied to electron microscopy, hold out the greatest hope for future advance in bacterial cytology, but it is also true that there is room for a great deal of technical improvement in the methods employed at present.

An interesting parallel to the use of a heavy metallic "stain" for electron microscopy occurs in the work of Yoshida *et al.* (*106*), who employed phosphotungstic acid to produce an electron dense complex on the cell wall. As described in Section 3, this chemical reacted preferentially with the protein component of the cell wall, and this information is of some interest insofar as the theory of action both of osmium tetroxide as an electron "stain", and of phosphomolybdic acid as a mordant for cell wall stains are concerned.

In view of the importance of this problem of the effect of fixatives and dehydration techniques [*cf.* also Bisset (*16, 21*)] the development of what may be termed biological methods for the demonstration of the cell envelopes is of great value. Such methods are very various. One of the most interesting is that of Newton (*74*) who employed a fluorescent derivative of polymyxin for the demonstration of the cell membrane of *Bacillus megaterium*. The original purpose of the experiment was the location of the site of action of the antibiotic; but the result was a demonstration in the living bacillus of the remarkable cellular complexity of the membrane (Plate II, Fig. 10).

However, one of the biggest advances in this field has been provided by the work of Tomcsik (*91*), Tomcsik and Guex-Holzer (*93, 94, 95, 96, 97, 98*) and Tomcsik and Grace (*92*). Tomcsik's methods arose in a similar manner to that of Newton, as part of an experiment designed to determine the location of certain chemically-defined antigens and the sites of action of the corresponding antibodies in the bacterial cell, as seen by phase-contrast microscopy. But it was immediately apparent that these reactions caused so striking an alteration in refractility of the cell envelopes that they provided a most important method for the demonstration of chemically definable constituents of the cell (Plate II, Fig. 17). The degree of complexity of the capsule revealed by this method has already been discussed, in Section 3, and can only be described as quite startling. Tomcsik and his collaborators also employed enzymes and other proteins, which they found to act similarly as "stains" for phase-contrast microscopy [Tomcsik and Guex-Holzer (*95, 97*)] by entering into combination, selectively, with cell components, at appropriate p_H values. Not only the capsule, but also the cell wall and cell membrane of bacteria have been thus demonstrated in the living cell [Tomcsik and Guex-Holzer (*94*)]. But the application of the method is by no means limited to bacteria; it is equally possible to investigate the immunochemical structure of living creatures of every type, and it is to be hoped that such applications of the method will soon become general. That they will enormously increase in our general knowledge of cytochemistry is quite indubitable.

6. Multicellularity in relation to the systematics of bacteria

In almost all branches of biology it is axiomatic that the foundation of any system of classification must be morphology. And whereas the assistance which may be given in the elucidation of natural relationships by other types of information, for example the blood groups of man, is often very great, it has never been suggested seriously that the morphological principle is at fault. The example just quoted, of blood groups in man, is indeed a very exceptional example of the situation which occurs commonly in the study of bacterial systematics, wherein a morphologically indeterminable group can be to some extent classified by the use of antigenic analysis.

Such errors as have in the past been acknowledged in the systematics of plants and animals have for the most part resulted not from any failure of morphology to provide adequate information, but rather from the failure of the systematists to utilise the information correctly.

The situation in bacterial systematology is much closer to that which has just been described than is commonly realised. Not only is the study of bacterial

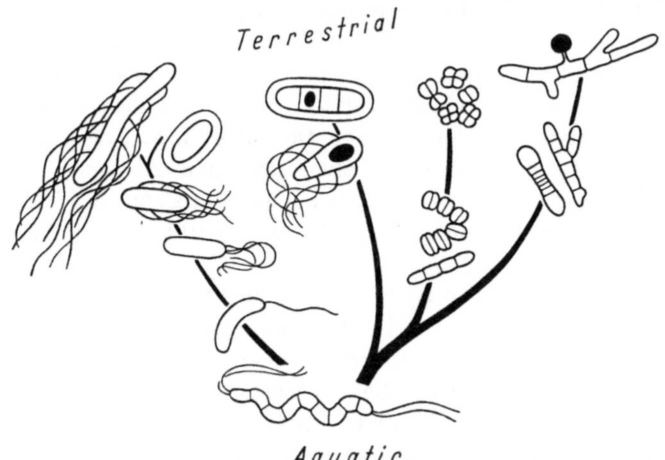

Text-fig. 11. A suggested scheme of evolution of the main bacterial types. As in other biological groups, the main evolutionary trend appears to be from an aquatic to a terrestrial mode of life. There are two main branches, the Gram-negative, unicellular (left) and the Gram-positive, multicellular (right). Both are presumed to derive from spiral, aquatic ancestors. The former evolve mainly by modification of flagellar pattern, the latter are much more diverse in their evolutionary trends. Most of them have lost their flagella and show a tendency to take advantage of the air as a measure of distribution

morphology capable of providing a very great deal of taxonomic information, but it is also capable of correcting errors which, in the past, have been founded upon inaccurate morphological studies.

In other branches of biology it has repeatedly been shown that even more important than the straightforward morphology of an organism is its life-history. To give an obvious example, the affiliations of a barnacle might, with very great difficulty, have been discerned from a study of the structure of its limbs. But they become immediately obvious when the nature and behaviour of the larva is realised. In the case of the ascidians, it is more than merely possible that their chordate relationships might never have been realised in the absence of any knowledge of their life-history. Thus it is all the more interesting that, of recent years, certain very distinct hints concerning the taxonomy of bacterial genera have been obtained from a study of their life-histories also.

In connection with the present study, it is important to attempt to discern whether the unicellular and multicellular bacteria are phylogenetically distinct, or whether closely related groups may contain examples of both. In fact, the former appears to be the case. Text-fig. 11 gives a diagrammatic representation of the relationships of the bacteria as the present writer [Bisset (*14*, *21*)] considers them to be. It will be observed that this scheme is markedly different from that of Kluyver and van Niel (*59*) which is well known, and which has been widely reproduced in text-books of systematic bacteriology. The reasons for this difference have been discussed at length elsewhere [Bisset (*22*)] and most of them are irrelevant to the present question of cell envelopes and multicellularity in bacteria. However, these questions are very cogent to the reason for the present rejection of Kluyver and van Niel's choice of a coccus, "the simplest form", as the ancestral morphological type for all bacteria. Simple unicellular cocci do indeed occur among the bacteria (Plate I, Figs. 3, 4) but they are not at all common in comparison with the complex, septate type of coccus, to which almost all Gram-positive forms belong (Plate I, Figs. 7, 8). Such septate cocci are very far from being either simple or primitive, and it is improbable that they are more closely related to the rod-shaped, flagellate, unicellular, Gram-negative *Bacteriaceae* than the latter are to the pseudomonads, from which they differ almost exclusively in the arrangement of their flagella. Yet it is a fundamental part of the scheme of Kluyver and van Niel that these two rod-like forms should be regarded as separately derived from such a coccus.

The flaw in their reasoning is, of course, due to their failure to recognise that the multicellularity of the cocci did, in fact, exist. And this is little or no reflection upon these experienced and competent authors, because, at the time when their work was written their beliefs were commonly shared.

It must be apparent to readers of this review that the symmetrical type of cell division found in these cocci is more probably indicative of a relationship with the symmetrically-dividing bacilli, the *Bacillaceae*, and that, granted the probability that terrestrial bacteria are derived from aquatic forms, both of these may be descended from a septate multicellular *Spirillum* (Plate II, Fig. 16). The evidence for this presumption is once more irrelevant to the major questions at present under discussion, but is presented in full in the papers already quoted. However, it may be appropriate at this point to remind readers of the fact that spirilla may be either Gram-positive or Gram-negative, septate or unicellular, and that according to Pijper (*78*), all rod-shaped bacteria are slightly spiral in form. It is thus a good deal more probable that a non-motile, slightly spiral bacillus should be descended from a motile, strongly spiral ancestor, by progressive loss of these characters on adaptation to a soil-dwelling existence, than that a soil-dwelling form should forecast an eventual return to an aquatic existence, on the part of its descendants, by the development of a slightly spiral morphology.

If, as seems probable, both Gram-negative unicellular, and Gram-positive multicellular bacteria are derived from spiral, aquatic ancestors, there is every reason to suppose that their origins were entirely separate, and that the two main groups of modern terrestrial bacteria are not closely related phylogenetically. It has already been suggested that the multicellular types have a cell wall which is much more readily stained by simple dyes than is that of the unicellular types, and it is known that considerable differences exist between the composition of these two types of cell wall (see Sections 3, 5).

Because the Gram stain is an arbitrary test, the correlation between Gram-positivity and multicellularity is not exact. But it is not difficult to discover whether apparently Gram-negative bacteria are truly of this phylogenetic group or whether they can be regarded as degenerate or specialised Gram-positives. And consideration of this possibility may help to explain a number of apparent anomalies. For example, GALE and TAYLOR (*43*) state that penicillin is highly effective against Gram-positive, but much less so against Gram-negative bacteria, but that the Gram-negative *Neisseria* are very sensitive. When this observation was recorded it was not yet generally realised that the resemblance between the cytological structure of a Gram-negative *Neisseria* and a Gram-positive *Staphylococcus* is very close indeed [BISSET (*14*)]. Both are subdivided by internal cross-walls, but in Gram stained preparations these are difficult to see (or at any rate, to notice) in the latter, and appear as gaps in the former, which accordingly is described as a "diplococcus". Both appearances are misleading, and both have the structure of septate cocci, as explained in this paper (Plate I, Figs. 7, 8; Text-fig. 9). It is thus apparent that the *Neisseria* are degenerate descendants of Gram-positive cocci, and their sensitivity to penicillin requires no further explanation. Indeed, it might be stated, with rather greater accuracy, that multicellular bacteria are penicillin sensitive, whereas unicellular bacteria are much less so.

A second problem of this type provides a remarkable illustration of the value of cytological studies in the elucidation of bacterial systematics. The three main groups of nitrogen fixing bacteria, apart from the autotrophic nitrifyers which appear to be pseudomonads [BISSET and GRACE (*24*)], are the aerobic *Rhizobium* and *Azotobacter*, and the anaerobic *Clostridium butyricum* group. According to published descriptions, the latter is a typical member of the multicellular, Gram-positive group, whereas *Rhizobium*, the symbiotic nitrogen-fixing bacterium, has normally been regarded as a rather aberrant, but perfectly definite member of the Gram-negative branch. *Azotobacter*, although Gram-negative, is sometimes likened to a yeast in appearance, and is notably protean in the main features of its morphology. These genera provide an example of the application to bacteria of the well-known biological principle that the most important evidence in the determination of the natural relationships of living organisms is obtainable from the life-cycle. A series of studies upon both *Rhizobium* and *Azotobacter* have made it apparent that these two genera, which are superficially unlike, and both of which are equally unlike *Clostridium* to the same superficial examination, have so much in common in their cytology and in the details of their life-cycles that it is apparent that all three of these major groups of nitrogen-fixing bacteria may well be derived from a relatively recent common ancestor [BISSET (*12, 14, 20*), BISSET & HALE (*25, 26*)]. It is not appropriate to this review to present all this evidence in detail, but in the photomicrographs of Plate II can be seen some suggestions of the structural characters which suggest that both *Rhizobium* and *Azotobacter* are really members of the multicellular group. Plate II, fig. 11, shows the cell walls of *Azotobacter* in very young culture, and Plate II, Fig. 13 illustrates the production of swarmers from the mother-cells of *Rhizobium*. The resemblance between these stages and such typical members of the symmetrical septate type as *Bacillus* (Plate II, Figs. 10, 12) or *Lactobacillus* (Plate II, Fig. 14) is very apparent. To which it need only be added that both of these genera show signs of Gram-positivity at different stages of their life-cycles, whereas *Clostridium* species, as is well known, are often weakly Gram-positive.

It is not only unsuspected resemblances between apparently dissimilar bacteria which can be revealed by this means. The converse is also true, in that bacterial types which are superficially alike may thus be distinguished and more accurately classified. It has already been shown that certain cocci may be unicellular and others septate [BISSET (*14*)] (Plate I, Figs. 3, 4, 7, 8), but more refined differences may also be discerned by a study of the cellular structure of septate forms. Plate II, Fig. 14 illustrates *Lactobacillus brevis*, which has just been quoted as an excellent example of a typically symmetrical septate bacillus. But not all species of this genus are morphologically alike. Some are very much shorter, with fewer cells, others filamentous with much longer cells [DAVIS, BISSET and HALE (*36*)]. These morphological distinctions are stable and diagnostic. They are of great assistance in the classification of the genus *Lactobacillus*. On the other hand the species which is miscalled "*Lactobacillus*" *bifidus* has an entirely different structure [HAYWARD, HALE and BISSET (*49*)] (Plate II, Fig. 15) and quite obviously bears a much closer resemblance to the asymmetrical group of septate bacteria, for example *Corynebacterium*. The strikingly multicellular bacillus from foot-rot in sheep (Plate 1, fig. 9) was originally described by THOMPSON (*89*) as a species of *Rhizobium*, but its remarkable structure, here illustrated, makes this improbable [BISSET and THOMPSON (*29*)]. In passing, it may be remarked that this organism (so far unclassified) is unique among bacillary forms in that its cells appear to divide both longitudinally and transversely, and that coccal elements actually bud off laterally from the rod. Whether or not this characteristic indicates an affinity with the septate cocci is dubious, however, since these elements are motile, which cocci very seldom are.

It will be observed from the suggested scheme of evolutionary relationship in Text-fig. 11 that the multicellular branch of the bacterial tree must be presumed to be much more complex in its philogenetic possibilities than the unicellular branch. The latter, figured on the right-hand side of the diagram, omits certain specialised sessile forms such as caulobacteria and chlamydobacteria. But these have much in common with vibrios and pseudomonads [BISSET and GRACE (*24*)]; the main line of descent of unicellular bacteria appears to be directly from *Vibrio*, through *Pseudomonas* and *Bacterium*, to such complex forms as *Proteus*. The major evolutionary change, so far as morphology is concerned, is from the polar, aquatic type of flagellation to the peritrichous flagellation or total loss of flagella associated with a terrestrial existence [BISSET (*22*)]. On the other hand the multicellular bacteria show every sign of being highly successful, and have diverged in all directions from the ancestral stock. Some have retained and some lost their flagella. In general form they range from spherical cocci to branched, mould-like *Actinomyces* and *Streptomyces*, and they have evolved a wide variety of different types of spore-like resting and distributive cells. Indeed, it is the present writer's opinion that both coccal morphology and the resistant endospore may well be adaptations to aerial distribution, like the more obvious distributive cells of *Streptomyces* [BISSET (*12*)].

We cannot determine the whole truth concerning to origin and nature of multicellularity in bacteria; but there is little doubt that those members of the group which possess this character have been both diverse and successful in their evolutionary development.

7. Summary

Although multicellular structure in bacteria can quite readily be discerned, when its existence is once realised, it has been and still is widely ignored, even among those workers who have an interest in other cytological structures in bacteria, such as nuclei. And it sometimes happens that for this reason their observations are less valuable than they would otherwise be. Attention is drawn especially to the misconceptions which can arise by too great a reliance upon a single technique. Many valuable techniques, such as nuclear stains and electron microscopy, give very little information concerning the bacterial cell envelopes. Where multicellularity exists unsuspected, components of more than one cell may be described as parts of a single cell, and structures which actually belong to those parts of the cell envelopes separating adjacent cells have often been confused with truly intracellular structures (Text-fig. 1).

Bacteria form part of the flagellate super-Kingdom, together with animals, plants and fungi, and are probably (in contrast with much accepted doctrine) less closely related to blue-green algae or viruses. Like the other members of the group, bacteria have their points of general resemblance with and specific difference from the remainder. In particular, the mode of division of the nucleus, although obeying fundamental rules, is in many respects atypical. Claims for a true mitosis in bacteria are not accepted. The blepharoplast is exceptional; and the multicellular forms produce cross-walls by centripetal ingrowth, unlike the majority either of plants and animals.

The degree of multicellularity found in bacteria varies enormously from genus to genus, and examples are illustrated in Plates I and II. Some cocci, actually the minority, are simple spheres, others are subdivided by cross-walls at right-angles into numerous tiny cells. Most rod-like bacteria have cross-walls only at right-angles to the long axis, but occasional types may have them also parallel to the main axis.

A second type of multicellularity is found in those bacteria which produce within themselves large numbers of tiny reproductive cells or gonidia. Some, for example *Rhizobium* show both types of multicellularity at one time.

The composition of the cell envelopes in typical multicellular bacteria, most of which are Gram-positive, differs considerably in its chemical nature from that in the unicellular genera. There is evidence of differences in nucleoprotein, lipid and carbohydrate composition. The cell wall and the underlying cell membrane both appear to be chemically complex; and recent researches have proved, rather surprisingly, that this is true of the capsule also, where it exists.

Cell division and fission of a multicellular micro-organism are not, of course synonymous. Fission may take place only when cell division has produced from two or four to an indefinite number of cells. However, the two are well co-ordinated in typically symmetrical septate bacilli and cocci, which usually have one complete cell division, an incipient point of fission, across the centre, and are subdivided in a regular manner by less mature cell divisions. The main development of the cell envelopes, including the capsule, takes place at the junctions of cell wall and cross-wall. An analogous condition occurs in the unicellular bacteria, where the main growth takes place at one or both poles. These poles and the growth areas at the cell wall junctions both stain well with basic dyes, and have been the cause of numerous misconceptions (Text-fig. 1). The active areas probably appertain to the cell membrane rather than the to cell wall proper, which is secreted by them.

Irregular fission, producing cells of varying sizes, and the lateral branching of bud-like daughter cells are found in some multicellular groups of bacteria. Permanent branching is rare, except among *Streptomyces*.

The cell envelopes may be demonstrated by a variety of methods, and the multicellular bacteria are easier than most unicellular types to stain in this manner. Certain simple dyes are capable of giving a direct demonstration of the cell wall, but most will do so only after mordanting with such agents as tannic or phosphomolybdic acid. But most of the well-known diagnostic staining procedures give only a very misleading picture of the cellular organisation of bacteria.

Electron microscopy is also unreliable as a method of determining the complexities of bacterial cell envelopes. Although both cross-walls and cell membranes may occasionally appear in electron micrographs, especially of disrupted material, their appearance is unpredictable and unreliable in the absence of controls.

Electron microscopy of ultra-thin sections of bacteria is a promising technique; the results achieved so far must be interpreted with great caution, because of the technically hazardous procedures to which the material is subjected, including treatment with protein precipitants.

By far the most valuable techniques which have recently been evolved are the biological staining methods employing fluorescent or phase-contrast microscopy. The cell envelopes have been demonstrated in the living state by the methods of Tomcsik and of Newton, and have proved in each case to be of an even greater degree of complexity than had been suspected. The methods of Tomcsik in particular, which enable immunochemically determinable structures of all kinds to be visualised, are of enormous potential value in all branches of cytology.

The cytological structure of multicellular bacteria is capable of providing much interesting information in respect of their taxonomy and natural relationships. It appears probable that most, if not all multicellular bacteria are members of the Gram-positive group, even in the case of those which have lost their power to retain the Gram stain; for example *Neisseria* and *Azotobacter*. Conversely, truly unicellular bacteria are almost if not quite invariably Gram-negative.

These two branches of the evolutionary tree of bacteria may be derived independently from spiral aquatic bacteria, which are intermediate between them in their main characters. The Gram-negative line leads directly, and with relatively little specialisation through *Vibrio*, *Pseudomonas* and *Bacterium* to such highly evolved terrestrial forms as *Proteus*. The Gram-positive line, possibly by virtue of its retention of the multicellular structure found in certain of the ancestral spirilla, has proved capable of very considerable variation and specialisation towards the evolutionary goal of terrestrial colonisation and aerial distribution.

References

1. Angelico, R., A. Calo, A. D'Amore, O. Mariani-Marelli e F. Scanga: Ricerche sull'idrazide dell'acido isonicotino. Ren. Istituto Superiore di Sanità **15**, 627 (1952).
2. Backus, R. C.: Adhesion partitioning: Intrasomatic observations on normal *Escherichia coli* and T 2 bacteriophage. J. Biophys. & Biochem. Cytology **1**, 99 (1955).
3. Bartholomew, J. W., and T. Mittwer: Cell structure in relation to the Gram reaction as shown during lysis of *Bacillus subtilis*. J. Gen. Microbiol. **5**, 39 (1951).
4. Bergersen, F. J.: Cytological changes induced in *Bacterium coli* by chloramphenicol. J. Gen. Microbiol. **9**, 353 (1953).
5. Birch-Andersen, A.: Reconstruction of the nuclear sites of *Salmonella typhimurium* from electron micrographs of serial sections. J. Gen. Microbiol. **13**, 327 (1955).
6. — O. Maaløe and F. S. Sjöstrand: High resolution electron micrographs of sections of *E. coli*. Biochim. et Biophysica Acta **12**, 395 (1953).
7. Bisset, K. A.: The structure of "rough" and "smooth" colonies. J. of Path. **47**, 223 (1938).
8. — The mode of growth of bacterial colonies. J. of Path. **48**, 427 (1939).
9. — The cytology of smooth and rough variation in bacteria. J. Gen. Microbiol. **2**, 83 (1948).
10. — Observation upon the cytology of corynebacteria and mycobacteria. J. Gen. Microbiol. **3**, 93 (1949).
11. — The cytology and life-history of bacteria, 1st edit. Edinburgh: Livingstone 1950.
12. — Evolution in bacteria and the significance of the bacterial spore. Nature (Lond.) **166**, 431 (1950).
13. — The development of surface structures in dividing bacteria. J. Gen. Microbiol. **5**, 155 (1951).
14. — Bacteria. Edinburgh: Livingstone 1952.
15. — Some relationships between morphology and antigenic structure in bacteria. Nature (Lond.) **171**, 1118 (1953).
16. — Do bacteria have mitotic spindles, fusion tubes and mitochondria? J. Gen. Microbiol. **8**, 50 (1953).
17. — Bacterial cell envelopes. Symp. Citologia Batterica. Rome: Paterno 1953.
18. — The production of stainable granules in bacteria by the adsorbtion of DNA upon the cell envelopes. Exper. Cell Res. **7**, 232 (1954).
19. — The cytology of *Micrococcus cryophilus*. J. Bacter. **67**, 41 (1954).
20. — Evidence from the cytology of *Azotobacter chroococcum* of a relationship with *Rhizobium* and the Bacillaceae. J. Gen. Microbiol. **13**, 442 (1955).
21. — The cytology and life-history of bacteria, 2nd edit. Edinburgh: Livingstone 1955.
22. — The value of cytological studies in elucidating natural relationships among bacteria. J. Gen. Microbiol. **12**, 325 (1955).
23. — Cellular organisation in bacteria. VI. Symp. Bacterial Anatomy, Soc. Gen. Microbiol. Cambridge Univ. Press 1956.
24. —, and J. Grace: The nature and relationships of autotrophic bacteria. IV. Symp. Autotrophic micro-organisms, Soc. Gen. Microbiol. Cambridge Univ. Press 1954.
25. —, and C. M. F. Hale: The production of swarmers in *Rhizobium* spp. J. Gen. Microbiol. **5**, 592 (1951).
26. — — The cytology and life-cycle of *Azotobacter chroococcum*. J. Gen. Microbiol. **8**, 442 (1953).
27. — — Complex cellular structure in bacteria. Exper. Cell. Res. **5**, 449 (1953).
28. —, and F. W. Moore: The relationship of certain branched bacterial genera. J. Gen. Microbiol. **3**, 387 (1949).
29. —, and R. E. M. Thompson: A further short note on the morphology of an organism isolated from strawberry foot-rot. J. of Path. (in the press).
30. Booij, H. L.: The protoplasmic membrane regarded as a lipo-protein complex. Discuss. Faraday Soc. **6**, 143 (1949).
31. Bradfield, J. R. G.: Organisation of bacterial cytoplasm. VI. Symp. Bacterial Anatomy, Soc. Gen. Microbiol. Cambridge Univ. Press 1956.

32. Brieger, E. M., and A. M. Glauert: The demonstration of tubercle bacilli in ultra-thin sections of infected tissue by electron microscopy. J. Brit. Tbc. Assoc. **35**, 80 (1954).
33. Burdon, K. L.: Fatty material in bacteria and fungi revealed by staining fixed slide preparations. J. Bacter. **52**, 665 (1946).
34. Chance, H. L.: A bacterial cell wall stain. Stain Technol. **28**, 205 (1953).
35. Chapman, G. B., and J. Hillier: Electron microscopy of ultra-thin sections of bacteria. J. Bacter. **66**, 362 (1953).
36. Davis, G. H. G., K. A. Bisset and C. M. F. Hale: Correlation between morphological and physiological characters in the classification of members of the genus *Lactobacillus*. J. Gen. Microbiol. **13**, 68 (1955).
37. Davis, J. C., and S. Mudd: The cytology of a strain of *Corynebacterium diphtheriae*. J. Bacter. **69**, 413 (1955).
38. Dawson, I. M., and H. Stern: Structure in the bacterial cell wall during cell division. Biochim. et Biophysica Acta **13**, 31 (1954).
39. DeLamater, E. D.: Cytology of bacteria. Part II. The bacterial nucleus. Annual Rev. Microbiol. **8**, 23 (1954).
40. —, and S. Mudd: The occurrence of mitosis in the vegetative phase of *Bacillus megatherium*. Exper. Cell Res. **11**, 499 (1951).
41. Eisenberg, P.: Weitere Methoden zur Darstellung des Ektoplasmas. Zbl. Bakter. I **53**, 481 (1910).
42. Fischer, R., and P. Larose: Contribution on the behaviour and structure of the cytoplasmic membrane of bacteria. Canad. J. Med. Sci. **30**, 86 (1952).
43. Gale, E. F., and S. Taylor: The assimilation of amino-acids by bacteria. J. Gen. Microbiol. **1**, 314 (1947).
44. Girbardt, M., and U. Taubeneck: Zur Frage der Zellwandfärbung bei Bakterien. Zbl. Bakter. I **162**, 310 (1955).
45. Grace, J. B.: Some observations on the flagella and blepharoplasts of *Spirillum* and *Vibrio* spp. J. Gen. Microbiol. **10**, 325 (1954).
46. Gutstein, M.: Das Ektoplasma der Bakterien. Zbl. Bakter. I **93**, 393 (1924).
47. Hale, C. M. F.: The use of phosphomolybdic acid in the mordanting of bacterial cell walls. Labor. Practice **2**, 115 (1953).
48. Harris, J. E.: Structure and function in the living cell. New Biology **5**, 26 (1948).
49. Hayward, A. C., C. M. F. Hale and K. A. Bisset: The morphology and relationships of "*Lactobacillus*" *bifidus*. J. Gen. Microbiol. **13**, 292 (1955).
50. Henry, H., and M. Stacey: Histochemistry of the Gram-staining reaction for micro-organisms. Nature (Lond.) **151**, 671 (1943).
51. Henry, H., M. Stacey and E. G. Teece: Nature of the Gram-positive complex in micro-organisms. Nature (Lond.) **156**, 720 (1945).
52. Hewitt, L. F.: Cell structure of *Corynebacterium diphtheriae*. J. Gen. Microbiol. **5**, 287 (1951).
53. Hoffman, H.: Localization of bacterial nucleic acids and mechanism of the Gram reaction. Nature (Lond.) **168**, 464 (1951).
54. Holdsworth, E. S.: The nature of the cell wall of *Corynebacterium diphtheriae*. Biochim. et Biophysica Acta 8, 110; **9**, 19 (1952).
55. Houwink, A. L.: A macromolecular mono-layer in the cell wall of *Spirillum* sp. Biochim. et Biophysica Acta **10**, 360 (1953).
56. Hunter, M. E.: Studies on the nucleus of giant cells of *Micrococcus cryophilus*. Exper. Cell Res. **9**, 221 (1955).
57. Hurst, H.: An electron diffraction study of the structure and orientation of the lipids in yeast and bacterial cell walls. J. of Exper. Biol. **29**, 30 (1952).
58. Järvi, O., and A. Levanto: On the structure of bacterial cells, as seen by the use of the histochemical polysaccharide tests. Acta path. scand. (Københ.) **27**, 473 (1950).
59. Kluyver, A. J., and C. B. van Niel: Prospects for a natural classification of bacteria. Zbl. Bakter. II **94**, 369 (1936).
60. Knaysi, G.: Cell structure and division of *Bacillus subtilis*. J. Bacter. **19**, 113 (1930).
61. Krieg, A.: Nachweis von Kernäquivalenten bei Bakterien in vivo. III. Mikrokokken. Z. Hyg. **139**, 61 (1954).

62. LABAW, L. W., and V. M. MOSLEY: Periodic structure in the flagella and cell walls of a bacterium. Biochim. et Biophysica Acta **15**, 325 (1954).
63. LARK, K. G., O. MAALØE and O. ROSTOCK: Cytological studies of nuclear division in *Salmonella typhimurium*. J. Gen. Microbiol. **13**, 318 (1955).
64. LEDERBERG, J.: Gene combinations and linked segregation in *Escherichia coli*. Genetics **32**, 505 (1947).
65. MAALØE, O., and A. BIRCH-ANDERSEN: On the organisation of the "nuclear material" in *Salmonella typhimurium*. VI. Symp. Bacterial Anatomy, Soc. Gen. Microbiol. Cambridge Univ. Press 1956.
66. MARSHAK, A.: Bacterial cytology. Internat. Rev. Cytology **4**, 103 (1955).
67. MCGREGOR, J. F.: Nuclear division and life-cycle in *Streptomyces* sp. J. Gen. Microbiol. **11**, 52 (1954).
68. MCQUILLEN, K.: Capabilities of bacterial protoplasts. VI. Symp. Bacterial Anatomy, Soc. Gen. Microbiol. Cambridge Univ. Press **1956**.
69. MITCHELL, P., and J. MOYLE: The Gram reaction and cell composition: nucleic acids and other phosphate fractions. J. Gen. Microbiol. **10**, 533 (1954).
70. MORRIS, E. O.: The life-cycle of *Actinomyces bovis*. J. of Hyg. **49**, 46 (1951).
71. MUDD, S.: Trends and perspectives of bacterial cytology. Symp. Citologia Batterica Rome: Paterno 1953.
72. —, and T. F. ANDERSON: Demonstration by the electron microscope of the combination of antibodies with flagellar and somatic antigens. J. of Immun. **42**, 251 (1941).
73. —, and L. C. WINTERSHEID: A note concerning bacterial cell walls mitochondria and nuclei. Exper. Cell Res. **5**, 251 (1953).
74. NEWTON, B. A.: A fluorescent derivative of polymyxin: its preparation and use in studying the site of action of the antibiotic. J. Gen. Microbiol. **12**, 226 (1955).
75. PEASE, P. E.: Some observations upon the development and mode of attachment of the flagella in *Vibrio* and *Spirillum* spp. Exper. Cell Res. **10**, 234 (1956).
76. — The gonidial stages in *Spirillum* and *Vibrio*. J. Gen. Microbiol. **14**, No 3 (1956).
77. PENNINGTON, D.: The use of periodate in microbiological staining. J. Bacter. **57**, 163 (1949).
78. PIJPER, A.: Shape and motility of bacteria. J. of Path. **58**, 325 (1946).
79. — Dark-ground studies of flagellar and somatic agglutination of *B. typhosus*. J. of Path. **47**, 1 (1938).
80. ROBINOW, C. F.: A study of the nuclear apparatus of bacteria. Proc. Roy. Soc. Lond., Ser. B **130**, 299 (1942).
81. — *Addendum to:* The Bacterial Cell. R. J. DUBOS. Harvard Univ. Press 1945.
82. —, and R. G. E. MURRAY: The differentiation of cell wall, cytoplasmic membrane and cytoplasm of Gram positive bacteria by selective staining. Exper. Cell Res. **4**, 390 (1953).
83. RYAN, F. J., and L. K. WAINWRIGHT: Nuclear segregation and the growth of clones of spontaneous mutants of bacteria. J. Gen. Microbiol. **11**, 364 (1954).
84. SALTON, M. R. J.: The nature of the cell walls of some Gram-positive and Gram-negative bacteria. Biochim. et Biophysica Acta **9**, 344 (1952).
85. —, and R. W. HORNE: Studies of the bacterial cell wall. Biochim. et Biophysica Acta **7**, 19, 177 (1951).
86. —, and R. C. WILLIAMS: Electron microscopy of the cell walls of *Bacillus megaterium* and *Rhodospirillum rubrum*. Biochim. et Biophysica Acta **14**, 455 (1954).
87. SCHULMAN, J. H., B. A. PETHICA, A. V. FEW and M. R. J. SALTON: The physical chemistry of haemolysis and bacteriolysis by surface active agents and antibiotics. Progr. Biophysics a. Biophysical Chem. **5**, 41 (1955).
88. STACEY, M.: Chemistry of the Gram-staining process. Nature (Lond.) **176**, 1145 (1955).
89. THOMPSON, P. E. M.: A species of *Rhizobium* isolated from strawberry foot rot in sheep. J. of Path. **48**, 445 (1954).
90. THORNTON, H. G., and N. GANGULEE: The life-cycle of the nodule organism. Proc. Roy. Soc. Lond., Ser. B, **99**, 427 (1926).
91. TOMCSIK, J.: Complex structure of the bacterial capsule in the genus *Bacillus*. Experientia (Basel) **7**, 459 (1951).

92. Tomcsik, J., and J. B. Grace: Bacterial cell walls as revealed by the specific cell wall reaction and by direct staining with Alcian blue. J. Gen. Microbiol. **13**, 105 (1955).
93. —, and S. Guex-Holzer: Anthrax-Polypeptide und andere spezies-spezifische Substanzen der Kapsel in der Bazillus-Gruppe. Schweiz. Z. Path. u. Bakter. **14**, 515 (1951).
94. — — Änderung der Struktur der Bakterienzelle im Verlauf der Lysozym-Einwirkung. Schweiz. Z. Path. u. Bakter. **15**, 516 (1952).
95. — — Ein neues Prinzip zur Färbung der Bakterienkapsel. Schweiz. Z. Path. u. Bakter. **16**, 882 (1953).
96. — — Genese der komplexen Kapselstruktur. Schweiz. Z. Path. u. Bakter. **17**, 221 (1954).
97. — — Demonstration of the bacterial capsule by means of a p_H-dependent, salt-like combination with proteins. J. Gen. Microbiol. **10**, 97 (1954).
98. — — A specific cell-wall reaction in *Bacillus* sp. J. Gen. Microbiol. **10**, 317 (1954).
99. Iterson, W. van: Some remarks on the present state of our knowledge of bacterial flagellation. Symp. Citologia Batterica. Rome: Paterno 1953.
100. Vendrèly, R.: Histochemistry of bacteria. Internat. Rev. Cytology **4**, 115 (1955).
101. Webb, R. B. and J. B. Clark: Cell division in *Micrococcus pyogenes*. J. Bacter. **67**, 94 (1954).
102. Weibull, C.: Observations on the staining of *Bacillus megaterium* with triphenyltetrazolium. J. Bacter. **66**, 137 (1953).
103. — The isolation of protoplasts from *Bacillus megaterium* by controlled treatment with lysozyme. J. Bacter. **66**, 688 (1953).
104. Witkin, E. M.: Nuclear segregation and the delayed appearance of induced mutants in *Escherichia coli*. Cold Spring Harbor Symp. Quant. Biol. **16**, 357 (1951).
105. Wong, S. C., and T. Tung: Further studies on the type-specific protein of *Corynebacterium diphtheriae*. Proc. Soc. Exper. Biol. a. Med. **43**, 749 (1940).
106. Yoshida, N., I. Fukuya, I. Kakutani, S. Tanaka, J. Tegawa and T. Hashimoto: Electron staining of the cell wall of *B. subtilis* by phosphotungstic acid. Tokushima J. Exper. Med. **1**, 8 (1954).
107. — — — — K. Takaishi, K. Nishino and J. Tegawa: Electron microscope studies of the formation of the bacterial septa. Jap. J. Bacter. **9**, 1059 (1954).
108. — — — — — — — Electron microscopic studies of the incompletely hydrolysed cell wall of *E. coli* B strain. Jap. J. Electron Micr. **6**, 5 (1954).
109. — — — — — — T. Hashimoto and J. Tegawa: Electron microscope studies of the cell wall of *Salmonella typhi* H-901 and the preliminary experiment for the morphological studies of the changes in the cell wall induced by the antigen-antibody reaction. Tokushima J. Exper. Med. **1**, 105 (1954).
110. — S. Tanaka, K. Nishino, I. Fukuya, K. Takaishi, I. Kakutani and T. Hashimoto: New techniques for staining the bacterial cell wall and septum. Tokushima J. Exper. Med. **1**, 153 (1954).
111. — — K. Takaishi, I. Fukuya, K. Nishino, I. Kakutani, S. Inoi, K. Fukui, A. Kono and T. Hashimoto: Studies on the role of the bacterial cell wall of *H. pertussis* an *V. cholerae* in immunochemical reactions. Tokushima J. Exper. Med. **2**, 11 (1955).

II. Bacterial flagella and motility

By

Adrianus Pijper-Pretoria[1]

With 38 figures

Table of contents

"There was, and still is, a tendency for morphologists to ascribe to organs and structures functional significance for which there was, or is, little observational evidence."

Gray, J.: Nature **173**, 227 (1954).

I. Introduction

Flagella, and especially bacterial flagella, have been thoroughly investigated during the past decade. Before that they were rather taken for granted. This awakening of interest was largely due to the electron microscope (hereafter

[1] Institute for Pathology, University of Pretoria, South Africa.

referred to as the EM) which made them more real, and revealed new features, without solving their mysteries. Perhaps also, as KNAYSI (*145*a) said, the doubts expressed by PIJPER (*233—251b*) as to the function and nature of bacterial flagella stimulated a good deal of much needed research. The entry of physicists like ASTBURY into this biological field radically affected the position and recalled D'ARCY THOMPSON's dictum (*302*) that the physicist will one day explain the properties of the flagellum.

This review attempts to present impartially all the available information, aiming at an unbiased but fearless assessment of the various contributions. Finality cannot be reached yet as many questions remain open.

New discoveries often invest the work of older authors with a new significance. Quoting their papers here is not a show of erudition but is essential for understanding new developments. There also is a warning in these pioneering efforts. For decades the study of bacterial flagella has been under the spell of staining methods. This time we must beware of it falling under the spell of the EM.

It should by now be generally recognised, but is not, that "flagella" in microbiology cannot be regarded as homologous. It is not even true anymore of bacterial flagella. The various categories of flagella will here be treated separately.

MILES (*197*) in 1949 wrote an essay on the *status* of some arguments about the bacterial surface, including flagella. According to him logic is not a strong point with otherwise capable bacteriologists, and they tend to regard observations and experiments as crucial, when they are not. Some points can be added to MILES' essay. Discussions and arguments about bacterial flagella often suffer from lack of statistical insight, too often there is no proper statistical approach. Observations and conclusions often lack adequate numerical support. This is "illustrated" by one and the same felicitous photomicrograph turning up year after year in a variety of lectures, journals and books, not, as might be acceptable, to elucidate a particular point, but as the sole and therefore quite inadequate support and supposed proof of the correctness of some particular view. This reverence for what has been seen only once is not uncommon with EM pictures, perhaps because the huge magnifications hardly allow more than one microbe per print. KNAYSI (*143*) had this in mind when he did not accept as evidence one particular EM photograph because it was "an isolated observation in an uncontrolled experiment, and one cannot build science on that". HILLIER (*117*) similarly warned in 1950 that studies of morphological changes-with-time or as the result of interactions between different entities must be carried out statistically. In this review there will be opportunities for this kind of criticism. Of similar unscientific nature are the naïve and futile arguments used by certain authors. ZETTNOW (*344*) in all seriousness insisted that to avoid faulty conclusions one should make sure that bacteria are motile before starting to stain their flagella. KINGMA BOLTJES (*140*) said that bacteria knew that they had flagella and knew how to use them! NEUMANN (*214*) admired their intelligence when they plaited their individual flagella to get more powerful action in viscous media!

The word "flagellum" is here used without prejudice as to meaning. A "little whip" suits both an active and a passive function (*128*). "Cilium" is better reserved for the usually short definite motor organs as a rule implanted in great numbers on the surface of animal cells. For designation of bacterial species BERGEY's Manual (*44*) has been followed.

II. Survey of methods of investigation

a) General

Any observer's views are influenced by the technique available to him, and its defects. COHN (*61, 62*) with his bright-field microscopy could only see the substantial polar flagella of large spirilla and regarded them as typical for all bacteria. KOCH (*147*) going on from there stained flagella of similar bacteria and confirmed their polar attachment. LOEFFLER (*172*) developed staining methods and applied them to a variety of bacteria, including *Salmonella typhosa* which surprised him by showing large numbers of spiral threads often not attached to bacterial bodies. He did not dare to call them flagella (although their shape was "correct") but imagined they were derived somehow from the outer coat of the bacteria! The following year (*173*) he found similar structures in many preparations of various bacteria and reluctantly decided they were flagella, although (again quite unexpectedly) they often seemed to be attached to the sides of the bacteria and not polarly. And thus the idea of *peri*trichousness arose simply as the result of staining methods. It was codified by MESSEA (*192*). ZETTNOW, the past master of flagella staining became the arbiter as to what was a flagellum. His mantle later fell on LEIFSON (*160—167*). The efforts by REICHERT (*261*), NEUMANN (*213, 214, 215*), NEUMÜLLER (*216*) and PIETSCHMANN (*225, 226*) to revert to the study of flagella on live bacteria remained isolated.

b) Staining methods

Staining techniques are given in textbooks, a critical review was written by LASSEUR *et al.* (*159*). The devastating drawback of staining is the drying of the drop of bacterial suspension on a glass surface. Such drying proceeds from the periphery inwards. Watched with sunlight dark-ground microscopy [PIJPER (*227, 233, 248*)] the centripetal tide breaks off flagella, disfigures, bends, stretches and dislocates them. The dried drop presents chaotic conditions with no resemblance to the original state. This unavoidable havoc reduces stained pictures to artefacts and explains the windswept or even bedraggled appearance of so many photomicrographs of stained flagella (*139*).

c) Electron microscopy

For technical information there are several textbooks, one by WYCKOFF of 1949 (*341*) being specially useful to microbiologists, and a very complete technical review was written by SELBY (*276g*). All techniques include drying of drops, and so apart from the disturbances caused by the vacuum, there are the same ravages as with flagella staining. A recent improvement is due to HILLIER, KNAYSI and BAKER (*118*) who grew bacteria on top of a film of collodium placed on an agar surface. This "near-dry" technique lessens disturbances from drying, but cannot exclude movements of bacteria on the spot or over short distances.

Early EM pictures show very thin flagella, looking rather too feeble to act as motor organs (examples in publications antedating 1945). Since Williams and Wyckoff's invention (*333*) of metallic shadow casting to get a tridimensional effect was applied to bacteria their flagella have acquired more body and look more impressive.

d) Dark-ground microscopy

The advantage of dark-ground microscopy (hereafter referred to as D-GM) is continuous observation of live bacteria in natural liquid surroundings. Started by Reichert in his classical work of 1909 (*261*) it was continued by Neumann (*213*, *214*, *215*), Neumüller (*216*) and Pietschmann (*225*, *226*). All these authors however, unwittingly dealt with live artefacts. They suspended bacteria in colloid solutions, (1) because they reasoned that increased viscosity would slow down movement thus making flagella easier to see, (2) because they believed that bacteria to counteract viscosity would plait their flagella, and (3) because they thought that such conditions would alter the refraction of the medium and increase the contrast of the flagella. The colloid solutions were gelatin, gums, agar and others. These authors, except Pietschmann, were *a priori* convinced of the reality of peritrichousness in their bacteria. They took great pains to interpret their observations from that point of view. Pietschmann (*225*, *226*) denied the existence of peritrichousness and reduced all flagella-equipment to a sub-polar flagellum. The error that made all these observations unacceptable was brought out by Pijper who showed that the unexpected thickness of the flagella of these authors, which they ascribed to plaiting of numerous thinner (peritrichous) flagella, was in reality due to a precipitate of colloid particles which in conformity with the laws of colloid chemistry had descended on flagella and their bodies. At first, in careful preparations, the separate granules can still be seen as in Fig. 1 which shows *Salmonella typhosa* in a solution of gum arabic. Fig. 2 shows how these granules coalesce into a complete sheath which these authors mistook for a plait. Fig. 3 taken much later shows how most of the sheath has disappeared, leaving the "naked" flagellum with a few particles of gum attached which can often be seen rushing up and down the helix like a man running up and down spiral stairs (*227*, *229*, *232*, *233*, *234*, *237*). Pijper showed at the same time that these and other "naked" flagella can only be made visible by using sunlight for D-GM, and that by this sunlight technique flagella devoid of colloid coatings could not only be made visible, but also photographed and filmed (*228*—*251a*). Different bacteria were found to have affinities for different colloids. The genus *Proteus* coats itself readily and thickly with agar when grown on this medium, but *Salmonella typhosa* hardly does. The degree of coating varies, some strains of bacteria may come to look like caricatures of themselves. An unusually suitable colloid is methylcellulose [Pijper (*234*)], this when supplied sparingly provides a minimal coating just sufficient to improve visibility without becoming deleterious. This may be due to its structure (*296*).

The tendency for particles to become attached to flagella is illustrated also by the work of Sykes and Reed (*298*) who added boric acid to *B. proteus* to control swarming and found the flagella covered with particles. Dautheribes (*74*) did not agree with their findings.

e) Phase contrast microscopy

This method shares with D-GM the advantages of continuous observation of live bacteria in liquid, but has been of little use in the study of bacterial flagella. "Naked" appendages do not become visible with it, except those of the large *Spirilla* (*239*), which can also often be seen with bright-field microscopy. "Giant-flagella", as shown by PELUFFO (*223*), and the big flagellum seen by RICHARDS (*262*) on *Ashbya gossypii*, are in a similar position. FLEMING *et al.* (*94, 95*) used it for their studies of *Proteus vulgaris* on agar, where, as mentioned in section IId, the flagella were thickened by a coating of agar. KVITTINGEN (*153, 154*) used the same technique as FLEMING *et al.* for his studies on *Proteus* bacteria as also did STOKES (*294*) with *Sphaerotilus natans*. POETSCHKE and BOMMER (*253*) saw the hefty flagella of protozoal *Flagellates* and sometimes of *Spirilla* but failed with other bacteria. BURCIK and PLANKENHORN (52) combined drying, mordanting and then phase contrast microscopy, thus doing away with the natural advantages of the method, and claimed to get recognisable mono-, lopho- and peritrichous preparations. Of more promise is the Anoptral microscope produced by WILSKA (*335, 336*) which, in the hands of its inventor, gave photomicrographs of undetermined oral bacteria with flagella. One of these probably is *Selenomonas* but the others resemble real bacteria. Here too there probably was a colloid coating.

f) Chemical and physical investigations

The application of X-ray diffraction to bacterial flagella by ASTBURY and WEIBULL (*12*) and by WEIBULL (*324*) placed them in the m-e-k-f group of fibrous proteins [ASTBURY (*7, 8, 9, 10, 11, 14*)]. WEIBULL's chemical analyses (*318, 319, 320, 322, 325, 326*) demonstrated chiefly proteins in flagella from two different bacteria.

III. Different kinds of flagella and cilia

To understand bacterial flagella one must know other flagella. The early studies of flagella and cilia concentrated on the mechanical aspects of animal flagella [GRAY (*102a*)]. KRIJGSMAN as early as 1925 (*151*), impressed by their various patterns of motility, predicted a complicated structure. LOWNDES in a series of papers (*176, 177*) explained his discovery, made largely by high speed cinematography, that the flagellum of protozoa generates its own energy on the spot and that its wave of distortion passes from base to tip, increasing in latitude during its course. This was confirmed in extensive studies, including the building of a model and using himself as a test object by BROWN (*48*). Still dealing with animal flagella LUCAS (*178*) in 1932 described in the infusorium *Cyathodinum piriforme* both during cell division and during periods of rejuvenescence intracellular "Anlages" in the form of cysts. These arose in the body of the cell and cilia were found growing into them, complete with basal bodies. The cysts then moved to the surface and "evaginated", providing a ready-made new locomotor apparatus. Von MIHÁLIK (*196*) in 1935 independently found the same phenomenon in the epithelium of the Fallopian tubes of humans and rabbits.

All this was before the advent of the EM. The EM revealed the amazing fact that a wide variety of animal and botanical flagella and cilia are all built in the same complicated way. As BRADFIELD (*41*) put it: "They all contain two similar central sub-fibrils surrounded by a ring of nine fibrils different in size and chemical composition from the central pair." In a few cases 18 plus 2 fibrils have been found instead of 9 plus 2. All authors agree that there is a liquid or semi-liquid mass around these fibrils and this mass is surrounded by a membrane, so the fibrils are properly ensheathed and bathed in a liquid directly communicating with the cell. GRIGG and HODGE (*105*) working with fowl spermatozoa thought that the 9 fibrils were the motor organ, and the 2 central ones a rudimentary nervous system. These complex organelles called flagella and cilia have now been described in human spermatozoa by HODGE (*120*), in fowl spermatozoa by GRIGG and HODGE (*105*), in fern spermatozoids by MANTON and CLARKE (*187*), in fungi and yellow-green algae by MANTON, CLARKE, GREENWOOD and FLINT (*188*), in green algae by MANTON (*189*), in mussels and in *Paramecium* by POTTS and TOMLIN (*254*), in mussels by BRADFIELD (*41*) and in molluscs, amphibians, mice and man in ciliated epithelium by FAWCETT and PORTER (*91*, *92*). The two central fibrils are more sensitive and disappear easily [HODGE (*120*)]. This explains ASTBURY's finding of only 9 fibrils in one of his algae (*13*), and BROWN and COX (*49*) giving the uncertain number of 9—12 for flagellates and WU and MCKENZIE (*340*) finding only 9 in ram spermatozoa after treatment with sodium-hydroxide. Similarly WOHLFARTH-BOTTERMANN (*337*) saw only 10 fibres in *Ciliata* and KRUGER and WOHLFARTH-BOTTERMANN (*152*) described 10—12 in *Paramecium*. A different technique might have given other and more definite results, the last two authors prefer dried preparations for the EM instead of the carefully fixed material or ultra-thin sections which other authors use.

All these complicated organelles have a basal granule or blepharoblast about which, again, there is no doubt, also from the statistical view. The fibrils can be followed through the blepharoblasts or are seen in close contact with them [SEDAR, BEAMS and JANNEY (*276f*), WOHLFARTH-BOTTERMANN and PFEFFERKORN (338), FAWCETT and PORTER (*91*, *92*)].

In contrast to all these marvellous structures *bacterial* flagella have never shown anything remotely as elaborate. In the EM bacteria generally just show wavy fibrils, varying in thickness but otherwise usually looking much the same. BRADFIELD (*41*) suggested that bacterial flagella might consist of or might be single fibrills comparable to one of the component fibrils of these complex organelles. There is a large field here for agreeable speculation, but the facts in their simplicity are not very helpful. The question of basal granules or blepharoblasts will be discussed in following sections. It is true that DE ROBERTIS and FRANCHI in the EM saw a tightly coiled double helix in a bacterial flagellar thread, which they throught was constituted by two interlacing filaments (*268*, *269*, *270*). STARR and WILLIAMS (*288*) in the flagella of a motile diphtheroid bacillus found a left-handed triple-threaded helix with constant period. They do not think this appearance is due to intertwisting of finer fibrils because there were no frayed ends nor thinner fibrils. BRAUN (*42a*) described similar appearances in *Escherichia coli* and *Proteus vulgaris* and ascribed them to twisting of thinner elements. Thinner elements were present, but superposition is not easily excluded in such cases. LABAW and MOSLEY (*156*) in an unidentified motile peritrichous bacterium saw in flagellar threads an external contour of a counter-clockwise double helix, and later on (*157*) in *Brucella bronchiseptica* the external contour of a left-handed triple helix. Such findings are exceptional, and anyhow still a long way from the complex organelles described above. An attempt at describing a sheath surrounding bacterial flagella was made by DE ROBERTIS and FRANCHI (*268*, *270*) but they merely show one line in one trypsin-digested preparation and at the same time claim that the sheath is digested by trypsin. Recently, VAN ITERSON (*130*) with her EM photographed what she regarded as sheaths round flagella of *Vibrio metschnikovii*. The sheath only appeared in preparations that had undergone autolysis. Autolysis naturally brings a quantity of colloid material from broken down cells into the medium, and, again following colloid laws, these materials precipitate onto the flagellar threads, and so the appearance of a sheath may originate. Material similar to that constituting VAN ITERSON'S sheaths can be seen in other autolysed bacterial preparations such as those of KELLENBERGER and WERNER (*136*), WERNER and KELLENBERGER (*331*), BONIFAS, KELLENBERGER and WERNER (*37*), and HAPPOLD, JOHNSTONE, ROGERS and YOUATT (*108*), sometimes attached to flagellar threads in somewhat sheathlike fashion, and sometimes lying loose in the field, as also happens in VAN ITERSON'S pictures. A very good picture of autolysed bacterial cells with flagella and particles lying about, the particles sometimes covering the flagella like VAN ITERSON'S sheath was given by WILLIAMS and WYCKOFF (*333*) but they did not call them sheaths. Particles like those of VAN ITERSON'S sheath also occur in a picture of *Rickettsiae* by WYCKOFF (*341*).

Must one imagine that whilst other microorganisms need elaborate organelles for motor function, the bacterial cell makes do with such very much simpler equipment?

IV. Different kinds of bacterial flagella

The fact that there are different kinds of bacterial flagella complicates this review. To proceed in an orderly manner the differences had better be explained first.

Starting with the smaller groups, it will simplify matters to state that *Spirochaetes* probably have no flagella at all (section XII). Of the flagellar arrangements in *Vibrio* too little is known to ally them with any other group and they will be treated the same as the *Spirochaetes*, in a separate section (XI). A few doubtful bacteria of which the flagellar position or even the taxonomic position is not clear will be discussed in section XIII.

This leaves most of the *Eubacteriales*, and here comes the great divide! The *Spirilla* differ so much as regards flagellar apparatus from all the other *Eubacteriales*, of which in this respect *Salmonellae* may be regarded as the type, that they must be discussed separately (section XIV). Neglect or ignorance of these fundamental differences causes confusion and has made many discussions and criticisms somewhat idle. Table 1 gives the points of difference and many of these points will have to be elaborated upon in this review.

Table 1. *Differences between flagella of Spirilla and other motile Eubacteriales*

	Spirilla	Other *Eubacteriales*
Appearance during normal swimming	permanent clear-cut definite shape with characteristic graceful curve and sharp pointed end (Fig. 4)	fuzzy-looking straight tail, varying in size, often stiffening into clear-cut helix or variously bent rod, point never sharp (Figs. 5 and 6)
Occurrence	rarely absent	often absent
Consistence	looks hard and horny	looks soft
Attachment	by narrow neck, normally at pole	indistinct as to way and place
Relation to cell wall	dark-ground, autolysis and staining show continuity with cell wall (Fig. 7)	no continuity with cel lwall, rather with mucous coat
Staining	with cell wall stains	difficult and with flagella stains only
Visibility	with simple dark-ground equipment, with phase contrast, sometimes with bright field	in „naked" state practically with sunlight dark-ground only
Drying	often keeps its shape well	does not, neither as straight tail nor as helix
Changes in shape	do not occur normally	transition of straight tail into stiff helix or rod is common, with "biplicity"
Splitting	probably only after death into similarly shaped structures all remaining attached at same spot	splitting of helices is common, leading to "peritrichousness"
Breaking off from body	not easily, and no linking" nor "packing" into "giant-flagella", and do not become attached to strange bodies	easily, and "linking" and "packing" often give "giant-flagella", and often become attached to strange bodies, adding to "peritrichous" appearance
Agglutination	do not take part in H agglutination	effect H agglutination

V. Flagella of Eubacteriales, except Spirilla

a) Sunlight dark-ground microscopy

D-GM, technically difficult, is not much used. EM is near to ousting it. EM has higher magnification and resolution, but photographs artefacts of dried bacteria *in vacuo*. D-GM watches live bacteria swimming freely in liquid under conditions particularly favourable to bring out appendages. For problems of motility and flagellation D-GM should serve as a basis for the interpretation of EM pictures.

D-GM is an exacting method and requires skill and the best of equipment, and this also refers to slides and coverslips (*244*). In strong sunlight all scratches and particles of dust show up and spoil the picture. A satisfactory slide is made by gluing to an ordinary, but thin microscope slide, a strip of clear white mica of the same shape, by means of canada-balsam. Before use a thin layer of mica is split off with a knife, leaving a new surface free from dust and scratches. Coverslips on the whole are badly made (*244*), and have as a rule innumerable round microscopic blemishes or bubbles on one surface. This surface if lying directly on the preparation speckles it badly. If the blemishes are on top they cause a diffuse grey hue. Only one brand was found free of these blemishes. Slides and coverglasses of fused quartz show sharp lines in bright D-GM and are unsuitable (*244*).

The light source must be brilliant and small, and with its collector lenses requires more careful manipulation than in bright-field microscopy. The pointolite type of lamp is suitable but has low brilliancy. Arclamps are brighter but unsteady and high-pressure mercury-lamps usually have an unsuitably shaped light source and a short life. A special water-cooled arclamp once used by SIEDENTOPF (*280*) which could take 50 Amps was less brilliant than South African sunlight.

Artificial light-sources do not suffice for viewing the finer structures of bacterial flagella. Sunlight D-GM has been used in South Africa for over 25 years and it has demonstrated, photographed and filmed many novel flagellar structures and many novel phenomena in the motility of bacteria [PIJPER (*228—251c*)]. This sunlight technique has remained unique and in books and journals little notice has been taken of the sunlight aspects. Yet sunlight is the most suitable source of light on account of shape, brilliancy and freedom from deleterious ultra-violet radiations, in fact some bacteria are attracted by it and foregather in its focus. It is of course not everywhere so readily available as in Pretoria which moreover is about 5,000 feet above sea-level. EINTHOVEN (*82*) has shown that with suitable arrangements the thinnest thread can be made visible, though not necessarily resolved, in D-GM.

Sunlight D-GM requires a heliostat (coelostat). The author uses a simple Zeiss instrument (*228*, *230*, *232a*, *235*, *242*). It directs the beam of sunlight direct into the microscope (Fig. 8). Minor adjustments are made by moving the microscope stand or by altering the tilt for which proviston is made as can be seen in the photograph. The beam is focussed into the dark-ground condenser by a 2 metre collector lens attached to the long arm at its far end. Behind this lens is a Chance heat-absorbing filter. The black metal screen in front of the microscope shields the observer's eyes. Provision for light-filters is made on the long arm and on the little table behind the screen. The microscope is BECK's Radial Research model. The film camera (Movikon) slides on the same optical bench as the microscope, both on saddles. The heavy construction prevents vibration even at high camera speeds. For visual work the camera is taken off and the microscope pushed up. The camera has a beamsplitter as seen in the photograph. For still photography a Kontax camera is attached to the microscope tube.

For D-GM objectives with an iris are preferable, and essential for higher magnifications with strong light such as sunlight. For low magnification the Zeiss × 20 ap. 0.40 combined with a Zeiss K 20 eyepiece is very good. For higher magnifications, photography and filming the Zeiss immersion objectives ×35, ×40 (with iris) and ×60 (with iris) are very good, combined with Zeiss projection or photographic eyepieces. The advantages of filming motility phenomena, not just for record purposes but as a research method have been detailed (*235*), they hardly need enumeration at this time. Films also provide an unrivalled method of communicating one's findings on motility (*230*, *232*, *232a*, *233*, *235*, *236*, *237*, *238*, *239*, *241*, *242*, *245*, *247*, *248*, *251a* and *b*).

Motility and flagella should always be studied primarily in broth or water. The deleterious effects of adding colloids (and this includes growth on agar-media) have been mentioned in section IId. Methylcellulose is the least harmful

(*234*). It was introduced into microbiology by MARSLAND (*190*) with the idea that viscosity would quieten down *Paramecium*. He found it non-toxic, of low tonicityand quite harmless. This was confirmed by STILES and HAWKINS (*291*) who preferred it to agar, gelatin, tragacanth and quince jelly as not interfering with normal physiology. We have used it but we have never presented photographic phenomena from methylcellulose preparations which had not previously repeatedly been seen in broth or water.

The fundamental novelty arising from sunlight D-GM was that in motile *Eubacteriales* (except *Spirilla*) normal motility is associated with a straight fuzzy-looking tail as in Figs. 5, 6 and 17. The body, moving in a straight line, seems to drag this straight tail, and that this tail has no definite attachment becomes obvious when the body wriggles as it often does.

The only bacteria in which this tail has not been seen so far are *Pseudomonas* and *Azotobacter* [PIJPER *et al.* (*251*)]. It was first seen in *Salmonella typhosa* in 1930 (*227*) and first filmed in 1940 (*230*). It is only visible as such in liquid preparations, as by the time a preparation reaches the EM or staining procedures, it has changed into the traditional wavy flagella. The same species of bacteria can exhibit a long substantial tail as in Fig. 6, even in plain broth, without any additions, or the more modest appearance of Figs. 5 and 17. In another batch of broth it may show an extremely long but very thin hardly visible tail, or a very short thickset one, or no tail at all, and yet the body swims in the same way every time. The tail curves when the bacterium circles, straightening again when a straight course is taken. How different batches of broth affect the appearance or presence of the tail is unknown. Good motility is usually associated with good tails like Fig. 6, but very good motility often occurs without any visible tails at all, or with very thin or short ones. The best tails do not always accompany the best motility. Efforts to associate particular ways of making broth (extracting the meats or meat preparations with water, HCl, NaOH, pepsin, trypsin, or papain) gave no consistent results. Variations in temperature or salt content have no definite effect (*233*). A final p_H of 6.3 is beneficial. Simple peptone water usually gave very good motility and no visible tails at all. Additions of growth factors, vitamins, carbohydrates and amino-acids make no difference. The unsatisfactory position is that occasionally a batch of broth will give excellent motility and excellent tails, but so far on chemical and physical analysis no special features have been discovered in such a broth (*233*).

Normal bacteria during normal activity swim in straight lines for long distances, at good speeds, varying from about 5 to perhaps 20 or 30 μ per second. All they possess in the way of external motor organs is the fuzzy-looking tail. It has been suggested from various sides that there are additional latent motor organs, not visible with the sunlight equipment [BISSET (*30*)]. But it has been demonstrated (*247*, *249*, *251*) that the fuzzy tail has the capacity of metamorphosing into all the structures commonly called and recognised and described as flagella. These metamorphoses have been witnessed with sunlight D-GM on innumerable occasions (*247*, *249*, *251*) and have been filmed (*251a*). Fig. 9 has diagrammatical drawings illustrating what can happen to the tail. Drawings seemed preferable to save space and facilitate understanding. All events drawn in Fig. 9 have been seen very often in "naked" tails.

"Rod-shaped" bacteria do not, as is generally assumed, move as straight rods. REICHERT (*261*) strangely enough always drew his quite straight. PIJPER (*233*, *234*) first described what he called gyrating undulating movement which makes the bacillary body look like Fig. 10. It looks like wriggling. A better term is three dimensional undulation [GRAY (*102*)]. It reveals that the tail is not definitely attached, as shown in Fig. 6b and c and in a recent film (*251a*).

The fuzzy straight tail takes part in a number of phenomena (*227*, *233*, *237*). The bacterial body often reverses as in Fig. 9, d to j. The tail then appears to drift back along the body and rearrange itself into a new tail at the other end. More dramatic is the semi-somersault by the body, as in Fig. 9, to n. This was first described in 1930 (*227*) and then called a polar turnabout. The tail seems to wait for the other pole to come round to resume its tail function (*227*, *236*, *237*, *251*a). Both reversal and semi-somersault take place at very high speed, a feature apparently not understood by those who deny their importance (*140*).

The metamorphosis into helices, usually called flagella, is easily observed with sunlight D-GM (*247*, *251*, *251a*). The tail may also stiffen into a stiff rod, as in Fig. 11, often bent at an angle, as in Fig. 9, x to y, but this is rare (*248*). As a rule the fuzzy tail suddenly becomes a clear-cut helix, illustrated by Fig. 9, n to o, and by Fig. 12 which shows two consecutive frames from a 16 mm film, with an interval of 1/12th second, the camera running at that speed. This helical stiffening of the tail often produces helical flagella of two different wavelengths in one and the same culture, and then the one wavelength is always twice the other (*247*, *249*, *251*, *251a*). (The only exception so far is a strain of *Bacillus megaterium* where the relation was 3 to 1.) Wavelength means the distance from crest to crest, it is also called period or pitch. The various ways in which straight tails change into helical structures are illustrated in Fig. 9, n to w.

The first bacterium on which this metamorphosis of the tail into two kinds of helices, one twice the wavelength of the other, was observed were motile *Sarcinae* [PIJPER and ABRAHAM (*249*)]. Hundreds of measurements on photomicrographs like Fig. 15, made from wet slide-coverslip preparations by means of sunlight D-GM established this phenomenon for both *Sarcina ureae* and *Sarcina agilis*. For this phenomenon the term "biplicity" was proposed (*247*). It is illustrated in Figs. 13, 14, 18 and 19. As a result of this discovery "biplicity" was searched for and met with in some older publications, unwittingly produced in photographs or drawings from wet preparations by their authors, including WEI (*317*), NEUMANN (*214*), PIETSCHMANN (*226*) and PIJPER (*229*, *234*, *237*). It is now established in many bacteria (*251*).

Further measurements on numerous bacteria of wavelengths of flagellar helices (*251*) established the fact, already hinted at by WEIBULL (*325*) who had measured two, that most bacteria have a characteristic wavelength of their own, which was found to vary between 0.6 μ and about 5 μ. Table 2 gives the wavelengths of the helices of a number of bacteria, abstracted from the paper by PIJPER, NESER and ABRAHAM (*251*). The table also states where biplicity was found. It is likely that with longer searching it would have been found in more bacteria. It must be added at once that measurements of wavelengths are not a simple proposition. Several factors have to be considered (*251*). Both p_H and temperature affect wavelength. The longer wavelengths are more susceptible to these factors than the short ones. Rises in temperature increase wavelength, necessitating sometimes warming and in hot Pretoria often cooling, the microscope stage (*251c*). Wavelength is shortest at p_H 7.0 and increases both on the alkaline and acid side. Colloid substances in the suspending solution also have an effect when compared with measurements of naked helices. Methylcellulose however hardly affects wavelength. Going rough seems to make no difference. Details of technique of these measurements which were always performed on large numbers are found in recent papers (*247, 248, 249, 251*). Peluffo (*223*) doing measurements on unknown numbers of what probably were giant-flagella, and not knowing of all the disturbing factors mentioned here, found figures deviating from ours in *Pseudomonaceae*, *Salmonellae* and *B. proteus*.

Table 2. *Wavelengths of bacterial helical flagella in microns*

Name	Wave-length	S. D.	Biplicity
Pseudomonas diminuta . . .	0.60		
Serratia marcescens 2302 . .	0.965	0.037	
Chromobacterium ianthinum .	1.115	0.035	
Pseudomonas X	1.342	0.075	
Pseudomonas aeruginosa . .	1.530	0.086	
Caryophanon latum.	1.831	0.054	c. 0.9
Proteus mirabilis.	1.962	0.081	
Azotobacter agile	2.060	0.074	
Proteus vulgaris	2.098	0.092	c. 1.0
Bacillus subtilis	2.186	0.103	c. 1.1
Salmonella dublin (local) . .	2.279	0.087	
Salmonella schottmuelleri . .	2.289	0.064	c. 1.13
Salmonella typhi.	2.293	0.061	c. 1.15
Salmonella enteritidis. . . .	2.335	0.088	
Aerobacter aerogenes	2.336	0.114	
Salmonella paratyphi	2.340	0.078	
Serratia marcescens 2446 . .	2.346	0.080	
Salmonella typhi-murium . .	2.350	0.091	c. 1.2
Bacillus pumilus.	2.353	0.112	c. 1.2
Salmonella dublin (German)	2.355	0.066	
Bacillus cereus.	2.356	0.164	c. 1.25
Escherichia coli	2.366	0.121	c. 1.15
Salmonella hirschfeldii . . .	2.400	0.084	
Nocardia turbata	2.500	0.125	
Serratia marcescens 8706 . .	2.591	0.108	
Serratia kilensis	2.649	0.089	
Sarcina ureae	3.055	0.100	c. 1.5
Bacillus megaterium	3.389	0.166	c. 1.1
Azotobacter insigne	5.058	0.157	

From Table 2 it is clear that there is no obvious pattern in the distribution of helical wavelengths over the different groups of bacteria. No correlation between wavelength and other attributes was found. Big *Caryophanon* has relatively short wavelengths. *Bacillus megaterium* and *Bacillus cereus* both are big bacteria, but their wavelengths differ considerably. In *Serratia marcescens* and in *Pseudomonas* three different wavelengths occur, and two in *Azotobacter*. *Salmonellae* tend to have wavelengths round about 2.3 μ, but the members of the group differ among themselves and two strains of *Salmonella dublin* treated in exactly the same way had different wavelengths. *Sarcina ureae* and *Sarcina agilis* had identical measurements, both showing marked biplicity (*248*). Biplicity proved a varying phenomenon, abundant in some strains and rare in others. It may one day become a clue to the structure of helices, but it cannot be evoked at will. Obviously flagellar wavelengths will be of limited use in classification (*251*). There is too much overlapping, there are the effects of heat, p_H and colloid additions, and the latent biplicity. It follows that stained preparations can never be used for this purpose.

A word about the helical shape may not be out of place. The helices can be seen as such with binocular stereoscopic microscopy. They conform to D'Arcy

THOMPSON's definition (*302*) that a spiral starts from a point of origin and continually diminishes its curvature as it recedes from that point, or in other words, it is a curve whose radius of curvature continually increases, whilst a (cylindrical) helix neither starts from a definite origin nor changes its curvature as it proceeds.

The metamorphoses from straight tail to clear-cut helix or rod do not always proceed in the same way (*247*). This is illustrated in Fig. 9, n to y. A straight tail can become a helix, either long or short wave, and then go back to the state of a long straight tail. Fig. 16 shows the same *Salmonella* photographed twice on the same piece of film with an interval of less than a second during which it changed back from a long wave helix to a straight tail. A long wavelength helix can become a short wavelength one and *vice versa*, as drawn in Fig. 9, s to w. Stiffening into a rod (Fig. 11) more often than not is immediately final (Fig. 9, x to y). Between straight rail and helix the changes can be rung a number of times, but once a first change has been made, finality is always reached within minutes. As a rule the change from straight tail to long wave helix is smooth, but the transition into a short wave helix is often accompanied by violent spasmodic lashings, contractions and shakings of the tail which are communicated to the body and may simulate a kind of erratic motility (*251a*). That the straight tails are fuzzy-looking and the helices clear-cut is well brought out in Fig. 17 showing numerous straight tails, all fuzzy, and one helix, clear-cut. Fig. 18 and 19 show that biplicity can occur on one bacterium and even on one flagellum.

At first these metamorphoses of the tail do not affect motility (*247*). Swimming goes on in normal fashion and even semi-somersaults, as drawn in Fig. 9, z to dd, take place. After a while the bodies that at first maintained their three dimensional undulations also stiffen, and this is the end of normal motility (*251a*).

Another conspicuous feature of the metamorphoses of the tail is that measurements on a large number of cases have shown that the over-all length of the tail does not alter when it becomes a helix (*247*). This is seen in Figs. 12 and 16, and refutes the suggestion that the straight tail is nothing but a long drawn-out helix which can become shortened and broadened. This erroneous idea, the result of casual observation (*228*, *233*) should not be perpetuated (*334a*). STOCKER (*293*) suggested that the straight tail as seen and photographed was just a blurred picture of a helix, the blurred look and fuzzy appearance being caused by its high speed. But then the tail in its fuzzy state would have to appear as broad as the resulting helix which is not the case, as seen in Figs. 12 and 16. Also the speeds in either picture of both Figs. 12 and 16 were practically the same.

The straight tail maintaining its *over-all* length when it changes into a helix, implies that its *total* length increases considerably, especially in the case of short wavelengths arising. In some cases this was calculated to amount to 50% (*247*), which is a very mysterious apparent gain in material.

Pseudomonas and *Azotobacter* so far have not been seen exhibiting straight fuzzy tails (*251*). This may one day explain why MAYR-HARTING (*190a*) could not differentiate between H and O agglutination in *Pseudomonas pyocyanea*. If however a preparation of such seemingly tail-less bacteria is left on the stage for a few hours a number of ordinary helices appear (*251*) (Fig. 20). It must be remembered that in other bacteria also the transition from straight tail to helix makes the appendage more clear-cut and more easily visible.

What causes the metamorphosis is unknown. Young cultures examined with sunlight D-GM at first show chiefly straight tails. Under the microscope the metamorphosis then starts on individuals and becomes general after several hours (*249*). In older cultures there are fewer straight tails and more helices to begin with, and the transition takes place quicker than in young cultures (*249*). If a culture were kept permanently fresh by adding new medium all the time, and removing used up medium at the same rate, it seems likely, if one could at the same time shelter the culture from light, that there would be nothing but straight tails *in perpetuo*. The straight tail certainly has the more normal appearance.

What underlies the metamorphosis is also unknown. The same material that eventually forms the helical flagella is of course present in the tail and presumably also around the body, be it in somewhat different molecular formation. It is thinkable that it is at that time amorphous and somewhat in the shape of a cloud. It would then, if present in sufficient quantity, form a tail which would be trailing behind at speed and be able to perform the phenomena described during sudden reversals and semi-somersaults. This assumption would imply that the bodies move themselves. The quantity of tail-material would decide the shape and appearance of the tail and even its presence. *Pseudomonades* and *Azotobacters* would have very little, *B. proteus* would have a great deal more. In previous publications (*233*, *234*, *236*, *237*, *240*) when on the authority of Stephenson (*290*) it was accepted that the outer covering of bacteria consisted largely of polysaccharides, it was accordingly suggested that this material was chiefly polysaccharide. This erroneous conception proved most unfortunate from a tactical point of view. When through the work of Weibull (*319*, *320*, *322*, *324*, *325*, *326*, *327*) it became likely flagellar material was largely protein, this finding was used as if it discredited all observations with sunlight D-GM. The transition of this material from straight tail to helix is a challenge to be met. There is a remote resemblance to the amorphous substance in cultures of *Acetobacter* becoming cellulose threads (*6*, *116*) through the activity of an extracellular enzyme.

In solutions of an amorphous collagen called ichthyol various substances can cause precipitation in the shape of typical collagen fibres (*106*, *217*). Perhaps the formation of fibrils by the silkworm (*191*) may be a closer analogy, because here motility also is a factor. It is true that the motility of many bacteria on which helical flagella have been seen, was limited, but movement on the spot and certainly cell wall movement was not excluded in such cases. The transition from fuzzy tail to clear-cut helix suggests a gel becoming coagulated. Van Herwerden (*113*) showed that protoplasm gels can coagulate by adding certain chemicals and that the process is reversible if the chemical is washed out in time. Seifritz (*276i*) does not think that this kind of coagulation is usually reversible, in the same way as one cannot unboil an egg, but that it might happen in living matter. It has been shown [Pijper (*237*)] that in acid broth of p_H 4.98 in which tails of *Salmonella typhosa* were absent or invisible, sudden addition of sodium carbonate to the preparation under the microscope brought them out, usually in the shape of a stiff rod, preceding the body and staying in front when the body reversed. Perhaps a chemical factor is only auxiliary and light may be the deciding factor, amoebae are known to coagulate in strong light. A gradual coagulation of the tail from the tip towards the body might explain continued normal motility during the process and also the continuation of semi-somersaults during the tail-stiffenng process, as happens (Fig. 9, z—dd).

The next stage in the evolution of the tail structures of these motile bacteria in sunlight D-GM is the splitting of the helices as in Figs. 21 and 10, first into two more or less equal thinner helices and then into several more, as seen in Figs. 22 and 23, and drawn in Fig. 9, ee and ff (*233*). Often the split helices remain around the bacterial body for a time (Fig. 9, gg) and give an impression of peritrichousness, and they may occasionally become dried and fixed in that position and appear in stained preparations and in the EM. It might be said that bacterial species like *Pseudomonas* and *Azotobacter* with so little material round their bodies that they cannot produce a visible straight tail, but can show one polar helix, would not have much material to split and thus appear monotrichous or at best lophotrichous. The difference between various types of flagellation might perhaps be just quantitative (*247*). In many cases the split flagella get loose and drift through the medium (Fig. 9, hh), as in Fig. 24 (*233*), and they then may become joined together by processes best described as linking and packing (*248*) and thus form giant-helices as seen in Fig. 25. It is at this stage that, as a rule, efforts at staining or photography with an EM are undertaken, and the drying drop then, as described in previous sections through its "receding tide" (Fig. 26), through dislocation, deformation and fortuitous attachment will produce the traditional pictures of "peritrichousness" (Fig. 9 ii).

The transition from straight tail to helix may be expected to set up tensions of a helical nature in the helices, the change over probably having something to do with molecular transformations (*247*), as indicated by its increase in total length. These tensions might then work themselves out by whirling movements in flagella that have remained attached at one end to bodies. Such whirling movements often occur and have been the cause of many misunderstandings. They are perhaps more pronounced in colloid suspensions, such as cultures on agar. If a body has stiffened and has become anchored to slide or coverslip, the whirlings of the attached helices give rise to local maelstroms (*247, 251a*). These can be made readily visible by adding indian ink or carmine. They made Rinker, Robinow and Koffler (*264*) and Mallett, Koffler and Rinker (*183*) and Stokes (for *Spherotilus natans*) (*294*) think that flagella had independent movement and showed themselves to be motor organs. This movement however obviously has nothing to do with motility as usually understood, although it has been used as an argument against the conception of a passive role of flagella in motility by Ørskov (*219*), by Kauffmann (*135*) and partly by Fleming *et al.* (*94, 95*). When this whirling occurs in helices attached to bodies that have stiffened but are not anchored down, it produces a kind of pseudomotility of the body, giving the impression of a liner pushed about by tug boats with their noses against its sides (*247, 251a*). Heat stimulates this kind of movement, followed by exhaustion [Fleming (*95*)] but this does not prove that it is a "live phenomenon". The cooperation between the different helices is not always good, which makes the movement of the body erratic. It sometimes happens that the whirling movement sets in before the body has stiffened. One then sees three dimensional undulation *combined* with tug boat pushing (*94, 95, 251a*).

B. proteus usually produces large numbers of split helical flagella surrounding the body. Cooperation here is often very good and the individual flagella seem to work harmoniously together. Taylor (*299, 300*) working with spermatozoa noticed that their tails waved in unison when the spermatozoa were close together and pointing the same way. His explanation was that if the waves of neighbouring tails are in phase much less energy is dissipated than when the waves are in opposite phase. If the phase in one tail lags behind the other there is a strong reaction due to the viscous stress in the fluid which forces them into phase. It would seem possible that this explains the harmonious working of the numerous helical flagella of *B. proteus*. Gray (*101*) described an apparently similar phenomenon in *Cristispira balbianii*. In *B. proteus* this helical whirling with its curious eventual harmony and pseudomotility also explains the horseshoe-shaped bacterial chains that were seen moving with the curved end in front by Kingma Boltjes (*140*), by Fleming *et al.* (*94, 95*), and by Ørskov (*219*). Pringsheim described it for *Lineola* (*258*). Whirling helices probably also explain the movement seen by Fleming *et al.* (*94, 95*) in spherical forms of *B. proteus*, and perhaps in old cultures of *Vibrio cholerae* by van Riemsdijk (*263*) and in penicillin-poisoned cultures of *Vibrio cholerae* by Bruce White (*50*).

b) Staining of flagella

Staining of bacterial flagella is notoriously difficult and erratic in its results, hence Henrici's dictum (*112a*) that the procedure is too laborious and uncertain for routine use in examining cultures. The technique, as shown in section IIb, inevitably leads to artefacts. Pijper from 1930 (*227*) insisted that the belief in peritrichousness originated in flagella-staining. This was supported by Pietschmann (226). Section V a showed that flagella-staining is usually undertaken at the stage when the straight tail has become a helix and has split into thinner helices. By artists like Zettnow (*344*), Levenson (*169, 170, 171*) and Leifson (*162, 164, 166, 167*) very impressive "peritrichous" stained pictures have been produced at this stage. In most instances however of flagella-staining the results have been adversely affected by the effects of drying as described in section IIb. Nevertheless, in such pictures *Pseudomonades* often have just one or just a few polar flagella, and *B. proteus* usually has large numbers of flagella round its body. These are the traditional standard examples, how they might be due to quantitative and not qualitative differences is suggested in section V a. In general it can be maintained that flagella-staining has caused more confusion than orderliness and for purposes of classification the differentiation between mono-, lopho- and peritrichousness breaks down when most needed. Knaysi (*143*) only recognises terminal and lateral flagella and maintains that when there are lateral ones, there are no terminal ones. Examples of the existing confusion are that Leifson and Palen (*166*) found they had to call the flagellation of *Bacterium (Listeria) monocytogenes* doubtful, and saw no correlation between flagellation and motility. Weeks (*316*) working with *Flavobacteria* could not establish a pattern of flagellation. Mueller and Stapp (*208*) found the flagellation of *Rhizobium* doubtful. Leifson and Hugh (*164*) stated that *Aeromonas* when young had only lateral, and when older only polar flagella, and that many cultures did show mixtures of these two types of flagellation. Leifson (*163*) declared that *Acetobacter* could be peritrichous but also "polar multitrichous". Knaysi stated (*143*) that mature cells had more flagella than young ones, but Neumann (*214*) gave as his (very extensive) experience that in so-called peritrichous bacteria the number of flagella unaccountably varies from none to very many. O'Toole (*220*) insisted that rapid transplantation preferably interspersed with animal passages was a sure method to create abundance of flagella. Hesselbrock and Foshay (*115*) denied the existence of flagella in *Pasteurella tularense* described by others. Brindle and Cowan (*46*) found that Whitmore's bacillus could appear with polar or with lateral flagella and Clark and Carr (*58*) stated that in *Cellulomonas* and in *Arthrobacter* both polar and lateral flagellation occurs and that the flagella are just randomly located. Conn and Elrod (*66*) in a general discussion of the problem declared that peritrichic and monotrichic *strains* could occur in the same species, and that probably monotrichic and peritrichic *cells* could appear in the same strain, leading to the inevitable conclusion that in such cases flagellation could not be used diagnostically. They admit that there are transitions from monotrichousness to peritrichousness, and at the same time maintain that some bacteria are truly the one or the other. This may be quite true but how and when does one decide which is which? In this publication the authors admit that they only photographed bacteria that had no loose flagella lying near, which sounds dangerously like selection and not fair sampling. Harrison and Hansen (*110*) discovered a motile lactobacillus (which is very unusual) and it had some threads round it that look very much like slime but they would make it peritrichous. Gray and Thornton (*103*) in *Mycoplana* found some that were polar and others that were peritrichous. Kingma Boltjes (*140*) stated that it was often very difficult to decide between polar or peritrichous flagellation and then somewhat surprisingly added that there would never be any confusion between bacteria with true polar and true peritrichous flagellation. Thjötta (*301*) described how to make pseudo-flagella. *Pseudomonas* is a special case, it is traditionally described as having polar flagella only. Bartholomew (*21*) however having examined six species claimed that flagella could originate both terminally and laterally. [Van Iterson (*129*) had pictures of *Pseudomonas* made with the EM that if it had been other bacteria would have been called peritrichous.] Pivnick (*252*) had 32 strains and they all stained polarly. A doubtful point is how many polar flagella *Pseudomonas* can have. Kjems (*141*) usually saw one and sometimes two. Crawford (*70*) on his unusual *Pseudomonas* that resembled *Escherichia coli* so much that he fully expected it to be peritrichous, got only one polar flagellum. Burckhardt (*53*) thought he could differentiate between species of *Pseudomonades* by the number of polar flagella but this was not accepted, and Caldwell and Ryerson (*54*) just found 1—5 polar flagella in various *Pseudomonades*.

Conn and Wolfe (*63*) made an effort to do away with difficulties about peritrichous and polar flagellation by introducing the conception of *degenerate* peritrichous flagellation. It would mean something that should have become peritrichous flagellation but did not. This proved useful as a term in doubtful cases but did not do away with the doubt and did not make classification on the basis of flagellation more efficient. Starkey (*287*) finding small forms of his *Vibrio desulfuricans* polarly, and large forms peritrichously flagellated, preferred to call them degenerate peritrichous. *Azotobacter* for a long time regarded as polarly flagellated when stained, was made peritrichous by HOFER (*123*) upon examination with an EM, and by DERX (*76*) after staining, but KYLE and EISENSTARK (*155*) want to make it degenerate peritrichous, adding that they found the difficulties in staining very great. HOFER (*122*) thought that *Bacterium radiobacter* which had been described as monotrichous and as peritrichous had better be called degenerate peritrichous. GRIFFIN and ROBBINS (*104*) think it better to call *Bacterium (Listeria) monocytogenes* which LEIFSON and PALEN (*166*) had called doubtful, degenerate peritrichous. Finally, CONN, WOLFE and FORD (*65*) described the flagellation of *Alcaligenes* as *doubtfully* degenerate peritrichous. There does not seem to be much diagnostic promise in CONN and WOLFE'S suggestion (*63*) that flagella-staining should become a routine procedure.

For classification purposes flagella-stained preparations obviously are unsuitable. The shape also of individual flagella can hardly escape deformation in the staining process as detailed in section IIb. The original shape may of course occasionally survive. As early as 1930 LASSEUR *et al.* (*159*) were struck by the differences in wavelengths of flagella exhibited in some of their stained preparations and started measuring them. They also noticed the constancy of wavelength of flagella in the pictures of NEUMANN (*214*). For *Pseudomonas aeruginosa* they found 1.5 μ and for *Bacillus subtilis* 2 μ, values very close to our own findings as given in Table 2. They did not think that it was worth while to measure amplitudes in their dried preparations. After this early start no attention was for many years given to flagellar wavelengths, probably because the irregular appearances of stained pictures did not suggest measuring. An exception was ARENA and SCHWARTZ (*4*) who saw "wider undulations" in smooth *Salmonella typhosa* and "narrower undulations" in the rough form; it probably was biplicity unrecognised. In 1951 LEIFSON (*162*) suggested that degree of curvature in a flagellum might be more important than total length. But the phenomenon of biplicity found by PIJPER (*247*, *248*, *249*) from observations on live bacteria with sunlight D-GM was not discovered by anybody in six decades of assiduous flagella staining. LEIFSON, CARHART and FULTON (*167*) measured one single wavelength of 10 flagella of each one of 75 strains of stained *B. proteus* and found large differences between otherwise identical strains. They also saw that *B. proteus* sometimes had both long and short wave flagella, these they called "normal" and "curly", but they did not grasp the "biplicity" suggested by their own pictures, nor did they ever describe it in so many words. Of course dried preparations do not show "biplicity" very definitely, but the indications were there. There is no doubt that because of the effect of drying, temperature, p_H and presence of colloid substances the only reliable way to measure wavelengths of bacterial flagella is by photographing and measuring very large numbers of them in wet preparations with all due precautions, as done by PIJPER, NESER and ABRAHAM (*251*). Only such measurements give a steep symmetrical frequency curve. What MIGULA (*195*) described in *Pseudomonas macroselmis* as alternating wavelengths looks more like an artefact than like biplicity. HILLIER (*117*) warned against placing biological significance on drying patterns, we have indicated some of the reasons, specially applying to helical wavelengths.

c) Anatomy and origin of flagella

This section is still concerned with *Eubacteriales* (excluding *Spirilla*) and, as it deals with the roots of flagellar problems, it is bound to be controversial. The main controversy is whether bacterial flagella go through the cell wall, and if so whether they are attached to blepharoblasts, also called basal granules. The argument assumes the existence of a tough cell wall and of at least a functional cytoplasmic membrane surrounding a cytoplasmic mass, as formulated by KNAYSI (*143*, *144*). PIEKARSKI and RUSKA in 1939 (*224*, *272*) first photographed bacterial flagella with an EM. The drying of the drop must have affected the results, the authors express surprise at the large number of flagella found. Shadowcasting of bacterial flagella was introduced by WILLIAMS and WYCKOFF (*333*) in 1945. The shadows cast differed in length and they concluded that flagella were of unequal thickness, and that some of them looked like ribbons. This is not the general view but it is obvious from the long series of magnificent pictures in the books by SCANGA (*276a*, *276b*). BARTHOLOMEW said that in some of his EM pictures the flagella looked tubular (*21*). The special technique of growing bacteria on a colloid film (*118*) with its "near-dry" conditions decreases the disturbances caused by drying, but even KNAYSI (*143*) admitted that the bacteria there had as much play as in a very thin slide-coverslip preparation. All EM pictures of flagella partake of artefacts, as described in section IIb and Vb for the case of stained pictures.

Some authors unhesitatingly state that flagella go through holes in the cell wall [HOUWINK and VAN ITERSON (*125*), PRINGSHEIM (*257*), BISSET (*30*, *32*), BISSET and HALE (*31*), SCANGA (*276b*), STOCKER (*293*), BRADFIELD (*41*)]. EM photographs often give this impression, as illustrated by Fig. 27 kindly supplied by Prof. SCANGA. Closer inspection reveals that many flagella go right through the body and come out at the other side, or turn round and come out at a different place on the same side. It is not always easy to see this because as long as the cytoplast is intact it is impervious to electrons and the flagella just seem to disappear in the dense mass, there is therefore no saying where they end. The amount of superimposition of flagella and bacterial bodies in all such pictures cannot be assessed. Superimposition would suffice to explain all the cases where penetration by flagella appears to take place. This danger of mistaking superimposition artefacts for reality was realized and warned against by HILLIER, KNAYSI and BAKER (*118*), KOFFLER and MALLETT (*149*), MUDD, POLEVITZKY and ANDERSON (*204*), RINKER and KOFFLER (*265*) and KNAYSI (*143*). Perhaps some authors too readily accept penetration, where it suits them, and overlook it in other instances. A picture of *Acetobacter xylinum* by VAN ITERSON (*130a*) shows cellulose fibres apparently piercing the cell wall and becoming attached to the cytoplast!

Several authors are convinced that bacterial flagella are attached to basal granules situated as a rule just underneath the cytoplasmic membrane and some of them illustrate their publications with drawings to that effect [BISSET (*30*, *30a*), TOMCSIK (*306*), BRADFIELD's idealized section (*41a*)]. Fig. 27 illustrates the kind of granules usually observed. Similar pictures were thought to be convincing by HOUWINK and VAN ITERSON (*125*) and VAN ITERSON (*129*, *130*). These granules however do not seem to be present every time, WEIBULL and

HEDVALL (*328*) could not find them, nor could MUDD, POLEVITZKY and ANDERSON (*204*). They are said only to become visible in bacteria that have undergone autolysis (*125, 276b*). Autolysis is apparently necessary because otherwise the cytoplasm is not sufficiently transparent to the electron beam, and the flagella cannot be followed in their course. The general structure of cytoplasm or protoplasm is not well known and autolysis of *non-motile* bacteria seometimes show similar granules, as shown by pictures taken by SCANGA (*276b*) and by WILLIAMS and WYCKOFF (*333*). WYCKOFF (*342*) found that the protoplasm of young bacteria is finely granular and becomes fibrous with normal ageing, this last feature being confirmed by KELLENBERGER and WERNER (*136*). But WYCKOFF (*341,342*) and HEDÉN and WYCKOFF (*111*) have also shown that heating and other damaging factors make the bacterial protoplasm go granular and these granules do not differ in appearance from the basal granules. WARREN and GRAY (*314*) produced similar granules in *Pseudomonas* with hyaluronidase. In any case these basal granules in most publications do not look uniform and although in many cases the flagella seem to stop inside them, in many others they appear to go right through them.

On the evidence brought up so far, the possibility that bacterial flagella go through cell walls and have basal granules cannot be denied, but it is far from proven. FISCHER and LAROSE (*93*) think that as the cytoplasmic membrane of bacteria has keratin-structure, and bacterial flagella grow from this membrane, and contain a protein of the keratin-myosin-epidermis-fibrinogen group, the analogy they see between wool and bacteria does not appear extraordinary. The reasoning is not so well-founded as it is ingenious.

Recent new techniques that might have furthered our knowledge of these relationships are the cutting of ultra-thin sections through bacteria which can then be looked at with an EM, and also the denuding of bacteria of their cell wall which gives direct access to the protoplast. Cutting ultra-thin sections was introduced by PEASE and BAKER (*222*), and BAKER and PEASE (*19*) sectioned *Bacillus megaterium.* Since then a number of authors have in such manner cut and examined by means of an EM several bacteria, some of which were at least potentially motile (*23, 24, 30, 42, 57, 97f, 107a, 137, 181, 307*) and even a *B. proteus* swarmer (*128a*) but none of them mentioned motility, flagella or holes in walls. The denuding technique was started by JOHNSON, ZWORYKIN and WARREN (*133*) by osmolysis through putting bacteria in liquids of lower tonicity, which produced naked protoplasts and empty ghosts of cell walls. They found flagella attached sometimes to the cell walls and sometimes to the protoplasts [not as STOCKER (*293*) quoted it, to the protoplasts only]. The technique was improved by TOMCSIK and GUEX-HOLZER (*305b*) who gently removed the cell wall by treatment with lysozym and again by TOMCSIK (*305c*) who found that the protoplasts could be preserved in Ringer solution. WEIBULL (*329*) used this technique on *Bacillus megaterium* and preserved the protoplasts with sucrose. He published one stained and one EM picture of a protoplast and these show a few threads attached to them which he regards as flagella. The threads however do not resemble flagella, they may easily have been washed up against the protoplasts during the drying process (there are a number of latex particles stuck to the protoplast in the EM picture), the protoplast is too thick to see whether and how far the flagella go into it, and there was complete absence of motility. Wiame *et al.* (*332*) in the same way procured protoplasts of *Bacillus subtilis* also showing supposed flagella in the EM picture which the authors describe as retained, but which may just as well have been washed up against them. There are a number of flagella lying loose in the field. There is no mention of motility though respiration was shown to have been going on. McQuillen (*211*) making protoplasts like Weibull (*329*) could see them divide but does not mention motility nor flagella.

The anatomy and function of flagella would be considerably furthered if they could be traced back to their origin. Pijper (*247—251a*) has brought evidence that they are derived from material outside the cell wall which through some physical or chemical change stiffens into helices. The opposite view is that flagella grow ready-made through the cell wall from the cytoplast and perhaps occasionally become twisted into a tail [Bisset (*30*)].

Here it seems indicated to dispose of a much quoted paper by van Tieghem (*304*) of 1879, written at a time when all flagella were believed to be terminal or polar. He used *Clostridium butyricum* and wrote that after cell division the motility of the two bodies, motility which he ascribed to protoplasmic contractions inside the bodies, would draw out the gelatinous capsule into a thread, this he admits he could only see after staining. The thread would then break and so produce two terminal flagella. It is not clear how much he saw and how much he visualized. Anyhow his reference to motility producing flagella did not go further than separation after cell division. One cannot help thinking that van Tieghem's thinking was inspired by the similar observations of Dallinger and Drysdale (*72*) of 1875 on *Bacterium termo* [according to Macé (*182*) this is not a well-defined microbe and may have been *Pseudomonas fluorescens*, *B. proteus* or *Bacillus subtilis*] where they had actually seen the two bacteria resulting from a cell division shake until separation followed. He may also have known of the similar observations by Ewart (*90*) in 1878, again on cell-division in *Bacterium termo*, which the author regarded as applicable to all motile bacilli. He saw the two "segments" as they separated slowly draw out a thin thread of "protoplasm", and then the one got fixed and the other one wriggled itself loose until the almost invisible thread gave way in the middle and two "cilia" were formed. Of all these authors only van Tieghem ascribed motility to protoplasmic contractions and this was no more than a suggestion.

The drawings made by Leifson in 1931 (*160*) supposed to represent the outgrowth of flagella through the cell wall are still quoted and accepted by several authors [Knaysi (*143*), van Iterson (*130*), Houwink and van Iterson (*125*)] although Leifson himself appears to have different views nowadays (*164*). Leifson let spores from three species of bacteria grow in broth and stained samples for flagella at short intervals. Curly flagella were drawn as growing out one by one from the cell wall all over its surface, at high speed, and whole fleeces apparently appeared in an hour or so. There was no effort at statistical analysis of the samples, or, as Houwink and van Iterson (*125*) nicely put it in connection with their own work: "the challenging task of correlating growth curve with flagellar development has not yet been undertaken by anybody". In Leifson's beautifully regular drawings the effect of drying and other disturbances were also not taken into account. BISSET and HALE (*31*) in more careful observations, starting with microcysts, planted on agar, with samples being taken at short intervals and examined both by staining and with an EM, found that flagellar development appeared to start by flagella piercing the wall in the polar region. Their pictures are too few to be convincing statistically and they add tracings from their EM pictures which appear rather arbitrary. In another paper (*32*) BISSET suggests that the polar region of a bacterial cell is a growth centre. The cell wall there would be soft and thin and there would be a local accumulation of stainable material (*29*). [This accumulation was confirmed by LANKFORD *et al.* (*158*), and by LUTTERINGER and LANKFORD (*180*) who thought O antigen was also localised there, but DARDANONI and CALVI (*73*) did not think the local granules were O antigen.] According to BISSET (*32*) a similar growth centre occurs at the point of cell division. Flagella make use of the thinness of the wall at the growth centre to grow through [BISSET (*31, 32*)]. Later on the wall expanding from the pole distributes the flagella in peritrichous fashion over the cell surface, at any rate in peritrichous bacteria. LEIFSON and HUGH (*164*) found this an attractive hypothesis, because they had come to the conclusion, in contradiction to LEIFSON's early paper (*160*), that ordinary cell wall was rather hard for the soft flagella to grow through. BISSET's hypothesis is in direct contradiction to the well-known solitary EM photograph of HOUWINK and VAN ITERSON (*125*) in which they show 6 straight short flagella "growing" out of the *lateral* cell wall. In 1953 VAN ITERSON still showed the identical picture as illustrating "juvenile" flagella (*130*). HOUWINK and VAN ITERSON (*125*) added that often in young bacteria they saw no flagella at the poles but only on the sides and they explained this as being caused by frequent and rapid cell division! A *reductio ad absurdum* of flagella freely piercing cell walls was provided in the observations of PRESTON and MAITLAND (*255*). They found *Pasteurella pseudotuberculosis*, when growing at 22 C, motile and in possession of flagella. Growing at 37 C there was no motility and no flagella. Cells grown at 37 C and then transferred to 22 C whilst measures were taken to prevent cell division, did "develop" flagella in a very short time. There must have been tremendous activity in hole-drilling during this short time.

KNAYSI (*143*) discussing the origin of flagella thinks that they arise singly and are irregularly distributed. Recent EM observations did show him that in *Bacillus mycoides* the flagella often approach the body tangentially and enter a fibrous layer that envelops the cell wall. Sometimes they run parallel with the fibres of that layer. KVITTINGEN (*153*) had a similar impression. SALTON and WILLIAMS (*275*) see no structure in the cell wall of *Bacillus megaterium* in their EM pictures, but isn't there a suggestion of reticular structure similar to what KNAYSI (*143*) thought he had seen in some of his own pictures? One of the EM pictures by SCANGA (*276a, 276b*), here reproduced with Professor SCANGA's kind permission (Fig. 28), shows an apparent transition of cell wall into fibres that might be flagella. MUDD *et al.* (*203*) after breaking up *Bacillus megaterium* through sonic oscillation found empty bags of cell wall with flagella still attached. KNAYSI (*143*) sums up his ideas and observations by saying that flagella are not slime but fibrous structures synthesized somewhere *in* the cells. The whole position is difficult to understand. It seems indicated to reserve judgment as to the anatomy and origin of flagella until more observations have been made.

d) Chemical analysis of flagella

The chemistry of bacterial flagella starts with WEIBULL's investigations. BOIVIN and MESROBEAU (*35*) in 1938 concluded from finding H and O agglutinins in rabbits injected with *Salmonella typhosa* which had been treated with trichloracetic acid, that flagella were of protein nature. In 1944 GARD (*97b*) worked out a method for collecting a large quantity of flagella of *Salmonella schottmuelleri* by shaking and differential centrifuging, similar to what CRAIGIE (*69*) had done with his *Salmonellae* in 1931. Using GARD's method WEIBULL and TISELIUS (*318*) showed that the flagella of *S. schottmuelleri* dissolve in acid. WEIBULL in a series of papers (*319—328*) first showed that the flagella of *B. proteus* behave similarly. He then started a chemical analysis of the flagella of *B. proteus* and *B. subtilis* and found (*320*) that they contained very little carbohydrate and that acid hydrolysis of flagella of *B. proteus* gave the following aminoacids: arginine, lysine, aspartic acid, glutamic acid, glycine, serine, alanine, threonine, valine, leucine, isoleucine, phenylalanine, tyrosine, methionine (very little) and proline (doubtful). Absent were histidine, tryptophane, cystine, hydroxyproline and norleucine. *B. subtilis* yielded the same aminoacids but there was more methionine and little proline. In his later papers WEIBULL confirmed that his two kinds of flagella (*B. proteus* and *subtilis*) contained no cystine and had 8% lysine. Myosin and actin are given by him as containing 1.5% cystine and 10—11% lysine. Keratins are given as containing 10—15% cystine and 2.3% lysine. A further differentiation is that flagella contained a smaller total number of aminoacids than keratin and myosin. An important fact is that cystine is recognised as an *essential* constituent of myosin. Flagella contain no cystine at all. Flagella therefore in their chemical constitution do not even resemble myosin nor keratin. WEIBULL did not find any energy-rich materials such as adenosine-triphosphate in his flagella (*325*) and their mineral content showed nothing striking. Fat-content was low, 0.7—0.8%, carbohydrate less than 0.2% and ashes about 1% (*326*). WEIBULL concluded that his flagella are almost entirely protein of constant chemical composition.

The protein nature of the flagella of *B. subtilis* was confirmed by KOFFLER and MALLETT (*148*) by means of ninhydrin and biuret reactions, and by HODGE (*121*) and by RINKER and KOFFLER (*265*) by specific staining with phosphotungstic acid. A strange finding is that of LURIA (*179*) who found flagella intact after cells had been lysed by bacteriophages.

e) Physical analysis of flagella

The X-ray diffraction investigations by ASTBURY (*7—12*) for which WEIBULL provided the material resulted in an entirely new conception of bacterial flagella. The conclusion reached on chemical grounds [WEIBULL (*319—328*)] that bacterial flagella are long protein (but not nucleoprotein) molecules invited X-ray diffraction studies. Dried small pools of flagellar material [for a different technique see WEIBULL (*324*)], subjected to X-ray diffraction by ASTBURY and WEIBULL (*12*) proved that the protein of these flagella belongs to the keratin-myosin-epidermis-fibrinogen group previously described by ASTBURY (*7*). Normally they showed the alpha configuration, squeezing brought on the beta form. They were therefore thought to be primitive hairs or muscles and their thinness suggested monomolecular hairs or muscles. The sharpness of the pictures indicated a structural simplification as compared with hair keratin (*10*). The absence of cystine makes them strikingly different from myosin [Astbury (*10*)] but they might still have the capacity of rhythmical change of form, making for motility (to be discussed in section VII). Astbury has grown more and more inclined to regard these flagella as monomolecular muscles (*11*) and recently Astbury, Beighton and Weibull (*14*) gave it as their view that bacterial flagella must be among the smallest organs of motility possible, embodying some mechanism of biological movement stripped to its barest minimals. The factual findings of Astbury cannot be criticized. It is different with the interpretation. There is already a growing tendency among bacteriologists to regard the matter as settled, by regarding flagellar protein practically as myosin. The new word "flagellin" proposed by Astbury *et al.* (*14*) for flagellar protein may curb this facile tendency. It would be worth while to subject to X-ray diffraction both long wave and short wave helices of flagella, if separate batches could be procured. Also, the straight tail from which Pijper (*226—251*) derives the helical flagella, invites examination by X-ray diffraction.

VI. Motility of non-flagellated bacteria

In this section will be discussed motility of *Myxobacteriales* and *Chlamydobacteriales* both as defined in Bergey's Manual (*44*), together with motility in representatives of what Pringsheim (*257*) calls *Myxophyceae (Cyanophyceae)*. Controversies of classification in these groups will thus be evaded, and conclusions as to systematics from appearances of motility avoided.

These groups contain large numbers of microbes in which motility occurs but flagella have never been found by any method, including efforts with an EM. [Tuffery (*308*) found peritrichous flagella in *Oscillospora guilliermondii* with an EM but this microbe together with its close relative *Caryophanon* most probably belongs to quite a different group.] The locomotion discussed here is usually referred to as creeping or crawling, to differentiate it from what in ordinary motile bacteria is designated as swimming. The problem of crawling or creeping is the mechanism. It is supposed to need a surface, and not to take place in the midst of liquid, but in many cases a water-air surface seems to suffice. Stanier (*286a*, *286c*) noticed in *Cytophaga* that it can move fast on an agar surface, as a mass of cells, with bending movements of the bodies, and obviously having no rigid cell walls. The vegetative cells showed extreme flexibility, greater than that of most other *Myxobacteria* and they also showed a swinging movement of their bodies, and often vibrating ends. Flexing movements of the bodies ccurred independently of linear movement, and often occurred as an interruption of progressive movement. He also described rotatory movements and slight displacements in wet mounts or hanging drops. Garnjobst (*97c*) described rotatory and waving movement of *Cytophaga* in water, with cells standing perpendicular, the upper end describing an incomplete circle at a regular rate, and long cells had a restless rotatory type of motion, some cells coiling up in circles. According to her, many of these movements seemed to need flagella which however were never found. Stanier (*286b*) had previously shown that *Chondrococcus exiguus* (now *coralloides*) planted on a sheet of agar that had been bent, and thereby slightly stretched, over a glass rod, produced swarms which travelled along the lines of tension of the agarsheet. HAROLD and STANIER (*109*) studying *Leucothrix* and *Thiothrix* saw a gliding movement similar to that of blue-green algae and *Myxobacteria*.

The point at issue in all these cases is whether these microbes get their motility from an asymmetric secretion of slime, which, by swelling, exerts pressure on one side of the cells, or from periodic alteration in cell form. The movement is usually rather slow. The point has been investigated by a number of workers, from VAHLE (*313*) in 1910 to SCHULZ in 1955 (*276e*), without finality being reached.

UELRICH (*309*) working with *Beggiatoa* and *Oscillatoria* did not believe in the excreted mucus theory and tended to accept a contraction and expansion theory, but not, as had been suggested previously, by parts of the body becoming bigger and smaller through water being taken up or changes in vacuoles. He thought of a wave-like motion running over a chain of cells. He actually saw that *Beggiatoa* had wrinkles running in spiral fashion around the threads. SCHULZ (*276e*) repeating this work, saw no waves running over the threads of *Oscillatoria*. He added carmin particles to moving threads of *Oscillatoria* and saw that this resulted in there being three kinds of particles. Some did not move, some moved spirally with the body and some moved spirally against the body. He concluded that all movement of the body was due to active participation of the excreted slime around the body. But all his efforts, including EM, even after ultrasonic shaking, to find holes or pores for the slime to come through proved negative. He also admitted that his theory met with great difficulties in explaining sudden reversals such as the bodies often performed. One feels tempted to suggest, forestalling on what will be brought up later, that by assuming a bulge to run in spiral fashion over the thread of cells against the direction in which the thread moves, the behaviour of the thread and the carmin particles becomes understandable. Particles outside the range of movement of course would lie still, others, sticking closely to the body, would be carried forward by the moving bulge in spiral fashion, and others again would be pushed backwards in spiral fashion by the thread pushing itself forward spirally against the surrounding slime. FRITSCH (*97a*) pointed out that perhaps there was no need for pores for the slime to pass through as the slime might arise by modification of the cell wall. He concluded however that there is increasing evidence that movement here is due to rhythmic longitudinal waves traversing the trichomes from end to end, and if such waves followed a spiral course, rotation would result, and that such contractile waves had been observed in *Beggiatoa* and in *Oscillatoria*. BEEBE (*22*) examining *Myxococcus xanthus* saw creeping motion but no wriggling in long flexible rods which had no flagella but whose shape however rather suggests gyrating movement. PRINGSHEIM (*257*) does not quite accept rhythmic waves of contraction for *Oscillatoria* and *Beggiatoa* although he admits they are commonly believed to be the chief cause of locomotion. According to him slime may have something to do with it, and *Myxobacters* also show this slime. KNAYSI (*143*) has no doubt that movement is by waves of contraction and BISSET (*30a*) simply says that in *Myxobacteria* the cell wall has been replaced by a muscular membrane whereby a worm-like progression is made possible and this was later repeated by BISSET and GRACE (*33*). JOHNSON and BAKER (*134*) ascribe motility in *Beggiatoa* to contractions which they observed taking place in the cytoplasm in rhythmic fashion.

It must not be thought that this kind of motility without flagella is limited to the special classes of non-flagellated microbes so far mentioned. MANDIA (*184*) described swarming colonies of *Clostridium perfringens* which had no flagella, and made the tentative suggestion that spiral movement of individual cells was not enough to produce motility and therefore individual cells showed none, but when a whole clump or colony worked together there would be enough power to move it, also because the total friction would be relatively less. PATOCKA and SEBEK (*221*) had a parallel case in a *Catenabacterium* where colonies moved though individual cells were motionless and there were no flagella whatever. The filaments were often seen to twist round one another and motility only occurred during very active growth. SACKS (*273*) noticed feeble motility in *Arthrobacter citreus*, spec. nov. but only before the fragmentation that breaks up the microbe into fragments. These fragments ultimately become cocci, without flagella ever becoming visible in an EM. TOMCSIK (*305a*) by chance got a mutant from *Bacillus anthracis* that was motile in a sort of creeping fashion, sometimes like a snake, and with rotation round the long axis. He was unable to demonstrate flagella with methods that proved successful with other bacteria. MACDONALD (*209*) and MACDONALD *et al.* (*210*) described a *Fusobacterium* which showed various kinds of movement with definite flexibility but no flagella even when they examined hundreds of cells with an EM. This is probably the same microbe PRÉVOT, GIUNTINI and THOUVENOT (*256*) described as *Fusocillus girans*, which had giratory motility and no flagella by any method. It had however an enormous reticulated mucilaginous expansion which the authors took to be a new organ. The position of some halophilic bacteria, one of which was investigated by SPRUIT and PIJPER (*285*) is doubtful. These authors described it as having a flat twisted shape, resembling a twisted ribbon. It rotated backwards and forwards along its long axis and had no flagella.

It does not seem excluded that the muscular coat which BISSET (*30a*) ascribed to *Myxobacteria* might be present in other bacteria underneath their cell wall and that it might be powerful enough to set the cell wall in helical motion. It has to be admitted that notwithstanding SEIFRITZ's thoughtful studies (*277*, *278*, *279*), very little is known of the streaming and moving that must be taking place in protoplasm as long as life is there.

VII. Mechanics of bacterial motility

Studying the mechanics of bacterial motility means entering the field of hydrodynamics [GRAY (*102*)]. This is a very difficult subject, which the average microbiologist cannot be expected to understand fully. It has recently been treated by GRAY (*102*) and by TAYLOR (*299*, *300*).

In orthodox bacteriology all flagella are helical or at least curly, and are motor organs [SALLE (*274*), BISSET (*30*)]. The straight tail described by PIJPER (*227—251a*) is often ignored. Investigations into the nature of motility usually start when the straight tail has metamorphosed into a helix as described in section V a (*247*, *248*, *249*, *251*, *251a*) and mechanical considerations then take their origin from this appearance. The phenomenon of biplicity (*247*) which must have a bearing on flagellar movements also has not been taken into account. For the sake of argument we shall, in most of this discussion, also overlook the straight tail and regard as flagella the helical shapes that metamorphose from the straight tail.

BISSET (*27*) described peritrichous bacteria, with all their flagella, as imperfect swimmers compared with monotrichous ones which have only one flagellum. PIETSCHMANN saw no difference (*226*).

In previous sections it has been shown that ASTBURY (*7—14*) has gradually evolved the theory that the (helical) bacterial flagella are effectively monomolecular muscles, with the capacity of rhythmic change of form, probably activated from a basal granule inside the body. He imagines a change in the state of folding passing along the polypeptide chains down one side of the flagellum and up the other side to produce a side-to-side waving movement. The granule is a sort of signal box. The arrangement would of course, as he admits, produce no more than a side-to-side effect but, he adds, that one would only have to invoke some kind of spiralisation of the structure to see also, fairly readily, how spiral waves might be propagated according to the same principle. Bacterial flagella would thus in ASTBURY's words become about the smallest possible organs of motility, some mechanism of biological movement stripped to its barest essentials. This attractive and highly ingenious theory is well on its way to become generally accepted [BISSET (*30*)]. Perhaps it is useful to bring up some criticisms. KNAYSI (*144*) has suggested, and this was supported by JAKOB and MAHL (*131*) that apart from their lack of structure which does not suggest motor organisation, the flagella look rather too feeble to possess enough power to push bacteria. One must think of them in their naked state, not as they appear in their metal armour after shadowing. The energy needed in moving bacteria was calculated by von ANGERER (*2*) and by MOROWITZ (*198*). VON ANGERER took more factors into account than MOROWITZ but neither went into the question as to whether flagella would be capable of the labour. It is generally accepted that flagella must generate their energy all along the helix, although the movement may be stimulated from a basal granule. In GRAY's words (*102*) this would necessitate internally generated bending couples derived from tensile and compression elements, of which there is no indication in the thin flagella. GRAY also discusses the need for a mechanism controlling the extent of contraction of the tensile elements and the phase difference between adjacent elements, for which he would postulate a neuroid transmission, of the existence of which again there is no indication. It might be thought that here the phenomenon of keeping in step by sheer propinquity as described by GRAY (*101*) and TAYLOR (*299, 300*) might enter into the picture, that is if we understand these authors correctly. But even so this would not suffice to explain the sudden reversals and sudden semi-somersaults bacteria execute, if one wants to explain these by flagellar activity. GRAY (*101*) finds it difficult to understand how a hairlike structure like a flagellum can have internal bending couples, maintain phase differences along the fibril and last, but not least, generate free mechanical energy at each short bending unit along the whole fibril. Nobody has ever even hinted at how such a source of energy might arise or act. ASTBURY's theory of monomolecular muscles going through cell walls implies definite points of attachment to the cell wall. It seems reasonable to think that during forward movement the flagella point away from the direction of movement [Bisset (*30*)]. This makes it difficult to imagine what happens to the firmly attached flagella during sudden reversal or semi-somersault. These events take place at very high speed, and would tangle up the numerous flagella on *B. proteus*. Pijper (*238*) insisted that motility was not impaired when vigorous shaking of *Salmonella typhosa* had removed the flagella, checked by sunlight D-GM. [The all-too vigorous ultrasonic shaking of *Salmonellae* by Caselitz (*56*) according to his own pictures damaged the bodies so thoroughly that it cannot be compared.] These shaking experiments by Pijper (*238*) would seem to prove that flagella are not essential for motility and also that the breaking off of numbers of flagella close to the cell wall at very tight exit holes did surprisingly little damage. Dubos (*78*) maintained that flagella are not essential to the life of the cell, but could this mean that the violent stripping off of numbers of penetrating fibres, leaving numbers of unemployed blepharoblasts with short bits of intracellular flagella still attached, would not upset the cell? Miles' remark (*197*) that there was no proof that *all* the flagella had been shaken off is not more than a good debating point. Loss of a large number of motor organs could never leave motility intact as it did, checked by sunlight D-GM. The shaking experiment was repeated by Mallett, Koffler and Rinker (*183*) on *Salmonella typhosa* and by means of an EM they still found flagella "attached" after shaking. There is no proof that the drying of the drop had not washed up flagella against bodies (there were plenty available after shaking) and there is no proof that the flagella found in contact with bodies really did "belong". Also, the authors do not say anything about the swimming of their bacteria after shaking.

Motility in many kinds of bacteria is not simple. It is not just a forward movement in the direction of the long axis. Many bacteria tumble by way of movement, especially *Sarcinae* (*248*), but others can also do it. Then there is movement on the spot described by Ellis (*85, 86*) as "trembling" movement and seen in other bacteria as movement of the cell wall [Kvittingen (*153, 154*), Pijper (*251a*)]. In the first kind of movement the body often takes on a helical structure [Pijper (*233*)], and when "coasting" or "freewheeling" is done in this shape, as often happens, it has been very confusing to some observers (*139, 140*). The suggestion first made by Pijper (*233*) that bacterial appendages result from bacterial movement, and are not its cause, should be considered with full awareness of all these types of movement.

Motility in bacteria is not an absolute quality, nor is speed. Stuart *et al.* (*295*) saw *Shigella alkalescens* and *Salmonella typhosa* (strain Bhatnagar) become motile after passages through soft agar, and bacterial motility has been improved or regained in many cases by this method. Clark and Carr (*58*) do not like this method because they found the percentage of agar of decisive importance. Passage through 0.2% may give motility and 0.3% may give negative results. They find that motility may occur spontaneously in unsuspected bacteria. Weil and Slafkowsky (*330*) found that the soft agar method did work but the results did not last. Occurrence of motility in bacteria is very much like their speed, a good deal depends on the method employed. The traditional figures given for bacterial speeds, usually measured in fluids [Knaysi (*144*), Gray (*102a*)], of up to 20 μ per second are very much higher than those given by CLOWES, FURNESS and ROWLEY (*60*) who made bacteria make their way through agar in glass capillaries.

Much has been made of the occasional finding of bacteria possessing flagella but showing no "motility" [HIRSCH (*119*), BÖE (*36*)] as if it refuted the idea expressed by PIJPER (*233*) that motility produced the bacterial appendages. Complete non-motility is too easily assumed. GRIFFIN and ROBBINS (*104*) among 43 strains of *Listeria monocytogenes* found 7 that were non-motile at 37 C and yet had stainable flagella. KAUFFMANN (*135*) said he had non-motile *Salmonellae* with stainable flagella. EDWARDS *et al.* (*81*) had two non-motile *Salmonellae* with stainable flagella, one of them however became motile after a passage through soft agar. FRIEWER and LEIFSON (*97*) also had stainable flagella on a non-motile variant of *Salmonella typhimurium*. A very little motility of some kind or other may have been at work here. The same kind of argument that a little may go a long way also applies to the objection that bacteria sometimes show appendages without having enough room to move in (*144*, *294*). Long distance movement may be essential to make a long straight tail but movement on the spot, such as would even be available with the "near-dry" technique of HILLIER *et al.* (*118*), might suffice to bring about the metamorphosis of "flagellin" into helices. It seems that production of stainable helical flagella is governed by a gen different from the one that governs directional movement [HIRSCH (*119*), STOCKER, ZINDER and LEDERBERG (*292*), MANNINGER and NOGRÁDI (*186*)].

The reasoning by SMITH (*282*) and similarly by DUGUID *et al.* (*79a*) that the close correlation between presence of flagella and motility *proves* that flagella move bacteria would seem bad logic, the argument works just as well the other way round. In any case, correlation however close never proves causality.

There seems to be no forceful argument to accept flagella as motor organs. There are fairly good reasons not to accept them as such, as just shown. It is becoming more and more obvious that, as PIJPER claimed (*233*), the bacterial body is capable of more change in shape than the dogma of rigid cell walls allows for. KVITTINGEN (*153*, *154*) using phase contrast microscopy saw bodily movements in *B. proteus* which he could not accept as due to flagella. His *Proteus* swarmers were highly flexuous and anything but rigid. Swarmer cells could bend double and then one leg get shorter and the other longer until the body appeared completely stretched. Swarmer cells moved only the end of their body or only its middle part showing definite local waves of the cell wall. There were undulating movements like fishes perform, and even short cells could bend their bodies, and the bodies all round showed rotating movement of their spiral shapes. He argued that such movements could not be done by means of flagella, which moreover, he reasoned, would break off at the fast sudden accelerations the bodies were capable of, if the flagella were as thin and easily broken off as usually described. Even KAUFFMANN (*135*), although strongly supporting the view that flagella are motor organs, admitted that from what he saw there are bodily movements in bacteria quite independent of flagellar activity! PIJPER (*233*) showed that *Salmonella typhosa* and other bacteria have a spiral, or rather helical shape, when swimming normally (*251a*) which becomes particularly clear in slow motion pictures or when movement was retarded in other ways, through lowering the temperature or by adding large quantities of methylcellulose (*234*). The cell walls appeared pliable and elastic. This kind of movement was also seen by RICHARDS (*262*). KNAYSI (*144*) admitted that there is considerable deformation of *Salmonella typhosa* whilst swimming and thought this might set the flagella in motion which would then contribute in a secondary manner to the motility. CONN and ELROD (*66*) in *Bacillus cereus* and *Escherichia coli* in methylcellulose noticed flexible rods, although most microbes to them appeared short and rigid. SMITH (*281*) made the observation that a large fusiform bacillus from the throat was motile when freshly taken, and also in early culture, and then changed into a slow worm-like spiral form.

The orthodox view that bacteria are rigid is still upheld by BISSET (*30*) who adds that any appearance of flexion is an optical illusion, although he admits spiral shape. DUBOS (*79*) only knows of cylindrical rods and STANIER and VAN NIEL in their scheme for classification of bacteria (*286*) regard cell wall rigidity as an important characteristic of bacteria. Other authors recognise (presumably rigid) spiral or helical shape in bacteria that are traditionally described as rods. UMBREIT and ANDERSON (*311*) saw it in *Thiobacillus*, GILLESPIE (*97e*) in *Bacillus megaterium* and HODGE (*121*) saw some degree of curvature in many bacteria although he could not see spiral shapes. KNAYSI (*143*) agreed that motile bacteria sometimes assume helical shape permanently. It might be pointed out that many textbook pictures have shown this for a long time. ORLA-JENSEN (*218*) knew this when he said very early on that the shape of the cell or the form of the body was not a fundamental character and that rods might be in the shape of screws. BISSET (*25, 26*) pictured colonies of bacteria with the bodies in loops and folds but he ascribed the kinking and coiling not to active movement, but to the elongating through growth, of bacterial chains in a limited area.

It is of course one thing to recognise as is gradually being done [BISSET (*30*)] that bacteria have a tendency to look like helices or spirals during movement and afterwards, and quite another thing to ascribe their motility to their assuming such shapes. According to HESSE (*114*) "Schlängelung" is the chosen method of propulsion for aquatic creatures from worms to whales, the fins merely having a steadying action. BREDER (*43*) said that fishes swim like eels, series of waves going caudad, the fins merely doing the steering. Fishes could reverse the direction of these waves and so come to a stop, or even swim backwards. An advantage of such movement according to GRAY (*100*) is that water flows easily past undulating surfaces. TAYLOR (*299*) is of the same opinion and adds that microscopic bodies have more stress from viscosity than from inertia. A leading article in "ENDEAVOUR" (*88*) once pointed out that dolphins have an undulatory propelling motion which is most economical because it minimizes eddies and the article suggested that submarines should be built with actively undulating surfaces as propelling forces. It is tempting to say that bacteria possessed this secret of economic propulsion long before dolphins and that they used it in more perfect manner by applying it in three dimensions and steadying it by means of a long tail. Swimming bacteria often give the impression of being propelled by the helical shape. This becomes particularly clear in bacteria which take on grotesque forms by being poisoned with penicillin [FLEMING *et al.* (*94, 95*)]. SHANAHAN (*279a*) spoke of the movements of *Escherichia coli* in such conditions as a sluggish swinging of the entire cell, resembling somewhat the movement of a tadpole. FLEMING *et al.* said that under such conditions the bodies of *B. proteus* undulated like snakes and we can confirm this from our own experience. Penicillin-poisoned *B. proteus* resembles nothing so much in its motility, except for a greater speed, as *Beggiatoa.* The three dimensional movement suggests a wriggling of the body. GRAY (*101*) discussing this kind of motility for bacteria concluded that effective propulsion is only possible if the body is long enough to exhibit one complete wave. In many short bacterial bodies this is not so. So neither the flagellar theory, with its complications of holes and impossibility of supplying energy where it is needed, nor the so much simpler theory which makes the whole bacterial body the motor organ, can be regarded as proven. For the theory however that the forces that move the bacterium are situated inside the body and have their effect on the cell wall there remains another possibility. There may be forces in the body that produce a bulge in the cell wall that runs in helical fashion around the body. If this spiral is of short enough wavelength, even the shorter bacteria might fulfil GRAY's postulate (*101*) for at least one complete wavelength. The helical shape of the whole bacterial body might then arise secondary to this kind of propulsion. There might be a parallel case with forces that seem to be at work in a similar way in the bodies of *Myxobacteria,* as discussed in section VI. Spiral bodies inside bacteria that might have something to do with such forces have been described by KRIEG (*150*) and by CLAUDE (*59*). In this connection it is of interest to follow BISSET (*27*, *30*) in some phylogenetic considerations. KLUYVER and VAN NIEL (*142*), with an attractive simple way of reasoning, derive all bacteria from the simplest form, both existent and conceivable, the sphere, in the shape of a coccus. RIPPEL and PIETSCHMANN (*266*) think that phylogenetic considerations are premature. BISSET (*27*, *30*) with as much conviction, reasons that the ancestral type of all bacteria must be sought under the aquatic bacteria and that the most completely aquatic form is the *Spirillum.* This ancestral quality incidentally is then thought to be confirmed by PIJPER (*233*) who showed that many rod-like bacteria are slightly spiral [BISSET (*30*)]. BISSET then traces evolution through flagellar patterns. From *Spirillum* and *Vibrio* he goes through *Pseudomonas* to *Bacterium,* which last one has found it necessary to become peritrichous (but still remembers its origin in that it always produces its terminal flagella first [BISSET (*27*)]. The most interesting aspect of this evolution is that the *Myxobacteria* have discarded flagella and move through muscular activity of the cell wall. This would be a suddenly newly acquired capacity. Would it not be less far fetched to assume that this muscular capacity of wriggling the cell wall has been present from the beginning? In the section on *Spirilla* it will be shown that their cell walls possess remarkable capacities of wave-like contraction.

VIII. Adenosinetriphosphate and bacterial motility

In muscle, adenosinetriphosphate (ATP) has to do with the conversion of chemical energy into mechanical energy. It seems to do the same to spermatozoa. MANN (*185*) found that ATP disappeared from semen as the motility of the spermatozoa became less. NELSON (*211a*) found an enzyme in the flagellum of bull spermatozoa that hydrolysed ATP. HOFFMANN-BERLING (*124*) made "models" of flagella by extracting flagella of spermatozoa and trypanosomes with glycerin-water and saw them contract readily if ATP was added, even if the heads of the spermatozoa had been taken off. Vorhaus and Deyrup (*312*) placed a piece of aluminium foil on the ciliated pharyngeal mucosa of a frog and saw that its rate of movement doubled when ATP was added. This effect was highly consistent, rapid and reversible. In all these experiments the organs affected by the ATP were cilia or similar other animal structures.

Goldacre and Lorch (*97g*) injected ATP into living amoebae and saw that the speed of protoplasma streaming increased several times. This effect lasted less than a minute and could be repeated. It seemed harmless to the protoplasm whilst control injections of distilled water or saline had no speeding up effect but seemed harmful.

De Robertis and Peluffo (*267*) undertook experiments with motile bacteria and ATP. They first anaesthetized proteus bacteria with chloral hydrate. They then revived them by washing out the chloral hydrate with various solutions containing phosphor and also ATP, every time noting the time elapsed before vibratory and translatory movement of the bacteria started. Apart from the obvious difficulty of making quite sure when such movements have properly started in a microscopic preparation, the experiments do not seem numerous enough to satisfy statistical requirements. It seemed that ATP worked about 74% faster than other solutions to re-establish translatory movement, but the authors admit that the times noted varied considerably from one experiment to another. They ascribe their results to an action of ATP on the bacterial flagella. This is not justified as flagella were not visible and the ATP might just as well have had an effect on the wriggling capacities of the bodies. These experiments have never been confirmed, we have repeated them in various ways without consistent results. De Robertis and Franchi (*268*, *270*) also tried to show by means of an EM the effect of ATP on detached flagella of *Bacillus brevis*. They claimed that their pictures showed a difference before and after treatment with ATP, the treated flagella showing contraction or even tightly packed tufts. The pictures do not look convincing, especially as the authors in a previous publication (*269*) admitted that untreated flagellar filaments of *Bacillus brevis* may show variations in pitch from one filament to another, and even in different regions of the same filament. Barlow and Blum (*20*) approached the same problem from a different angle. They made a suspension of bacterial flagella in the way Gard (*97b*) did and then found that in light-scattering experiments ATP caused no change in turbidity. This implied that there was no change in state of aggregation or shape. Myosin under such conditions had shown them very drastic changes. Snellman *et al.* (*283*, *284*) showed that ATP contracts actomyosin. It would appear that ATP has no effect on bacterial flagella, but has a clear-cut definite effect on other flagella and motor structures.

IX. Flagellar agglutination

The widespread belief that specific H serum added to motile bacteria causes *immediate paralysis* of the flagella which stops motility, is difficult to trace to its origin. Addition of H serum to sensitive motile bacteria interferes with motility, but the effect is not dramatically immediate. H agglutination should be studied with sunlight D-GM which makes flagella visible before and after [PIJPER (*229* — *232*)]. The erroneous belief of sudden paralysis probably started with ARKWRIGHT who studied H agglutination under the microscope without being able to see flagella (*5*). He nevertheless stated that flagella become hampered in their movement and that the bodies in the clumps may, at any rate for some time, dance at the end of their tether and occasionally show violent movements to and fro and up and down. This of course is really a description of what is visible in already formed clumps. He had passed over what happened before the formation of clumps. Adding a drop of H serum to a broth culture of *Salmonella typhosa* under a coverslip in sunlight D-GM has, as its first effect, a slowing down of motility but this is not universal (*229*, *232*). Tails still in full movement begin to show localised thickenings in the shape of granular deposits and so the tails become thickened and stiffened into helices and eventually become covered with a complete sheath. A similar precipitate descends on the bodies and makes them appear much thicker than is normal, they also become covered by a sheath. All this hampers movement. The precipitate is very similar to that caused by the colloid substances described in section II d (Figs. 1 and 2). Motility decreases gradually. Spasmodic sudden movements happen in between quieter periods. Often bodies keep on turning like windmills at the end of a fixed flagellum. This last phenomenon may be due to the whirling of flagella caused by tensions in the helical structure as described in section V a. Agglutination takes place through fortuitous entanglement of the thickened corkscrew-like flagella as shown previously and filmed (Fig. 29) (*229*, *232*).

Addition of H serum in the absence of *salt* has no effect, swimming goes on normally for hours with straight tails and no precipitates appear. All this indicates that there is no such thing as nervous paralysis but only a hampering precipitate. This precipitate was also seen with an EM by MUDD and ANDERSON (*202*); HEIDELBERGER and KABAT (*112*) said long ago that agglutination is precipitin reaction at bacterial surfaces. A forgotten observation was made many years ago, before H and O had been discovered, by PROCA (*259*). He added specific serum to *Salmonella typhosa* and saw them becoming non-motile and he observed thickened flagella in his dark-ground microscope. He added however that he got a similar effect with normal horse serum in strong concentration. He evidently was dealing with partly specific and parthy non-specific, just colloid precipitates. Pseudo-agglutination is common in colloid solutions.

Similar agglutination without definite loss of motility is apparently also performed by a soil amoeba that feeds on bacteria (*260*).

The effect of H serum appears to be similar in effect to that of precipitating colloids. But it is specific, takes place in lower dilutions and probably also acts quicker. Adding H serum to motile *Sarcina ureae* also resulted in thickening of tails by granular precipitates and steadily decreasing motility. The resulting thickened helices and stiff rods also caused agglutination through fortuitous entanglement (Fig. 30) [PIJPER *et al.* (*248*)].

H serum thus has an *indirect* effect on motility. A *direct* effect on motility can be seen when Vi serum is added to *Salmonella typhosa*. The flagella are not visibly affected but the bodies show erratic movement which, through very much increased frequency of bumping, leads to agglutination (*231*).

A difference between bacterial flagella and definite motor organelles was again demonstrated by Senterfit (*276 h*). He added an immune serum against *Schistosoma mansoni* to live miracidia and saw them becoming immobilized and their cilia becoming paralysed. In low serum dilutions the cilia were only affected in limited areas on the miracidial surface.

X. Motility and flagella of Sarcinae and other cocci

Ellis (*83*, *84*) claimed that he could produce motility and stainable flagella in all cocci, including *Sarcinae*, simply by rapid transplantation. Apart from what he described as "trembling" motility-on-the-spot his description of motility is not convincing, as pointed out by van Niel (*212*). Van Niel condemned Ellis' flagella on account of their shapelessness. They resemble what Zettnow (*344*) produced as attached to staphylococci and other supposedly non-motile bacteria. In some instances however Zettnow also achieved some quite shapely appendages. Flagella staining is a notoriously unreliable and deceptive method.

There is no doubt that motility exists in some *Sarcinae* and other cocci. Pijper, Crocker and Savage (*248*) studied motility and flagellation in *Sarcina ureae* and *Sarcina agilis* with sunlight D-GM and found many points of resemblance in their flagella with those of other *Eubacteriales* like *Salmonella typhosa*. The kind of motility however was different. Packets of *Sarcinae* move by tumbling. This is probably the same as what Sames (*276*) called "waltzing" movement. To see this tumbling, agar cultures are very suitable and the microculture technique on agar described by Knaysi (*145*) was of great help (*248*). Tumbling motility occurs in other *Eubacteriales* but only when motility is not very good. A semi-somersault in *Salmonella typhosa* as described in section V a and as filmed (*251a*) might be regarded as an occasional tumbling movement interrupting a course of steady forward movement. In *Sarcinae* during forward movement by tumbling some packets are seen to be followed by a straight fuzzy-looking tail (Figs. 31 and 32). It seems to be so loosely attached that it is not affected by tumbling. It gives the impression of being dragged along without taking an active part in motility. Apart from tumbling, trembling movement occurs. It is possible that Ellis (*83*, *84*) saw this in otherwise non-motile *Sarcinae*, motility may be a relative thing as suggested in section VII.

In *Sarcinae* the fuzzy-looking tail can stiffen into a rod (Fig. 11), straight or angular, or, more usually, into a helix, either short or long wave. The *Sarcinae* were the bacteria in which this doubling of wavelengths (biplicity) was first seen [Pijper and Abraham (*249*)]. The metamorphoses of the tail follow the same pattern as described for other *Eubacteriales* in section V a and tumbling goes on undisturbed. Splitting takes place before or after stiffening. Splitting and chance distribution of washed up broken off flagella cause peritrichous appearances, as in the beautiful stained pictures of Levenson (*170*, *171*). *Sarcina ureae* and *Sarcina agilis* showed flagella of exactly identical wavelengths (*249*) but *Sarcina agilis* showed relatively more short waves. There always is a good deal of linking and packing leading to formation of giant-flagella as in Fig. 25.

Pijper *et al.* concluded from their observations with sunlight D-GM (*248*, *249*) that the tails of the packets are structures very similar to the tails of other

Eubacteriales, that they are not motor organs and that motility is due to cytoplasmic movements inside the cell wall which cause tumbling, trembling being a minor form of this kind of movement. H agglutinating serum was seen to work in exactly the same way as described for *Salmonella typhosa* in section IX (Fig. 30).

Motile streptococci have been described by KOBLMUELLER (*146*), AUERBACH and FELSENFELD (*15*) and by GRAUDAL (*99*) but these authors say very little about the kind of motility and flagella.

XI. Motility and flagella of Vibrios

Vibrios are commonly supposed to move by means of one or sometimes two or three polar flagella [BISSET (*30a*), KNAYSI (*143*)]. MUDD, POLEVITZKY and ANDERSON (*204*) suggested continuity between wavy threads and protoplasm in *Vibrio schuylkiliensis* in an EM picture but saw no basal granule. KNAYSI'S refusal (*143*) to accept a basal granule that had appeared in an EM picture of *Vibrio cholerae* by MUDD and ANDERSON (*205*) because it was an isolated finding, is mentioned above. VAN ITERSON (*129*) had the impression that in one of her EM pictures of *Vibrio metschnikovii* a flagellum originated from a basal granule. Later (*130*) she printed several EM photographs of the same bacterium, intending to show basal granules and flagellar sheaths. It is very tempting to accept these ready solutions to long standing problems. One can also be critical and wonder whether it is not merely a question of chance localisation whether in these pictures a small body is regarded as a blepharoblast, part of a sheath or just an odd particle on the screen. GRACE (*98*) showed an EM picture of *Vibrio cholerae*, later used again by BISSET (*30*), where a flagellum apparently ends at a roundish body inside the cell, but there are four similar, unattached, bodies quite near inside the same cell.

A study by PIJPER and NUNN (*241*) by means of dark-ground microscopy showed that *Vibrio metschnikovii* moves with three dimensional undulation, rather like *Salmonella typhosa*. Instead of a fuzzy tail however there often is a stiff rod in front, looking like a snout or a bowsprit. On closer examination it often appears as a fine coil or helix. Sudden reversals were common, and then, contrary to what happens in *S. typhosa*, where the tail would flow to the other end, the bowsprit as such stayed where it was and appeared as a tail. With the next reversal it became a snout again. Some individuals had a bowsprit and a tail. Bowsprits were thicker than tails and thus easier to see, creating an impression of a majority of bowsprits. Probably bowsprits became compressed (and thicker) through the movement, and tails elongated (and thinner). When movement stopped the coil-like nature of the structure became more obvious. All these appendages were visible with simple D-GM, no sunlight being needed. Addition of colloid substances made little difference to visibility. The appendage sometimes split into two or three helices, which never became distributed over the body. There are therefore similarities and differences between the appendages of *Vibrios* and *S. typhosa*, and it seems reasonable to suspend judgment as to the nature of these appendages until they have been further investigated. There are indications in the observations of VAN RIEMSDIJK (*263*) and BRUCE WHITE (*50*) that these appendages may develop the same kind of whirling movement as described in section Va for other *Eubacteriales*.

XII. Flagella of Spirochaetales

Here too the classification of BERGEY's Manual (*44*) is followed. The motility of the *Spirochaetales* has not been studied thoroughly, they are traditionally regarded as flexuous.

Reports on their flagellation are confused. In the days when workers had to rely on staining, ZETTNOW (*343*) pictured numerous flagella on *Borrelia recurrentis* from Africa, and FRAENKEL (*96*) similar ones on other *Borreliae*. LEIFSON (*161*) very much later (1950) still made stained pictures of flagella on *Borrelia novyi*. The supposed body of the microbe however looks strange and large and the attached flagella resemble *Borreliae*. LOFGREN and SOULE (*174, 175*) applying an EM to non-shadowed *Borrelia novyi* found long thin flagella-like fibres and a terminal filament but decided these were shredded periplasts. BABUDIERI and BOCCIARELLI (*17, 18*) never saw real flagella at the sides of *Borreliae* with an EM, and no axostyle, but they did see an undulating membrane which macerated into flagella-like fibres. They added that there might be tufts of real flagella at a rounded end of the body. WATSON *et al.* (*315*) also using an EM on a *Borrelia* saw what looks like peeled off threads and called them flagella.

As to *Leptospirae* there is general agreement that there are no flagella. MORTON and ANDERSON (*200*) saw no flagella-like structures nor any internal structures in *L. icterohaemorrhagiae* and *L. canicola*. VAN THIEL and VAN ITERSON (*303*) said the same about *L. biflexa*. BABUDIERI (*16*) said about the whole genus that they were long thin protoplasmic cylinders enclosed by a membrane, and wound round a thin axial filament, without flagella or cilia. BREESE, GOCHENOUR and YAGER (*45*) examined seven kinds of *Leptospirae* and do not mention flagella, but they saw a fine filament entwined about the main structure, often extending beyond one or both ends. CZEKALOWSKI (*71*) described an axostyle which was both backbone and locomotion organ.

There is no such unanimity about *Treponema*. LEIFSON (*162*) using his staining method described *Treponema pallidum* as polar monotrichous, or polar (perhaps subpolar) multitrichous. NEUMANN (*215*) found no flagella nor an axostyle in *T. pallidum*. MORTON and ANDERSON (*199*) using an EM on *T. pallidum* without shadowing saw terminal filaments which they ascribed to cell division and also flagella-like filaments on the sides. MUDD, POLEVITZKY and ANDERSON (*206*) with an EM saw flagella on all their strains of *T. pallidum*, and also on *T. macrodentium* but not on *T. microdentium*. HAMPP, SCOTT and WYCKOFF (*107*) using an EM on *T. pallidum* and oral spirochaetes found terminal filaments and flagella on the sides but make no mention of an axial filament although it seems to be visible in some of their pictures. Similar pictures of oral spirochaetes appear in WYCKOFF's book (*341*). MORTON, RAKE and ROSE (*201*) produced magnificent EM pictures of treponemes with flagella which they take to be definite structures and not fragmentations. ANGULO *et al.* (*3*) and WATSON *et al.* (*315*) found flagella in most of the EM pictures of the several *Treponemata* they examined. DYAR (*80*) in a free-living *Spirochaeta* not further defined, examined with D-GM, phase contrast and EM saw no flagella and although there seemed to be an axial filament he took this to be an artefact.

A study by BRADFIELD and CATER (*40*) of *Leptospira*, *Treponema* and *Cristispira* by means of an EM brought them to the conclusion that these microbes all have one or more fibrils wound spirally round their body from end to end, sometimes apparently outside the cell membrane and sometimes inside. They deny the existence of free fibrils or flagella. SWAIN (*297*) in a similar study of a large variety of *Spirochaetales* came to similar conclusions, denying the existence of true flagella.

It would appear that the majority of workers say that *Spirochaetales* have no real flagella. If their structure is as BRADFIELD and CATER and later on SWAIN described, this structure may easily give rise to artefacts resembling flagella. The complicated structure of *Spirochaetales* so different from the usual *Eubacteriales* raises doubts as to whether they can be regarded as bacteria.

XIII. Some other flagellated bacteria

It is doubtful whether *Selenomonas* is a bacterium. WOODCOCK and LAPAGE (*339*) found this microbe, somewhat resembling the sickle of the moon, in the rumen, but noticed that it could also adopt an oval shape. In either shape it moved by tumbling although in the crescent shape, when it sometimes but not always showed a flagellum, the flagellum might act as a pulsellum. BOSKAMP (*38*) described this microbe and ascribed all motility to flagella, but admitted that movement sometimes resembled tumbling. He saw no flexibility in the body. MACDONALD (*209*) probably had the same microbe and called it *Spirillum sputogenum*, and he saw an erratic tumbling movement. WILSKA (*336*) apparently saw it in the mouth and photographed its flagella with his Anoptral microscope. When present the flagellar bundle is always attached to the concave side of the half moon. LESSEL and BREED (*168*) also saw it moving with a tumbling kind of motility. They suggest that MACDONALD has shown that it is a bacterium and that it may well be intermediate between *Spirillum* and *Spirochaeta*, because it has a tuft of flagella but can also move independent of flagella. If this were correct it would confirm that bacteria can move by changing the shape of their body from the inside. However, JEYNES (*132*) has given a whole list of diagnostic points (nuclear structure, blepharoblast, mode of division, no bacterial cell wall and others) which would place the affinities of *Selenomonas* with the *Protozoa*.

A similar uncertainty surrounds *Caulobacter*. It forms stalks in a manner that is not clear [HOUWINK (*127*), BOWERS *et al.* (*39*)]. Filaments often radiate from the bases of the stalk whilst a new young cell is formed at the other end and shows a flagellum in the EM. It seems advisable to wait for the position to clear before these appearances are applied to bacterial physiology.

In *Nocardiae* motility was recently described by TOPPING (*307a*) and by ØRSKOV (*219a*). Topping mentioned peritrichous flagella, ØRSKOV saw no flagella at all. Previous reports on motility of *Nocardiae* had better be ignored as not being made on assuredly pure cultures. JENSEN (*131a*) in *Nocardia citrea* received from ØRSKOV saw motile hyphae with undetermined pattern of flagellation. ERIKSON (*89*) described what she has called *Nocardia turbata* n. sp., also received from ØRSKOV, as showing motile cells under appropriate conditions, but she saw no flagella. Examined with sunlight D-GM numerous wriggling motile cells were seen in this *Nocardia*, also branching forms, and they showed helical flagella irregularly attached and often broken off (*251*). This *Nocardia* only showed motility under special conditions. This confirms the opinion of JENSEN (*131b*) both for *Nocardiae* and *Corynebacteria* that finality as to motility cannot be reached until special methods of cultivation have been applied in all cases where motility is in doubt.

A new genus of the *Actinomycetales*, *Actinoplanes*, was described by COUCH (*67*, *68*) as having motile flagellated spores, it is perhaps better regarded as not a bacterium.

XIV. Motility and flagella of Spirilla

In section IV and Table 1 the reasons were given as to why the flagella of *Spirilla* must be regarded as entirely different from those of the other *Eubacteriales* and why they must be discussed separately. Neglect of this differentiation is still apparent in recent publications [SALLE (*274*), BISSET (*30*)] and is causing unnecessary confusion and holding up research (*245*, *246*, *250*). The total number of recognised *Spirilla* is not high, nearly all of them can be found, very well described, in GIESBERGER'S book (*97d*). New ones have been added since by DIMITROFF (*77*) and especially by WILLIAMS (*334*).

They are traditionally regarded as spiral shaped rigid bodies, just as the *Spirochaetes* are traditionally regarded as flexuous [BERGEY'S Manual (*44*)]. Most textbooks and writers insist upon this rigidity [GIESBERGER (*97d*), DIMITROFF (*77*)] except perhaps BISSET (*30*) who said the cell wall might be soft or rigid. GIESBERGER (*97d*) pointed out that older laboratory cultures often show an increasing number of straight and less motile individuals. From the observations of PIJPER *et al.* (*245*) there is no doubt that *Spirilla* are very flexuous and elastic. At rest they appear as broadly wound coils, as in the centre of Fig. 33. When getting up speed they become stretched, and at high speed they look perfectly straight as at the periphery of Fig. 33. As soon as they stop they become broad coils again. In the publication mentioned (*245*) a series of pictures copied from a cinemicrographic film (*251b*) show these consecutive changes in the shape of the body. The elasticity of the body is so great that even the total length of the body is not constant.

Most *Spirilla* carry flagella, usually one at each end (Figs. 4 and 7). These can be seen, in wet preparations, occasionally in bright field microscopy [HOUWINK (*126*), PIJPER (*239*)], better in phase contrast microscopy (*240*), quite easily in simple D-GM (*239*), and best in sunlight D-GM (*239*, *245*, *246*, *250*, *251b*). EM shows them well but by the time they reach it they usually have split (*129*).

The thin neck that connects the flagellum to the body occurs in every good picture (*97d*, *140*) but has never been especially noted except by PIJPER (*239*, *245*, *250*) and is of importance because it so definitely localizes the place of attachment at the pole (Figs. 4 and 7).

It has been known for a long time [METZNER (*193*, *194*), ULEHLA (*310*)] that in D-GM one can see during forward movement the frontal flagellum slung round the body and the rear one rotating freely in the liquid (*245*, *250*, *251b*). When the body reverses the two flagella change their function, the rear one becoming slung round the body and the front one rotating freely in the liquid. Spirilla with only one flagellum go through the same exercises but it is difficult to understand if motility is dependent on the flagellum how movement can be equally fast to and fro, or even how it comes about at all in one of the two directions. A flagellum swung round the body cannot exert much propelling power. Equally good motility can be seen in *Spirilla* without flagella, whether they are taken off by shaking or naturally absent (*239*, *251b*).

It is generally assumed by those who believe that motility is due to flagellar action that the rear flagellum rotates by itself. This would then maintain a rotation of the spiral body which through its spiral shape would screw itself through the water [Metzner (*193*)]. Metzner studied this kind of motility by means of models and a stroboscope (*194*). In a study of live *Spirilla* also with a stroboscope (*193*) he claimed that the rear flagellum on an average rotated three to four times faster than the body. As *Spirilla* do a good deal of freewheeling (*251b*) these figures need not be taken at face value. In sunlight D-GM the bodies of *Spirilla* can often be seen rotating at high speed whilst their rear flagellum somewhat languidly performs an occasional revolution (*251b*). Metzner (*193*) insisted that the rear flagellum always rotated in a direction opposite to that of the body. If this were true the thin neck of the flagellum could not possibly last. Metzner (*193*) is concerned as to how the coupled system of two flagella works in harmony especially at sudden reversals, how the plasma conducts impulses from the one flagellum to the other. He does not like the idea of there being blepharoblasts as kinetic centres. His difficulty would have resolved itself if he had accepted the view that it is the body that through stretching and coiling is responsible for all movement. It is an easy matter to imagine the whole body reversing the direction of its convolutions to go into reverse. The flagella would then automatically be slung round the front end and turn freely at the rear. Obviously, if the rear flagellum did push the body, the coils of the body, as a result of the pressure encountered, would become broader. It is an observational fact (Fig. 33) that they stretch (*245*). It is strange that Metzner (*193*) *had* noticed that the body was not always rigid and that the "diameter" (amplitude) of its windings got less during rapid swimming, sometimes making the body look quite straight! But he ascribed this in very complicated fashion to a "centripetal pull" that became effective particularly during very fast rotation and of which he saw the cause in the distribution of the resistance of the moved body.

The flagellum of *Spirilla* can split into a number of flagella of exactly the same shape of the original one, illustrating the toughness of the material. Splitting is common at death and probably is pathological, it is not seen during normal activity (*239*, *245*). These sub-flagella remain attached through the same thin neck. They often keep on performing whip-like movements which do not move the body. Each subflagellum acts independently, some may keep still whilst others in the same bunch lash out, each having its own tempo (*251b*). This does not suggest a common blepharoblast, as claimed by Bisset (*30*).

Spirilla readily undergo autolysis, resulting in a completely empty cell wall. The flagellum or subflagella are then seen as a continuation of the cell wall (Fig. 7). They have also been found to stain like cell wall, not like flagella (*239*, *245*). There are EM pictures by Houwink (*126*) and by Bisset (*30*) where the flagellum or subflagella appear to be continuous with the fibres of the striated cell wall. DIMITROFF (*77*) from what he saw with ordinary microscopy placed the origin of the flagellum in the cell membrane (wall) because both reflected light with the same intensity. The absence of refractive material in the cytoplasm opposite the place of attachment of the flagellum proved to him that the flagellum did not go through the cell wall. Many authors still believe that the flagellum pierces the cell wall. Most of this belief is based on one solitary EM photograph of VAN ITERSON (*129*), which keeps on appearing in various publications (*30, 140, 274*), always as "proof" that flagella pierce cell walls. It proves nothing of the kind as shown by PIJPER *et al.* (*245*). A cytoplast pulling a flagellum or subflagella by a thin neck through a cell wall would show only one piercing of the cell wall, not several, as in the photograph. The correct explanation of the picture is that *Spirilla* are apt to blow out their cell wall in bubbles as in Fig. 34. If such a bubble eventuates at the pole it displaces the place of attachment of the thin neck. Looking at Fig. 34 in the plane of the paper would give a picture in which it would appear as if there were flagella going through the wall, and this is the explanation of the VAN ITERSON picture.

Linked up with this question of penetrating flagella is the question of blepharoblasts. VAN ITERSON (*130*) has abandoned her belief in blepharoblasts in *Spirilla*. At the same time GRACE (*98*) and BISSET (*30*) have produced a few EM pictures with supposedly visible blepharoblasts and BISSET (*30*) is sure there is only one blepharoblast for each polar bunch of flagella. To make these pictures convincing there should be fewer similar bodies lying about, inside and outside the cells. The possibility of blepharoblasts cannot be ruled out but in view of the improbability that the flagella go through the wall a sceptic might ascribe the presence of these bodies as due to chance. The solitary EM picture by BRADFIELD (*41a*) where one *Spirillum* is equipped with a whole bunch of supposed blepharoblasts rather goes against BISSET's conception.

Flagella thus do not appear to be essential nor very suitable for moving *Spirilla*. The body can manage quite well for itself by stretching its coils to get propulsion. Periods of freewheeling in between the stretchings would give an opportunity to regain the broadly coiled state. These features have been filmed (*251b*) and some idea of them can be gained from published pictures (*245*). KING and BEAMS (*138*) centrifuged *Spirillum volutans* at very high speed until the distribution of the material inside the bodies was very much upset and this brought motility to a stop. The disturbances of material however were in the "troughs" of the body, not at the poles where blepharoblasts would be situated.

The origin of the flagella of *Spirilla* is unknown. The terminal flagellum may be supposed to arise from cell division as described by DALLINGER and DRYSDALE (*72*) and by CARPENTER (*55*). The separating daughter cells may draw out a cell wall filament between themselves which eventually snaps. PIJPER (*250, 251b*) working with two new *Spirilla* isolated by and received from Miss MARION WILLIAMS (*334*) found that as cultures aged they showed an increasing number of additional flagella of various shapes distributed all over the body and creating a sort of peritrichous appearance (Fig. 35). This was followed by a stage when these additional flagella began to look abnormal in appearance and the cell wall became surrounded by large numbers of fibrillae of various shapes and sizes (Fig. 36). These additional appendages soon got loose and drifted through the medium (Fig. 37). They were explained as excessive efforts at cell wall growth, being obviously derived from the cell wall. BÜNNING and GÖSSEL (*51*) found similar structures after treating *Spirillum volutans* with chloral hydrate. They also occur in other *Spirilla* and may be the explanation of the fibre lying on top of the solitary picture by BRADFIELD (*41a*) of *Spirillum serpens*. It seems worth while to investigate these structures further, chemically and physically. BLANK (*34*) found chitin in the cell wall of dermatophytes, the hardness of these flagellar structures suggests something similar.

XV. Bacterial fimbriae

ANDERSON (*1*) had an EM picture of a non-motile strain of *Escherichia coli* with a fringe of curiously shaped filaments round the bacterial body. He called them "streamers" and supposed them to be mucous threads of capsular material. A good illustration of such structures is Fig. 38 made by SCHREIL and SCHLEICH (*276c* and *d*) and kindly presented by SCHREIL. Since ANDERSON first showed them in 1939 they have been photographed, always with an EM, by numerous observers. HOUWINK and VAN ITERSON (*125*) thought at first that they were only formed on solid surfaces. VAN DEN ENDE *et al.* (*87*) found them on *Pseudomonas* and thought it was difficult to differentiate them from real flagella. DE *et al.* (*75*) photographed them on a variant of *E. coli*. MUDD *et al.* (*207*) saw them as filiform extensions of the cell wall in *E. coli* and called them pseudopod-like. ROHRER (*271*) photographed similar structures in much coarser form. WEIBULL and HEDVALL (*328*) stated that on *B. proteus* they did not disintegrate at low p_H like flagella did, and that they were not destroyed by either 1 N HCl nor 1 N NaOH, nor by trypsin, pepsin, ribonuclease or desoxyribonuclease. BRINTON *et al.* (*47*) found that in a strain of *E. coli* there were two kinds of bacteria. In electrophoresis one travelled fast and the other slowly. There were no intermediate forms. Only the slow ones possessed filaments. In viscosity experiments those with filaments were slower to settle. The filaments came off in a Waring Blendor and this converted the slow ones into fast ones. This experiment also showed that although in the EM pictures before shaking the filaments appeared to go through the cell wall, in the same way as flagella often seem to do, after shaking the filaments were lying about and the cell walls looked intact. SMITH (*282*) also photographed such structures on *E. coli* and later DUGUID, SMITH, DEMPSTER and EDMUNDS (*79a*) produced very clear pictures of them on *E. coli*. They suggested the name of "fimbriae" for them and this sounds acceptable. They find them to be more numerous than flagella, and to measure 0.01 μ when flagella measure 0.02 μ. SCHREIL and SCHLEICH (*276d*) give similar measurements. BRAUN (*42a*) found similar things, which he called fibrils, coming out of bacterial bodies of *B. proteus* and *E. coli*. His fibrils often show a structure like a chain of pearls, somewhat resembling what VAN ITERSON (*130*) described as a sheathed flagellum. Whilst other authors saw the "fimbriae" on non-motile bacteria, BRAUN (*42a*) is sure they only occur on motile ones. He thinks that in *E. coli* the flagella are built out of smaller units and that these smaller units might be these fimbriae. He has pictures showing fimbriae coming out of flagella but as always in EM pictures superposition is not excluded. On the whole ANDERSON'S (*1*) explanation seems the most acceptable.

References

1. ANDERSON, T. F.: On the mechanism of adsorption of bacteriophages on host cells. In: The nature of the bacterial surface ed. A. A. MILES and N. W. PIRIE. Oxford: Blackwell Scientific Publications 1949.
2. ANGERER, K. v.: Über die Arbeitsleistung eigenbeweglicher Bakterien. Arch. f. Hyg. **88**, 139 (1919).
3. ANGULO, J. J., J. H. L. WATSON, C. COURTNEY WEDDERBURN, F. LEON-BLANCO and G. VARELA: Electronmicrography of treponemas from cases of Yaws, Pinta and the co-called Cuban form of Pinta. Amer. J. Trop. Med. **31**, 458 (1951).
4. ARENA, A., et R. SCHWARTZ: Coloration des cils par la méthode de Cesares-Gil. C. r. Soc. Biol. Paris **124**, 500 (1937).
5. ARKWRIGHT, J. A.: Microscopic evidence of the different manner of clumping of motile bacteria with somatic and flagellar agglutinins. J. of Path. **30**, 566 (1927).
6. ASCHNER, M., and S. HESTRIN: Fibrillar structure of cellulose of bacterial and animal origin. Nature (Lond.) **157**, 659 (1946).
7. ASTBURY, W. T.: The forms of biological molecules. In: Essays on growth and form. Oxford: Clarendon Press 1945.
8. — Some recent adventures among proteins. Publ. Staz. zool. Napoli **23**, 1 (1951).
9. — Flagella. Sci. Amer. **184**, 21 (1951).
10. — Hairs, muscles and bacterial flagella. Nature (Lond.) **167**, 880 (1951).
11. — Structure of algal flagella. Nature (Lond.) **171**, 280 (1953).
12. —, and C. WEIBULL: X-ray diffraction study of the structure of bacterial flagella. Nature (Lond.) **163**, 280 (1949).
13. —, and N. N. SAHA: Structure of algal flagella. Nature (Lond.) **171**, 280 (1953).
14. — E. BEIGHTON and C. WEIBULL: In: Fibrous proteins and their biological significance. Symposia of the Soc. exper. Biology. Cambridge: University Press 1955.
15. AUERBACH, H., and O. FELSENFELD: An unusual strain of streptococcus isolated from subacute bacterial endocarditis. J. Bacter. **56**, 587 (1948).
16. BABUDIERI, B.: Ricerche di microscopia elettronica. Studio del genere *Leptospira*. Rend. Inst. super. Sanità **11**, 1046 (1948).
17. —, and D. BOCCIARELLI: Ricerche di microscopia elettronica. Studio morfologico del genere *Spironema*. Rend. Inst. super. Sanità **6**, 305 (1943).
18. — — Electron-microscope studies on relapsing fever spirochaetes. J. of Hyg. **46**, 438 (1948).
19. BAKER, R. F., and D. C. PEASE: Sectioning of the bacterial cell for the electron microscope. Nature (Lond.) **163**, 282 (1949).
20. BARLOW, G. H., and J. J. BLUM: On the "contractility" of bacterial flagellae. Science (Lancaster, Pa.) **116**, 572 (1952).
21. BARTHOLOMEW, J. W.: Flagellation of certain species of *Pseudomonas* as seen with the electron microscope. J. Gen. Microbiol. **3**, 340 (1949).
22. BEEBE, J. M.: Morphology and cytology of *Myxococcus xanthus*. J. Bacter. **42**, 193 (1941).
23. BIRCH-ANDERSEN, A.: Reconstruction of the nuclear sites of *Salmonella typhimurium* from electron micrographs of serial sections. J. Gen. Microbiol. **13**, 327 (1955).
24. — O. MAALØE and F. S. SJÖSTRAND: High-resolution electron micrographs of sections of *E. coli*. Biochim. et Biophysica Acta **12**, 395 (1953).
25. BISSET, K. A.: The structure of "rough" and "smooth" colonies. J. of Path. **47**, 223 (1938).
26. — The mode of growth of bacterial colonies. J. of Path. **48**, 427 (1939).
27. — Evolution in bacteria and the significance of the bacterial spore. Nature (Lond.) **166**, 431 (1950).
28. — The morphology and cytology of bacteria. Annual Rev. Microbiol. **5**, 1 (1951).

29. Bisset, K. A.: Some relationships between morphology and antigenic structure in bacteria. Nature (Lond.) **171**, 1118 (1953).

30. — The cytology and life-history of bacteria, 2nd ed. Edinburgh and London: E. & S. Livingstone Ltd. 1955.

30a. — Bacteria. Edinburgh and London: E. & S. Livingstone Ltd. 1952.

31. —, and C. M. F. Hale: The development of bacterial flagella in the germinating microcyst. J. Gen. Microbiol. **5**, 150 (1951).

32. — The development of the surface structures in dividing bacteria. J. Gen. Microbiol. **5**, 155 (1951).

33. —, and J. B. Grace: The nature and relationships of autotrophic bacteria. In: Autotrophic Micro-organisms. Cambridge: The University Press 1954.

34. Blank, F.: The chemical composition of the cell walls of dermatophytes. Biochim. et Biophysica Acta **10**, 110 (1953).

35. Boivin, A., et L. Mesrobeau: Sur la resistance à l'acide trichloracétique de l'antigène flagellaire du bacille typhique. C. r. Soc. Biol. Paris **129**, 136 (1938).

36. Böe, J.: On the motility of fusobacterium. Acta path. scand. (Copenh.) **20**, 573 (1943).

37. Bonifas, V., E. Kellenberger et G. Werner: L'application de la microscopie électronique aux recherches biologiques. J. de Radiol. et Électrol. **30**, 286 (1949).

38. Boskamp, E.: Über Bau, Lebensweise und systematische Stellung von *Selenomonas palpitans* (Simons). Zbl. Bakter. I Orig. **88**, 58 (1922).

39. Bowers, L. E., R. H. Weaver, E. A. Grula and O. F. Edwards: Studies on a strain of caulobacter from water. J. Bacter. **68**, 194 (1954).

40. Bradfield, J. R. G., and D. B. Cater: Electron-microscopic evidence on the structure of spirochaetes. Nature (Lond.) **169**, 944 (1952).

41. — New features of protoplasmic structure observed in recent electron microscope studies. Quart. J. Microsc. Sci. **94**, 351 (1953).

41a. — Organization of bacterial cytoplasm. In: Bacterial anatomy, ed E. T. C. Spooren and B. A. D. Stocker. Cambridge, England: Cambridge University Press 1956.

42. — Electron microscopic observations on bacterial nuclei. Nature (Lond.) **173**, 184 (1954).

42a. Braun, H.: Beiträge zur Feinstruktur der Bakteriengeißeln. Arch. Mikrobiol. **24**, 1 (1956).

43. Breder, C. M.: On the locomotion of fishes. Zoologica **4**, 159 (1926).

44. Breed, R. S., E. G. D. Murray and A. Parker Hitchens: Bergey's Manual of determinative bacteriology. London: Baillière, Tindall & Cox 1948.

45. Breese, S. S., W. S. Gochenour and R. H. Yager: Electron microscopy of leptospiral strains. Proc. Soc. Exper. Biol. a. Med. **80**, 185 (1932).

46. Brindle, C. S., and S. T. Cowan: Flagellation and taxonomy of Whitmore's bacillus. J. of Path. **63**, 571 (1951).

47. Brinton, C. C., A. Buzzell and M. A. Lauffer: Electrophoresis and phage susceptibility studies on a filament-producing variant of the *E. coli* B bacterium. Biochim. et Biophysica Acta **15**, 533 (1954).

48. Brown, H. P.: On the structure and mechanics of the protozoan flagellum. Ohio J. Sci. **45**, 247 (1945).

49. —, and A. Cox: An electron microscope study of protozoan flagella. Amer. Midl. Naturalist **52**, 106 (1954).

50. Bruce White, P.: A note on the globular forms of *Vibrio cholerae*. J. Gen. Microbiol. **4**, 36 (1950).

51. Bünning, E., u. I. Gössel: Eine Wirkung von Chloralhydrat auf Spirillen. Arch. Mikrobiol. **21**, 411 (1955).

52. Burcik, E., u. B. Plankenhorn: Darstellung von Bakteriengeißeln mittels Phasenkontrast. Arch. Mikrobiol. **19**, 435 (1953).

53. Burckhardt, J. L.: Die Begeißelung als differentialdiagnostisches Merkmal in der Fluorescenz-Gruppe. Zbl. Bakter. I Orig. **79**, 321 (1917).

54. Caldwell, M. E., and D. L. Ryerson: A new species of the genus *Pseudomonas* pathogenic for certain reptiles. J. Bacter. **39**, 323 (1940).

55. Carpenter, W. B.: The microscope and its revelations. Revised and edited by W. H. Dallinger. London: Churchill 1891.

56. Caselitz, F. H.: Ultraschall-Wirkung bei Bakterien der *Salmonella*-Gruppe. Z. Hyg. **133**, 113 (1951).

57. Chapman, G. B., and J. Hillier: Electron microscopy of ultra-thin sections of bacteria. J. Bacter. **66**, 362 (1953).

58. Clark, F. E., and P. H. Carr: Motility and flagellation of the soil coryne bacteria. J. Bacter. **62**, 1 (1951).

59. Claude, A.: Electron microscope studies of cells by the method of replicas. J. of exper. Med. **89**, 425 (1949).

60. Clowes, R. C., G. Furness and D. Rowley: The measurements of speeds of motility in *Escherichia coli*. J. Gen. Microbiol. **13**, ii (1955).

61. Cohn, F.: Untersuchungen über Bakterien. I. Beitr. Biol. Pflanz. **1**, 146 (1872).

62. — Untersuchungen über Bakterien. II. Beitr. Biol. Pflanz. **1**, 141 (1875).

63. Conn, H. J., and G. E. Wolfe: Flagella staining as a routine test for bacteria. J. Bacter. **36**, 517 (1938).

64. — — The flagellation of bacteria. Science (Lancaster, Pa.) **87**, 283 (1938).

65. — — and M. Ford: Taxonomic relationships of *Alcaligenes* sp. to certain soil saprophytes and plant parasites. J. Bacter. **39**, 207 (1940).

66. —, and R. P. Elrod: Concerning flagella and motility. J. Bacter. **54**, 681 (1947).

67. Couch, J. N.: A new group of organisms related to *Actinomyces*. J. Elisha Mitchell Sci. Soc. **65**, 315 (1949).

68. — *Actinoplanes*, a new genus of the *Actinomycetales*. J. Elisha Mitchell Sci. Soc. **66**, 87 (1950).

69. Craigie, J.: Studies on the serological reactions of the flagella of *B. typhosus*. J. of Immun. **21**, 417 (1931).

70. Crawford, I. P.: A new fermentative pseudomonad, *Pseudomonas formicans*, n. sp. J. Bacter. **68**, 734 (1954).

71. Czekalowski, J. W., and G. Eaves: The structure of *Leptospirae* as revealed by electron microscopy. J. of Path. **69**, 129 (1955).

72. Dallinger, W. H., and J. J. Drysdale: On the existence of flagella in *Bacterium termo*. Monthly Microsc. J. **14**, 105 (1875).

73. Dardanoni, L., and P. Calvi: Colorazione di polisaccaridi intracellulari in alcune specie di *Enterobatteri*. Riv. Ist. Sieroterapico ital. **29**, 51 (1954).

74. Dautheribes, P.: L'action inhibitrice de l'acide borique sur le developpement des *Proteus*. Ann. Biol. clin. **8**, 607 (1950).

75. De, M. L., A. Guha and N. N. das Gupta: A variant of *Bacterium coli*. Nature (Lond.) **171**, 879 (1953).

76. Derx, H. G.: *Azotobacter insigne* spec. nov., fixateur d'azote à flagellation polaire. Proc. Kon. ned. Akad. v. Wetensch. C **54**, 342 (1951).

77. Dimitroff, V. T.: *Spirillum virginianum* nov. spec. J. Bacter. **12**, 19 (1926).

78. Dubos, R. J.: The bacterial cell. Cambridge, Mass.: Harvard University Press 1945.

79. — Bacterial and mycotic infections in man. Philadelphia-London-Montreal: Lippincott Co. 1948.

79a. Duguid, J. P., I. W. Smith, G. Dempster and P. N. Edmunds: Non-flagellar filamentous appendages ("fimbriae") and haemagglutinating activity in *Bacterium coli*. J. of Path. **70**, 335 (1955).

80. Dyar, M. T.: Isolation and cytological study of a free-living spirochete. J. Bacter. **54**, 483 (1947).

81. Edwards, P. R., A. B. Moran and D. W. Bruner: Flagella and flagellar antigens in "non-motile" *Salmonella* cultures. Proc. Soc. Exper. Biol. a. Med. **62**, 296 (1946).

82. Einthoven, W.: Über die Beobachtung und Abbildung dünner Fäden. Pflügers Arch. **191**, 60 (1921).

83. Ellis, D.: Der Nachweis der Geißeln bei allen *Coccaceen*. Zbl. Bakter. II **9**, 546 (1902).

84. — Untersuchungen über *Sarcina*, *Streptococcus* und *Spirillum*. Zbl. Bakter. I Orig. **33**, 1, 81, 161 (1903).

85. — On the discovery of cilia in the genus *Bacterium*. Zbl. Bakter. II **11**, 241 (1903).

86. — Sulphur bacteria. London-New York-Toronto: Longmans, Green & Co. 1932.

87. Ende, M. van den, P. A. Don, W. J. Elford, C. E. Challice, I. M. Dawson and J. E. Hotchin: The bacteriophages of *Pseudomonas aeruginosa*, filtration, measurements and electron microscopy. J. of Hyg. **50**, 12 (1952).

88. Endeavour: Natural and mechanical locomotion. Leading article, April, 1944.

89. Erikson, D.: Factors promoting cell division in a „soft" mycelial type of Nocardia: *Nocardia turbata* n. sp. J. Gen. Microbiol. **11**, 198 (1954).

90. Ewart, J. C.: The life-history of *Bacterium termo* and *Micrococcus*, with further observations on *Bacillus*. Proc. Roy. Soc. Lond. **27**, 474 (1878).

91. Fawcett, D. W., and K. R. Porter: A study of the fine structure of ciliated epithelial cells with the electron microscope. Anat. Rec. **113**, 539 (1952).

92. — — A study of the fine structure of ciliated epithelia. J. of Morph. **94**, 221 (1954).

93. Fischer, R., and P. Larose: Contribution on the behaviour and structure of the cytoplasmic membrane of bacteria. Canad. J. Med. Sci. **30**, 86 (1952).

94. Fleming, A., A. Voureka, I. R. H. Kramer and W. H. Hughes: The morphology and motility of *Proteus vulgaris* and other organisms cultured in the presence of penicillin. J. Gen. Microbiol. **4**, 257 (1950).

95. — Further observations on the motility of *Proteus vulgaris* grown on penicillin agar. J. Gen. Microbiol. **4**, 457 (1950).

96. Fraenkel, C.: Geißelfäden an den Spirillen des Rekurrens- und des Zeckenfiebers. Zbl. Bakter. I Orig. **47**, 471 (1908).

97. Friewer, F. I., and E. Leifson: Non-motile flagellated variants of *Salmonella typhimurium*. J. of Path. **64**, 223 (1952).

97a. Fritsch, F. E.: The structure and reproduction of the algae, vol. II. Cambridge: Univ. Press 1945.

97b. Gard, S.: Preparation of bacterial flagellae. Ark. Kemi, Mineral. Geol. A **19**, No21 (1944).

97c. Garnjobst, L.: *Cytophaga columnaris* (Davis) impure culture: a myxobacterium pathogenic to fish. J. Bacter. **49**, 113 (1945).

97d. Giesberger, G.: Beiträge zur Kenntnis der Gattung *Spirillum* Ehrenberg. Doctor's Thesis, Univ. of Utrecht 1936.

97e. Gillespie, H. B., and L. F. Rettger: Bacterial variation: formation and fate of certain variant cells of *Bacillus megatherium*. J. Bacter. **38**, 41 (1939).

97f. Giuntini, J., et E. Edlinger: Obtention de coupes très minces pour la microscopie électronique par une méthode utilisant la dilatation thermique. Ann. Inst. Pasteur **86**, 671 (1954).

97g. Goldacre, R. J., and I. J. Lorch: Folding and unfolding of protein molecules in relation to cytoplasmic streaming, amoeboid movement and osmotic work. Nature (Lond.) **166**, 497 (1950).

98. Grace, J. B.: Some observations on the flagella and blepharoblasts of *Spirillum* and *Vibrio* spp. J. Gen. Microbiol. **10**, 325 (1954).

99. Graudal, H.: Motile streptococci. Acta path. scand. (Copenh.) **31**, 46 (1952).

100. Gray, J.: Aspects of the locomotion of whales. Nature (Lond.) **161**, 199 (1948).

101. — Undulatory propulsion in small organisms. Nature (Lond.) **168**, 929 (1951).

102. Gray, J.: Undulatory propulsion. Quart. J. Microsc. Sci. **94**, 551 (1953).

102a. — Ciliary movement. Cambridge: University Press 1928.

103. Gray, P. H., and H. G. Thornton: Soil bacteria that decompose certain aromatic compounds. Zbl. Bakter. II **73**, 74 (1928).

104. Griffin, A. M., and M. L. Robbins: Flagellation of *Listeria monocytogenes*. J. Bacter. **48**, 114 (1944).

105. Grigg, G. W., and A. J. Hodge: Electron microscopic studies of spermatozoa. Austral. J. Sci. Res. B **2**, 271 (1949).

106. Gross, J., J. H. Highberger and F. O. Schmitt: Some factors involved in the fibrogenesis of collagen in vitro. Soc. Exper. Biol. a. Med. **80**, 462 (1952).

107. Hampp, E. G., D. B. Scott and R. W. G. Wyckoff: Morphologic characteristics of certain cultured strains of oral spirochaetes and *Treponema pallidum* as revealed by the electron microscope. J. Bacter. **56**, 755 (1948).

107a. Hannay, C. L.: Inclusions in bacteria. In: Bacterial anatomy, ed E. T. C. Spooner and B. A. D. Stocker. Cambridge, England: Cambridge University Press 1956.

108. Happold, F. C., K. I. Johnstone, H. J. Rogers and J. B. Youatt: The isolation and characteristics of an organism oxidizing thiocyanate. J. Gen. Microbiol. **10**, 261 (1954).

109. Harold, R., and R. Y. Stanier: The genera *Leucothrix* and *Thiotrix*. Bacter. Rev. **19**, 49 (1955).

110. Harrison, A. P., and P. A. Hansen: A motile lactobacillus from the cecal feces of turkeys. J. Bacter. **59**, 444 (1949).

111. Hedén, C. G., and R. W. G. Wyckoff: The electron microscopy of heated bacteria. J. Bacter. **58**, 153 (1949).

112. Heidelberger, M., and E. A. Kabat: Chemical studies on bacterial agglutination. J. of Exper. Med. **60**, 643 (1934).

112a. Henrici, A. T.: The biology of bacteria, 3rd ed., revised by E. J. Ordal. Boston-New York-Chicago: Heath & Co. 1948.

113. Herwerden, M. A. van: Reversible gelation and fixation of tissues. Proc. Kon. Ned. Akad. v. Wetensch. **29**, 975 (1926).

114. Hesse, R.: Tierbau und Tierleben. Jena: Gustav Fischer 1935.

115. Hesselbrock, W., and L. Foshay: The morphology of *Bacterium tularense*. J. Bacter. **49**, 209 (1945).

116. Hestrin, S., M. Aschner and J. Mager: Synthesis of cellulose by resting cells of *Acetobacter xylinum*. Nature (Lond.) **159**, 64 (1947).

117. Hillier, J.: Electron microscopy of microorganisms and viruses. Annual Rev. Microbiol. **4** (1950).

118. — G. Knaysi and R. F. Baker: New preparation techniques for the electron microscopy of bacteria. J. Bacter. **56**, 569 (1948).

119. Hirsch, W.: A new bacterial variant: the non-motile H form. J. of Hyg. **45**, 417 (1947).

120. Hodge, A. J.: Electron microscopic studies of spermatozoa. Austral. J. Sci. Res. B **2**, 368 (1949).

121. — The motility and flagellation of bacteria. Austral. J. Sci. **11**, 115 (1949).

122. Hofer, A. W.: A characterisation of *Bacterium radiobacter* (Beyerinck and van Delden) Löhnis. J. Bacter. **41**, 193 (1941).

123. — Flagellation of *Azotobacter*. J. Bacter. **48**, 697 (1944).

124. Hoffmann-Berling, H.: Geißelmodelle und Adenosinphosphat. Biochim. et Biophysica Acta **16**, 146 (1955).

125. Houwink, A. L., and W. van Iterson: A study on flagellation. Biochim. et Biophysica Acta **5**, 10 (1950).

126. — A macromolecular mono-layer in the cell wall of *Spirillum* spec. Biochim. et Biophysica Acta **10**, 360 (1953).

127. Houwink, A. L.: *Caulobacter*, its morphogenesis, taxonomy and parasitism. Ant. v. Leeuwenhoek **21**, 49 (1955).

128. Hughes, C. G.: The motility of bacteria. Austral. J. Sci. **10**, 149 (1948).

128a. Hughes, W. H.: The structure and development of the induced long forms of bacteria. In: Bacterial anatomy, ed E. T. C. Spooner and B. A. D. Stocker. Cambridge, England: Cambridge University Press 1956.

129. Iterson, W. van: Some electron-microscopical observations on bacterial cytology. Biochim. et Biophysica Acta **1**, 527 (1947).

130. — Some remarks on the present state of our knowledge of bacterial flagellation. Symp. Citologia batterica Rome, Sept. 24, 1953.

130a. — Discussion in: The nature of the bacterial surface ed A. A. Miles and N. W. Pirie. Oxford: Blackwell Scientific Publications 1949.

131. Jakob, A., u. H. Mahl: Über die Adsorption von Metallkolloiden an Bakterien. Z. Naturforsch. **3**b, 26 (1948).

131a. Jensen, H. L.: The genus *Nocardia.* In: *Actinomycetales.* Roma: International Union of Biological Sciences Ser. B No. 14, 1953.

131b. — The coryneform bacteria. In: Annual review of microbiology. Stanford, California: Annual Reviews, Inc. 1952.

132. Jeynes, M. H.: Taxonomic position of the genus *Selenomonas* (von Prowazek). Nature (Lond.) **176**, 1077 (1955).

133. Johnson, F. H., N. Zworykin and G. Warren: A study of luminous bacterial cells and cytolysates with the electron microscope. J. Bacter. **46**, 167 (1943).

134. —, and R. F. Baker: The electron and light microscopy of *Beggiatoa.* J. Cellul. a. Comp. Physiol. **30**, 131 (1947).

135. Kauffmann, F.: Über Bakteriengeißeln als aktive Bewegungsorgane. Schweiz. Z. Path. u. Bakter. **11**, 378 (1948).

136. Kellenberger, E., et G. H. Werner: Microscopie électronique et cytologie microbienne. Schweiz. Z. Path. u. Bakter. **12**, 500 (1949).

137. —, et A. Ryter: Contribution á l'étude du noyau bactérien. Schweiz. Z. Path. u. Bakter. **18**, 122 (1955).

138. King, R. L., and H. W. Beams: Ultracentrifugation and cytology of *Spirillum volutans.* J. Bacter. **44**, 597 (1942).

139. Kingma Boltjes, T. Y.: Some remarks on microphotography. Ant. van Leeuwenhoek **12**, 232 (1947).

140. — Function and arrangement of bacterial flagella. J. of Path. **60**, 275 (1948).

141. Kjems, E.: Studies on five bacterial strains of the genus *Pseudomonas.* Acta path. scand. (Copenh.) **36**, 531 (1955).

142. Kluyver, A. J., and C. B. van Niel: Prospects for a natural system of classification of bacteria. Zbl. Bakter. II **94**, 369 (1936).

143. Knaysi, G.: Elements of bacterial cytology. Ithaca, New York: Comstock Publishing Co. Inc. 1951.

144. — The structure of the bacterial cell. In: Bacterial physiology. New York: Academic Press Inc. 1951.

145. — The structure, division and fusion of nuclei observed in living cells of *Mycobacterium thamnopheos.* J. Bacter. **64**, 859 (1952).

145a. — The present status of bacterial cytology. 6th Intern. Congr. Microbiol., Rome 14, 1954.

146. Koblmueller, L. O.: Untersuchungen über Streptokokken. Zbl. Bakter, I Orig. **133**, 311 (1935).

147. Koch, R.: Untersuchungen über Bakterien. Beitr. Biol. Pflanz. **2**, 399 (1877).

148. Koffler, H., and G. E. Mallett: Evidence that the flagella and slime material of *Bacillus subtilis* differ chemically. Bacter. Proceed. G **9** (1952).

149. Koffler, H., and G. E. Mallett: Sind Bakterien-Geißeln aktive Bewegungsorgane oder nur Fäden der Außenschicht? Zbl. Bakter. I Orig. **158**, 357 (1952).

150. Krieg, A.: Zur Zytologie und Enzymatik der Bakterien. Z. Naturforsch. **9** b, 342 (1954).

151. Krijgsman, B. J.: Beiträge zum Problem der Geißelbewegung. Arch. Protistenkde **52**, 478 (1925).

152. Krueger, F., u. K. E. Wohlfarth-Bottermann: Elektronenoptische Beobachtungen an Ciliatenorganellen. Mikroskopie (Wien) **7**, 121 (1952).

153. Kvittingen, J.: Some observations on the nature and significance of bacterial flagella. Acta path. scand (Copenh.) **37**, 89 (1955).

154. — Studies of the life-cycle of *Proteus Hauser*. Acta path. scand. (Copenh.) **32**, 363 (1953).

155. Kyle, T. S., and A. Eisenstark: The genus *Azotobacter*. Bull. Oklahoma Agricult. Mech. College **48**, No 16 (1951).

156. Labaw, L. W., and V. M. Mosley: Periodic structure in the flagella and cell walls of a bacterium. Biochim. et Biophysica Acta **15**, 325 (1954).

157. — — Periodic structure in the flagella of *Brucella bronchiseptica*. Biochim. et Biophysica Acta **17**, 322 (1955).

158. Lankford, C. E., H. Hoyo and J. F. Lutteringer: Intracellular polysaccharide of *Enterobacteriaceae*. J. Bacter. **62**, 621 (1951).

159. Lasseur, P., P. Vernier, A. Dupaix et S. Zahl: Observations sur les fouets de quelques bactéries. Trav. Labor. Microbiol. Fac. Pharmacie Nancy **3**, 75 (1930).

160. Leifson, E.: Development of flagella on germinating spores. J. Bacter. **21**, 357 (1931).

161. — The flagellation of spirochaetes. J. Bacter. **60**, 678 (1950).

162. — Staining, shape, and arrangement of bacterial flagella. J. Bacter. **62**, 377 (1951).

163. — The flagellation and taxonomy of species of *Acetobacter*. Ant. van Leeuwenhoek **20**, 102 (1954).

164. —, and R. Hugh: Variation in shape and arrangement of bacterial flagella. J. Bacter. **65**, 263 (1953).

165. — — A new type of polar monotrichous flagellum. J. Gen. Microbiol. **10**, 68 (1954).

166. —, and A. I. Palen: Variations and spontaneous mutations in the genus *Listeria* in respect to flagellation and motility. J. Bacter. **70**, 233 (1955).

167. — S. R. Carhart and Mac D. Fulton: Morphological characteristics of flagella of proteus and related bacteria. J. Bacter. **69**, 73 (1955).

168. Lessel, E. F., and R. S. Breed: *Selenomonas* Boskamp, 1922. — A genus that includes species showing an unusual type of flagellation. Bacter. Rev. **18**, 165 (1954).

169. Levenson, S.: Coloration des cils bactériens par un procédé simple. Ann. Inst. Pasteur **56**, 634 (1936).

170. — Entérocoques mobiles. Ann. Inst. Pasteur **60**, 99 (1938).

171. — Mobilité et Cils des Bactéries. Thèse de Paris 1938.

172. Loeffler, F.: Eine neue Methode zum Färben der Mikroorganismen, im besonderen ihrer Wimperhaare und Geißeln. Zbl. Bakter. **6**, 209 (1889).

173. — Weitere Untersuchungen über die Beizung und Färbung der Geißeln bei den Bakterien. Zbl. Bakter. **7**, 625 (1890).

174. Lofgren, R.: The structure of *Spirillum rubrum*. Bacter. Proceed. **1**, 15 (1948).

175. —, and M. H. Soule: The structure of *Spirochaeta novyi* as revealed by the electron microscope. J. Bacter. **50**, 679 (1945).

176. Lowndes, A. G.: Flagella movement. Nature (Lond.) **138**, 210 (1936).

177. — The term tractellum in flagellate organisms. Nature (Lond.) **152**, 51 (1943).

178. Lucas, M. S.: The cytoplasmic phases of rejuvenescence and fission in *Cyathodinum piriforme*. Arch. Protistenkde **77**, 407 (1932).

179. Luria, S. E.: Electron microscope studies of bacteriophage. J. Bacter. **45**, 73 (1943).

180. Lutteringer, J. F., and C. E. Lankford: Relation of O antigen to polysaccharide polar bodies of *Enterobacteriaceae.* J. Bacter. **65**, 746 (1953).

181. Maaløe, O., A. Birch-Andersen and F. S. Sjöstrand: Electron micrographs of sections of *E. coli* cells infected with the bacteriophage T . Biochim. et Biophysica Acta **15**, 12 (1954).

182. Macé, E.: Traité pratique de bactériologie. Paris: Baillière & fils 1913.

183. Mallett, G. E., H. Koffler and J. N. Rinker: The effect of shaking on bacterial flagella and motility. J. Bacter. **61**, 703 (1951).

184. Mandia, J. W.: The migration of cultures of *Clostridium perfringens* in semisolid medium. J. Bacter. **60**, 275 (1950).

185. Mann, T.: Studies on the metabolism of semen. Biochemic. J. **39**, 451 (1945).

186. Manninger, R., u. A. Nográdi: Über induzierte Mutationserscheinungen an Milzbrand und Kartoffelbazillen. Experientia (Basel) **4**, 276 (1948).

187. Manton, I., and B. Clarke: Demontration of compound cilia in a fern spermatozoid with the electron microscope. J. of Exper. Bot. **2**, 125 (1951).

188. — — A. D. Greenwood and E. A. Flint: Further observations on the structure of plant cilia, by a combination of visual and electron microscopy. J. of Exper. Bot. **3**, 204 (1952).

189. — Number of fibrils in the cilia of the green algae. Nature (Lond.) **171**, 485 (1953).

190. Marsland, D. A.: Quieting *Paramecium* for the elementary student. Science (Lancaster, Pa.) **98**, 414 (1943).

190a. Mayr-Harting, A.: The serology of *Pseudomonas pyocyanea.* J. Gen. Microbiol. **2**, 31 (1948).

191. Mercer, E. H.: Formation of silk fibre by the silkworm. Nature (Lond.) **168**, 792 (1951).

192. Messea, A.: Contribuzione allo studio delle ciglia dei batteri. Riv. Igiene e Sanita publ. **1890**, 513 (quoted by Knaysi, 143).

193. Metzner, P.: Die Bewegung und Reizbeantwortung der bipolar begeißelten Spirillen. Jb. wiss. Bot. **59**, 325 (1920).

194. — Zur Mechanik der Geißelbewegung. Biol. Zbl. **40**, 49 (1920).

195. Migula, W.: System der Bakterien. Jena: Gustav Fischer 1897.

196. Mihálik, P. v.: Über die Bildung des Flimmerapparates im Eileiterepithel. Anat. Anz. **79**, 259 (1935).

197. Miles, A. A.: The status of some arguments about the bacterial surface. In: The nature of the bacterial surface ed. A. A. Miles and N. W. Pirie. Oxford: Blackwell scientific publications 1949.

198. Morowitz, H. J.: The energy requirements for bacterial motility. Science (Lancaster, Pa.) **119**, 286 (1954).

199. Morton, H. E., and T. F. Anderson: Some morphological features of the Nichols strain of *Treponema pallidum* as revealed by the electron microscope. Amer. J. Syph. **26**, 565 (1942).

200. — — The morphology of *Leptospira icterohaemorrhagiae* and *L. canicola* as revealed by the electron microscope. J. Bacter. **45**, 143 (1943).

201. — G. Rake and N. Rose: Electron microscope studies of treponemes. III Flagella. Amer. J. Syph. **35**, 503 (1951).

202. Mudd, S., and T. F. Anderson: Demonstration by the electron microscope of the combination of antibodies with flagellar and somatic antigens. J. of Immun. **42**, 251 (1941).

203. — K. Polevitzky, T. F. Anderson and L. A. Chambers: The bacterial cell wall in the genus *Bacillus.* J. Bacter. **42**, 251 (1941).

204. — — — Bacterial morphology as shown by the electron microscope. Arch. of Path. **34**, 199 (1942).

205. —, Mudd, S., and T. F. Anderson: Selective "staining" for electron photography. J. of Exper. Med. **76**, 103 (1942).

206. — K. Polevitzky and T. F. Anderson: Bacterial morphology as shown by the electron microscope. J. Bacter. **46**, 15 (1943).

207. — J. Hillier, E. H. Beutner and P. Hartman: Light and electron microscope studies of *Escherichia coli*-coliphage interactions. Biochim. et Biophysica Acta **10**, 153 (1953).

208. Mueller, A., u. C. Stapp: Beiträge zur Biologie der Leguminosenknöllchenbakterien mit besonderer Berücksichtigung ihrer Artverschiedenheit. Arb. Biol. Reichsanst. Land- und Forstw. **14**, 26 (1926).

209. Macdonald, J. B.: The motile non-sporulating anaerobic rods of the oral cavity. Toronto: University of Toronto Press 1953.

210. — M. L. Knoll and R. M. Sutton: Motility in a species of non-flagellated bacteria. Proc. Soc. Exper. Biol. a. Med. **84**, 459 (1953).

211. McQuillen, K.: Bacterial protoplasts: growth and division of protoplasts of *Bacillus megaterium*. Biochim. et Biophysica Acta **18**, 458 (1955).

211a. Nelson, L.: Enzyme distribution in fragmented bull spermatozoa. Biochim. et Biophysica Acta **14**, 312 (1954).

212. Niel, C. B. van: Über die Beweglichkeit und das Vorkommen von Geißeln bei einigen Sarcina-Arten. Zbl. Bakter. II **60**, 289 (1923).

213. Neumann, F.: Die Sichtbarmachung von Bakteriengeißeln am lebenden Objekt im Dunkelfeld. Zbl. Bakter. I Orig. **96**, 250 (1925).

214. — Die Sichtbarmachung von Bakteriengeißeln am lebenden Objekt im Dunkelfeld. Zbl. Bakter. I Orig. **109**, 143 (1928).

215. — Bewegungsvorgänge beweglicher Mikroorganismen, insbesondere von Spirochäten festgehalten mit dem Kinematograph. Klin. Wschr. **1929**, 2081.

216. Neumüller, O.: Bemerkung zu der Arbeit von Dr. Neumann: Die Sichtbarmachung von Bakteriengeißeln am lebenden Objekt im Dunkelfeld. Zbl. Bakter. I Orig. **102**, 90 (1927).

217. Noda, H., and R. W. G. Wyckoff: The electron microscopy of reprecipitated collagen. Biochim. et Biophysica Acta **7**, 494 (1951).

218. Orla-Jensen, S.: The main lines of the natural bacterial system. J. Bacter. **6**, 263 (1921).

219. Ørskov, J.: Method for the demonstration of bacterial flagellum activity. Acta path. scand. (Copenh.) **2**, 181 (1947).

219a. — Untersuchungen über Strahlenpilze. Zbl. Bakter. II **98**, 344 (1938).

220. O'Toole, E.: Flagella staining of anaerobic bacilli. Stain Technol. **17**, 33 (1942).

221. Patocka, F., and V. Sebek: Moving colonies in anaerobic microbe isolated from vaginal discharge. Bull. internat. Acad. tchèque sci. **61**, 1, 12 (1951).

222. Pease, D. C., and R. F. Baker: Preliminary investigations of chromosomes and genes with the electron microscope. Science (Lancaster, Pa.) **109**, 8 (1949).

223. Peluffo, C. A.: Significance of the spiral period of bacterial flagella. VI. Congr. Internat. Microbiol. Roma, 1, 370, 1953.

224. Piekarski, G., and H. Ruska: Übermikroskopische Darstellung von Bakteriengeißeln. Klin. Wschr. **1939**, 383.

225. Pietschmann, K.: Entspricht der Ausdruck „peritrich" begeißelt bei Bakterien den tatsächlichen Verhältnissen? Arch. Mikrobiol. **10**, 133 (1939).

226. — Über die Begeißelung der Bakterien. Arch. Mikrobiol. **12**, 377 (1942).

227. Pijper, A.: Begeißelung von Typhus- und Proteusbazillen. Zbl. Bakter. I Orig. **118**, 113 (1930).

228. — Nochmals über die Begeißelung von Typhus- und Proteusbazillen. Zbl. Bakter. I Orig. **123**, 13 (1931).

229. Pijper, A.: Dark-ground studies of flagellar and somatic agglutination of *B. typhosus*. J. of Path. **47**, 1 (1938).

230. — Microcinematography of the motile organs of typhoid bacilli. J. Biol. Photogr. Assoc. **8**, 158 (1940).

231. — Dark-ground studies of Vi agglutination of *B. typhosus*. J. of Path. **53**, 431 (1941).

232. — Microcinematography of the agglutination of typhoid bacilli. J. Bacter. **42**, 395 (1941).

232a. — Filming Vi agglutination. J. Biol. Photogr. Assoc. **11**, 65 (1942).

233. — Shape and motility of bacteria. J. of Path. **58**, 325 (1946).

234. — Methylcellulose and bacterial motility. J. Bacter. **53**, 257 (1947).

235. — Filming as a method of research in microbiology. Ant. van Leeuwenhoek **12**, 26 (1947).

236. — Bacterial flagella and motility. Nature (Lond.) **161**, 200 (1948).

237. — Bacterial surface, flagella and motility. In: The nature of the bacterial surface ed. A. A. Miles and N. W. Pirie. Oxford: Blackwell scientific publications 1949.

238. — Evidence that amputation of bacterial flagella does not affect motility. Science (Lancaster, Pa.) **109**, 379 (1949).

239. — The flagella of *Spirillum volutans*. J. Bacter. **57**, 111 (1949).

240. — Zur Frage der Bakterien-Geißeln. Schweiz. Z. Path. u. Bakter. **12**, 681 (1949).

241. —, and A. J. Nunn: Flagella and motility of *Vibrio metschnikovii*. J. Roy. Microsc. Soc. **69**, 138 (1949).

242. — Cinemicrography with sunlight. S. Afric. J. Sci. **46**, 296 (1950).

243. — Bacterial flagella. Nature (Lond.) **168**, 749 (1951).

244. —, and A. Kistner: Slides and coverglasses for dark-ground microscopy. J. Roy. Microsc. Soc. **71**, 176 (1951).

245. — C. G. Crocker, J. P. van der Walt and N. Savage: Flagellum and motility of *Spirillum serpens*. J. Bacter. **65**, 628 (1953).

246. — — — — Specific agglutination of *Spirillum* serpens. J. Bacter. **65**, 636 (1953).

247. — Shape of bacterial flagella. Nature (Lond.) **175**, 214 (1954).

248. — C. G. Crocker and N. Savage: *Sarcinae*; motility, kind of flagella, and specific agglutination. J. Bacter. **69**, 151 (1954).

249. —, and G. Abraham: Wavelengths of bacterial flagella. J. Gen. Microbiol. **10**, 452 (1954).

250. — Flagella and cell wall of spirilla. J. Roy. microsc. Soc. **75**, 38 (1955).

251. — M. L. Neser and G. Abraham: Wavelengths of bacterial flagella. J. Gen. Microbiol. **14**, 371 (1956).

251a. — Genesis of bacterial flagella, sunlight dark-ground cinemicrographic film. Bacter. Proceed. **5**, 16 (1956).

251b. — Flagella of spirilla, sunlight dark-ground cinemicrographic film. Bacter. Proceed. **6**, 16 (1956).

251c. — Heating and cooling the microscope stage. Amer. J. Clin. Path. **25**, 980 (1955).

252. Pivnick, H.: *Pseudomonas rubescens*, a new species from soluble oil emulsions. J. Bacter. **70**, 1 (1955).

253. Poetschke, G., and W. Bommer: Die Technik der Lebensbeobachtung von Bakterien im Phasenkontrastmikroskop. Zbl. Bakter. I Orig. **156**, 127 (1950).

254. Potts, B. P., and S. G. Tomlin: The structure of cilia. Biochim. et Biophysica Acta **16**, 66 (1955).

255. Preston, N. W., and H. B. Maitland: The influence of temperature on the motility of *Pasteurella pseudotuberculosis*. J. Gen. Microbiol. **7**, 117 (1952).

256. Prévot, A. R., J. Giuntini et H. Thouvenot: Recherches sur l'organe giratoire de *Fusocillus girans*. Ann. Inst. Pasteur **86**, 774 (1954).

257. PRINGSHEIM, E. G.: The relationship between bacteria and myxophyceae. Bacter Rev. **12**, 47 (1948).

258. — The bacterial genus *Lineola*. J. Gen. Microbiol. **4**, 198 (1950).

259. PROCA, G.: Action des sérums agglutinants sur les cils. C. r. Soc. Biol. Paris **72**, 73 (1912).

260. RAY, D. L.: Agglutination of bacteria: a feeding method in the soil ameba *Hartmanella* sp. J. of Exper. Zool. **118**, 443 (1951).

261. REICHERT, K.: Über die Sichtbarmachung der Geißeln und die Geißelbewegung der Bakterien. Zbl. Bakter. I Orig. **51**, 1 (1909).

262. RICHARDS, O. W.: Phase microscopy in bacteriology. Stain Technol. **23**, 55 (1948).

263. RIEMSDIJK, M. VAN: Der Einfluß des Sauerstoffs auf die Beweglichkeit und Form der Choleravibrionen. Zbl. Bakter. **113**, 161 (1929).

264. RINKER, J. N., C. ROBINOW and H. KOFFLER: Flagellar activity of immobilized cells of a motile bacterium. Bacteriol. Proceed. **32** (1950).

265. —, and H. KOFFLER: Preliminary evidence that bacterial flagella are not "polysaccharide twirls". J. Bacter. **61**, 421 (1951).

266. RIPPEL, A., and K. PIETSCHMANN: Die Bedeutung der subpolaren Begeißelung der Bakterien für einige Fragen der Bakteriensystematik. Nachr. Akad. Wiss. Göttingen **1941**.

267. ROBERTIS, E. DE, and C. A. PELUFFO: Chemical stimulation and inhibition of bacterial motility studied with a new method. Proc. Soc. Exper. Biol. a. Med. **78**, 584 (1948).

268. — W. W. NOWINSKI and F. A. SAEZ: General cytology. Philadelphia and London: W. B. Saunders Company 1954.

269. —, and C. M. FRANCHI: Electron microscope observations of the fine structure of bacterial flagella. Exper. Cell Res. **2**, 295 (1951).

270. — — Macromolecular structure of the contractile protein of bacterial flagella. J. Appl. Physics **23**, 161 (1952).

271. ROHRER, E.: Elektronenmikroskopische Studien an Bakteriengeißeln. Schweiz. Z. Path. u. Bakter. **15**, 297 (1952).

272. RUSKA, H.: Übermikroskopische Darstellung organischer Struktur. Arch. exper. Zellforsch. **22**, 673 (1939).

273. SACKS, L. E.: Observations on the morphogenesis of *Arthrobacter citreus*, spec. nov. J. Bacter. **67**, 345 (1954).

274. SALLE, A. J.: Fundamental principles of bacteriology. New York-London-Toronto: McCraw-Hill Publishing Co. Ltd. 1954.

275. SALTON, M. R. J., and R. C. WILLIAMS: Electron microscopy of the cell walls of *Bacillus megatherium* and *Rhodospirillum rubrum*. Biochim. et Biophysica Acta **14**, 455 (1954).

276. SAMES, T.: Eine bewegliche Sarcine. Zbl. Bakter. **4**, 664 (1898).

276a. SCANGA, F.: Microorganismi al microscopio elettronico. Milano: Sagdos 1953.

276b. — La Cellula batterica Roma: Il pensiero scientifico 1954.

276c. SCHREIL, W.: Ein präparativer Beitrag zur durchstrahlenden Elektronenmikroskopie von Bakterien. Mikroskopie (Wien) **10**, 38 (1955).

276d. —, u. F. SCHLEICH: Elektronenmikroskopische Untersuchungen an fixierten und lebenden Mikroorganismen unter Anwendung der Abdruckverfahren. Z. Hyg. **141**, 576 (1955).

276e. SCHULZ, G.: Bewegungsstudien sowie elektronenmikroskopische Membranuntersuchungen an Cyanophyceen. Arch. Mikrobiol. **21**, 335 (1955).

276f. SEDAR, A. W., H. W. BEAMS and C. D. JANNEY: Electron microscopic studies of the ciliary apparatus of the gill cells of *Mya arenaria*. Proc. Soc. Exper. Biol. a.Med. **79**, 303 (1952).

276g. SELBY, C. C.: Electron microscopy, a review. Cancer Res. **13**, 753 (1953).

276h. SENTERFIT, L. B.: Immobilization of *Schistosoma mansoni* by immune serum. Proc. Soc. Exper. Biol. a. Med. **84**, 5 (1953).

276i. Seifriz, W.: Protoplasm. New York and London: McGraw-Hill Book Co. Ltd. 1936.

277. — A theory of protoplasmic streaming. Science (Lancaster, Pa.) **86**, 397 (1937).

278. — Torsion in protoplasm. J. Colloid. Sci. **1**, 27 (1945).

279. — Mechanism of protoplasmic movement. Nature (Lond.) **171**, 1136 (1953).

279a. Shanahan, A. J., and F. W. Tanner: Further studies on the morphology of *E. coli* exposed to penicillin. J. Bacter. **55**, 537 (1948).

280. Siedentopf, H.: Personal communication.

281. Smith, D. T.: Oral spirochaetes and related organisms in fuso-spirochetal disease. London: Bailliere, Tindall & Cox 1932.

282. Smith, I. W.: Flagellation and motility in *Aerobacter cloacae* and *Escherich. coli*. Biochim. et Biophysica Acta **15**, 20 (1954).

283. Snellman, O., and T. Erdös: An electron microscope study of myosin, actin and actomyosin. Biochim. et Biophysica Acta **2**, 660 (1948).

284. —, and B. Gelotte: An investigation of the physical chemistry of the contractile proteins. Exper. Cell Res. **1**, 234 (1950).

285. Spruit, C. J. P., and A. Pijper: An obligate halophilic bacterium from solar salt. Ant. van Leeuwenhoek **18**, 190 (1952).

286. Stanier, R. Y., and C. B. van Niel: The main outlines of bacterial classification. J. Bacter. **42**, 437 (1941).

286a. — Studies on the cytophagas. J. Bacter. **40**, 619 (1940).

286b. — A note on elastotaxis in myxobacteria. J. Bacter. **44**, 405 (1942).

286c. — The cytophaga group. Bacter. Rev. **6**, 143 (1942).

287. Starkey, R. L.: A study of spore formation and other morphological characteristics of *Vibrio desulfuricans*. Arch. Mikrobiol. **9**, 268 (1938).

288. Starr, M. P., and R. C. Williams: Helical fine structure of flagella of a motile diphtheroid. J. Bacter. **63**, 701 (1952).

289. Stempen, H.: Demonstration of a cell wall in the large bodies of *Proteus vulgaris*. J. Bacter. **70**, 177 (1955).

290. Stephenson, M.: Bacterial metabolism. London: Longmans, Green & Co. 1943.

291. Stiles, K. A., and D. A. Hawkins: Immobilizing paramaecia for study in the introductory course. Science (Lancaster, Pa.) **105**, 101 (1947).

292. Stocker, B. A. D., N. D. Zinder and A. J. Lederberg: Transduction of flagellar characters in *Salmonella*. J. Gen. Microbiol. **9**, 410 (1953).

293. — Bacterial flagella; morphology, constitution and inheritance. In: Bacterial anatomy, ed. E. T. C. Spooner and B. A. D. Stocker. Cambridge, England: Cambridge University Press 1956.

294. Stokes, J. L.: Studies on the filamentous sheathed iron bacterium *Sphaerotilus natans*. J. Bacter. **67**, 278 (1954).

295. Stuart, C. A., K. M. Wheeler, V. McGann and I. Howard: Motility and swarming of some *Enterobacteriaceae*. J. Bacter. **52**, 519 (1946).

296. Svedberg, T., and K. O. Pedersen: The ultracentrifuge. Oxford: Clarendon Press 1940.

297. Swain, R. H. A.: Electron microscopic studies of the morphology of pathogenic spirochaetes. J. of Path. **69**, 128 (1955).

298. Sykes, J. A., and R. Reed: The control of the swarming of *Proteus vulgaris* by boric acid. J. Gen. Microbiol. **3**, 117 (1949).

299. Taylor, G.: Analysis of the swimming of microscopic organisms. Proc. Roy. Soc. Lond., Ser. A **209**, 447 (1951).

300. — The action of waving cylindrical tails in propelling microscopic organisms. Proc. Roy. Soc. Lond., Ser. A **211**, 225 (1952).

301. Thjötta, T., and E. Käss: Pseudoflagella in bacteria. Acta path. scand. (Copenh.) **23**, 215 (1946).

302. THOMPSON, D'ARCY W.: On growth and form. Cambridge: University Press 1942.

303. THIEL, P. H. VAN, and W. VAN ITERSON: An electronmicroscopical study of *Leptospira biflexa.* Proc. Kon. Ned. Akad. v. Wetensch. **50**, 976 (1947).

304. TIEGHEM, P. VAN: Sur les prétendus cils des bactéries. Société botanique de France, Séance du 14 février 1879.

305a. TOMCSIK, J.: Über eine bewegliche Mutante des *B. anthracis.* Schweiz. Z. Path. u. Bakter. **13**, 616 (1950).

305b. —, u. S. GUEX-HOLZER: Änderung der Struktur der Bakterienzelle im Verlauf der Lysozym-Einwirkung. Schweiz. Z. Path. u. Bakter. **15**, 517 (1952).

305c. — Effect of disinfectants and of surface active agents on bacterial protoplasts. Proc. Soc. Exper. Biol. a. Med. **89**, 459 (1955).

306. — Die Struktur der Bakteriengrenzflächen. Ergebnisse der medizinischen Grundlagenforschung 1956.

307. TOMLIN, S. G., and J. W. MAY: Electron microscopy of sectioned bacteria. Austral. J. Exper. Biol. a. Med. Sci. **33**, 249 (1955).

307a. TOPPING, L.: The predominant microorganisms in soil. Zbl. Bakter. II **97**, 289 (1937).

308. TUFFERY, A. A.: Systematic position of the genus *Oscillospora.* Nature (Lond.) **174**, 838 (1954).

309. UELRICH, H.: Über die Bewegungen von *Beggiatoa mirabilis* und *Oscillatoria jenensis.* Z. wiss. Biol. Abt. D, Planta **2**, 295 (1926).

310. ULEHLA, V.: Ultramikroskopische Studien über Geißelbewegung. Biol. Zbl. **31**, 645, 721 (1911).

311. UMBREIT, W. W., and T. F. ANDERSON: A study of *Thiobacillus thiooxidans* with the electron microscope. J. Bacter. **44**, 317 (1942).

312. VORHAUS, E. F., and I. J. DEYRUP: The effect of adenosine-triphosphate on the cilia of the pharyngeal mucosa of the frog. Science (Lancaster, Pa.) **118**, 553 (1953).

313. VAHLE, C.: Untersuchungen über die Myxobakteroizeen. Zbl. Bakter. II **25**, 178 (1910).

314. WARREN, G. H., and J. GRAY: Studies on the properties of a polysaccharide constituent produced by *Pseudomonas aeruginosa.* J. Bacter. **70**, 152 (1955).

315. WATSON, J. H. L., J. J. ANGULO, F. LEON-BLANCO, G. VARELA and C. C. WEDDERBURN: Electron microscopic observations of flagellation in some species of the genus *Treponema schaudinn.* J. Bacter. **61**, 455 (1951).

316. WEEKS, O. B.: *Flavobacterium aquatile* (FRANKLAND and FRANKLAND) BERGEY *et al.*, type species of the genus *Flavobacterium.* J. Bacter. **69**, 649 (1955).

317. WEI, H.: A study of bacterial flagella by dark field examination. Chinese Med. J. Suppl. **1**, 135 (1936).

318. WEIBULL, C., and A. TISELIUS: Note on the acid hydrolysis of bacterial flagellae. Ark. Kemi, Mineral. Geol. B **20**, No 3 (1945).

319. — Some chemical and physico-chemical properties of the flagella of *Proteus vulgaris.* Biochim. et Biophysica Acta **2**, 351 (1948).

320. — Chemical and physico-chemical properties of the flagella of *Proteus vulgaris* and *Bacillus subtilis*, a comparison. Biochim. et Biophysica Acta **3**, 378 (1949).

321. — Morphological studies on salt precipitated bacterial flagella. Ark. Kemi (Stockh.) **1**, 21 (1949).

322. — Electrophoretic and titrimetric measurements on bacterial flagella. Acta chem. scand. (Copenh.) **4**, 260 (1950).

323. — Ordered aggregation of salted out and dried bacterial flagella. Ark. Kemi (Stockh.) **1**, No 65, 573 (1950).

324. — X-ray diffraction pattern given by bacterial flagella. Nature (Lond.) **165**, 482 (1950).

325. — Investigations on bacterial flagella. Acta chem. scand. (Copenh.) **4**, 268 (1950).

326. — Some analytical evidence for the purity of proteus flagella protein. Acta chem. scand. (Copenh.) **5**, 529 (1951).

327. Weibull, C.: Movement of bacterial flagella. Nature (Lond.) **167**, 511 (1951).

328. —, and J. Hedvall: Some observations of fractions of disintegrated bacterial cells obtained by differential centrifugations. Biochim. et Biophysica Acta **10**, 35 (1953).

329. — Isolation of protoplasts from *B. megaterium*. J. Bacter. **66**, 688 (1953).

330. Weil, A. J., and H. Slafkovsky: The position of the Wakefield bacilli (Berger). J. of Immun. **60**, 173 (1948).

331. Werner, G. H., et E. Kellenberger: Le rôle du microscope électronique dans l'étude du mode d'action des antibiotiques. Bull. Acad. suisse Sci. Méd. **4**, 263 (1948).

332. Wiame, J. M., R. Storck et E. Vanderwinkel: Biosynthèse induite d'arabokinase dans les protoplastes de *Bacillus subtilis*. Biochim. et Biophysica Acta **18**, 353 (1955).

333. Williams, R. C., and R. W. G. Wyckoff: Shadowed electron micrographs of bacteria. Proc. Soc. Exper. Biol. a. Med. **59**, 265 (1945).

334. Williams, M. A.: A study of the genus *Spirillum* Ehrenberg. Master's thesis, University of Southern California, Los Angeles, California 1952.

*334*a. Winkler, A.: Die Bakterienzelle. Ein Überblick über den gegenwärtigen Stand unseres Wissens. Stuttgart: Gustav Fischer 1956.

335. Wilska, A.: A new method of light microscopy. Nature (Lond.) **171**, 353 (1953).

336. — Observations with the anoptral microscope. Mikroskopie (Wien) **9**, 1 (1954).

337. Wohlfarth-Bottermann, K. E.: Weitere Untersuchungen an Ciliatencilien. Verh. dtsch. zool. Ges. (Freiburg) **1952.**

338. —, u. G. Pfefferkorn: Protisten-Studien. V. Zur Struktur des Wimperapparates. Protoplasma (Berl.) **42**, 227 (1953).

339. Woodcock, H. M., and G. Lapage: On a remarkable type of protisten parasite. Quart. J. Microsc. Sci., N. S. **59**, 431 (1913).

340. Wu, S. H., and F. F. McKenzie: Electron microscopic study of spermatozoa. Science (Lancaster, Pa.) **119**, 213 (1954).

341. Wyckoff, R. W. G.: Electron microscopy. New York: Interscience Publishers Inc. 1949.

342. — The fine structure of protoplasm in healthy and virus-diseased cells. Faraday Soc. Disc. **1951**, No 11.

343. Zettnow, E.: Geißeln bei Hühner- und Recurrensspirochäten. Dtsch. med. Wschr. **1906**, 376.

344. — Über Schleimgeißeln. Z. Hyg. **86**, 25 (1918).

Plate I

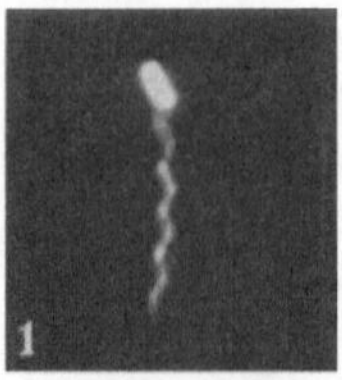

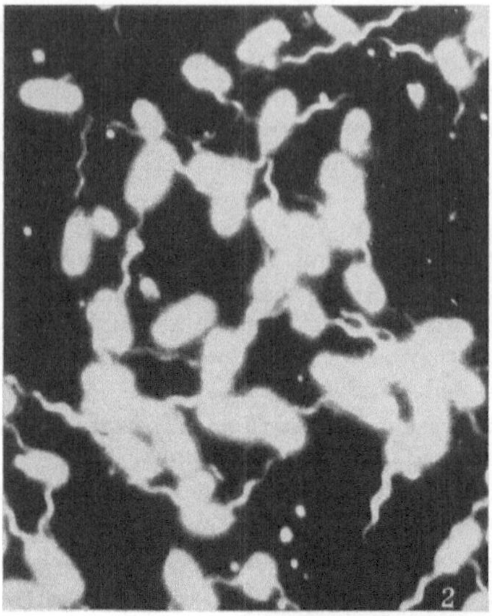

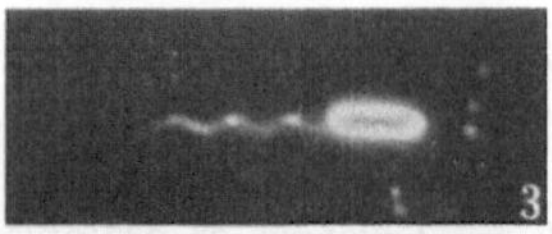

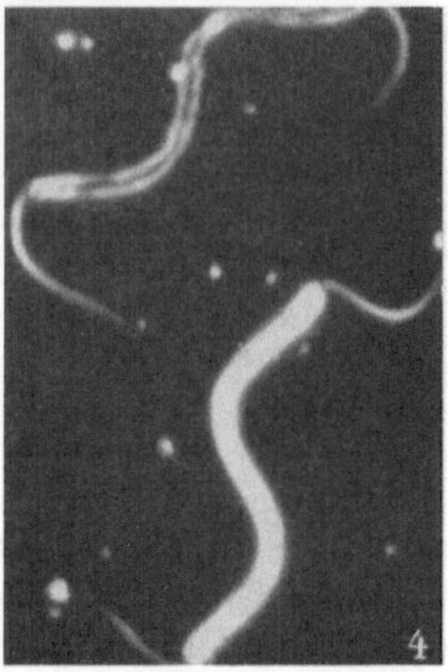

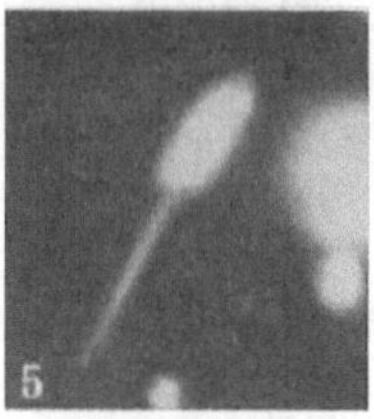

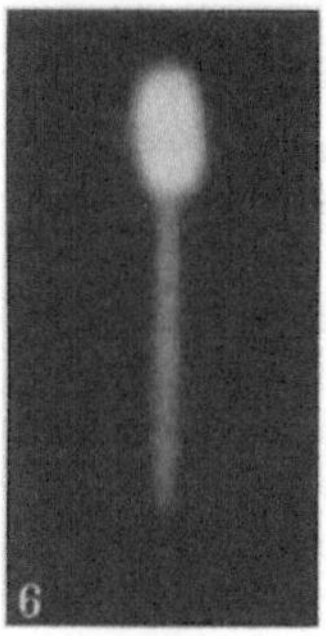

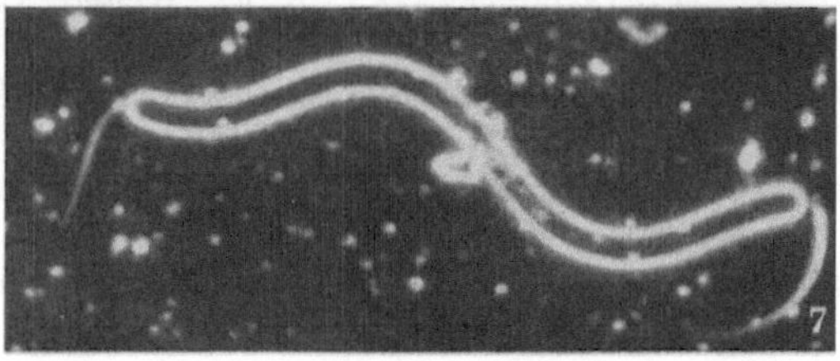

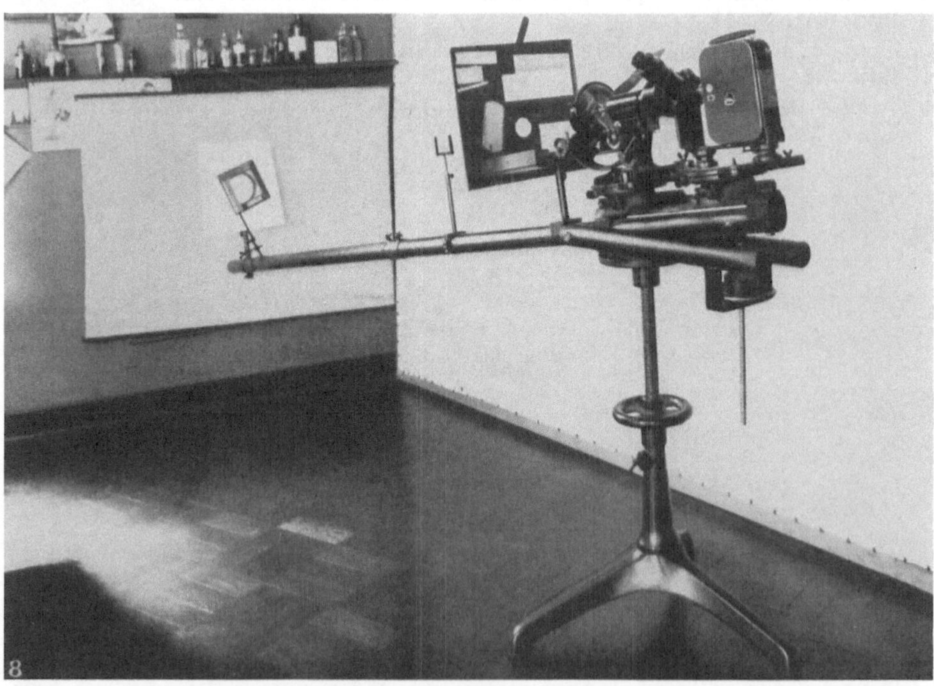

Fig. 1. *Salmonella typhosa*, granular precipitate on flagellum. ×1200. Fig. 2. *S. typhosa*, colloid sheath round flagella. ×1500. Fig. 3. *S. typhosa*, odd granules on flagellum. ×1500. Fig. 4. *Spirillum volutans*. ×1000. Fig. 5. *S. typhosa* (broth). ×2000. Fig. 6. *S. typhosa* (methylcellulose). ×2000. Fig. 7. *Sp. volutans*, autolysed. ×1500. Fig. 8. Sunlight dark-ground microscope with film camera

Plate II

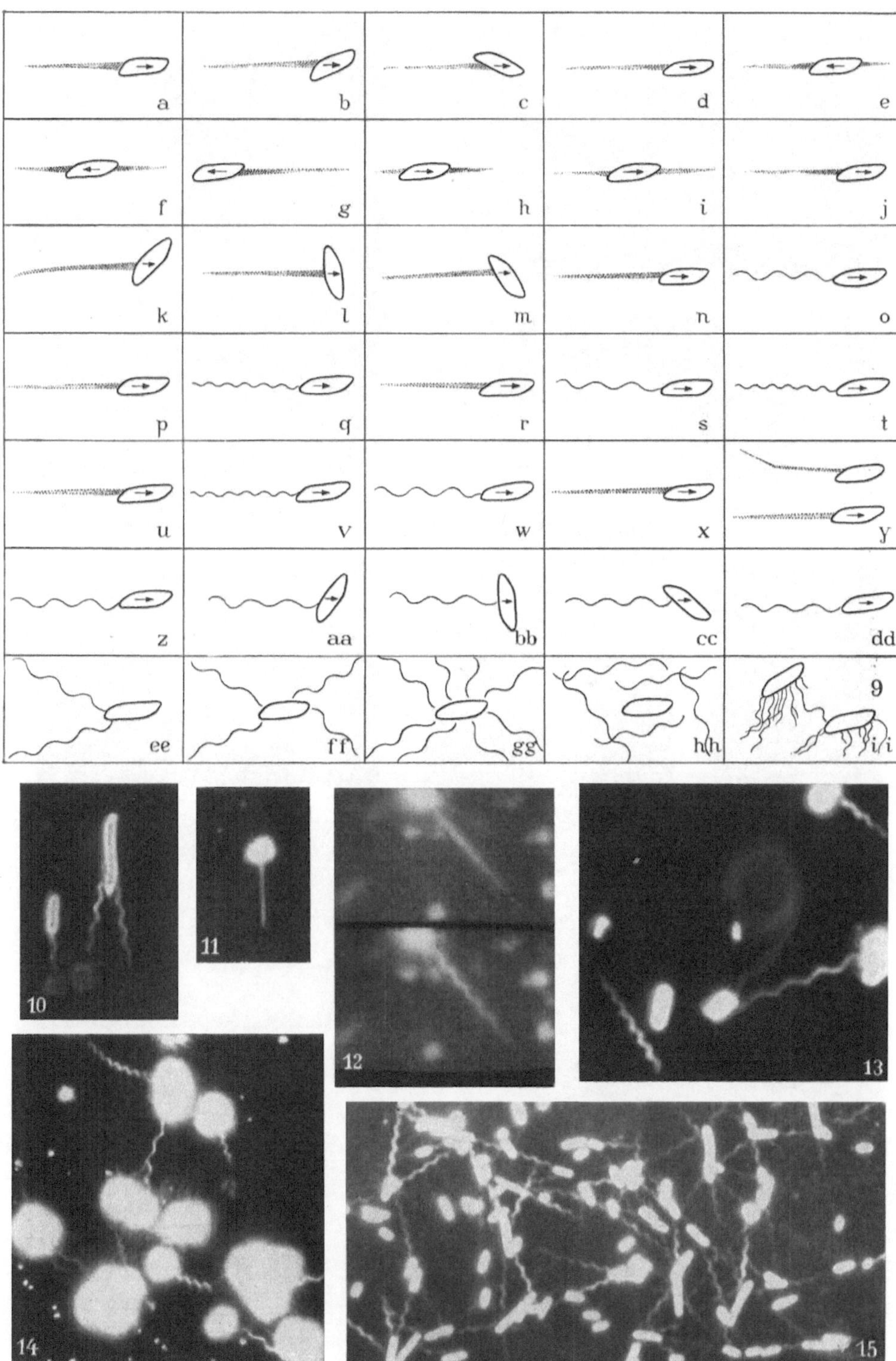

Fig. 9. Diagrams of metamorphoses of tail in *S. typhosa.* **Fig. 10.** *S. typhosa,* helical shape of body, flagellum split. × 1200. **Fig. 11.** *Sarcina ureae,* tail as stiff rod. × 1000. **Fig. 12.** *S. schottmuelleri,* 16 mm film, transition from straight tail into long wave helix. × 1200. **Fig. 13.** Biplicity in *S. schottmuelleri.* × 2000. **Fig. 14.** Biplicity in *S. ureae.* × 1000 **Fig. 15.** *B. proteus.* × 1000.

Plate III

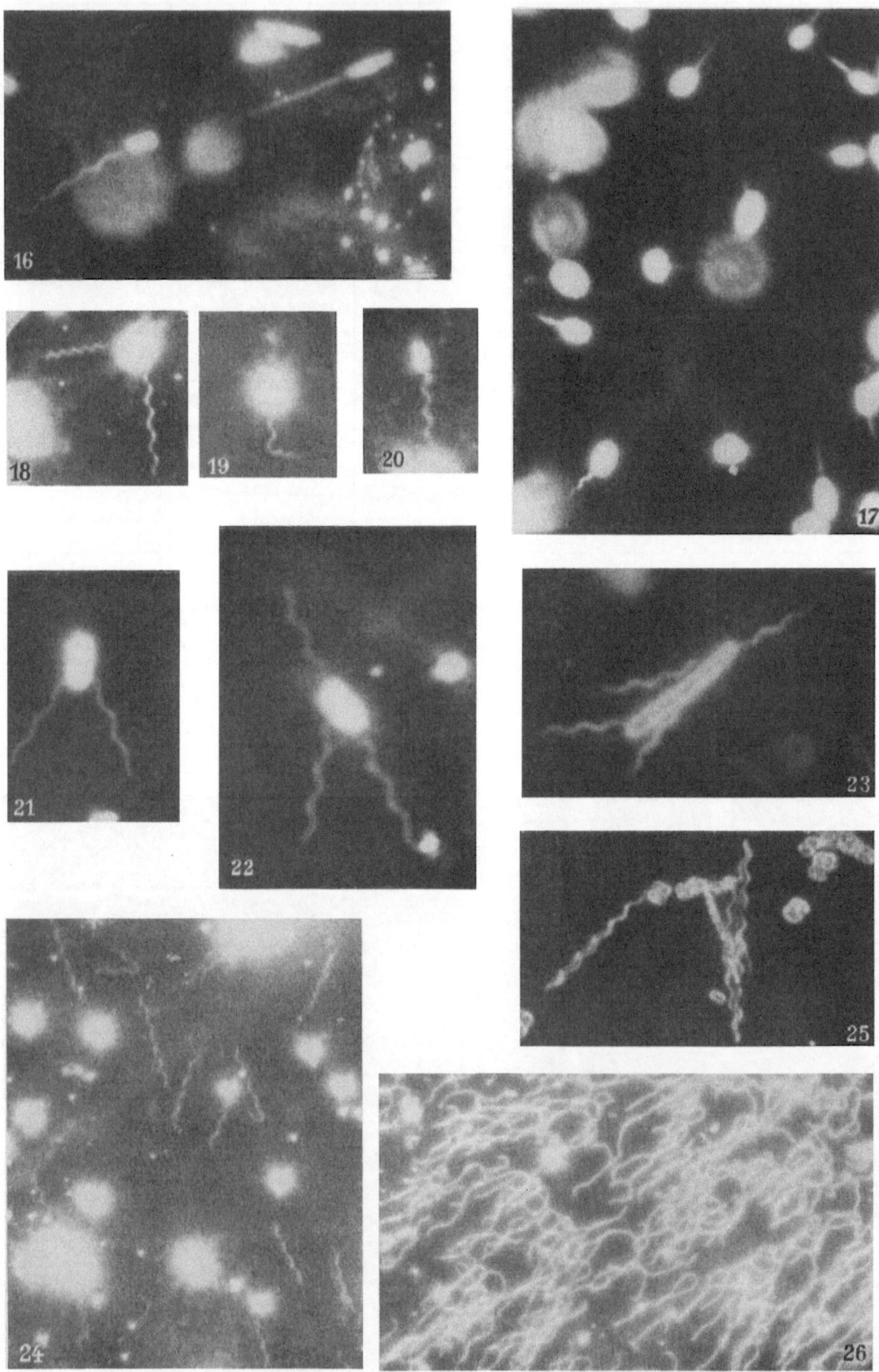

Fig. 16. *S. schottmuelleri*, transition from helix to straight tail. × 1500. Fig. 17. *S. typhosa*, fuzzy tails, clear-cut helix. × 500. Fig. 18. Biplicity, *S. ureae*. × 1000. Fig. 19. Biplicity, *S. ureae*. × 1000. Fig. 20. *Pseudomonas aeruginosa*. × 2000. Fig. 21. *S. typhosa*, flagellum split. × 2000. Fig. 22. *S. typhosa*, flagellum split further. × 2500. Fig. 23. *B. proteus*, flagellum split. × 2000. Fig. 24. *S. typhosa*, loose flagella. × 1000. Fig. 25. *S. uraee*, giant flagellum. × 1000. Fig. 26. *S. typhosa*, drying drop. × 1000

Plate IV

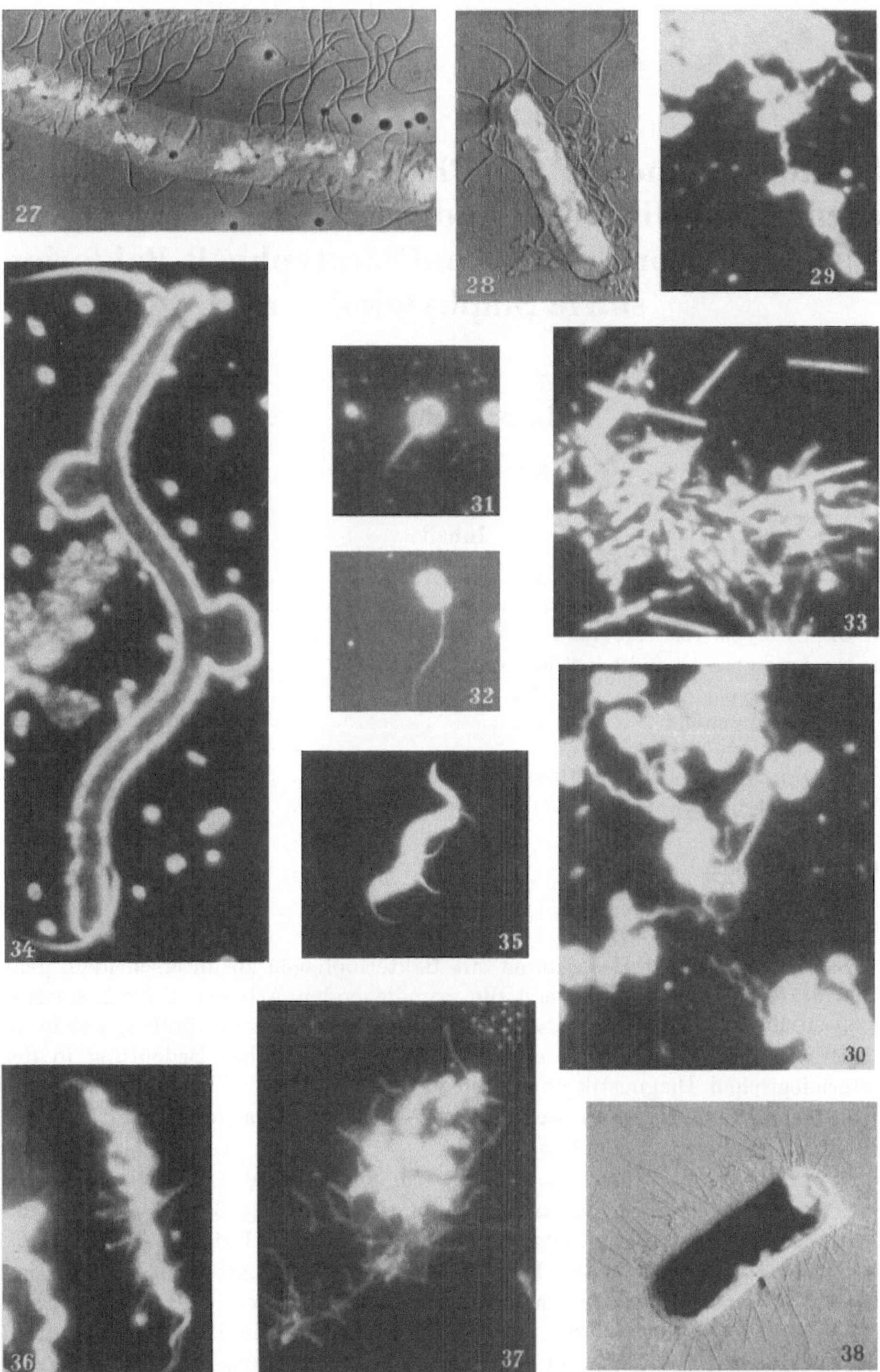

Fig. 27. *Clostridium tetani.* × 6000. From Prof. SCANGA. Fig. 28. *Clostridium botulinum.* × 8000. From Prof. SCANGA. Fig. 29. *S. typhosa*, H agglutination. × 800. Fig. 30. *S. ureae*, H agglutination. × 1000. Fig. 31. *S. ureae*, on straight course. × 1000. Fig. 32. *S. ureae*, on curved course. × 1000. Fig. 33. *Sp. serpens.* × 600. Fig. 34. *Sp. volutans*, bubbles of cell wall. × 4000. Fig. 35. *Sp. cohnii*, various flagella. × 1000. Fig. 36. *Sp. cohnii*, flagella and fibrils. × 1000. Fig. 37. *Sp. cohnii*, fibrils. × 1000. Fig. 38. *Escherichia coli*, with fimbriae. × 10.000. From SCHREIL

III. Die Anwendung von Phagen in der bakteriologischen Diagnostik mit besonderer Berücksichtigung der Typisierung von Typhus- und Paratyphus B-Bakterien sowie Staphylokokken[1]

Von

HENNING BRANDIS

Mit 10 Abbildungen

Inhalt

Einleitung

Der Gedanke, Bakterienstämme mit Bakteriophagen zu unterscheiden, geht wohl auf BAIL zurück. Aber erst die grundlegenden Arbeiten SONNENSCHEINS (1925—1929) haben die Voraussetzung für eine Methodik geschaffen, welche in den letzten Jahren in zunehmendem Maße an praktischer Bedeutung in der bakteriologischen Diagnostik gewonnen hat.

Während SONNENSCHEIN sich hauptsächlich mit Phagen beschäftigte, die eine Gruppen- bzw. Artzugehörigkeit der entsprechenden Bakterienstämme anzeigten, so liegt aus epidemiologischen Gründen heute das Hauptgewicht auf den Phagen, die es ermöglichen, innerhalb einer Art sog. Typen aufzustellen. Es hat sich gezeigt, daß die Phagen außerordentlich feine Indicatoren darstellen, mit deren Hilfe zwischen den Bakterienstämmen einer Art noch Divergenzen aufzudecken sind, welche mit anderen Nachweismethoden, z. B. serologischen Verfahren, unerkannt bleiben. Da diese Unterschiede in der Phagenempfindlichkeit erblich fixiert sind und die einzelnen Bakterienphagentypen, die nach

[1] Aus dem Hygiene-Institut der Stadt und Universität Frankfurt a. M. (Direktor: Professor Dr. med. K. HERZBERG).

NICOLLE zweckmäßig Lysotypen genannt werden, z. B. bei den Typhus- und Paratyphus B-Bakterien eine hohe Konstanz besitzen, stellt die Typenbestimmung von Bakterien mit Bakteriophagen (Lysotypie) nicht nur ein theoretisch interessantes Problem dar, sondern bedeutet für die Klärung epidemiologischer Fragen auch eine wertvolle Unterstützung.

Grundsätzlich gibt es drei verschiedene Verfahren zur Typisierung von Bakterien mit Bakteriophagen. Man kann einmal durch Empirie ausgewählte Phagen mit verschiedenartiger Wirkungsweise benutzen oder aber solche, die sich von einem einzelnen Phagenstamm durch Anpassung herleiten. Schließlich ist es möglich, auf Grund der Zahl und Art der nachweisbaren Lysogenitätsphagen bei der zu prüfenden Bakterienkultur indirekt auf einen bestimmten Bakterientyp zu schließen. Da das zuletzt genannte Verfahren technisch umständlich ist, werden vor allem die beiden zuerst erwähnten Methoden in der Routinepraxis verwendet. Auf Einzelheiten wird im speziellen Teil eingegangen. Nach ANDERSON und WILLIAMS (1956) sind an eine Typisierungsmethode mit Bakteriophagen folgende Anforderungen zu stellen: 1. Die Typen müssen eine hohe „praktische“ Konstanz besitzen. 2. Es sollte eine nicht zu große aber auch nicht zu kleine Zahl an unterscheidbaren Typen feststellbar sein. 3. Die Technik muß einfach sein und reproduzierbare sowie vor allem eindeutige Ergebnisse liefern. 4. Die zum Typisieren verwendeten Phagen sollen stabil sein. 5. Die Methode soll sich für eine Standardisierung eignen und vor der allgemeinen Anwendung auf ihre Brauchbarkeit in ausgedehnten epidemiologischen Untersuchungen geprüft sein.

Besonders gute Erfolge sind bisher mit der Lysotypie von Typhus- und Paratyphus B-Bakterien erzielt worden. Es seien daher zunächst diese Bakterienarten besprochen.

I. Die Lysotypie von Typhusbakterien

Erstmalig hat sich SONNENSCHEIN (1925—1929) mit der Diagnostik von Typhusbakterien durch Phagen beschäftigt. Es gelang diesem Autor, einen Phagen zu isolieren, der eine hohe spezifische Wirkung auf Typhusstämme besaß. So blieben von 512 geprüften verschiedenartigen Bakterienkulturen 403 Nichttyphusstämme unbeeinflußt, während von 109 Typhuskulturen 101 (92,7%) gelöst wurden. Der Phage erwies sich daher als brauchbares Hilfsmittel für die Artdiagnose. Darüber hinaus stellte SONNENSCHEIN auch fest, daß sein Phage ein gewisses Anpassungsvermögen besaß. Der Phage, der zunächst einige Typhusstämme nur schwach angriff, zeigte nach 1—2 Bouillonpassagen mit diesen Stämmen eine wesentlich verstärkte Wirksamkeit. Auch MARCUSE (1934) hat von den Diagnostikphagen SONNENSCHEINS ausgehend Anpassungsversuche bei Typhusstämmen vorgenommen. Er erhielt so vier neue Phagenrassen, mit denen er einschließlich des Originalphagen die von ihm geprüften 469 Typhuskulturen in 5 Gruppen mit unterschiedlichem Phagenverhalten aufteilen konnte. Diese Untersuchungen MARCUSES stellten den Beginn einer Typenaufstellung durch Phagen bei den Typhusbakterien dar. MARCUSE erwähnte bereits, daß Typhusstämme von ein und derselben Person, die zu verschiedenen Zeiten gezüchtet worden waren, zur gleichen Gruppe gehörten. Allerdings beobachtete MARCUSE auch Ausnahmen.

Zwei Jahre nach der Beschreibung des Vi-Antigens durch FELIX und PITT (1934) konnte durch CRAIGIE und BRANDON (1936) und unabhängig von ihnen durch SCHOLTENS (1936) sowie SERTIC und BOULGAKOV (1936) die Existenz von Phagen, die spezifisch auf das Vi-Antigen eingestellt sind (Vi-Phagen), nachgewiesen werden. Bald darauf zeigten CRAIGIE und YEN (1938), daß die Vi-Phagen sich in besonders guter Weise für die Typisierung der Typhusbakterien eignen, und das von ihnen angegebene Verfahren wird auch heute noch — allerdings in erheblich erweiterter Form — angewendet. CRAIGIE und YEN berichteten über 4 Vi-Phagen, die sich hinsichtlich ihrer Lochgröße auf den Wirtsstämmen, der Thermoresistenz, dem Verhalten auf einer Reihe von Typhusteststämmen und vor allem auch serologisch voneinander unterschieden. Diese Vi-Phagen wurden mit Typ I, Typ II, Typ III und Typ IV bezeichnet. Der Vi-Phage II besaß im Gegensatz zu den übrigen Phagen eine große Anpassungsfähigkeit. Hierdurch gelang es CRAIGIE und YEN, diesen Vi-Phagen II einer Reihe von Typhusstämmen anzupassen und typenspezifische Varianten des Vi-Phagen II zu erhalten. Diese adaptierten Vi-Phagen griffen jeweils ihren homologen Züchtungsstamm und die mit diesem Stamm übereinstimmenden Kulturen weitaus stärker an als andere Typhusstämme. Bei Verwendung der Phagen in der kritischen Testverdünnung, d. h. derjenigen Phagenverdünnung, die beim Auftropfen auf den homologen Stamm noch eine konfluierende Lysis ergibt, erwiesen sich die angepaßten Vi-Phagen als typenspezifisch. Sie lösten nur noch ihren homologen Stamm und die Kulturen, welche diesem entsprachen, während andere Stämme unbeeinflußt blieben. CRAIGIE und YEN stellten eine Serie von elf derartigen vom Vi-Phagen II sich herleitenden angepaßten Vi-Phagen auf, mit denen sie elf verschiedene Lysotypen unterschieden, die mit großen Buchstaben (A, B_1, B_2, C, D_1, D_2, E, F, G, H, J) bezeichnet wurden, und von denen auch die entsprechenden Phagen jeweils ihre Benennung erhielten (Tabelle 1).

Tabelle 1. *Reaktionen der Typenstämme von S. typhi mit den kritischen Testverdünnungen der Vi-Phagen II-Anpassungen.* (Nach CRAIGIE und YEN)

Typen-stämme	Phagen										
	A	B 1	B 2	C	D 1	D 2	E	F	G	H	J
A	cL	cL	cL	cL	cL	cL	cL	cL	cL	cL	cL
B 1	—	cL	+++	±	—	—	±	+++	+	—	+++
B 2	±	—	cL	—	+	—	—	+	—	—	—
C	—	—	±	cL	±	—	±	±	±	—	±
D 1	—	—	—	±	cL	+++	—	—	—	—	—
D 2	—	—	—	—	±	cL	—	—	—	—	—
E	—	—	—	—	—	—	cL	—	—	—	—
F	—	—	—	—	—	—	—	cL	—	—	—
G	—	—	—	—	—	—	—	—	cL	—	—
H	—	—	—	—	—	—	—	—	—	cL	—
J	—	—	—	—	—	—	—	—	—	—	cL

cL = konfluierende Lysis, +++ = zahlreiche einzelne Löcher, + = vereinzelte Löcher.

Aus der Tabelle 1 geht das typenspezifische Verhalten der angepaßten Vi-Phagen deutlich hervor. Übergreifende Phagenreaktionen, die jedoch wesentlich schwächer sind als mit dem homologen Stamm, kommen nur bei den Typen B_1,

B_2 und C vor, außerdem wird der Typenstamm A von allen Phagenanpassungen angegriffen. Für jeden Lysotyp besteht demnach ein typisches und für praktische epidemiologische Zwecke vor allem auch konstantes und reproduzierbares Reaktionsbild, so daß es berechtigt ist, von wohl charakterisierten Typhusbakterientypen zu sprechen. Die verschiedenartige Reaktionsweise der einzelnen Lysotypen deutet darauf hin, daß bei ihnen das als Phagenreceptor dienende Vi-Antigen nicht völlig übereinstimmend sein kann. Vielmehr müssen geringe bis jetzt serologisch und chemisch noch nicht faßbare Unterschiede bestehen, welche nur mit Hilfe der Phagen erkennbar sind.

Das Schema von CRAIGIE und YEN hat im Lauf der Jahre eine ständige Ausweitung erfahren. Neue Phagenanpassungen wurden von CRAIGIE [(1939), Unterteilung des Typs E in E_1 und E_2, des Typs F in F_1 und F_2], HELMER und Mitarbeitern (Typ M), FELIX [(1943—1947) D_4, D_5, D_6, L_2, N, O, T], YEN [(1939) K, L_1] hinzugefügt, so daß im Jahre 1947 der Phagensatz 24 Phagenanpassungen umfaßte, die alle vom Vi-Phagen II abstammten. Es waren dies die Phagen A, B_1, B_2, B_3, C, D_1, D_2, D_4, D_5, D_6, E_1, E_2, F_1, F_2, G, H, J, K, L_1, L_2, M, N, O und T. In dem gleichen Jahr machten CRAIGIE und FELIX (1947) Vorschläge zur Standardisierung der Lysotypie, um die Methoden und Arbeitsweisen der an der Lysotypie in den einzelnen Ländern interessierten Laboratorien zu koordinieren. Außerdem wurde während des Internationalen Mikrobiologenkongresses in Kopenhagen (1947) das „International Committee for Enteric Phage Typing" gegründet. Diesem Komitee gehören heute die nationalen und auch die regionalen Institute aller der Länder an, in denen die Lysotypie ausgeführt wird. Nach den Vorschlägen von CRAIGIE und FELIX arbeiten sämtliche Untersuchungsstellen mit der gleichen Methodik, und vor allem werden ausschließlich die von der Internationalen Lysotypiezentrale in London zur Verfügung gestellten Typen-Vi-Phagen benutzt. Eigene Phagenanreicherungen finden keine Verwendung. Auf diese Weise hat es sich ermöglichen lassen, einheitliche verläßliche und vor allem vergleichbare Resultate aus den Lysotypielaboratorien der verschiedensten Länder zu erhalten und falsche Typendiagnosen auf ein Minimum zu beschränken.

Seit dem Jahre 1947 sind bis zu dem gegenwärtigen Zeitpunkt (1956) folgende neuangepaßte Vi-Phagen zu dem Phagensatz hinzugekommen und von dem „International Committee for Enteric Phage Typing" auf den internationalen Kongressen für Mikrobiologie in Rio de Janeiro 1950 bzw. Rom 1953 anerkannt worden, soweit sie bis dahin schon nachgewiesen waren: 25 (LIE KIAN JOE), 26 (CLARK), 27 und 28 (SCHOLTENS), 29 (BORMAN), 30 (ANDERSON), 31 (= E_8, EDWARDS), 32 (FELIX und FRASER), 33 (= C_2, DESRANLEAU), C_4 (WILSON und EDWARDS), C_5 (SCHOLTENS), 34 (WILSON und EDWARDS), 35 (WILSON und EDWARDS), 36 (SCHOLTENS), 37 (NICOLLE und BRAULT)[1]. Ferner ist die E-Gruppe, die 1947 im offiziellen Schema nur die Lysotypen E_1 und E_2 umfaßte, in den letzten Jahren stark erweitert worden. DESRANLEAU (1947) beschrieb die Lysotypen E_3 und E_4. Hierbei ist zu bemerken, daß der ursprünglich von DESRANLEAU benutzte Vi-Phage E_4 nicht vom Vi-Phagen II CRAIGIES

[1] ANDERSON (1956) beschrieb außerdem einen neuen Typ T 4904. Das Wirtsspektrum des adaptierten Vi-Phagen T 4904 umfaßt die Lysotypen A, D_1, D_5, D_6, E_7, 29 und den homologen Typ.

abstammte, sondern von einem anderen Vi-Phagen (Q 1467—43), den DESRANLEAU von einer E_1-Kultur isoliert hatte, und der sich serologisch als eng verwandt

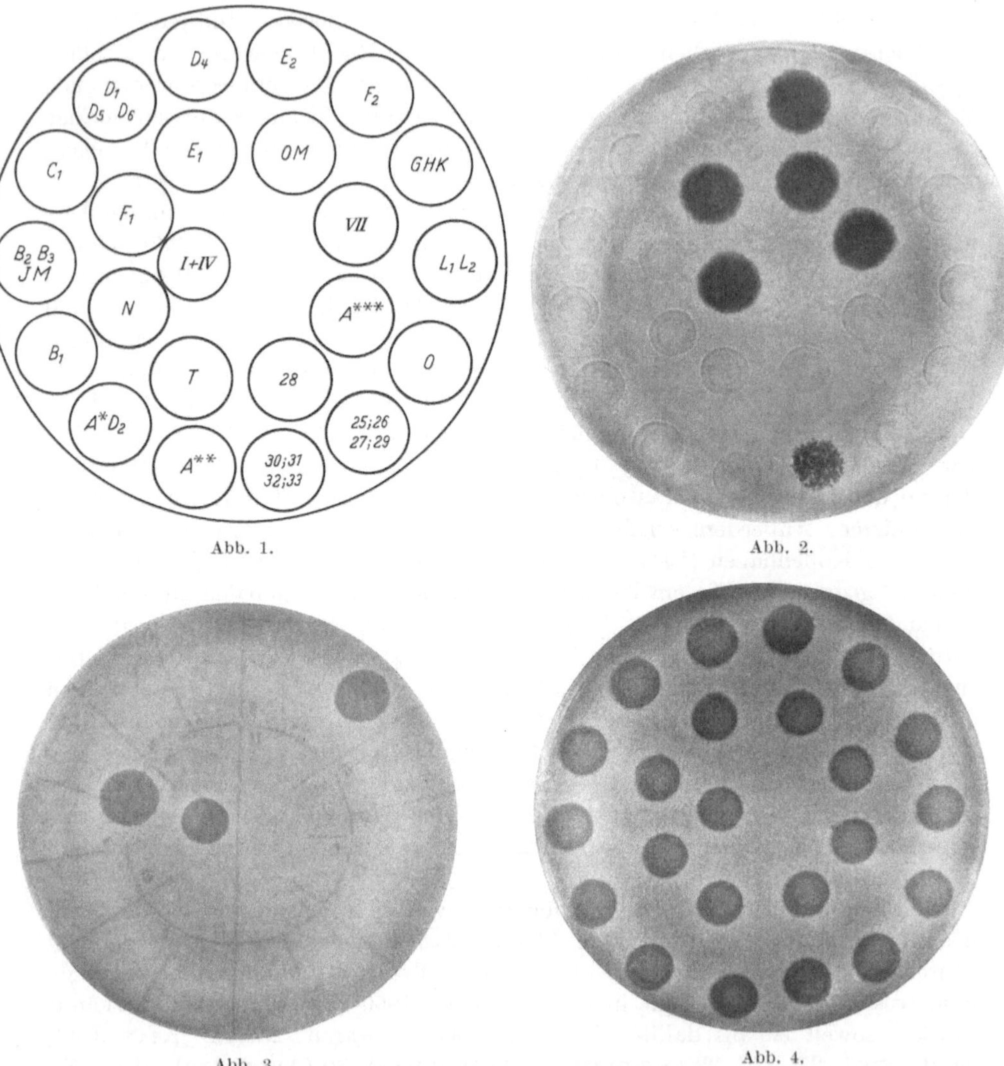

Abb. 1. Auftropfschema der Vi-Typenphagen. *A-Phage in RTD, **A-Phage in 20fach stärkerer und ***A-Phage in 100fach stärkerer Konzentration als die der RTD. OM = Gemisch von Salmonella-Anti-O-Phagen

Abb. 2. Lysotyp E_1. Lysis mit den Phagen E_1, I + IV, E_2, Anti-O, Vi VII und 31 (= E_8), (vgl. Schema Abb. 1). [Die Feldereinteilung für die einzelnen Phagen auf dem Boden der Petri-Schale ist für die photographische Aufnahme fortgelassen.]

Abb. 3. Lysotyp F_1. Lysis mit den Phagen F_1, I + IV, F_2 (vgl. Schema Abb. 1)

Abb. 4. Lysotyp A. Lysis mit allen Phagen (s. [ ] Abb. 2)

mit dem Vi-Phagen II erwies. Neuerdings sind aber SCHOLTENS sowie EDWARDS auch Anpassungen des Vi-Phagen II an den Typenstamm E_4 von DESRANLEAU gelungen, so daß auch hier die Regel wieder eingehalten ist, daß sämtliche für die

Lysotypie gebrauchten Vi-Phagen Anpassungen vom Vi-Phagen II CRAIGIEs sein sollen. Im Jahre 1954 sind ferner noch die Lysotypen E_5 (SCHOLTENS), E_6 (WILSON und EDWARDS) und E_7 (vgl. Anmerkung 3 Tabelle 2) aufgefunden worden (s. SCHOLTENS 1955). Da der Lysotyp 31 nach seiner Reaktionsweise auch zur E-Gruppe zu zählen ist, hat FELIX vorgeschlagen, diesen in E_8 umzubenennen, so daß die E-Gruppe zur Zeit die Lysotypen E_1—E_8 enthält.

Insgesamt umfaßt das Typisierungsschema jetzt (1956) einen Satz von 44 angepaßten Vi-Phagen, mit denen also 44 Typhuslysotypen unterschieden werden können (s. Tabelle 2). Wie die Tabelle 2 zeigt, ist trotz der großen Zunahme der Typen seit der ersten Veröffentlichung von CRAIGIE und YEN im Jahre 1938 das Prinzip der Methode gewahrt geblieben. Es besteht eine deutliche Spezifität der einzelnen Typenphagen für ihre homologen Lysotypen. Daß Lysotypen von mehreren Phagen angegriffen werden, kommt außer bei dem Typ A, der durch seine Empfindlichkeit für alle Phagen eine Ausnahme macht, und dem Typ C praktisch nur bei verwandten Lysotypen (B-, C-, D- und E-Gruppe) vor. Aber auch diese Lysotypen lassen sich an ihren verschiedenartigen charakteristischen Lysisreaktionen sicher voneinander unterscheiden.

Die technische Durchführung der Lysotypiemethode geht so vor sich, daß auf die ganze Oberfläche einer Nähragarplatte (brauchbar ist Nutrient Broth-Agar (Difco) s. CRAIGIE und FELIX 1947) entweder eine 2stündige Bouillonkultur des zu prüfenden Typhusstammes gleichmäßig verteilt wird, oder daß man die Bouillonkultur in einzelnen kreisrunden Flächen von etwa 1 cm Durchmesser ausimpft. Nach Antrocknen der Kultur erfolgt das Auftropfen der Typenphagen, die im Kühlschrank bei $+4^0$ C aufbewahrt werden und bei dieser Temperatur lange haltbar sind, in ihrer kritischen Routinetestverdünnung (RTD), d. h. derjenigen Phagenverdünnung, die auf dem homologen Stamm noch mit einem Ösentropfen eine konfluierende Lysis ergibt. Als Kontrollphagen für das Vorhandensein des Vi-Antigens dienen die nichtadaptierten Vi-Phagen I und IV von CRAIGIE, die man gemischt verwendet. Zusätzlich können auch noch andere Vi-Phagen [z. B. V und VI (DESRANLEAU), VII (BRANDIS) oder Salmonella-Anti-O-Phagen (FELIX und CALLOW)] mitgeführt werden (s. S. 111). Wie schon erwähnt, handelt es sich bei den Testphagen um Präparationen, welche die internationale Lysotypiezentrale in London zur Verfügung stellt. Im Gegensatz zu den Staphylokokkenphagen (s. S. 134) werden die Typhus-Vi-Phagen also nicht selbst angereichert, um Abweichungen in der Spezifität der Reaktionen zu vermeiden. ANDERSON und WILLIAMS (1956) haben Einzelheiten über die Züchtung der Vi-Phagen, wie sie im Central Enteric Reference Laboratory vorgenommen wird, mitgeteilt. Das Verfahren besteht darin, daß die Typenphagen auf ihrem homologen Typenstamm in Bouillon vermehrt werden, wobei als Ausgang gleiche Mengen von Phagen und Bakterien meist den höchsten Titer ergeben. Nach Bebrütung bis zu einem Maximum von $7^1/_2$ Std und Abtöten der Wirtsbakterien durch Erhitzen auf 57^0 C für 40 min sowie Zentrifugieren wird die RTD durch Austitrieren auf dem Typenstamm A und dem homologen Stamm festgestellt. Danach prüft man den Phagen in der RTD und in einer 100fach stärkeren Konzentration auf allen Typhustypenstämmen. Zur Gewinnung hochtitriger Phagen verwenden ANDERSON und WILLIAMS eine Agarplattenmethode. Sie züchten hierbei die Phagen mit den Wirtsbakterien in einer dünnen Schicht 0,45%igen Nähragars, die auf Nutrient-Broth-Agar aufliegt. Nach 16 Std Bebrütung erfolgt mit 6 cm^3 Bouillon die Abschwemmung der Platte. Die Befreiung der Phagensuspension von Wirtsbakterien und Agarpartikeln geschieht durch Zentrifugieren und Erhitzen auf 57^0 C für 40 min.

Die Bebrütung der mit den zu prüfenden Typhuskulturen beimpften und den Testphagen beschickten Platten erfolgt 7 Std bei 37^0 C. Bei der Ablesung, die nach einer weiteren Bebrütung von 12 Std noch ein zweites Mal stattfindet und mit bloßem Auge sowie einer 10fachen Lupe bei schräg durchfallendem Licht geschieht, werden alle Grade der Lysis aufgezeichnet (s. Tab. 2 u. 7). Vereinfachungen der Methode durch Zusammenfassung geeigneter Phagen in Mischungen oder Auswahl von Phagen, die den häufig vorkommenden Typen

Tabelle 2[1]. *Provisorisches erweitertes Typisierungsschema von Salmonella*

Typenstämme																		Vi-
	A	B 1	B 2	B 3	C	30	C 2	C 4	C 5	D 1	D 2	D 4	D 5	D 6	E 1	E 2	E 3	E 4
A	CL	CL	CL	CL	CL	CL	CL	CL	CL	CL	CL	CL	CL	CL	CL	CL	CL	CL
B 1	++	CL	++	+	+	++	±	++	–×	–	–	–	+	+	+	+	–	±
B 2	+	–	CL	–	–	++	++	–	–×	–	–	+	–	–	+	–	++	±
B 3	++	–	++	CL	++	++	++	++	×	++	++	–	–	–	++	++	++	++
C [2]	±	±	±	±	CL	CL	CL	CL	CL	±	±	±	±	±	±	±	±	±
30 (Anderson) [2]	–	–	–	–	±	CL	CL	±	±	–	–	–	–	±	–	–	–	–
C 2 (33, Desranleau) [2]	–	–	–	–	–	+	CL	±	–	–	–	–	–	±	–	–	–	–
C 4 (Wilson u. Edwards)	–	–	–	–	–	–	–	CL	–	–	–	–	–	–	–	–	–	–
C 5 (Scholtens)	–	–	–	–	–	+m	–	–	CL	–	–	–	–	–	–	–	–	–
D 1	–	–	–	–	–	–	–	–	–	CL	CL	CL	++	++	–	–	++m	++m
D 2	–	–	–	–	–	–	–	–	–	–	CL	–	–	–	–	–	–	–
D 4	–	–	–	–	–	–	–	–	–	–	–	CL	–	–	–	–	–	–
D 5	–	–	–	–	–	–	+++	–	–	–	–	–	CL	CL	–	–	–	–
D 6	–	–	–	–	–	–	CL	–	–	–	–	–	–	CL	–	–	–	–
E 1	–	–	–	–	–	–	–	–	–	–	–	–	–	–	CL	CL	CL	CL
E 2	–	–	–	–	–	–	–	–	–	–	–	–	–	–	–	CL	–	–
E 3 (Desranleau)	–	–	–	–	–	–	–	–	–	–	–	–	–	–	–	–	CL	CL
E 4 (Desranleau)	–	–	–	–	–	–	–	–	–	–	–	–	–	–	–	–	–	CL
E 5 (Scholtens)	–	–	–	–	–	–	–	–	–	–	–	–	–	–	–	–	–	–
E 6 (Wilson u. Edwards)	–	–	–	–	–	–	–	–	–	–	–	–	–	–	–	–	–	–
E 7 [3]	–	–	–	–	–	–	–	–	–	–	–	–	–	–	–	–	–	–
E 8 (31, Wilson u. Edwards)	–	–	–	–	–	–	–	–	–	–	–	–	–	–	–	–	–	–
F 1	–	–	–	–	–	–	–	–	–	–	–	–	–	–	–	–	–	–
F 2	–	–	–	–	–	–	–	–	–	–	–	–	–	–	–	–	–	–
G	–	–	–	–	+ n	–	–	–	–	–	–	–	–	–	–	–	–	–
H	–	–	–	–	–	–	–	–	–×	–	–	–	–	–	± s	–	–	–
J	–	–	–	–	–	–	–	–	–	–	–	–	–	–	–	–	–	–
K	–	–	–	–	–	–	–	–	–	–	–	–	–	–	–	–	–	–
L 1	–	–	–	–	–	–	–	–	–	–	–	–	–	–	–	–	–	–
L 2	–	–	–	–	–	–	–	–	–	–	–	–	–	–	–	–	–	–
M	–	–	–	–	–	–	–	–	–	–	–	–	–	–	–	–	–	–
N	–	–	–	–	–	–	–	–	–×	–	–	–	–	–	–	–	–	–
O	–	–	–	–	–	–	–	±	–×	–	–	–	–	–	–	–	–	–
T	–	–	–	–	–	–	–	±	–×	–	–	–	–	±m	–	–	–	–
25 (Lie Kian Joe)	–	–	–	–	–	–	–	–	–	–	–	–	–	–	–	–	–	–
26 (Clark)	–	–	–	–	–	–	–	–	–	–	–	–	–	–	–	–	–	–
27 (Scholtens)	–	–	–	–	–	–	–	–	–	–	–	–	–	–	–	–	–	–
28 (Scholtens)	–	–	–	–	–	–	–	–	–	–	–	–	–	–	–	–	–	–
29 (Borman)	–	–	–	–	–	CL	CL	±	–	–	–	–	–	CL	–	–	±m	–
32 (Felix u. Fraser)	–	–	–	–	–	–	–	–	–	–	–	–	–	–	–	–	–	–
34 (Wilson u. Edwards)	–	–	–	–	–	–	–	–	–	–	–	–	–	–	–	–	–	–
35 (Wilson u. Edwards)	–	–	–	–	–	–	–	–	–	–	–	–	–	–	–	–	–	–
36 (Scholtens)	–	–	–	–	–	–	–	–	–	–	–	–	–	–	–	–	–	–
37 (Nicolle u. Brault)	–	–	–	–	–	–	–	–	–	–	–	–	–	–	–	–	–	–

CL = konfluierende Lysis; n = normale Phagenlöcher; x = der verfügbare Typenstamm
Löcher; + = gewöhnlich einige wenige Löcher; +, ++, +++

[1] An dieser Stelle sei den Herren Dr. EDWARDS, Dr. NICOLLE und Dr. SCHOLTENS für die Genehmigung gedankt, das Reaktionsbild der von ihnen neu entdeckten aber noch nicht beschriebenen Typhuslysotypen in die Tabelle aufzunehmen.

[2] Von FELIX (1955) sind für die Typen C, 30 und C_2 die noch nicht vom I.C.E.P.T. anerkannten Bezeichnungen „C_1, C_2 und C_3" vorgeschlagen worden.

[3] Der Typ E_7 wurde zuerst von ANDERSON experimentell in vitro erzeugt und später von SCHOLTENS auch unter natürlichen Bedingungen gefunden.

entsprechen, sind natürlich möglich. Die Typhuskulturen sollen durch Abimpfung von mehreren auf den Originalplatten gewachsenen Kolonien gewonnen und sobald wie möglich nach der Isolierung typisiert werden, da bei älteren Kulturen schlechtere Ergebnisse auftreten. Zur Aufbewahrung der Kulturen (bei Zimmertemperatur) eignet sich vor allem der DORSET-Eiernährboden (CRAIGIE und FELIX 1947). Über den Einfluß der Zusammensetzung des Nährbodens auf die Lysotypieergebnisse liegen Untersuchungen von FELIX und ANDERSON (1951) sowie RISCHE und SCHNEIDER (1955) vor. Durch Zusatz von Glycerin zum Nährboden werden die Phagenlöcher größer (DESRANLEAU 1947).

typhi nach der Standardmethode von CRAIGIE *und* FELIX, *November 1955*

Phagen																										Typen-stämme
E 5	E 6	E 7	E 8	F 1	F 2	G	H	J	K	L 1	L 2	M	N	O	T	25	26	27	28	29	32	34	35	36	37	
CL	CL	CL	CL	CL	CL	CL	CL	CL	CL	CL	CL	CL	CL	CL	CL	CL	CL	CL	CL	CL	CL	CL	CL	CL	CL	A
−	−	+	±	+	+	+	+	−	+	+	+	+	+	++	+	−	−	+	+	++	+	−	+	×	++	B 1
++	++	−	++	+	−	+	+	+	−	−	±	±	+	−	+	++	++	++	++	−	++	−	+	−×	++	B 2
++	+	+	++	++	++	++	++	++	++	++	++	++	++	++	++	++	++	++	++	−	++	−	++	×	++	B 3
±	±	±	+	+	±	+	±	±	±	±	±	±	±	±	±	±	±	±	±	±	±	±	+	±	±	C
−	−	±	−	−	±	−	−	−	−	−	−	−	−	−	−	−	−	−	−	±	−	−	−	±	−	30
−	−	−	−	−	−	−	−	−	−	−	−	−	−	−	−	−	−	−	−	−	−	−	−	±	−	C 2
−	−	−	−	−	−	−	−	−	−	−	−	−	−	−	−	−	−	−	−	−	−	−	−	−	−	C 4
−	−	−	−	−	−	−	−	−	−	−	−	−	−	−	−	−	−	−	−	−	−	−	−	−	−	C 5
−	−	−	−	−	−	−	−	−	−	−	−	−	−	−	−	−	−	−	−	−	−	−	−	−	−	D 1
−	−	−	−	−	−	−	−	−	−	−	−	−	−	−	−	−	−	−	−	−	−	−	−	−	−	D 2
−	−	−	−	−	−	−	−	−	−	−	−	−	−	−	−	−	−	−	−	−	−	−	−	−	−	D 4
−	−	−	−	−	−	−	−	−	−	−	−	−	−	−	−	−	−	−	−	−	−	−	−	−	−	D 5
−	−	−	−	−	±m	−	−	−	−	−	−	−	−	−	−	−	−	−	−	−	−	−	−	CL	−	D 6
CL	CL	CL	CL	−	−	−	−	−	−	−	−	−	−	−	−	−	−	−	−	−	−	−	−	−	−	E 1
−	−	−	−	−	−	−	−	−	−	−	−	−	−	−	−	−	−	−	−	−	−	−	−	−	−	E 2
−	−	−	−	−	−	−	−	−	−	−	−	−	−	−	−	−	−	−	−	−	−	−	−	−	−	E 3
−	−	−	−	−	−	−	−	−	−	−	−	−	−	−	−	−	−	−	−	−	−	−	−	−	−	E 4
CL	−	−	−	−	−	−	−	−	−	−	−	−	−	−	−	−	−	−	−	−	−	−	−	−	−	E 5
−	CL	−	−	−	−	−	−	−	−	−	−	−	−	−	−	−	−	−	−	−	−	−	−	−	−	E 6
−	−	CL	−	−	−	−	−	−	−	−	−	−	−	−	−	−	−	−	−	−	−	−	−	−	−	E 7
−	−	−	CL	−	−	−	−	−	−	−	−	−	−	−	−	−	−	−	−	−	−	−	−	−	−	E 8
−	−	−	−	CL	CL	−	−	−	−	−	−	−	−	−	−	−	−	−	−	−	−	−	−	−	−	F 1
−	−	−	−	±	CL	−	−	−	−	−	−	−	−	−	−	−	−	−	−	−	−	−	−	−	−	F 2
−	−	−	−	±	−	CL	−	−	−	−	−	−	−	−	−	−	−	−	−	−	−	−	±	−	−	G
−	−	−	−	±	±	−	CL	−	−	−	−	−	±	±	±	−	−	−	−	−	−	−	±	×	−	H
−	−	−	−	−	−	−	−	CL	−	−	−	−	−	−	−	−	−	−	−	−	−	−	−	−	−	J
−	−	−	−	−	−	−	−	−	CL	−	−	±m	−	−	−	−	−	−	−	−	−	−	−	−	−	K
−	−	−	−	−	−	−	−	−	−	CL	CL	−	−	−	−	−	−	−	−	−	−	−	−	−	−	L 1
−	−	−	−	−	−	−	−	−	−	±	CL	−	−	−	−	−	−	−	−	−	−	−	−	−	−	L 2
−	−	−	−	−	−	−	−	−	−	−	−	CL	−	−	−	−	−	−	−	−	−	−	−	−	−	M
−	−	−	−	−	−	−	−	−	−	−	−	−	CL	−	±m	−	−	−	−	−	−	−	−	−×	−	N
−	−	−	−	−	±m	−	−	−	−	−	−	−	−	CL	−	−	−	−	−	−	−	−	−	×	−	O
−	−	−	−	−	−	−	−	−	−	−	−	−	−	−	CL	−	−	−	CL	−	−	−	−	×	−	T
−	−	−	−	−	−	−	−	−	−	−	−	−	−	−	−	CL	−	−	−	−	−	−	−	−	−	25
−	−	−	CL	−	−	−	−	−	−	−	−	−	−	−	−	−	CL	−	−	−	−	−	−	−	−	26
−	−	−	−	−	−	−	−	−	−	−	−	−	−	−	−	−	−	CL	−	−	−	−	−	−	−	27
−	−	−	−	−	−	−	−	−	−	−	−	−	−	−	−	−	−	−	CL	−	−	−	−	−	−	28
−	−	CL	−	−	CL	−	−	−	−	−	−	−	−	−	−	−	−	−	−	CL	−	−	±	CL	−	29
CL	−	−	−	−	−	−	−	−	−	−	−	−	−	−	−	−	−	−	−	−	CL	−	−	−	−	32
−	−	−	−	−	−	−	−	−	−	−	−	−	−	−	−	−	−	−	−	−	−	CL	−	−	−	34
−	−	−	−	−	−	−	−	−	−	−	−	−	−	−	−	−	−	−	−	−	−	−	CL	−	−	35
−	−	−	−	−	−	−	−	−	−	−	−	−	−	−	−	−	−	−	−	−	−	−	−	CL	−	36
−	−	−	−	−	−	−	−	−	−	−	−	−	−	−	−	−	−	−	−	−	−	−	−	−	CL	37

war degradiert (FELIX); — = keine Phagenlöcher; s = kleine mit bloßem Auge noch sichtbare = ansteigende Zahl von Löchern; m = Mikrolöcher.

Nicht alle Typhuskulturen können mit den zur Verfügung stehenden angepaßten Vi-Phagen typisiert werden. Diese unbestimmbaren Stämme zerfallen in 3 Gruppen: 1. Die sog. unbestimmbaren Vi-Stämme, das sind solche Kulturen, die zwar Vi-Antigen enthalten, aber von den Typenphagen nicht angegriffen werden. Dagegen findet eine Lysis mit anderen nichtangepaßten Vi-Phagen, z. B. den Phagen I und IV von CRAIGIE, statt. Mit Hilfe derartiger nichtadaptierter Vi-Phagen sowie auch mit Nicht-Vi-Phagen (s. S. 110) ist es möglich, bei den unbestimmbaren Vi-Stämmen auf Grund verschiedenartiger Reaktionsweise Unterteilungen vorzunehmen. Dies hat u. U. für epidemiologische Fragestellungen einen Nutzen. Die Unempfindlichkeit der unbestimmbaren Vi-Stämme für den Vi-Phagen II kann nach ANDERSON und WILLIAMS (1956) auf einer von vornherein vorhandenen Resistenz beruhen. In diesem Fall verlaufen auch Anpassungsversuche mit dem Vi-Phagen II negativ. Oder es handelt sich um

Tabelle 2a. *Salmonella typhi; Liste der Typenstämme A — 33*
(Zusammenstellung von Dr. A. FELIX; vgl. CRAIGIE und FELIX 1947)[1]

Lysotyp	Aufgestellt von	Bezeichnung des Typenstammes	Isoliert	Literaturangabe
A	CRAIGIE u. YEN 1938	A	1937 Canada	CRAIGIE, J., and C. H. YEN: Canad. Publ. Health J. **29**, 448, 484 (1938).
B 1	CRAIGIE u. YEN 1938	B 1	1937 Canada	CRAIGIE, J., and C. H. YEN: Canad. Publ. Health J. **29**, 448, 484 (1938).
B 2	CRAIGIE u. YEN 1938	B 2	1937 Canada	CRAIGIE, J., and C. H. YEN: Canad. Publ. Health J. **29**, 448, 484 (1938).
B 3	CRAIGIE 1940	B 3	1938 Schottland	CRAIGIE, J.: Canad. Publ. Health J. **31**, 4 (1940). (Abstract.) — PIERSOL and BORTZ: Cyclopedia of Medicine, Section of Bacteriology, S. 5. 1941.
C[3]	CRAIGIE u. YEN 1938	C	1937 Canada	CRAIGIE, J., and C. H. YEN: Canad. Publ. Health J. **29**, 448, 484 (1938).
D 1	CRAIGIE u. YEN 1938	D 1	1937 Canada	CRAIGIE, J., and C. H. YEN: Canad. Publ. Health J. **29**, 448, 484 (1938).
D 2	CRAIGIE u. YEN 1938	D 2	1936 England	CRAIGIE, J., and C. H. YEN: Canad. Publ. Health J. **29**, 448, 484 (1938).
D 4	FELIX 1941	T 107	1941 England	FELIX, A.: Brit. Med. J. **1943**, **I**, 435.
D 5	FELIX 1943	T 5077	1951 Madagaskar	FELIX, A.: Brit. Med. J. **1943**, **I**, 435.
D 6	FELIX 1946	T 4274	1950 England	FELIX, A.: First Report International Committee for Enteric Phage Typing, Rio de Janeiro, 1950.
E 1	CRAIGIE u. YEN 1938	E 1	1918 Rußland	CRAIGIE, J., and C. H. YEN: Canad. Publ. Health J. **29**, 448, 484 (1938).
E 2	CRAIGIE 1939	T 84	1940 England	CRAIGIE, J.: Canad. Publ. Health J. **30**, 37 (1939). (Abstract.)
E 3	DESRANLEAU 1945		1945 Canada	DESRANLEAU, J. M.: Canad. J. Publ. Health **88**, 343 (1947).
E 4	DESRANLEAU 1947		1947 Canada	DESRANLEAU, J. M.: Canad. J. Publ. Health **88**, 343 (1947).
E 5	SCHOLTENS 1954	4316	1954 Holland	SCHOLTENS, R. TH.: Ann. Inst. Pasteur **89**, 216 (1955).
E 6	WILSON u. EDWARDS 1954	640	USA	SCHOLTENS, R. TH.: Ann. Inst. Pasteur **89**, 216 (1955).
E 7	ANDERSON[2], SCHOLTENS 1954	563	1954 Holland	SCHOLTENS, R. TH.: Ann. Inst. Pasteur **89**, 216 (1955).
F 1	CRAIGIE u. YEN 1938	F 1	1900 England	CRAIGIE, J., and C. H. YEN: Canad. Publ. Health J. **29**, 448, 484 (1938).
F 2	CRAIGIE 1939	F 2	1938 Canada	CRAIGIE, J.: Canad. Publ. Health J. **30**, 37 (1939). (Abstract.)

G	Craigie u. Yen 1938	G	1937 Canada	Craigie, J., and C. H. Yen: Canad. Publ. Health J. **29**, 448, 484 (1938).
H H	Craigie u. Yen 1938	H (1) H (3)	1937 Canada	Craigie, J., and C. H. Yen: Canad. Publ. Health J. **29**, 448, 484 (1938).
J	Craigie u. Yen 1938	T 5105	1953 Tunis	Craigie, J., and C. H. Yen: Canad. Publ. Health J. **29**, 448, 484 (1938).
K	Yen 1939	T 2078	1943 England (von Indien)	Yen, C. H.: Proc. Soc. Exper. Biol. a. Med. **41**, 162 (1939).
L 1	Yen 1939	L 1	1938 China (Stamm P 16)	Yen, C. H.: Proc. Soc. Exper. Biol. a. Med. **41**, 162 (1939).
L 2	Felix 1943	T 131	1940 England	Felix, A.: Brit. Med. J. **1943, I,** 435.
M	Craigie 1940	M	1939 Canada	Craigie, J.: Canad. Publ. Health J. **31**, 4 (1940). (Abstract.) — Piersol and Bortz: Cyclopedia of Medicine, Section of Bacteriology, p. 5.
N N	Felix 1943	T 1388 T 2658	1945 England 1948 England	Felix, A.: Brit. Med. J. **1943, I,** 435.
O O	Felix 1943	T 858 T 1190	1943 England 1944 England	Felix, A.: Brit. Med. J. **1943, I,** 435.
T T	Felix 1943	T 820 T 879	1943 England (von S. Afrika)	Felix, A.: Brit. Med. J. **1943, I,** 435.
25 (Lie Kian Joe)	Lie Kian Joe 1948	T 2957 a T 2957 d	1948 Indonesien	Lie Kian Joe: Documenta neerl. et indones. morbis trop. **1**, No 2 (1949).
26 (Clark)	~~Eina M. Clark 1948~~	To 359	1948 Puerto Rico	Felix, A.: First Report Internat. Comm. for Enteric Phage Typing, Rio de Janeiro, 1950.
27 (Scholtens)	Scholtens 1948	T 2958	1948 Holland	Scholtens, R. Th.: Antonie van Leeuwenhoek **16**, 245 (1950).
28 (Scholtens)	Scholtens 1948	T 2960	1948 Holland	Scholtens, R. Th.: Antonie van Leeuwenhoek **16**, 245 (1950).
29 (Borman)	Borman 1950	T 4019	1950 USA	Felix, A.: First Report Internat. Comm. for Enteric Phage Typing, Rio de Janeiro, 1950. — Felix, A., and E. S. Anderson: J. of Hyg. **49**, 349 (1951).
30 (Anderson)[3]	Anderson 1951	T 4677	1951 England	Anderson, E. S., and A. Felix: J. Gen. Microbiol. **9**, 65 (1953).
31 = E 8 (Edwards)	Edwards 1951	T 4931	1951 USA	Felix, A.: Second Report Internat. Comm. for Enteric Phage Typing, Rome, 1953.
32 (Felix u. Fraser)	Felix u. Fraser 1953	T 3122	1953 England	Felix, A.: Second Report Internat. Comm. for Enteric Phage Typing, Rome, 1953.
33 (Desranleau C_2)[3]	Desranleau 1946	T 5051	1946 Canada	Desranleau, J. M., and I. Martin: Canad. Publ. Health J. **41**, 128 (1950). — Felix, A.: Second Report Internat. Comm. for Enteric Phage Typing, Rome, 1953.

[1] Die Liste ist durch folgende neu aufgestellte Typen C4 (Wilson und Edwards), C5 (Scholtens), 34 (Wilson und Edwards), 35 (Wilson und Edwards), 36 (Scholtens), 37 (Nicolle und Brault) und 38 (Wilson und Edwards) zu ergänzen.

[2] Anderson hat den Typ E_7 erstmalig experimentell in vitro erzeugt. Scholtens fand ihn später unter natürlichen Bedingungen.

[3] Vgl. Anmerkung 2 Tabelle 2.

einen Typhustyp, für den noch kein angepaßter Typenphage zur Verfügung steht. 2. Die „degradierten" Vi-Stämme, die zwar mit den Typenphagen reagieren, aber atypische und oft auch inkonstante Reaktionen geben. Es handelt sich um Kulturen, die nach ANDERSON und FELIX (1953) als Stadien zwischen dem spezifischen Typ und dem Typ A, der für alle Phagen empfindlich ist, zu betrachten sind, und 3. die Vi-negativen Stämme (W-Formen nach KAUFFMANN), denen das Vi-Antigen fehlt und die infolgedessen von den Vi-Phagen nicht beeinflußt werden können. Über die Häufigkeit des Vorkommens derartiger unbestimmbarer Typhuskulturen gibt Tabelle 6 Auskunft. Vi-negative Stämme sind bei Dauerausscheidern häufig zu finden (RISCHE 1955, BRANDIS 1955 u. a.). Gelegentlich kann die Resistenz gegen die Typenphagen durch eine Kontamination der zu prüfenden Kultur mit exogenen Phagen bewirkt sein. In diesen Fällen gelingt es nach DUNBAR (1948), den Stamm durch Züchtung in einem Anti-O-Phagenserum in eine typisierbare Form zu überführen.

Zusammenfassend läßt sich sagen, daß die Methode von CRAIGIE und YEN sich auf dem Vorhandensein des Vi-Antigens der Typhusbakterien aufbaut, dessen Uneinheitlichkeit durch eine Serie spezifisch von dem Vi-Phagen II abstammender angepaßter Vi-Phagen erkannt werden kann, wodurch die Typhusbakterien in eine Reihe von Lysotypen aufgeteilt werden. Über die Entstehung der Anpassungsformen des Vi-Phagen II gibt es mehrere Anschauungen. Nach CRAIGIE und YEN müssen sie als Mutanten (host range mutants) angesehen werden. Andererseits wird nach BERTANI und WEIGLE die Spezifität der Phagen weitgehend von dem jeweiligen Wirtsstamm bestimmt. Auf Grund experimenteller Versuche bei degradierten Vi-Stämmen neigen FELIX sowie ANDERSON und FELIX (1953) zu dieser Anschauung und nehmen an, daß die Phagenanpassungen zum Teil phänotypische Variationen darstellen. Offenbar sind nach ANDERSON (1955) bei einer Reihe von Typenphagen Mutanten und bei anderen dagegen phänotypische Adaptationen oder aber beide Arten dieser Änderungen anzunehmen. Wie ANDERSON und WILLIAMS (1956) ausführen, ist beispielsweise der Phage E_1 eine phänotypische Modifikation des Vi-Phagen II. Bei Züchtung des Phagen E_1 auf dem Typenstamm A gehen nämlich sämtliche Phagenpartikel in den Phagen A über, welcher nach ANDERSON und FRASER (1955) die Wildform des Vi-Phagen II darstellt. Demgegenüber behalten diejenigen Typenphagen, die Abkömmlinge von host range mutants sind, bei Züchtung auf dem Typenstamm A ihre Eigenschaften unverändert bei. So unterliegt z. B. der Phage D_1 bei Vermehrung auf dem Typenstamm A keinerlei Änderung. Nach Versuchen von ANDERSON und FRASER (1956) muß man annehmen, daß der Phage A von vornherein in geringer Anzahl mutierte Phagenpartikel mit verschiedenem host range enthält. In serologischer Hinsicht tritt beim Vi-Phagen II durch die Anpassungen keine Abwandlung ein.

Die Spezifität der einzelnen Lysotypen wird indessen nicht allein durch das Vi-Antigen bewirkt. Wie CRAIGIE (1942, 1946), FELIX und ANDERSON (1951), ANDERSON (1951), ANDERSON und FELIX (1953) sowie FERGUSON, JUENKER und FERGUSON (1955) nachwiesen, spielen auch bei mehreren Lysotypen latente Phagen[1] für das Zustandekommen der Typenreaktion eine determinierende Rolle. Das Vi-Antigen dient bei den lysogenen Typhusstämmen nur

[1] In dieser Arbeit wird für Prophagen die Bezeichnung latente Phagen verwendet.

zur Adsorption der Vi-Phagen. Ob es aber zu einer Weiterentwicklung der Phagen kommt, hängt bei diesen Kulturen weitgehend von deren latenten Phagen ab, die wie ANDERSON und FELIX (1953) zeigten, im freien Zustand am O-Antigen angreifen und demnach keine Vi-Phagen darstellen. In Tabelle 3 sind die von ANDERSON und FELIX bei verschiedenen Typhuslysotypen nachgewiesenen temperierten Phagen aufgeführt.

Tabelle 3. *Lysogene Typhustypen*
(Nach ANDERSON und FELIX, ergänzt nach ANDERSON)

Lysotyp	Bezeichnung der temperierten Phagen	Lysotyp	Bezeichnung der temperierten Phagen
B 3	b 3	T	t
D 1	d 1	25	25′
D 4	d 1	26	26′
D 6	d 6	28	28′
F 2	f 2	29	(29′ =) f 2
K	k	30	(30′ =) f 2
		31	26′
		33	d 6

FERGUSON und Mitarbeiter (1955) haben die Ergebnisse von ANDERSON und FELIX weitgehend bestätigen können. Sie bezogen in ihre Versuche noch die temperierten Phagen b_2, d_5, a-phi (isoliert von DESRANLEAU), b_3 (isoliert von EDWARDS) und a-TH-25 ein. Aus 16 verschiedenen Lysotypen gelang es den Autoren bisher nicht, temperierte Phagen zu gewinnen.

Wenn man nun einen bestimmten Lysotyp (besonders geeignet ist der Typ A) gegen den aus einem anderen Lysotyp stammenden Phagen resistent werden läßt, so wird die Kultur dabei lysogen und ändert ihren Typ. Dies hat zuerst CRAIGIE (1942) nachgewiesen. Wenn also beispielsweise eine Kultur des Typs A mit dem temperierten typenbestimmenden Phagen d_1 behandelt wird, so nimmt die Kultur die Reaktionsweise des Lysotyps D_1 an, d. h. sie hat ihre Empfindlichkeit, mit allen Vi-Phagen eine Lysis zu geben, verloren und ist nur noch für die Phagen der D-Gruppe empfindlich (CRAIGIE 1942, ANDERSON und FELIX 1953). Eingehend hat ANDERSON (1951) diese Verhältnisse bei den Lysotypen F_1 und F_2 studiert. So wechselt der Typ F_1 beim Resistentwerden gegenüber dem vom Typ F_2 stammenden temperierten Phagen f_2 seine Typeneigenschaften und verhält sich wie der Typ F_2. Beim Verlust des latenten Phagen f_2 geht der

Tabelle 4. *Künstliche Beeinflussung der Bakteriophagenreaktionen bei S. typhi.*
(Nach ANDERSON und FELIX, geändert nach ANDERSON sowie FERGUSON und Mitarbeitern)

Verwendeter typentransformierender Phage	Ausgangslysotyp	Künstlich entstandener Lysotyp
d 1	A	D 1
	E 1	unbestimmbarer Vi-Stamm
d 5	A	D 1
d 6	A	D 6
	C	33
	D 1	D 6
	F 1	F 2
f 2	A	29
	C	30
	D 1	D 6
	D 4	unbestimmbarer Vi-Stamm
	E 1	Typ der E-Gruppe (E 7)
	F 1	F 2
t	A	T
25′	A	25
26′	A	26
30′	A	29
	C	30
	F 1	F 2
31′ (=26′)	A	26
	B 3	26
	E 3	31
	G	unbestimmbarer Vi-Stamm
	L 1	unbestimmbarer Vi-Stamm
	M	unbestimmbarer Vi-Stamm
	O	unbestimmbarer Vi-Stamm
33′ (=d 6)	A	D 6
	C	33
	B 1	unbestimmbarer Vi-Stamm
	J	unbestimmbarer Vi-Stamm
	L 1	unbestimmbarer Vi-Stamm

künstliche Typ F_2 dann wieder in den Typ F_1 über. Wird dagegen der Typ A mit dem Phagen f_2 behandelt, so entsteht der Typ 29. Wie FERGUSON und Mitarbeiter (1955) mitteilen, kann jedoch auch bei den Umwandlungsversuchen eine nichtspezifisch reagierende Kultur entstehen, die demnach einem degradierten Vi-Stamm entsprechen würde. Schließlich treten gelegentlich auch unbestimmbare Vi-Stämme auf, die sich resistent gegenüber allen Vi-II-Phagenanpassungen verhalten. Hinsichtlich der serologischen Verwandtschaftsverhältnisse und anderer Eigenschaften der genannten temperierten Phagen muß auf ANDERSON und FELIX sowie FERGUSON und Mitarbeiter verwiesen werden, die zu weitgehend übereinstimmenden Ergebnissen kamen. In Tabelle 4 sind die Änderungen im Phagenverhalten durch Einwirkung temperierter Phagen auf eine Reihe von Lysotypen nach den Versuchen von ANDERSON und FELIX sowie FERGUSON und Mitarbeitern wiedergegeben.

Tabelle 5. *Klassifizierung einiger Lysotypen von S. typhi nach ihren Strukturformeln* (Nach ANDERSON und FRASER)

Gruppe	Bezeichnung des Typs	Strukturformel
A	A	A
	D 1	A (d 1)
	D 6	A (d 6)
	29	A (f 2) oder A (30′)
	T	A (t)
	25	A (25′)
	26	A (26′)
C	C	C
	30	C (f 2) oder C (30′)
	33 (= C 2 Desranleau)	C (d 6)
E	E 1	E 1
	T 4904	E 1 (d 6)
	31	E 1 (26′)
	künstliche Typen	E 1 (d 1) E 1 (f 2)
F	F 1	F 1
	F 2	F 1 (f 2)

Die künstlich erzeugten Lysotypen entsprechen zum Teil den natürlich vorkommenden Lysotypen, zum Teil verhalten sie sich aber in ihrer Reaktion gegenüber nicht vom Vi-Phagen II abstammenden Vi-Phagen unterschiedlich. So ist nach ANDERSON und FELIX der künstliche Lysotyp 29 gegenüber dem Phagen IV von CRAIGIE voll empfindlich, während der natürlich vorkommende Lysotyp 29 nur gering auf den Phagen IV anspricht. Nach ANDERSON (1955) kann man die lysogenen Typhusbakterientypen nach ihrer mutmaßlichen Entstehungsweise bezeichnen, d. h. nach dem nichtlysogenen[1] Ursprungsstamm und dem typendeterminierenden Phagen. Der Typ F_2 würde demnach die Formel F_1 (f_2) oder der Typ D_1 die Formel A (d_1) haben. ANDERSON sowie ANDERSON und FRASER (1955) betrachten diese Strukturformeln als nützlich für die Aufdeckung der phylogenetischen Verwandtschaft der Lysotypen und haben eine Gruppierung der lysogenen Typhustypen nach der Abstammung von den gemeinsamen Ursprungsstämmen vorgenommen, die in Tabelle 5 wiedergegeben ist. Aus der Tabelle geht hervor, daß, wie ANDERSON ausführt, die Typenspezifität der mit typdeterminierenden latenten Phagen behafteten Lysotypen von der Natur des nichtlysogenen Ausgangsstammes (A, C, E_1 usw.) und zugleich von der Art der typbestimmenden Phagen abhängt.

Die Entstehung degradierter Typhusstämme ist schwierig zu erklären. In einer Reihe von Fällen dürften auch hier latente Phagen und vor allem der Verlust der ursprünglich typisch reagierenden Kulturen an bestimmten latenten Phagen eine Rolle spielen, wodurch diese Kulturen ihre Spezifität verlieren und

[1] Hinsichtlich typenbestimmender temperierter Phagen.

empfindlich für den Angriff mehrerer Typen-Vi-Phagen werden können (ANDERSON und FELIX 1953). Andererseits ist es auch denkbar, daß Veränderungen am Receptorenapparat der Bakterienoberfläche das Erscheinungsbild der degradierten Typhusstämme bewirkt. Die Bezeichnung „degradiert" bezieht sich, worauf ANDERSON hinweist, nur auf die Reaktion des betreffenden Typhusstammes mit den Typen-Vi-Phagen und sagt nichts über die antigenen Verhältnisse und die Virulenz der Kultur aus.

Wie ANDERSON weiter betont, bedeutet eine „Degradation" nicht immer einen Verlust, d. h. den Übergang von einer hohen Spezifität zu einer niederen, vielmehr können Stämme des Lysotyps A nach Lysogenisation mit geeigneten Phagen das Erscheinungsbild von degradierten Typhusstämmen zeigen. Dies würde dann im Gegensatz zu der zuvor genannten Möglichkeit den umgekehrten Weg darstellen. Der Nachweis von temperierten Phagen und deren Eigenschaften läßt sich u. U. zur Charakterisierung degradierter Typhusstämme heranziehen (vgl. RISCHE 1956). In diesem Zusammenhang sei noch bemerkt, daß wir gegenwärtig noch nicht endgültig sagen können, welche Typhuslysotypen nicht lysogen sind, da der Nachweis der temperierten Phagen zum Teil sehr schwierig ist. Ferner muß betont werden, daß auch nicht allen aus den verschiedenen Typhuslysotypen gewonnenen temperierten Phagen (z. B. a-TH25, b_2, b_3, k, 28' [FERGUSON und Mitarbeiter]) ein typendeterminierender Wert zukommt.

Die Stämme mancher Lysotypen verhalten sich der Serie der angepaßten Vi-Phagen gegenüber nicht völlig übereinstimmend, so daß hier innerhalb von Lysotypen Varietäten zu unterscheiden sind. Wie DESRANLEAU (1951) sowie NICOLLE, HAMON und EDLINGER (1953) nachwiesen, tritt neben dem üblicherweise vorkommenden Typ N, der nur von seinem homologen Phagen und dem Phagen E_4 von DESRANLEAU gelöst wird (Varietät Richmond), noch eine Spielart auf, welche eine zusätzliche Lysis mit den Phagen der D-Gruppe ergibt (Varietät Chattam). Diese zuletzt genannte Varietät, die auch als N + D_1 bezeichnet wird und in Frankreich selten, dagegen in der Türkei, in Nordafrika und in Österreich häufiger vorkommt, ist indessen als offizieller Typ nicht anerkannt. Ferner treten Stämme des Typs A auf, die mit dem Typenphagen B_1 oder aber mit anderen Vi-Typenphagen keine Reaktion geben. Wie NICOLLE, HAMON und EDLINGER (1953) feststellten, stimmt die atypische Reaktionsweise meist bei allen Kulturen aus einem gleichen Infektionsherd überein, so daß derartige abweichende Formen, die möglicherweise den ersten Schritt zu den „degradierten" Vi-Stämmen darstellen, für epidemiologische Zwecke verwendbar sind. Ferner zeigten NICOLLE, VAN OYE, CROCKER und BRAULT (1955), daß beim Lysotyp C eine Varietät auftritt, welche sich von dem üblichen Typ C durch eine fehlende Lysis mit den Typenphagen D_1, D_2, D_4, D_5 und D_6 und eine fast völlige Unempfindlichkeit gegenüber dem Typenphagen 30 auszeichnet. Während die typische Form ubiquitär verbreitet ist, beschränkt sich die C-Varietät vor allem auf das äquatoriale Afrika und auf Madagaskar.

Außer der Möglichkeit der Typisierung von Typhusstämmen mit Hilfe der angepaßten Vi-Phagen, können noch zusätzliche Unterscheidungsmerkmale herangezogen werden, die vor allem bei den häufiger vorkommenden Lysotypen und auch bei der Gruppe der unbestimmbaren Vi-Stämme eine praktisch epidemiologische Bedeutung haben. Hier ist zunächst das Verhalten gegenüber Xylose und Arabinose zu nennen. Es zeigte sich nach den Untersuchungen von PAVLATOU und NICOLLE (1953) sowie BRANDIS und MAURER (1954), daß gewisse Beziehungen zwischen den Lysotypen und ihren fermentativen Eigenschaften bestehen. So gehören abgesehen vom Typ A die Typen C—G vorwiegend dem Vergärungstyp I (Xylose +) an, während bei den übrigen Typen ein größerer Prozentsatz von xylosenegativen Formen vorkommt. Beim Lysotyp A waren von den aus Deutschland von 139 verschiedenen Ausbrüchen geprüften Stämmen 57,6% xylosepositiv und 42,4% xylosenegativ. Da der Lysotyp A zu den häufig auftretenden Typen gehört, erweist sich gerade hier die zusätzliche biochemische Differenzierung als wertvoll. Auch in der Gruppe der unbestimmbaren Vi-Stämme und der degradierten Stämme findet sich ein größerer Prozentsatz xylosenegativer Kulturen. Die Stämme des Lysotyps E_1 sind in Europa offenbar fast ausschließlich xylosepositiv. NICOLLE und Mitarbeiter (1956) unterscheiden hinsichtlich des Xylose- und Arabinoseverhaltens zwischen heterogenen und homogenen bzw. fast homogenen Lysotypen. Zu den homogenen Lysotypen zählen nach diesen Autoren die Typen C, 30, C_4, D_2, D_4, E_1, E_3, E_4, F_1, G, L_1 und 35, welche fast immer den Biotyp I ergeben, ferner der Lysotyp M, von dem die bisher geprüften 163 von verschiedenen Gegenden stammenden Kulturen alle den Biotyp II aufwiesen. Die heterogenen Lysotypen demgegenüber setzen sich aus einem variablen Prozentsatz der Biotypen I—III zusammen.

Für den Lysotyp A haben NICOLLE, PAVLATOU und DIVERNEAU (1953, 1954) ferner eine Unterteilung in 9 Subtypen (Coquilhatville, Montreal, Tananarive, Chamblee, Welshpool, Duala, Oswestry, Maracaibo, Léopoldville) mit Hilfe von 7 Nicht-Vi-Phagen vorgeschlagen. Die Autoren fanden mit ihrer „lysotypie auxiliaire" (oder besser „complementaire") eine gute Übereinstimmung zwischen Subtyp, Kulturtyp und epidemiologischen Gegebenheiten. 4 von den benutzten 7 Phagen stammen von lysogenen Kulturen des Typs A, und zwar sind die Subtypen Coquilhatville, Chamblee, Welshpool, Maracaibo und Léopoldville immer und die Subtypen Montréal und Duala zum Teil lysogen. Als nichtlysogen erweisen sich die Subtypen Tananarive und Oswestry (s. NICOLLE und DIVERNEAU 1955). In Deutschland scheinen vor allem nur die Subtypen Tananarive und Chamblee vorzukommen (BRANDIS 1955). Auch die Gruppe der unbestimmbaren Vi-Stämme kann mit geeigneten Nicht-Vi-Phagen unterteilt werden (NICOLLE, PAVLATOU und DIVERNEAU); praktisch verwertbare Ergebnisse liegen aber hierüber noch nicht vor.

	Vi-Phagen			
	I	III	V	VI
Lysotyp A	cL	cL	cL	cL
Subtyp phi	—	cL	—	cL
Subtyp psi	cL	—	cL	—

cL = konfluierende Lysis.

Während NICOLLE und Mitarbeiter für ihre Lysotypie auxiliaire vorwiegend Nicht-Vi-Phagen heranzogen, zeigten DESRANLEAU und MARTIN (1950), daß der Lysotyp A bei Prüfung mit den nichtangepaßten Vi-Phagen I, III, IV von CRAIGIE sowie zwei von ihnen neu gezüchteten Vi-Phagen V und VI nach vorstehendem Schema unterteilt werden kann.

Ferner ergab sich auch die Möglichkeit, den Lysotyp E_1 mit Hilfe eines nichtangepaßten Vi-Phagen, der in Fortführung der Nomenklatur von CRAIGIE mit VII bezeichnet worden ist (BRANDIS und IMAMURA 1956), in 2 Subtypen zu zerlegen (BRANDIS 1955)

	Vi-Phagen	
	E 1	VII
E 1a	cL	cL
E 1b	cL	—

In Deutschland macht der Subtyp E_1b etwa $^1/_4$—$^1/_3$ aller E_1-Stämme aus (s. auch RISCHE 1955) und ist für praktische epidemiologische Zwecke infolge seiner Konstanz verwertbar. In anderen Ländern (z. B. USA, Frankreich) wird er dagegen auffälligerweise nur selten beobachtet.

Mit 13 aus Oberflächenwasser isolierten nichtadaptierten Vi-Phagen („Utrecht-System") gelang es SCHOLTENS (1950),die Gruppe der unbestimmbaren Vi-Stämme in eine Reihe von „Typen“ aufzuteilen. In diesem Zusammenhang sei vor allem die Beobachtung von SCHOLTENS hervorgehoben, daß er mit der Serie dieser nichtangepaßten Vi-Phagen auch bei den typisierbaren Stämmen, also den einzelnen Lysotypen, jeweils charakteristische Reaktionsbilder feststellte. Die Lysotypen geben demnach mit den nichtangepaßten Vi-Phagen typische Lysebilder, d. h. sie werden von den einzelnen Phagen in bestimmter und konstanter Weise gelöst. Diese Beobachtungen wurden von BRANDIS und IMAMURA (1956) bestätigt und unterstreichen die strukturellen Unterschiede, die im Vi-Antigen der verschiedenen Lysotypen vorhanden sein müssen. Es zeigte sich ferner, daß 4 Vi-Phagen von SCHOLTENS (22 Utrecht, 23 Utrecht, 25 Utrecht und 26 Utrecht) eine mit dem Vi-Phagen II fast übereinstimmende Anpassungsfähigkeit besaßen. Auch DESRANLEAU (1947) berichtete über einen Vi-Phagen Q 1467—43, den er im Jahre 1943 von einem Typhusstamm des Typs E_1 isoliert hatte. Dieser Phage erwies sich als serologisch eng mit dem Vi-Phagen II von CRAIGIE verwandt und hatte ebenfalls eine gute Adaptationsfähigkeit (s. S. 99). In dieser Eigenschaft verhalten sich die Vi-Phagen offenbar sehr unterschiedlich, da es auch Vi-Phagen gibt, mit denen Anpassungsversuche nicht gelingen.

Schließlich sei noch erwähnt, daß die von FELIX und CALLOW 1943 beschriebenen Salmonella-Anti-O-Phagen zur Unterscheidung von Typhusstämmen verwendet werden können (FELIX und ANDERSON 1951). So lassen sich die unbestimmbaren Vi-Stämme in 2 Gruppen zerlegen: In solche Stämme, die von den Anti-O-Phagen in der RTD[1] gelöst werden, und in andere, die unbeeinflußt bleiben. Alle diese zuvor geschilderten Methoden können in geeigneten Fällen das Lysotypieverfahren von CRAIGIE und FELIX vervollständigen, aber keinesfalls ersetzen.

Bevor auf die Ergebnisse der Lysotypie von Salmonella typhi, die in den verschiedenen Ländern erhalten worden sind, eingegangen wird, seien zunächst noch einige Ergänzungen zu den vorstehenden Ausführungen gemacht. Hinsichtlich der Beziehungen zwischen Vi-Antigen und Vi-Phagen konnte CRAIGIE zeigen, daß Vi-Phagen auch an Vi-antigenhaltige Typhusstämme adsorbiert werden, welche keine Lysis mit diesen Vi-Phagen geben. Die Adsorption ist nach ANDERSON und FRASER (1956) für die Bakterienzellen letal. Beim Auftropfen eines konzentrierten Phagen kann infolgedessen im Bakterienrasen ein wachstumsfreier Bezirk entstehen, der eine konfluierende Lysis, welche durch eine Vermehrung von Phagen hervorgerufen ist, vortäuscht. Werden Erythrocyten mit Vi-Antigen beladen, so vermögen diese vorbehandelten Erythrocyten ebenfalls die Vi-Bakteriophagen zu fixieren (EDLINGER und VIEUCHANGE 1955). Auch Vi-antigenhaltige Kulturen von S. paratyphi C, aus der Ballerup- und Coligruppe sind für die Wirkung von Vi-Phagen, z. B. des Vi-Phagen I von CRAIGIE, empfindlich (NICOLLE, JUDE und DIVERNEAU 1953). Allerdings besteht offenbar keine absolute Korrelation zwischen Vi-Phagen und Vi-Antigen, da CHERRY und Mitarbeiter (1954) nachwiesen, daß einige Stämme aus der E. freundii- und Ballerup-Gruppe auch ohne nachweisbares Vi-Antigen von dem Vi-Phagen I gelöst werden. Es wäre sicher lohnend, die chemische Natur dieser die Vi-Phagen bindenden Substanz aufzuklären. Vi-antigenhaltige Typhusstämme ohne den O-Antigenfaktor XII_2 sind in gleicher Weise typisierbar wie andere Typhus-

[1] „routine test dilution“.

kulturen (ROSCHKA 1953). Elektronenmikroskopische Beobachtungen über die Vi-Phagen I, II und IV haben GIUNTINI, EDLINGER und NICOLLE (1953) angestellt. Hiernach besitzt der Vi-Phage I einen kugeligen Kopf mit einem Durchmesser von 100 mμ, der Fortsatz ist 65 mμ lang und 12 mμ breit. Der Vi-Phage II hat ebenfalls einen sphärischen Kopf mit einem Durchmesser von 40—50 mμ, die Länge des Fortsatzes beträgt 100 mμ. Der Vi-Phage IV schließlich hat einen Kopfdurchmesser von 50—60 mμ, der Fortsatz ist 100 mμ lang. Temperierte Phagen von Typhuslysotypen sind elektronenmikroskopisch durch FERGUSON und Mitarbeiter (1955) dargestellt worden. Über vergleichende serologische Untersuchungen bei einer Reihe von nichtangepaßten Vi-Phagen haben BRANDIS und IMAMURA (1956) berichtet.

Über die Verteilung der Lysotypen auf der Erde liegen mehrere zusammenfassende Veröffentlichungen vor. Es sei besonders auf diejenigen von NICOLLE, HAMON und EDLINGER (1953), NICOLLE und HAMON (1954), EDLINGER, NICOLLE und HAMON (1954), und vor allem von FELIX (1955) verwiesen. Tabelle 6, die der Übersicht von FELIX entnommen ist, faßt unsere gegenwärtigen Kenntnisse über die geographische Verteilung der Lysotypen zusammen. Es geht aus der Tabelle hervor, daß der Typ E_1 in den meisten Ländern weitaus am häufigsten beobachtet wird, an zweiter Stelle steht der Typ A. Bemerkenswert ist, daß in Vietnam der Typ M, der in Europa ganz fehlt, die erste Stelle einnimmt und anscheinend in Ostasien nicht selten vertreten ist (NICOLLE und Mitarbeiter 1953, NICOLLE und Mitarbeiter 1955). Offenbar muß man in Ostasien mit einer anderen Verteilung und Häufigkeit der Lysotypen rechnen. Der Typ G tritt in Europa und Nordamerika kaum auf, dagegen findet er sich in den Ländern um den Indischen Ozean (VAN OYE und NICOLLE 1953). Der Typ A kommt in Südafrika in 50—82% aller Typhusstämme vor, während dort die Typen D_1 und F_1 selten sind. Dagegen wurde der Typ A in Jamaika nur in 3,7% der geprüften Kulturen gefunden. Nach NICOLLE und Mitarbeitern (1956) kann zwischen 1. ubiquitären, in Europa häufig auftretenden Lysotypen, 2. exotischen, in bestimmten Gegenden der Erde vermehrt vorkommenden und 3. seltenen sowie in ihrer Verbreitung noch ungenügend bekannten Lysotypen unterschieden werden. Ubiquitär sind die Typen A, C, D_1, E_1, F_1. Zu den exotischen in Europa nur vereinzelt angetroffenen Typen zählen nach den genannten Autoren die Typen 30 (Nordafrika, Sardinien), D_2 (Indonesien, Vietnam, Japan, Nordafrika), D_6 (Nordafrika, Zentralafrika, Vietnam, Peru), G (östlich Belgisch-Kongo, Madagaskar, Iran, Indien, Indonesien, Vietnam, Cuba), H (Vietnam, Japan), J (Nordafrika, Ägypten, Französisch-Westafrika, Indonesien, Indien, Vietnam, Japan), K (Nordafrika, Indien, Vietnam, USA), L_1 und L_2 (Nordafrika, Marokko), M (Iran, Indien, Indonesien, Vietnam, Japan, Chile, Peru, Cuba). Auch einige Subtypen des Typs A (s. S. 110) sind auf bestimmte Gegenden beschränkt, so z. B. der Subtyp Maracaibo, der bisher von NICOLLE nur in Venezuela, Französisch-Antillen, Französisch-Guyana und in Französisch-Westafrika und Südafrika nachgewiesen wurde. Was schließlich die 3. Gruppe betrifft, so sind wir bei einer Reihe von Lysotypen infolge des seltenen und sporadischen Auftretens über ihre eigentliche Verbreitung noch unvollkommen orientiert.

Der Nutzen, welchen die Lysotypie von S. typhi für die Aufklärung epidemiologischer Fragen besitzt, dürfte heute kaum mehr zu bezweifeln sein. Wertvoll

Tabelle 6. *Geographische Verteilung der Typhuslysotypen*[1]. [Nach Berichten der Lysotypielaboratorien, zusammengestellt von Dr. A. Felix (1953) und ergänzt (1955)]

	Gewöhnlich vorkommende Typen, nach der Häufigkeit angeordnet. Zahlenangaben in %	Seltene oder neue Typen, die in dem betreffenden Land gefunden wurden	Durchschnittlicher Prozentsatz der		Gesamtzahl der geprüften Patienten und Ausscheider
			unbestimmbaren Vi-Stämme	degradierten Vi-Stämme	
Afrika:					
Marokko, Algerien, Tunesien	E 1, A, T, D 1 31,4 28 5,7 5,4	D 6, L 2, O L 1 2,4 1,5 0,6 0,3	4,83	8,15	331
Franz.-West- u. Ostafrika, Belg.-Kongo	E 1, A, C 45,8 35,4 6,5	D 6, G, O 1,4 0,87 0,29	3,79	4,38	342
Madagaskar	A, E 1 46,6 42,6	D 5, G 2,6 1,3	1,33	3,99	75
Südafrika, Transvaal 1952/53	A, E 1 50 12,1	D 1, F 1 3,7 1,9	22,4	4	356
Südafrika, Natal u. Zululand 1952/53	A, 82,6	E 1 1,9	7,7	5,5	334
Amerika:					
Kanada 1950 bis 1953[2]	C, E 1, A, D 1 21 20 11,7 9	E 4, F 2, E 2 8,4 2 0,33	1,3	1	450
Jamaika[2]	E 1, T, C, A 45,5 19,8 9,3 3,7		9	0,4	329
USA[2] 1. 7. 51 bis 30. 6. 52	E 1, C, A, F 1 19,8 13,8 7,7 7,7	G, H, J, K, L 2, C 2 Typ 26 u. Typ 29	9,9	8,5	1263
Venezuela	A, T, E 1 69,8 11,1 3,1	D 6 1,58	3,16	11,1	63
Asien:					
Indien (Bombay) 1948 bis 1954	A, E 1, O, D 1 38,9 10,9 9,1 5	D 6, J, K, M 0,3 0,4 2,1 0,07	13	13,2	1028
Indonesien a) Djakarta	D 6, A, E 1, D 2 28, 14,5 10,7 13,4 B 1 8	25, M, D 1, J 4,7 3,6 1,3 0,4	8,6	3,5	910
b) Medan	E 1, B 1, E 2, A 26,5 23 12,9 6,6	D 1, J, 25 8,9 1,1 5,9	6	2	441
Iran	F 1, A 36,7 7,6	B 1, B 2, M, N, T	7,6	32,5	223
Vietnam	M, A, N, T 18,9 13,9 10,3 8,5	B 2, D 6, G, J, 25	12,8	8,9	280
Japan		M, H, J, D 6	.	.	
Türkei	T, D 1, A, E 1, F 1 29 13 8 9 9	B 2, C, N 3 2 1	4	21	100
Australien	E 1, T, A	J	1,4	4,6	74
Europa:					
Österreich	E 1, D 1, A, F 1 42,6 15,3 10,8 9,3	B 2 O, T, 28, 29 1,2 0,2 1,2	4,9	4,5	480
Tschechoslowakei 1948 bis 1950[2]	E 1, A, D 1, F 1	D 6, E 2, G, 28	10,3	11,8	1389
Dänemark[2]	A, F 1, E 1, C 33,3 24,2 21,2 9	T 3	3	—	59

Tabelle 6. Fortsetzung

	Gewöhnlich vorkommende Typen, nach der Häufigkeit angeordnet. Zahlenangaben in %	Seltene oder neue Typen, die in dem betreffenden Land gefunden wurden	Durchschnittlicher Prozentsatz der unbestimmbaren Vi-Stämme	Durchschnittlicher Prozentsatz der degradierten Vi-Stämme	Gesamtzahl der geprüften Patienten und Ausscheider
Frankreich	C, E 1, A, D1 28 16,5 15,7 6,1	B 2, O, T, M 3,1 1,9 3 0,1	8,6	7,4	416
England 1950 bis 1952[2]	E 1, A, C, D 1 24,8 19,2 10,1 7,1 F 1 2,7	G, F 2, H, J, L 2, O 0,6 0,6 Typ 29, 30, 31, 32	12,7	10,7	612
Westdeutschland[2]	E 1, F 1, A, D 1 31,18 10 9,9 7,9 C 7,1	B 1, F 2, J, T 0,1 0,2 0,4 0,3 E 2, L 1, K	12,6	5,8	1807
Mitteldeutschland[2]	E 1, F 1, A, D 1, C		8,5	10,5	542
Irland[2]	E 1, A, D 1, C		21,3	—	61 Foci
Mittelitalien 1950—1952[2]	C, E 1, D 1, A 26,9 15,1 13,9 12,9	D 6, F 2, J, H 0,5 2,3 0,5 0,5	14,6	2,86	372
Holland 1952[2]	E 1, A, D 1, C 28,4 12 9,9 7,1	B 2, D 6, F 2, L 2 0,7 0,7 0,7 0,7 Typ 25, 27, 28 1,4 1,4 3,5	5,7	17,7	206
Norwegen[2]	E 1, C, F 1	J	10,7	7,1	28 Foci
Polen 1948 bis 1950	E 1, F 1, A, D 1	D 5, D 6, E 2, G, H, J, L 1, M			2228
Portugal[2]	B 3, A, E 1, T, D 1	O	2,1	5,4	185
Rumänien[2]	A, E 1, F 1, D 1, C	D 6, E 2	16,2	14,8	4172
Jugoslawien (Kroatien) (1952)[2]	A, E 1, F 1, D 1, 34,1 19,1 8 6,4 C 5,7	D 6, F 2, O 6,4 0,3 0,3	4,7	3,3	611

[1] Es sind in der Liste nicht alle Lysotypen aufgeführt, die in den einzelnen Ländern gefunden wurden.

[2] Berechnung der prozentualen Häufigkeit nach „Foci"; in den übrigen Ländern nach der Zahl der Patienten und Ausscheider.

ist vor allem die Lysotypie bei dem Ausschluß des Zusammenhanges von Typhusinfektionen. Erkrankungen, bei denen verschiedene Lysotypen gezüchtet werden, muß man auf getrennte Infektionsquellen zurückführen. Die hohe praktische Konstanz der Lysotypen hat sich durch eine große Zahl an Einzelbeobachtungen von Autoren aus den verschiedensten Ländern immer wieder bestätigt. Auch bei Dauerausscheidern bleibt selbst über Jahrzehnte hinweg der gleiche Lysotyp nachweisbar. Allerdings können gelegentlich bei Dauerausscheidern wie auch bei Kulturen, die in vitro längere Zeit aufbewahrt werden, Übergänge in den Typ A stattfinden (CRAIGIE und FELIX 1947, SCHOLTENS 1950).

Es ist im Rahmen dieses Überblickes, der vor allem die bakteriologische Seite der Lysotypieverfahren beleuchten soll, nicht möglich, die vielen Veröffentlichungen über die Ergebnisse und epidemiologische Bedeutung der Typisierung von Typhusbakterien im einzelnen aufzuzeigen. Es sollen aber im folgenden die Autoren einiger Berichte, in denen die Resultate und Erfahrungen aus verschiedenen Ländern zusammengefaßt sind, angeführt werden:

Australien (BUCKLE 1946), Belgischer Kongo (VAN OYE und NICOLLE 1953), Canada (DESRANLEAU 1942, DESRANLEAU und MARTIN 1950), Deutschland (BRANDIS 1955, RISCHE 1955, RISCHE, ROHNE und SCHNEIDER 1954, TRÜB und SAUER 1955, FREYTAG und PLOCHMANN 1955), England (FELIX 1943, 1951), FRANKREICH (NICOLLE und HAMON 1954), Holland (SCHOLTENS 1950), Indien (Bombay) (DHAYAGUDE und BANKER 1951), Irland (MCCORMACK 1954), Italien (DE BLASI und BUOGO 1952, BUONOMINI und D'AMELIO 1954), Jugoslawien (TOMASIC und BAJZER 1954), Nordafrika (NICOLLE und DIVERNEAU 1954), Norwegen (HENRIKSEN 1952), Österreich (EDLINGER, NICOLLE und HAMON 1954), Palästina (OLITZKI und Mitarbeiter 1945), Polen (HIRSZFELD und Mitarbeiter 1948, LACHOWICZ und BUCZOWSKI 1950, MACIEREWICZ 1950, PRAZMOWSKI und STEMPIEŃ 1950, WIZA 1952), Puerto Rico (SOTO und OTERO 1948), Rumänien (COMBIESCO und POPVICI 1948), Sumatra (ERBER und DE MOOR 1954), Tschechoslowakei (RASKA und Mitarbeiter 1950, LIBIKOVA 1950, ŠEFČOVICOVÁ 1953), Ungarn (EÖRSI 1956), USA (MORRIS und Mitarbeiter 1945, HENDERSON und FERGUSON 1949), Vietnam (NICOLLE und Mitarbeiter 1953 [a]).

Für epidemiologische Zwecke hat es sich als sehr brauchbar erwiesen, die Lysotypiebefunde auf Karten einzutragen, wodurch ein anschauliches Bild über die in einem bestimmten Bezirk vorhandene örtliche Verteilung der Typhuserkrankungen und Dauerausscheider mit den jeweiligen Lysotypen erhalten wird. Derartige epidemiologische Karten sind unter anderem von FELIX (1951), MCCORMACK (1954), TOMASIC und BAJZER 1954, BRANDIS (1955)), WÜSTENBERG und KORELL (1955) sowie FREYTAG und PLOCHMANN (1955) veröffentlicht worden.

II. Die Lysotypie von S. paratyphi B.

Im Jahre 1943 teilten FELIX und CALLOW eine Methode zur Typisierung von S. paratyphi B mit, welche sich eng an das Verfahren von CRAIGIE und YEN anlehnte. Mit Hilfe von angepaßten Phagen war es den Autoren möglich, die Paratyphus B-Stämme in 4 Typen (1, 2, 3a und 3b) zu unterteilen. In ihrem Schema entsprach der Lysotyp 1 dem Lysotyp A bei den Typhusbakterien, da der Typ 1 von allen 4 Phagen angegriffen wurde. Der Typ 2 zeigte dagegen nur eine Lysis mit seinem homologen Phagen. Der Typ 3a wurde von den Phagen 3a und 3b und der Typ 3b nur von dem 3b-Phagen angegriffen. Der Prozentsatz der untypisierbaren Stämme betrug nur 12%. Epidemiologisch zusammenhängende Kulturen ergaben stets den gleichen Typ, so daß sich die Methode auch für praktische Zwecke als brauchbar erwies. FELIX und CALLOW nahmen zunächst an, daß die von ihnen verwendeten Phagen auf ein nur bei den Paratyphus B-Bakterien vorkommendes spezifisches Vi-Antigen eingestellt seien und nannten ihre Phagen daher Paratyphus B-Vi-Phagen. Es hat sich aber später gezeigt (FELIX, CHERRY und Mitarbeiter 1953, SICCA und D'AMELIO 1953), daß diese Phagen auch auf andere Salmonellen (z. B. S. pullorum und S. gallinarum u. a.) wirken können. Die Bezeichnung „Vi"-Phagen wurde daher fallen gelassen, wie überhaupt das Vorkommen eines Vi-Antigens bei den Paratyphus B-Bakterien von KAUFFMANN (1947) abgelehnt wird. Der Angriffspunkt der Paratyphus B-Phagen bei den Paratyphus B-Bakterien ist noch nicht genau bekannt. Es dürfte sich hierbei wahrscheinlich nicht um das V-Antigen handeln, da auch

V-freie Kulturen (Var. Odense) typisierbar sind (NICOLLE, L. und S. LE MINOR und DIVERNEAU 1952, ROSCHKA 1953, BRANDIS 1955, RISCHE 1955).

FELIX und CALLOW konnten bis zum Jahr 1951 die Zahl der Paratyphus B-Phagen auf 10 erhöhen. Das von ihnen in dem gleichen Jahr veröffentlichte Typisierungsschema ist noch heute in unveränderter Form gültig (Tabelle 7). Es umfaßt die Paratyphus B-Phagen und die entsprechenden Lysotypen 1, 2, 3a, 3a I, 3b, Jersey, Beccles, Taunton, B. A. O. R. und Dundee. Die technische Ausführung der Methode entspricht derjenigen bei S. typhi. Die Phagen werden ebenfalls von der internationalen Lysotypiezentrale in London verteilt und in ihrer kritischen Routinetestverdünnung verwendet. Natürlich treten auch bei

Tabelle 7. *Reaktionsschema der Salmonella paratyphi B-Stämme mit der Serie der Routinetest-Paratyphus B-Phagen.* (Nach FELIX und CALLOW)

S. paratyphi B-Stämme; Typ	Test-Phagen in RTD									
	1	2	3a	3a I	3b	Jersey	Beccles	Taunton	B.A.O.R.	Dundee
1	cl	cl	cl	cl	scl	cl	scl	< cl	ol	< cl
2	—	cl	—	—	—	±	—	—	—	scl
3a	—	—	< cl	< cl	ol	±	< cl	< cl	< cl	< cl
3a I	—	—	< cl	< cl	—	—	—	—	—	< cl
3b	—	—	—	±	ol	±	< cl	< cl	< cl	< cl
Jersey	—	±	±	±	—	< cl	< cl	< cl	< cl	< cl
Beccles	—	±	—	—	—	—	< cl	< cl	< cl	< cl
Taunton	—	—	—	—	—	—	—	< cl	—	< cl
B.A.O.R.	—	—	—	—	±	—	—	—	< cl	—
Dundee	—	—	—	—	—	—	—	—	—	< cl

cL = vollständige Lyse; < cl = fast vollständige Lyse; scl = halbkonfluierende Lyse; ol = opake Lyse mit Inseleffekt; ± = gewöhnlich wenige Löcher; — = keine Löcher.

den Paratyphus B-Stämmen unbestimmbare Kulturen auf. Der Prozentsatz ist aber klein. Vielfach handelt es sich dabei um rauhe oder halbrauhe Stämme. Atypisch reagierende Kulturen kommen ebenfalls nur selten vor. Derartige abweichende Reaktionen können darauf beruhen, daß ein Stamm vorliegt, der zu einem noch nicht bekannten Typ gehört (vgl. NICOLLE und Mitarbeiter 1952), oder aber daß die Kultur durch einen fremden Phagen lysogenisiert und hierdurch ihre Reaktionsfähigkeit gegenüber den Typenphagen verändert worden ist (s. S. 124). Nach FELIX und CALLOW sowie eigenen Beobachtungen scheint unter den Paratyphus B-Typen einzig der Typ Beccles nicht völlig stabil zu sein und zu einer Veränderlichkeit zu neigen, die sich in einer vermehrten Empfindlichkeit gegenüber den Paratyphus B-Typenphagen äußert.

Über die Herkunft der Paratyphus B-Typenphagen ist zu sagen, daß sie von FELIX und CALLOW mit Ausnahme der Phagen Beccles und B. A. O. R. durch Anpassung des Phagen 1 an die verschiedenen Paratyphus B-Lysotypen gewonnen wurden. Der Beccles-Phage stammt von einem anderen Typ 1-Phagen ab, und der Phage B. A. O. R. wurde von einer Bakterienkultur des Lysotyps 1 isoliert. Serologische Untersuchungen der 8 vom Typ 1 angepaßten Phagen ergaben nun aber, daß diese nur zum Teil übereinstimmten. Vom Antiserum 1 werden nur die Phagen 1, 2, 3a, 3a I und Jersey neutralisiert, die Phagen 3b, Beccles, Taunton und Dundee dagegen nicht. Die zuletzt genannten Phagen zerfallen wiederum in zwei serologisch verschiedene Gruppen, wie Tabelle 8 zeigt.

Tabelle 8. *Serologisches Verhalten der Paratyphus B-Phagen.* (Nach Felix und Callow 1951)

Antiserum gegen B-Phagen	Neutralisierte Typen der B-Phagen
1. Typ 1	1, 2, 3a, 3a I, Jersey (unvollständig)
2. Typ 3b	3b, B.A.O.R.
3. Typ Beccles	Beccles, Taunton, Dundee (unvollständig)

Es ist daher unwahrscheinlich, daß die Paratyphus B-Phagen eine gemeinsame Wurzel haben. Dies bewiesen Felix und Callow auch dadurch, daß es ihnen ein einziges Mal gelang nachzuweisen, daß im Stammphagen 1 ganz vereinzelte Partikelchen des Phagen 3b enthalten waren. Man muß auf Grund dieser Feststellung mit Felix und

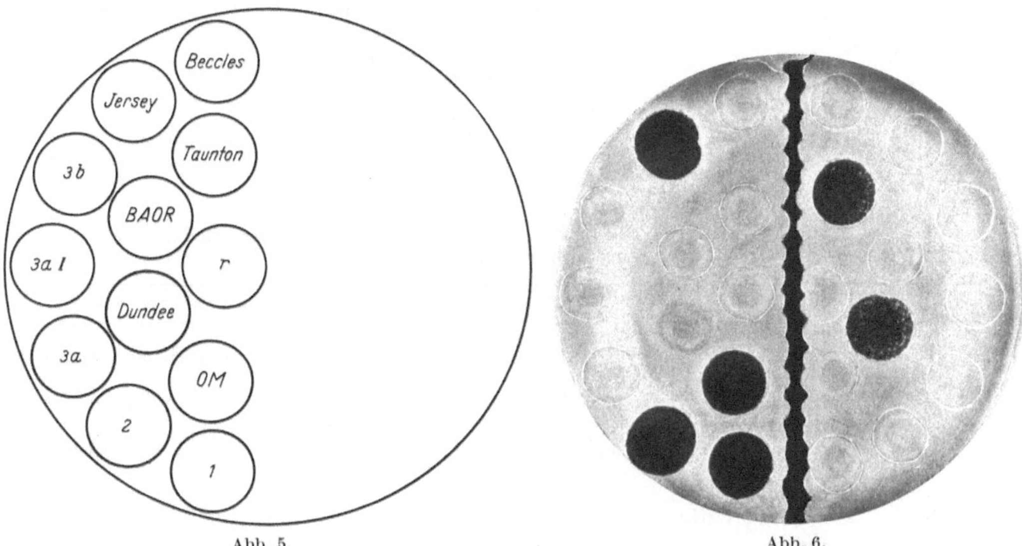

Abb. 5. Abb. 6.

Abb. 5. Auftropfschema der Paratyphus B-Phagen. Für die rechte Seite ist das Schema um 180° zu drehen. *r* = Anti-R-Phage. *OM* = Gemisch von Salmonella-Anti-O-Phagen

Abb. 6. Agarplatte mit 2 verschiedenen Paratyphus B-Stämmen. Links Typ 1 (Lysis mit den Phagen 1, 2, Jersey, Anti-O, außerdem mit der Lupe (Originalplatte) erkennbare Phagenlöcher mit den Phagen 3a, 3a I und Dundee); rechts Typ B.A.O.R. (Lysis mit den Phagen B.A.O.R. und Anti-O) (s. [ ] Abb. 2)

Callow annehmen, daß die serologisch mit dem Phagen 1 nicht übereinstimmenden Phagen von lysogenen Kulturen abstammen, die zur Züchtung der Phagen benutzt worden sind (s. auch bei den Staphylokokkenphagen, S. 133). Hieraus erklärt sich auch die Tatsache, daß eine Reihe von Paratyphus B-Lysotypen nicht nur — wie es bei den Typhusbakterien der Fall ist — mit dem homologen Phagen reagiert, sondern mit mehreren Typenphagen, und die Paratyphus B-Lysotypen also durch ein bestimmtes, allerdings konstantes Reaktionsbild gekennzeichnet sind. Das Zustandekommen dieser Lysebilder wird wesentlich durch die bei allen Paratyphus B-Stämmen vorkommenden latenten Phagen verursacht. In Tabelle 9 sind die Angaben von Felix und Callow über die latenten Phagen bei den einzelnen Paratyphus B-Lysotypen wiedergegeben.

Liste der Typenstämme der 10 *Paratyphus B-Lysotypen.* (Nach FELIX 1954)

Lysotyp	Aufgefunden von	Typen-stamm Nr.	Isoliert in	Literaturangaben
1	FELIX and CALLOW 1943	B 76	England 1940	FELIX, A., and B. R. CALLOW: Brit. med. J. **1943 II,** 127
1		B 309	England 1941	
2	„ „ 1943	B 228	England 1941	
2		B 300	England 1941	
3a	„ „ 1943	B 62	England 1940	
3a		B 365	England 1941	
3a I	„ „ 1944	B 624	England 1941	FELIX, A., and B. R. CALLOW: Lancet **1951 II,** 10
3a I		B 1305	England 1945	
3b	„ „ 1943	B 97	England 1941	FELIX, A., and B.R. CALLOW: Brit. med. J. **1943 II,** 127.
3b		B 656	England 1942	
Jersey	„ „ 1950	B 4182	Jersey 1950	FELIX, A., and B.R. CALLOW: Lancet **1951 II,** 10.
Beccles	„ „ 1947	B 1742	England 1944	
Taunton	„ „ 1947	B 2253	England 1947	
B.A.O.R.	„ „ 1948	B 2227	Deutschland 1947	
Dundee	„ „ 1948	B 2590	Schottland 1947	

Eingehend haben sich auch NICOLLE und Mitarbeiter (1951 [a]) mit der Lysogenität bei den verschiedenen Paratyphus B-Lysotypen beschäftigt. Die Befunde

Tabelle 9. *Bisher nachgewiesene temperierte Phagen bei Paratyphus B-Lysotypen.* (Nach FELIX und CALLOW)

Lysotypen	Typen der latenten („natürlichen“)-Phagen, die in den Kulturen nachzuweisen sind
1	Lytische Wirkung für den Typ B.A.O.R., teilweise neutralisiert durch Beccles-Serum
2	Lytische Wirkung für die meisten Typenstämme, nicht neutralisiert mit verfügbaren Antiseren
3a	Schwache Wirkung auf die Typen 3a I und Taunton, teilweise neutralisiert durch Anti-Beccles-Serum
3a I	Phage 3b
3b	Lytische Wirkung für Beccles, teilweise neutralisiert durch Anti-Beccles-Serum
Jersey	Phage 3b, ein zweiter Phage ist noch nicht näher bestimmt
Beccles	Phage 3b, ein zweiter Phage ist noch nicht näher bestimmt
Taunton	Phage 3b und Phage Beccles
B.A.O.R.	Phage Beccles und ein zweiter Phage, der auf den Typ 2 wirkt
Dundee	Phage 3b und ein zweiter Phage, der auf den Typ 2 wirkt

sind in der Tabelle 10 wiedergegeben, die zeigt, daß die Autoren zwischen typdeterminierenden, typhalbdeterminierenden und indifferenten temperierten Phagen unterscheiden.

Die typdeterminierenden Phagen sind, wie durch die Bezeichnung ausgedrückt ist, für das charakteristische Lysisbild ihrer Trägerstämme mit den Paratyphus B-Phagen verantwortlich. Der Typ Beccles enthält nach NICOLLE und Mitarbeitern mindestens zwei verschiedene latente Phagen, davon ist einer typbestimmend. Wenn nämlich dieser temperierte Phage auf einen Stamm des Lysotyps 3b einwirkt, so geht die Kultur unter Resistentwerden gegen diesen Phagen in den Lysotyp Beccles über. Der zweite temperierte Phage vom Lysotyp Beccles vermag dagegen beim Typ 3b keine Änderung im Verhalten gegenüber den B-Phagen zu bewirken, er ist daher indifferent. Demnach können mit den

Tabelle 10. *Temperierte Phagen bei Paratyphus B-Lysotypen.* (Nach NICOLLE und Mitarbeitern)

Lysotyp	Zahl der Eigenphagen	Lytische Wirkung auf die Typen
1	1	S. java (44 231) (typindifferent)
2	1	Alle Lysotypen außer Typ 2 (typdeterminierend)
3a	1	Lytische Wirkung auf Taunton und 3a I (Phage konnte nicht isoliert werden)[1]
3a I (B 624)	3	1. 1, 3a, 3b, B.A.O.R. (typhalbdeterminierend) 2. ..... (typhalbdeterminierend) 3. 1, 2, 3a, 3a I (1305), 3b, Beccles, Taunton, B.A.O.R., Dundee
3b	1	1, 2, 3a, Taunton (typindifferent)[2]
Jersey	3	.....
Beccles (Gruppe B_2)	1	1, ... (typdeterminierend)
Beccles (Gruppen B_1 u. B_0)	2	1. 1, ... (typdeterminierend) 2. 1, ... (typindifferent)
Taunton	4	1. 1, 3a, 3b, Beccles, B.A.O.R. (typhalbdeterminierend) 2. 1, ... (typhalbdeterminierend) 3. 1, ... (typindifferent) 4. 1, ... (typindifferent)
B.A.O.R.	1	1, 2, 3a (minus 3b), 3a I (1305), 3b, Beccles (typdeterminierend)
Dundee	4	1. 1, ... (typhalbdeterminierend) 2. 1, ... (typhalbdeterminierend) 3. 1, ... (typindifferent) 4. 1, ... (typindifferent)

Zu dem Zeitpunkt der Untersuchungen von NICOLLE und Mitarbeitern (1951[a]) rechnete man die jetzt als S. java bezeichneten Kulturen (d-tartratpositiv) noch zu S. paratyphi B. Die Versuche der Autoren erstreckten sich daher auf d-tartratpositive Stämme wie d-tartratnegative Paratyphus B-Kulturen.

[1] In Mischkultur mit S. java (44231) wurde ein Phage isoliert, der die Lysotypen 1 und 2 sowie eine Rauhkultur des Typs Beccles angriff.

[2] Der Phage stammt aus einer Mischkultur mit S. java (44231).

typdeterminierenden temperierten Phagen Typenumwandlungen bei den Paratyphus B-Bakterien vorgenommen werden, wobei die transformierten Kulturen den betreffenden Phagen aufgenommen haben und im Prophagenstadium beherbergen. Wenn der typdeterminierende Phage des Typs 2 auf den Typ 1 einwirkt, so tritt auch hier eine Änderung im Phagenverhalten ein, und es entsteht eine Kultur mit der Reaktionsweise des Typs 2.

Für die Typenumwandlung mit halbdeterminierenden Phagen sind zwei verschiedene Phagenstämme verantwortlich. So lassen sich nach NICOLLE und Mitarbeitern aus dem Lysotyp Taunton zwei halbdeterminierende Phagen gewinnen. Beide Phagen für sich allein vermögen den Lysotyp 3b nicht in den Typ Taunton zu überführen. Sie ergeben entweder den Typ Beccles (1. Phage) oder eine Reaktionsweise (2. Phage), die im Schema von FELIX und CALLOW nicht enthalten ist (Typ 3b ohne Reaktion mit dem Beccles-Phagen). Der 2. Phage führt aber den durch die Wirkung des 1. Phagen entstandenen Typ Beccles in den Typ Taunton über. Auch wenn beide Phagen direkt auf den Lysotyp 3b einwirken, entsteht der Lysotyp Taunton. Tabelle 11 gibt über die von NICOLLE und Mitarbeitern vorgenommenen Typenumwandlungen Auskunft. Es bestehen zweifellos gewisse Übereinstimmungen mit dem Schema, das ANDERSON bei den Typhuslysotypen aufgestellt hat (s. S. 108).

Tabelle 11. *Typenumwandlungen von Paratyphus B-Lysotypen mit temperierten typdeterminierenden Phagen.* (Nach HAMON und NICOLLE 1951)

Ausgangstyp	Verwendeter temperierter Phage	Künstlich erzeugter Typ
1	Phage aus dem Typ 2	2
3a	1. Phage aus dem Typ 3a I	3a I
3b	Phage aus dem Typ B.A.O.R.	B.A.O.R.
	1. Phage aus dem Typ Beccles	Beccles
	1. und 2. Phage aus dem Typ Taunton	Taunton
	1. und 2. Phage aus dem Typ Dundee	Dundee

Unterschiede in der Zahl der Eigenphagen bei Stämmen eines Lysotyps bewirken, daß diese Stämme sich den Typenphagen gegenüber nicht gleich verhalten. FELIX und CALLOW (1951) beobachteten so bei einer Reihe von Lysotypen neben der üblicherweise vorkommenden Reaktionsform Abweichungen im Lysisbild. Die Autoren fanden bei dem Lysotyp 1 4 Varianten, beim Lysotyp 3a 4 Varianten, beim Lysotyp 3a I 3 Varianten und beim Lysotyp 3b 4 Varianten[1]. Die meisten dieser Varianten werden aber nur selten beobachtet. Über ihre epidemiologische Bedeutung und Konstanz ist daher nur wenig bekannt. NEWELL (1955) beschrieb Paratyphus B-Infektionen in England, die wahrscheinlich von gefrorenem chinesischen Vollei ihren Ausgang nahmen und durch den Typ 3a, var. 2 verursacht waren. Die Variante 3a I, var. 1, die sich vom Typ 3a I in einer zusätzlichen Reaktion mit den Typenphagen Beccles und Taunton unterscheidet, scheint nach unseren Beobachtungen in Deutschland häufiger aufzutreten. Ob aber alle bisher nachgewiesenen Varianten auf den Einfluß von latenten Phagen zurückzuführen sind, dürfte nach NICOLLE und Mitarbeitern fraglich sein. Beim Lysotyp 1 ist nur eine latente Phagenform bekannt, trotzdem kennt man, wie zuvor erwähnt, bei diesem Typ 5 Variationen.

Die Stämme des Lysotyps Beccles besitzen eine verschiedene Zahl an latenten Phagen (NICOLLE und Mitarbeiter 1953 [b]). Allen Stämmen gemeinsam ist der typdeterminierende Phage „a". Die Gruppe Beccles „B_2" hat nur diesen Phagen.

Subtyp mit Angabe der nachweisbaren Phagen	Phagen		
	a	b	c
B_2 (a)	—	+	+
B_1 (a, c)	—	+	—
B_0 (a, b)	—	—	—

+ = Lysis

Der Beccles-Subtyp B_1 weist darüber hinaus noch einen latenten Phagen „c" auf. Der Subtyp B_0 schließlich enthält die latenten Phagen „b" und „a". Die Kulturen des Subtyps B_2 sind gegen die Phagen b und c und die des Subtyps B_1 gegen den Phagen b empfindlich, während die Stämme der Untergruppe B_0 gegen die Phagen b und c resistent sind. Auf diese Weise kann nach NICOLLE und Mitarbeitern bei Verwendung der drei genannten Phagen der Lysotyp Beccles in 3 Subtypen zerlegt werden.

Beccles-Subtyp	Phagen				
	Beccles	Taunton	B.A.O.R.	Dundee	B_0
B_2	cL	cL	—	cL	cL
B_1	cL	cL	—	cL	—
B_0	cL	cL	cL	—	cL

Diese 3 Subtypen konnten NICOLLE und Mitarbeiter auch durch die verschiedenartigen Reaktionen mit den Typenphagen bei zusätzlicher Ver-

[1] Nach CALLOW (s. ANDERSON und WILLIAMS 1956) sind inzwischen beim Typ 1 5, beim Typ 3a 6, beim Typ 3a I 7, beim Typ 3b 5 Variationen und bei den Typen 2, Beccles und Dundee je eine Variation aufgefunden worden.

Tabelle 12. *Natürliche Gruppen der Paratyphus B-Lysotypen.* (Nach SCHOLTENS, gekürzt)

Gruppe	Typen nach FELIX und CALLOW	Neue Typen nach SCHOLTENS	Phagen in RTD													
			Gruppenreaktion										Typenreaktion			
			1	2	3a 3a I	Jersey	Beccles-Meppel	e	d (Dundee)	f	h	b	I (3b)	IVb (Beccles)	IVa (Taunton)	VI
A	Beccles	„22“	—	—	—	—	—	—	cl	—	cl	—	—	cl	cl	cl
	3b, Var. 2		—	—	—	—	—	—	cl	—	cl	—	cl	—	—	—
	Taunton	„the Hague“	—	—	—	—	—	—	cl	—	cl	—	—	—	cl	—
M	3b, Var. 1		—	—	—	—	—	cl	—	—	cl	—	cl	cl	cl	cl
	Beccles	„Midwoud“	—	—	—	—	—	cl	—	—	cl	—	—	cl	cl	cl
S	3a	„54“	—	—	cl	—	—	cl	cl	cl	cl	+1	cl	cl	cl	cl
	Beccles ? 3a I ?	„Sittard“	—	—	(cl)	—	—	cl	cl	cl	cl	+1	—	cl	cl	cl
		„18“	—	—	cl	—	—	cl	cl	cl	cl	+1	cl	—	—	—
J	3a, Jersey ?	„60“	—	—	cl	cl	cl	cl	cl	—	scl	cl	cl	cl	cl	cl
	Jersey		—	—	(cl)	cl	cl	cl	cl	—	scl	cl	—	cl	cl	cl
	3a, Var. 3		—	—	cl	cl	cl	cl	cl	—	scl	cl	cl	—	—	—
B.M.	Beccles	„Meppel“	—	—	—	—	cl	cl	cl	—	—	cl	—	cl	cl	cl
B	3a I, Var. 1,2		—	—	cl	—	—	cl	cl	+1 m^4	cl	cl	—	cl	cl	cl
	3a I	„Schiedam“	—	—	cl	—	—	cl	cl	—	cl	cl	—	—	—	cl
	3a I	„Leeuwarden“	—	—	cl	—	—	cl	cl	—	cl	cl	—	—	—	—
	Taunton	„Kampen“	—	—	(cl)	—	—	cl	cl	—	cl	cl	—	—	cl	—

(cl) = Reaktion fehlt oft; cl = konfluierende Lysis; + 1 = geringe Zahl von Einzellöchern; m^4 = staubähnliche sehr kleine Löcher.

wendung eines mit B_0 bezeichneten Phagen nachweisen, welcher eine Mutante des B.A.O.R.-Phagen darstellt und auf einem Stamm des Beccles-Subtyps B_0 gezüchtet worden war.

Stämme aus einem Herd erwiesen sich in ihren Reaktionen stets übereinstimmend. Allerdings hat die Unterteilung des Lysotyps Beccles infolge seiner geringen Häufigkeit nur eine untergeordnete Bedeutung. NICOLLE und Mitarbeiter konnten ebenfalls die Lysotypen Taunton und Dundee mit den Phagen b, c und B_0 in jeweils 3 Subtypen zerlegen, von denen zwei jedoch so selten vorkommen, daß sie keinen praktischen Wert besitzen.

Wie aus den vorstehenden Ausführungen hervorgeht, hat sich durch die Untersuchungen von FELIX und CALLOW sowie NICOLLE und Mitarbeitern ergeben, daß ein enger Zusammenhang zwischen Lysotyp und der Art der vorhandenen latenten Phagen besteht. Diese Feststellung ist ebenfalls von SCHOLTENS gemacht worden, der darüber hinaus zeigte, daß die verschiedenartige Lysogenität der Paratyphus B-Stämme als Grundlage für eine Typeneinteilung dienen kann. Für seine Versuche hat SCHOLTENS (1950, 1952, 1955, 1956) folgende Methoden verwendet: 1. Nachweis der von den Paratyphus B-Stämmen spontan abgegebenen Phagen, 2. Gewinnung von Phagen aus Mischkulturen zweier Paratyphus B-Stämme, 3. Lysisreaktionen der Paratyphus B-Stämme mit den Spontanphagen, den Phagen aus Mischkulturen sowie den vom Phagen 1 sich herleitenden Typenphagen von FELIX und CALLOW (das sind 1, 2, 3a, 3a I, Jersey und Beccles-Meppel[1]).

[1] Der Beccles-Meppel-Phage wurde durch Anpassung des Typenphagen Jersey an einen Beccles-Stamm erhalten.

Die direkt nachweisbaren Spontanphagen wurden von SCHOLTENS aus Bouillonkulturen isoliert oder aber aus Mischkulturen des zu prüfenden Stammes mit dem Typenstamm 3a (B 62). Bei diesem letztgenannten Verfahren lassen sich nach SCHOLTENS ebenfalls nur die spontan freiwerdenden Phagen feststellen. SCHOLTENS hat folgende spontan abgegebene Phagen gefunden (s. Tabelle 14): I (entspricht dem Typenphagen 3b), II, IV und VI. Beim Phagen IV wurden 2 Varianten IVa und IVb beobachtet; der Phage IVb erwies sich mit dem Typenphagen Beccles und der Phage IVa mit dem Typenphagen Taunton identisch.

Außer diesen direkt nachweisbaren Phagen, die zum Teil den typendeterminierenden Phagen von NICOLLE und Mitarbeitern gleichzusetzen sind, gewann SCHOLTENS auch Phagen aus Mischbouillonkulturen zweier Paratyphus B-Stämme, welche eine Woche lang bebrütet wurden. Die hierbei entstandenen Phagen, die die Bezeichnung b, c, d[1], e, f, g, h erhielten, waren niemals bei einem Wachstum der Stämme für sich allein festzustellen. Die Herkunft dieser Phagen ist in der Tabelle 13 wiedergegeben (hinsichtlich der in dieser Tabelle aufgeführten Gruppen s. weiter unten).

Tabelle 13. *Phagen aus Mischkulturen verschiedener Lysotypen von S. paratyphi B.* (Nach SCHOLTENS)

Phagen mit dem lytischen Spektrum	Gewonnen aus Mischkulturen von Stämmen der Gruppen	Stämme aus folgenden Gruppen werden angegriffen
b	A + J A + B.M. A + B	S, J, B.M., B
c	Aus allen Mischkulturen, bei denen Stämme der Gruppe M beteiligt sind	Alle Gruppen
d	Aus allen Mischkulturen, bei denen Stämme der Gruppe M beteiligt sind	Alle Gruppen außer M
e (577)	A + M	Alle Gruppen außer A
f (572)	A + S A + Typ 1	S, Typ 1
h (580)	B.M. + B.A.O.R.	Alle Gruppen außer B.M.
g	Typ 1 + Sittard	....

SCHOLTENS nimmt an, daß es sich bei den aus den Mischkulturen erhaltenen Phagen um genetische Rekombinationen von Phagenelementen, die in beiden Stämmen vorhanden sind, handelt.

Bei Verwendung der adaptierten Typenphagen von FELIX und CALLOW, den direkt nachweisbaren Spontanphagen I, IVb, IVa und VI, sowie den Phagen aus Mischkulturen fand SCHOLTENS eine Reihe neuer Typen, die mit der Methode von FELIX und CALLOW nicht zu erkennen sind (s. Tabelle 12). BRANDIS und STORCH (1956) haben die von SCHOLTENS angegebenen Reaktionsbilder und das Vorkommen der Typen Taunton „The Hague", Taunton „Kampen", Beccles-Meppel, „22", Midwoud, Sittard, „87" in Deutschland bestätigen können.

SCHOLTENS nahm ferner eine neue Gruppierung der Paratyphus B-Lysotypen vor, welche die natürlichen Verwandtschaftsverhältnisse zum Ausdruck bringen soll. Diejenigen Lysotypen, die mit den angepaßten Phagen von FELIX und

[1] Der Phage mit dem lytischen Spektrum d entspricht dem Typenphagen Dundee.

Callow (1, 2, 3a, 3a I, Jersey, Beccles-Meppel) sowie den Phagen aus Mischkulturen (b, c, d, e, f, h) die gleichen übereinstimmenden Lysisreaktionen geben, werden in einer Gruppe zusammengefaßt (Tabelle 12). Außerdem bilden alle Stämme einer Gruppe in Mischkulturen mit Stämmen anderer Gruppen eine bestimmte Phagenart aus. So entsteht in Mischkulturen von Stämmen der Gruppe A mit Typen der Gruppe M stets eine Phagenrasse mit dem lytischen Spektrum e oder aber bei Verwendung von Kulturen der Gruppe S stets eine solche mit dem lytischen Spektrum f usw. Scholtens stellte insgesamt die Gruppen A, M, S, J, B.M. und B auf. Innerhalb jeder Gruppe unterscheiden sich die in ihnen vorkommenden einzelnen Lysotypen durch ihren verschiedenartigen Gehalt an direkt nachweisbaren Spontanphagen sowie durch ihre Lysisreaktion mit diesen Phagen (Tabelle 14).

Tabelle 14. *Natürliche Gruppierung der Lysotypen von S. paratyphi B.* (Nach Scholtens)

Gruppen der Lysotypen						Lysisreaktionen mit den direkt nachweisbaren Spontanphagen				In den Lysotypen vorhandene typendeterminierende Phagen
A	M	S	J	B.M.	B	I	IV b	IV a	VI	
3b (m. c.)	3b, Var. 1	„54“	„60“		3a (m. c.)	cL	cL	cL	cL	—
„22“	Midwoud	Sittard	Jersey	Meppel	3a I, Var. 1,2	—	cL	cL	cL	I
					3a I (Schiedam)	—	—	—	cL	I
3b, Var. 2	B.A.O.R.	„18“	3a, Var. 3		3a, Var. 4	cL	—	—	—	VI
Dundee		„87“			3a I, (Leeuwarden)	—	—	—	—	I, VI
Taunton (The Hague)						—	—	cL	—	IV
					Taunton (Kampen)	—	—	cL	—	I, IV

m. c. = „most common“.

Es geben also beispielsweise alle Lysotypen der Gruppe A (3b (m. c.), „22“, 3b, var. 2, Dundee, Taunton (The Hague)) mit den vom Phagen 1 angepaßten Phagen 1, 2, 3a, 3a I, Jersey, Beccles-Meppel sowie den Phagen aus Mischkulturen e, d, (= Dundee), c, f, h, b das gleiche Lysisbild. Eine Differenzierung der einzelnen Typen dieser Gruppe ist möglich nach den Lysisreaktionen mit den Spontanphagen I, IVb (Beccles), IVa (Taunton) und VI. Die Anordnung der Lysotypen innerhalb der Gruppen ist so vorgenommen, daß zunächst die Typen aufgezählt sind, die keine Spontanphagen abgeben, dann folgen Typen, die nur den Phagen I haben, solche die den Phagen VI aufweisen, dann andere mit den Phagen I und VI und schließlich Typen, welche die Phagen I und IV oder nur IV besitzen. Die Gruppe A ist bisher am vollständigsten. Es ist aber durchaus möglich, daß auch noch bei den anderen Gruppen korrespondierende Typen gefunden werden.

Bemerkenswerterweise bestehen bei den Paratyphus B-Lysotypen gewisse Beziehungen zwischen Lysotyp und Kulturtyp. So haben Brandis und Thomsen (1956) in Übereinstimmung mit den Befunden von Rische (1955) nachgewiesen, daß die in Deutschland vorkommenden Stämme des Lysotyps Taunton in der Mehrzahl inositnegativ sind. Ferner gehörten von den auf Simmons-Agar die

Rhamnose nicht vergärenden Stämme 87,5% zum Lysotyp Jersey, 7,5% zum Lysotyp Beccles und 5% zum Lysotyp 3a I, var. 1. Diese Ergebnisse stehen in Übereinstimmung mit denen von SCHOLTENS (1956), der ebenfalls fand, daß die Stämme seiner Gruppen J (Jersey) und B.M. (Beccles-Meppel) Rhamnose nur langsam vergären. ANDERSON (1955) stellte fest, daß bemerkenswerterweise alle von ihm untersuchten Stämme des Typs Jersey monophasisch waren. Nach diesem Autor dürfte die Verbindung von Lysotyp und monophasischem Charakter ohne innere Abhängigkeit voneinander sein.

Die Paratyphus B-Lysotypen sind für die praktische Epidemiologie von gleichem Wert wie die Typhuslysotypen. Auch bei den B-Typen besteht unter natürlichen Bedingungen eine große Konstanz, selbst bei experimentellen ungünstigen Verhältnissen wie Austrocknung, kurzem Erhitzen auf 60^0, Einwirkung von Streptomycin oder Formol bleibt der Lysotyp unverändert erhalten (NICOLLE und Mitarbeiter 1950). Durch Kontamination mit fremden („exogenen") Phagen können Paratyphus B-Stämme gelegentlich ein atypisches Verhalten oder eine vollständige Resistenz gegenüber den Typenphagen zeigen. Bei Prüfung einer Reihe von Einzelkolonien der betreffenden Kultur oder durch Einwirkung von Formol (NICOLLE) gelingt es aber meist, typisierbare Subkulturen zu erhalten. BRANDIS (1953, 1955) beschrieb einen mit exogenen Phagen infizierten B.A.O.R.-Stamm, der mit den Typenphagen 3a und 3a I eine Lysis gab. Es stellte sich heraus, daß die Ursache hierfür in einer Rekombination der in der Kultur vorhandenen exogenen Phagen mit den beim Typisieren aufgetropften Typenphagen 3a bzw. 3a I zu suchen ist, wodurch neue Phagenrassen entstehen, welche das beobachtete atypische Lysisbild bewirken (BRANDIS und STORCH 1956).

Die Verteilung der Paratyphus B-Lysotypen in den einzelnen Ländern ist recht verschieden (Tabelle 15). Während beispielsweise in Deutschland der Typ Taunton mit 32,5% weitaus der häufigste Typ ist, so steht in England der Typ 1 und in Kanada der Typ 3a an erster Stelle. In Frankreich wird der Typ Dundee in 23,7% beobachtet; in Deutschland kommt er dagegen nur in 1,68% vor. Bemerkenswert ist auch das häufige Auftreten dieses Typs im Saarland; GÄRTNER (1956) vermutet, daß dies nicht nur auf die engen Verkehrskontakte mit Frankreich zurückzuführen ist, sondern daß darüber hinaus der Typ Dundee im Saarland und in Lothringen endemisch verbreitet ist. Der Typ 2 wurde bisher nur in England (8,8%) beobachtet, während er auf dem europäischen Kontinent fehlte. Im Sommer 1956 stellten wir diesen Typ kurz hintereinander in mehreren deutschen Städten (Darmstadt, Frankfurt, Osnabrück, Stade, Berlin) fest. Epidemiologische Zusammenhänge ließen sich nicht aufdecken. Auch blieb die Ursache der Infektionen unbekannt. In Frankreich wurde der Typ 2 aus gefrorenem chinesischen Vollei isoliert (NICOLLE). Da in England Paratyphus B-Infektionen mit den Typen 3a und wahrscheinlich 3a, var. 2 durch gefrorenes Ei vorgekommen sind (HOBBS und SMITH 1955, NEWELL 1955), darf vielleicht angenommen werden, daß die Typ 2-Infektionen in Deutschland durch das gleiche Produkt verursacht wurden.

Über das Vorkommen der Paratyphus B-Lysotypen in den verschiedenen Ländern geben neben einer allgemeinen Übersicht von FELIX (1955) folgende Arbeiten Auskunft: Belgien (BEUMER und VANDEWEYER-DEVEEN 1954), Kanada

(Desranleau 1947), Deutschland (Rische, Rohne und Schneider 1954, Rische 1955, Brandis 1955), England (Felix und Callow 1943, 1951, Felix 1951), Frankreich (Nicolle, Hamon und Edlinger 1953, Nicolle und Hamon 1954), Mittelitalien (de Blasi und Buogo 1952, Buonomini und d'Amelio 1954), Norwegen (Henriksen 1952), Österreich (Edlinger 1951, Edlinger, Nicolle und Hamon 1954), USA (Cherry und Mitarbeiter 1953).

Tabelle 15. *Geographische Verteilung der Paratyphus B-Lysotypen.* (Nach Angaben der Lysotypielaboratorien, zusammengestellt von Dr. A. Felix 1953)

	Die gewöhnlich vorkommenden Typen, angeordnet nach der Häufigkeit. Prozentsatz[1] der gefundenen Lysotypen	Prozentsatz[1] der unbestimmbaren Stämme	Gesamtzahl der geprüften Patienten und Ausscheider
Österreich	Taunton, Typ 1, 3a I, B.A.O.R. 50,4 12,1 8,6 8,4	1,21	1163
Belgien	Taunton, B.A.O.R., 3a, Beccles, 1 28,1 28,1 12,5 10,9 7,8	0	68
Dänemark	Taunton, 3a, Typ 1, Beccles 32,6 28,1 23,6 7,9	5,6	153
Frankreich[2]	Taunton, Dundee, Jersey 1, 3a I 41,5 23,7 22,3 4 2,4	1,56	1365
Mitteldeutschland	Taunton, Typ 1, 3a I, B.A.O.R. 53 14,4 10,4 3,2	12,1	222
Westdeutschland	Taunton, 3a I, 3a, Typ 1, Beccles 32,5 17,9 12,23 10,83 6,5	11,12	2130
England[3]	Typ 1, Taunton, Typ 2, 3a 46,6 11 8,8 7,2	10,6	2409
Mittelitalien[4]	Taunton, 3a I, 3a, Typ 1 49,7 16,5 14,7 6,5	6,5	366
Holland 1952	Taunton, 3a I, Typ 1, Beccles, Jersey 28,4 16,2 11,4 12,4 7,3	4,1	293
Norwegen[5]	Typ 1, 3a I, Dundee	17,4	60
Kanada[6]	3a, Typ 1, 3b, 3a I, Taunton 70,3 10,9 7,8 4,7 1,6	4,7	109

[1] Prozentsatz berechnet nach „Foci" in allen Ländern außer Österreich und Frankreich.
[2] Die Typen B.A.O.R. und 2 wurden nicht gefunden.
[3] Angaben für die Periode 1950—1952.
[4] Kein Nachweis der Lysotypen Jersey, Beccles und Dundee.
[5] Die Phagen Beccles und Taunton waren nicht verfügbar.
[6] Angaben für die Periode 1. Juli 1950 bis 1. Juli 1953.

III. Die Lysotypie bei anderen Salmonellen

1. S. paratyphi A

In Fortführung gemeinsamer Versuche mit Felix gelang es Banker (1955), einen Paratyphus A-Phagen, der 1943 von Dhayagude aus Abwasser in Bombay isoliert worden war, an verschiedene Paratyphus A-Stämme nach dem Vorgehen von Craigie und Yen anzupassen und somit typenspezifisch zu machen. Banker erhielt auf diese Weise 4 Phagenpräparationen, mit denen die Mehrzahl der von ihm geprüften 636 Paratyphus A-Stämme typisierbar war (Tabelle 15a). Inzwischen hat sich die Zahl der Typenphagen auf 5 erhöht (Felix).

Zweifellos dürfte der Lysotypie von Paratyphus A-Bakterien in den Ländern, in welchen der Paratyphus A endemisch verbreitet ist, der gleiche epidemiologische Wert zukommen wie der Typenbestimmung der Typhus- und Paratyphus B-Bakterien.

Tabelle 15a. *Lysisschema der Paratyphus A-Phagen.* (Nach BANKER)

Typ	Isoliert in	Phagen (in RTD)					Häufigkeit in %
		1	2	3	4	5	
1	Bombay	cL	cL	cL	cL	cL	59
2	Palästina	—	cL	—	—	—	14
3	Indien	—	—	cL	—	—	2,2
4	Ägypten	—	—	—	cL	—	7,9
5	Indonesien	—	—	—	—	cL	.
Unbestimmbar							13,8
Degradiert							3,1

cL = konfluierende Lysis.

2. S. typhi-murium

SCHMIDT (1931) paßte den Paratyphus B-Diagnostikphagen von SONNENSCHEIN an S. typhi-murium an und erhielt so einen spezifisch auf diese Bakterienart wirkenden Phagen. Mit einer Typendifferenzierung der Breslau-Bakterien hat sich vor allem LILLEENGEN (1948) beschäftigt. Dieser Autor stellte mit Hilfe von 12 in der RTD angewandten Anti-O-Phagen, die von 44 Phagen verschiedener Herkunft (Abwasser, Dung, Hühner- und Hundekot, lysogene Kulturen) ausgewählt worden waren, 24 Typen auf, mit denen 667 (= 94,9%) der untersuchten 703 Breslaukulturen erfaßt werden konnten. Die für die einzelnen Typen charakteristischen Lysisbilder mit den 12 Testphagen erwiesen sich in vitro und, soweit geprüft, auch unter natürlichen Bedingungen als stabil. Einige Typen fanden sich besonders bei bestimmten Tierarten, so der Typ 11 bei Ratten und die eng verwandten Typen 6 A und 6 B beim Schwein. Die Typen 3 und 9 kamen wesentlich häufiger bei Vögeln als bei anderen Tierarten vor.

FELIX und CALLOW erwähnten bereits 1943 die Unterteilung von S. typhimurium mit spezifischen Typhi-murium-Phagen in 2 Typen. Von dieser Feststellung ausgehend hat dann FELIX (1956) ein Typenschema für S. typhi-murium veröffentlicht, mit dem 12 Typen (1, 1a, 1b, 2, 2a, 2b, 2c, 2d, 3, 3a, 3b[1] und 4) mit einem Satz von 12 korrespondierenden Phagen unterschieden werden können. Die Typenphagen gehören drei verschiedenen serologischen Gruppen an. In der gleichen Veröffentlichung hat FELIX ferner ein Typenschema bekannt gegeben, das sich auf der Verwendung von adaptierten Phagen aufbaut, die sich alle von einem einzelnen Phagenstamm (dem 3b-Paratyphus B-Typenphagen) herleiten und daher serologisch einheitlich sind. Die Typisierung der Breslau-Bakterien besitzt infolge des sporadischen Auftretens von Breslau-Infektionen bei Menschen nicht die große epidemiologische Bedeutung wie diejenige der Typhus- und Paratyphus B-Bakterien, sie kann jedoch die Aufklärung von Nahrungsmittelvergiftungen oder aber von Schmierinfektionen in Schlachthöfen und dergleichen unterstützen (ROHNE 1956). Stämme eines Ausbruchs gehören nach FELIX, der eine Reihe von Beispielen anführt, stets einheitlich zu einem bestimmten Typ.

[1] Der Typ 3b ist neuerdings in 1a, var. 1 umbenannt worden (s. ANDERSON und WILLIAMS 1956).

Einen anderen Weg der Typisierung von Breslau-Stämmen hat BOYD (1950, 1951) vorgeschlagen. Dieser Autor zeigte, daß Kulturen von S. typhi-murium nach der Art oder der Kombination der nachweisbaren temperierten (symbiontischen) Phagen in Typen einzuteilen sind, da die Lysogenität ein permanenter Charakter der Kulturen ist. Es erfolgt also bei dem Vorgehen von BOYD nicht die Prüfung der Breslau-Stämme mit einer Serie von Testphagen, sondern man untersucht ihren Besitz an Lysogenitätsphagen, deren Eigenschaften auf 2 Indicatorstämmen und falls nötig auch serologisch festgestellt werden.

3. Übrige Salmonellen

Bei einer Reihe von Salmonellen ist von verschiedenen Autoren eine Unterteilung in Lysotypen teils durch Verwendung von empirisch ausgewählten Anti-O-Phagen, teils auf Grund verschiedenartiger Lysogenitätsverhältnisse vorgenommen worden.

Mit Hilfe von Anti-O-Phagen (aus Abwasser u. dgl.) hat LILLEENGEN (1950) bei S. dublin 6 Lysotypen (mit 5 Testphagen) und bei S. enteritidis 8 Typen (mit 4 Testphagen) nachgewiesen. In ähnlicher Weise stellte dieser Autor (1952) verschiedene Lysotypen bei S. pullorum und S. gallinarum auf, nachdem schon THUMES (1933), NAIDU (1935) sowie NOBREGA (1935, 1936) gezeigt hatten, daß diese beiden Salmonellatypen mit Phagen unterschieden werden können. WILLIAMS SMITH (1951) benutzte 7 Phagenpräparationen von lysogenen Salmonella thompson-Stämmen und konnte mit der lytischen Wirkung dieser Phagen 11 Lysotypen bei S. thompson feststellen. Der gleiche Autor (1951) fand bei S. dublin mit 6 Phagen von lysogenen Dublin-Kulturen insgesamt 11 Lysotypen. 294 (94%) von 306 Stämmen waren typisierbar. ATKINSON und Mitarbeiter (1952, 1953) unterteilten, ebenfalls mit Lysogenitätsphagen, S. adelaide, S. waycross und S. bovis-morbificans in eine Reihe von Lysotypen. BRANDIS und THOMSEN (1955) zeigten, daß Stämme von S. java mit den Paratyphus B-Typenphagen von FELIX und CALLOW Reaktionsbilder ergeben, welche den Paratyphus B-Lysotypen entsprechen.

Kurz sei noch auf Salmonellaphagen mit einem größeren Aktionsbereich eingegangen. Auf die speciesspezifischen Diagnostikphagen von SONNENSCHEIN (1925—1929), die diesem Autor gute Dienste bei der Erkennung von Typhus- und Paratyphus B-Bakterien leisteten, wurde schon kurz hingewiesen. WASSERMANN und SAPHRA (1955) verwendeten Anti-O-Phagen, die spezifisch auf die Salmonella-O-Gruppen B, C_1, C_2, D und E eingestellt sind. Die Autoren fanden eine gute Übereinstimmung der Phagenreaktionen mit den Ergebnissen der Agglutinationsprobe mit O-Seren. Dies unterstreicht die von BURNET u. a. hervorgehobene Beziehung der Salmonellaphagen zu der O-Antigenstruktur. CHERRY und Mitarbeiter (1954 [a]) zeigten, daß der O_1-Phage von FELIX und CALLOW (1943) eine spezifische Wirkung auf Bakterienkulturen der Salmonellagruppe hat. Dieser Phage greift bei unverdünnter Anwendung die Salmonellatypen aus allen bisher bekannten O-Gruppen an, lediglich die Angehörigen der O-Gruppe 35 erwiesen sich infolge des Vorhandenseins eines K-Antigens resistent. Bei 427 verschiedenen Salmonellakulturen fanden CHERRY und Mitarbeiter bei 415 Stämmen (97%) eine Lysis. Von 195 Shigellen wurde dagegen nur eine einzelne Kultur angegriffen. Überhaupt keine Lysis trat mit Stämmen aus den Gattungen

Proteus, Serratia, Klebsiella und Aerobacter auf. Von 93 geprüften Colistämmen wurde nur eine Kultur gelöst und alle untersuchten 195 Bethesda-Stämme blieben unbeeinflußt. Dagegen wirkte der O_1-Phage bei Arizonastämmen, die eng verwandt mit den Salmonellen sind, lytisch. Thal und Kallings haben ebenfalls mit dem O_1-Phagen gute Erfahrungen gesammelt. Wir selbst wiesen unter 511 Kulturen aus der Coligruppe bisher nur 3 Stämme nach, die eine Lysis mit dem O_1-Phagen gaben. Es kommen aber auch gelegentlich Salmonellen vor, die gegen den O_1-Phagen resistent sind. So sahen wir keine Wirkung bei einigen Typhus- und Paratyphus-B-Stämmen sowie bei S. ahuza. Hofmann fand, daß S. kaltenhausen nicht gelöst wird. Wichtig ist, um brauchbare Ergebnisse zu erzielen, daß der O_1-Phage weder einen zu hohen noch einen zu niederen Titer hat. Ferner sei noch angefügt, daß nach Felix und Anderson (1951) Typhusstämme der Typen D_6, F_1, F_2, H, J, K, L_1 und L_2 und M durch den 1:3000 verdünnten O_1-Phagen beim Auftropftest nicht beeinflußt werden. Die unbestimmbaren Vi-Typhusstämme zeigen sich teils empfindlich, teils resistent gegenüber dem verdünnten O_1-Phagen. Dies kann zur Charakterisierung derartiger Stämme mitverwendet werden. Zusammenfassend läßt sich über den O_1-Phagen sagen, daß er die serologischen Standardmethoden zwar nicht zu ersetzen vermag aber andererseits zweifellos eine Bereicherung der diagnostischen Möglichkeiten zur Erkennung der Gattung Salmonella darstellt.

IV. Lysotypie von Staphylococcus aureus[1]

Die ersten eingehenden Untersuchungen über eine Reihe von Staphylokokkenphagen stammen von Burnet und Lush (1935). Die Autoren arbeiteten mit 8 Phagenstämmen, die von Staphylokokkenkulturen verschiedener Herkunft stammten und unterschieden nach der Art der Vermehrungsfähigkeit in Bouillon- bzw. auf Agarkulturen stark (Au 1, Au 2, Au 3 und Au 4), schwach (Au 11, Au 12 und Au 13) sowie mittelstark wirkende Phagen (Au 21). Es zeigte sich, daß Staphylokokkenstämme von diesen Phagen in ungleicher Weise angegriffen wurden (Tabelle 16), wobei hervorzuheben ist, daß, je aktiver ein Phage wirkte, desto größer die Zahl der gelösten Stämme war.

Die von Burnet und Lush erhaltenen Resultate weisen auf die Möglichkeit einer Unterscheidung von Staphylokokkenstämmen durch Phagen hin. Diese Frage wurde jedoch von den Autoren, die an anderen Problemen interessiert waren, nicht weiter verfolgt. Williams und Timmins (1938) haben dagegen zur Aufklärung von Infektionsquellen bei Erkrankungen an akuter Osteomyelitis versucht, Staphylokokkenkulturen mit den 4 starken Phagen Au 1 bis Au 4 zu differenzieren. Sie verwendeten das Bouillonverfahren und fanden mit den geprüften Staphylokokkenstämmen folgende Verhaltensweisen: A (völlige Klärung der Bouillon mit allen Phagen), B (bei Gegenwart aller Phagen Wachstum), C (Klärung der Bouillon mit den Phagen Au 1, Au 2, Au 4), D (Klärung mit den Phagen Au 2, Au 3 und Au 4), E (Klärung mit den Phagen Au 1, Au 3 und Au 4), F (Klärung mit den Phagen Au 1 und Au 3). In der Nase und im Rachen der von ihnen untersuchten Kinder waren vor allem Staphylokokken der Gruppen A und B vorherrschend.

---

[1] Unter Mitarbeit von Dr. H.-Ph. Pöhn.

Tabelle 16. *Empfindlichkeit von Staphylococcus aureus-Kulturen gegenüber Staphylokokkenbakteriophagen.* (Nach BURNET und LUSH)

Staphylokokkenstämme	Zahl der gleichsinnig reagierenden Kulturen	Phagen							
		Au 1	Au 2	Au 3	Au 4	Au 21	Au 11	Au 12	Au 13
Eggleston	5	(+++)	(+++)	(+++)	(+++)	—	—	—	—
Krueger	25	+++	+++	+++	+++	—	—	—	—
Latham	5	+++	+++	+++	+++	+++	+++	—	—
Copland	8	+++	+++	+++	+++	+++	+++	+++	+++
Ebdon	2	+++	+++	+++	+++	+++	—	+++	+++
Pyle	1	+	+	+++	+++	—	—	—	—
Wauer	1	+++	+++	—	—	—	—	—	—
SF (Sta. albus)	.	+++	±±±	±±±	±±±	—	—	—	—

Zeichenerklärung: +++ = konfluierende Lysis auf Agar; (+++) = konfluierende Lysis auf Agar, aber keine Vermehrung in Bouillon; ±±± = konfluierende Lysis mit sekundärem Bakterienwachstum; + = vereinzelte Löcher; — = keine Lysis oder Wachstumshemmung in der Bouillon.

Einen wesentlichen Fortschritt bedeuteten die Arbeiten von FISK (1942 [1]). FISK wies nach, daß lysogene Stämme bei Staphylokokken nicht ungewöhnlich sind. Der Prozentsatz betrug bei 43 Kulturen 44,2%. Von diesen lysogenen Staphylokokkenstämmen isolierte FISK mit Hilfe der Kreuzkulturmethode eine Reihe von verschiedenen Phagen, die er nach ihrer Wirkung auf einer Serie von Teststämmen in 24 Gruppen einteilte. Die Phagen glichen in ihrem Verhalten den „schwachen Phagen" Au 11, Au 12 und Au 13 von BURNET und LUSH, da eine Anreicherung ebenfalls wie bei diesen nur auf Agarplatten gelang und mit Bouillonkulturen nicht möglich war.

Um sich einen Überblick über die Differenzierungsmöglichkeit von Staphylokokkenstämmen zu verschaffen, untersuchte FISK (1942 [2]) in einer anderen Versuchsreihe 95 Kulturen mit 27 Phagen, die er mit großen Buchstaben (A, B, C, D usw.) bezeichnete. Die Technik bestand darin, daß eine 18—24stündige Tryptose-Bouillonkultur des zu prüfenden Stammes gleichmäßig auf einer Tryptose-Agarplatte ausgeimpft wurde. Nach Trockenwerden der Platte erfolgte das Auftropfen der verschiedenen Phagen und die Bebrütung für 3 Std bei 37° C. Vor der Ablesung wurden die Platten über Nacht bei Zimmertemperatur gehalten. Da es FISK nur auf qualitative Unterschiede ankam, verwendete er die Phagen unverdünnt und nahm keine Titration der Konzentrationen vor. Dieses Verfahren erwies sich als ein brauchbares Mittel zur Unterscheidung von Staphylokokkenstämmen. Identische Kulturen gaben das gleiche Lysisbild, während nichtverwandte Stämme unterschiedliche Reaktionen zeigten (Tabelle 17) [vgl. auch FISK und MORDVIN (1944)].

Aus Tabelle 17 geht hervor, daß die Empfindlichkeit der Staphylokokkenstämme gegenüber Staphylokokkenphagen sehr verschieden ist. Teils werden die Stämme von vielen Phagen angegriffen, teils aber auch nur von einer geringeren Anzahl. Je nach ihrem charakteristischen Lysisbild bei Verwendung sämtlicher Phagen konnten demnach die einzelnen Staphylokokkenstämme voneinander unterschieden werden.

WILSON und ATKINSON haben im Jahre 1945 die Methode von FISK weiter ausgebaut und wesentlich vervollkommnet. Die Autoren gewannen ihre Phagen

Tabelle 17. *Reaktionen von Staphylokokkenstämmen mit 27 verschiedenen Phagen.* (Nach Fisk, stark gekürzt)

Patient	Herkunft der Kultur	Datum der Isolierung	Stamm Nr.	Hämolyse	Pigment	Mannit	Coagulase	Lysis mit den Phagen
1. Vater	Nasenabstrich	10. 8. 1940	8	—	O	+	+	A L Q
	Nasenabstrich	16. 2. 1941	80	+	O	+	+	A L Q
	Nasenabstrich	4. 7. 1941	N 55	+	O	+	+	A L Q
2. Sohn	Nasenabstrich	10. 8. 1940	9	+	O	+	+	A L Q
	Nasenabstrich	4. 7. 1941	N 56	+	O	+	+	A L Q
Osteomyelitis	Blutkultur	10. 2. 1941	72	—	O	+	+	I N U Y
	Urinkultur	15. 4. 1941	118	+	O	+	+	I N U Y
Osteomyelitis	Blutkultur	11. 3. 1941	86	+	Y	+	+	I M N O U V W X Y
	Blutkultur	14. 3. 1941	87	+	Y	+	+	I M N O U V W X Y
	Knochenfistel	11. 3. 1941	95	+	O	+	+	I M N O U V W X Y

Zeichenerklärung: O = Orange; Y = gelb; — = keine Hämolyse.

nach der Fiskschen Kreuzkulturmethode, die darin besteht, daß zunächst mit einem Staphylokokkenstamm die ganze Oberfläche einer 0,2% Glucose enthaltenden Agarplatte gleichmäßig beimpft wird. Nach Eintrocknen werden dann auf diesen „Basalstamm" Tropfen 18stündiger Bouillonkulturen anderer Staphylokokkenstämme aufgesetzt. In anschließenden Versuchen dienen diese aufgetropften Kulturen als Basalstämme. Die Bebrütung dauert 6—8 Std bei 37° C, und dann werden die Platten über Nacht bei Zimmertemperatur aufbewahrt. Falls eine Phagenwirkung vorhanden ist, läßt sie sich an den entstandenen Löchern oder einer Wachstumshemmungszone am Rand der aufgetropften Bouillonkulturen erkennen. Vom Ort der Phagenwirkung wird mit der Öse etwas Material entnommen und zuerst geprüft, gegen welchen Stamm (Basalstamm oder aufgetropfter Stamm) der Phage wirksam ist. Wilson und Atkinson haben die auf die vorstehend geschilderte Weise isolierten Phagen in Bouillonpassagen mit den jeweiligen empfindlichen Stämmen angereichert und dann von Einzelphagenlöchern ausgehend in mehrfachen Überimpfungen reine Phagenstämme gewonnen. Die Abtrennung von den Bakterien erfolgte durch Seitz-Filtration. Danach wurde die „kritische Testkonzentration" festgestellt, d. h. diejenige höchste Phagenverdünnung, welche mit einem Tropfen beim Agarplattentest noch eine konfluierende Lysis ergab. Diese Phagenverdünnung kam bei den Typisierungsversuchen zur Anwendung.

Wilson und Atkinson isolierten nach dem Fiskschen Verfahren insgesamt sieben verschiedene Phagenstämme, hinzu kamen elf andere Phagenpräparationen, die ausgehend von den Originalphagen nach dem Vorgehen von Craigie und Yen bei den Typhus Vi-Phagen durch Anpassung an geeignete Staphylokokkenkulturen erhalten worden waren. Mit diesen Phagen stellten die Autoren 21 Staphylokokkentypen bzw. Subtypen auf. Schwache Lysisreaktionen fanden für die Typeneinteilung keine Berücksichtigung. Jeder Staphylokokkentyp sollte also, ähnlich wie bei den Typhusbakterien, durch eine bestimmte konstante Reaktionsweise mit den Typenphagen charakterisiert sein (Tabelle 18).

Wilson und Atkinson prüften insgesamt 460 Staphylokokkenkulturen, von diesen waren 278 (= 60,4%) typisierbar, 104 (= 22,6%) Stämme wurden durch

Tabelle 18. *Bezeichnungen der Staphylokokkenlysotypen.* (Nach WILSON und ATKINSON)

Typ	Bakteriophagenfiltrate																					
	3/284 (3 A)	3/211 (3 B)	51/145 (51)	6/3 (6)	7/4 (7)	42/1163 (42 B)	47/36 (47)	47/1163 (47 C)	29/33 (29)	31/18 (31)	52/144 (52)	3/284 (3 A)	3 B/1339 (3C)	42 A/1307 (42 C)	44/18 (44)	44/373 (44 A)	47/761 (47)	47/987 (47 B)	47/1163 (47 C)	51/145 (51)	52/144 (52)	52/925 (52 A)
1 A	+	—	+																			
1 B	—	+	—																			
1 C	±	+	+																			
2 A				+	—	+	+	+														
2 B				±	+	+	+	+														
2 C				—	—	+	—	+														
2 D				—	—	—	+	+														
3 A									+	+	—											
3 B									—	+	—											
3 C									—	+	+											
4												+										
5													+									
6														+								
7															+							
8																+						
9																	+					
10																		+				
11																			+			
12																				+		
13																					+	
14																						+

+ = konfluierende Lyse; ± = halbkonfluierende Lyse; — = geringe Lysisgrade oder keine Lyse.

Phagen zwar beeinflußt, waren aber nicht eindeutig einem der aufgestellten 21 Typen zuzuordnen, und 78 (= 17,0%) Stämme erwiesen sich gegenüber allen Phagen als unempfindlich. Es stellte sich heraus, daß mit Ausnahme der Typen 10 und 14 kein Typ mehr als 10% der Gesamtzahl der typisierbaren Kulturen ausmachte. In epidemiologischer Hinsicht erzielten WILSON und ATKINSON bei Nahrungsmittelvergiftungen durch Staphylokokken und Pemphigusausbrüchen auf geburtshilflichen Stationen mit ihrem Verfahren gute Erfolge.

Die ursprünglichen Phagenbezeichnungen sind später von WILSON und ATKINSON geändert worden. Tabelle 18 gibt eine Gegenüberstellung der Benennungsweisen. Außerdem fügten diese Autoren dem Phagensatz noch drei neue Phagen (29 A, 42 D und 42 E) hinzu (WILLIAMS SMITH 1948).

Während WILSON und ATKINSON, wie schon kurz erwähnt, nur die Phagenreaktionen der Staphylokokkenstämme mit einer starken Lysis registrierten und darauf ihr Typenschema aufbauten, berücksichtigten WAHL und LAPEYRE-MENSIGNAC (1950), die mit den gleichen Phagen arbeiteten, auch schwächere Lysegrade. Außerdem unterschieden diese Autoren zwischen sog. „phages majeurs“ und „phages mineurs“. Hierdurch kamen WAHL und LAPEYRE-MENSIGNAC zu einer Aufteilung der Staphylokokkenstämme in 5 Gruppen und einer Anzahl von Subgruppen, wobei die Stämme der einzelnen Staphylokokkengruppen jeweils durch eine kräftige Lysis mit einem oder mehreren der „phages

Tabelle 19. *Gruppeneinteilung der Staphylokokkenstämme mit „phages majeurs“.* (Nach WAHL und LAPEYRE-MENSIGNAC)

Gruppe	Phages majeurs	Prozentsatz der in die Gruppen fallenden Stämme
A	68[1]	29
B	52 A	18
C	68, 52, 52 A	5
D	3 C	26
E	47 A, 68	5
Nicht angegriffene Stämme		15,5
Nicht klassifizierbare Stämme		8

[1] Der Phage 68 wurde von WAHL und LAPEYRE-MENSIGNAC dem Satz der 21 Phagen von WILSON und ATKINSON hinzugefügt.

Tabelle 20. (Nach WILLIAMS und RIPPON)

Labor-Nr.		Nr. der lysinogenen Staphylokokkenstämme	Phagenstamm, von dem angepaßt wurde	Serologische Gruppe des Phagen
Phagen	Züchtungsstämme			
3 A	284	—	3	A
3 B	211	—	3	A
3 C	1339	—	3 B	A
6	3	42 ?	—	A
7	4	5	—	A
29	33	21	—	B
29 A	1351	—	29	B
31	18	24	—	B
31 A	R 48/2329	—	31	B
42 B	1163	—	42	A
42 C	1307	—	42 A	B
42 D	1363	—	42 C	B ? F
42 E	1670	—	42	A
44	18	35	—	B
44 A	373	—	44	B
47	36	17	—	A
47 A	761	—	47	A
47 B	987	—	47	A
47 C	1163	—	47	A
51	145	40	—	A
52	144	Rad. 2	—	B
52 A	925	—	52	B
53	R 48/3292	R 48/2311	—	B
54	R 48/R 3303	R 48/3298	—	A

Die Züchtungsstämme 18 und 1163 erscheinen 2mal, da sie zur Züchtung zweier verschiedener Phagen dienten.

majeurs“ charakterisiert waren (Tabelle 19).

Dieses Schema wurde später (1952) von WAHL und FOUACE vor allem unter Fortlassung des Phagen 68, der sich als nicht genügend spezifisch erwies, modifiziert. Die Autoren ordneten nun die Phagen in 4 Serien (s. S. 141). Jede Serie umfaßt diejenigen Phagen, die miteinander am häufigsten auf einen gleichen Staphylokokkenstamm einwirken:

1. Serie: 52, 52 A, 29, 29 A, 31,
2. Serie: 3 C, 3 A, 3 B, 51,
3. Serie: 47, 53, 54, 44, 44 A, 6, 7,
4. Serie: 42 B, 42 C, 42 D, 42 E, 47 A, 47 B, 47 C.

Jeder dieser Phagenserien stellten WAHL und FOUACE eine entsprechende Staphylokokkengruppe gegenüber. Mit dieser Einteilung näherten sich die Autoren den von WILLIAMS und RIPPON (1952) erhobenen Feststellungen, die grundlegende Bedeutung für die Methode der Staphylokokkentypisierung erlangt haben und aus diesem Grunde hier ausführlicher besprochen werden sollen. WILLIAMS und RIPPON verwendeten einen Satz von 24 Phagen. Es handelte sich hierbei um die 21 Phagen von WILSON und ATKINSON, ergänzt durch 3 Phagen (31 A, 53 und 54) von ALLISON. Zehn dieser Phagenstämme sind mit Hilfe der Kreuzkulturmethode von FISK isoliert worden und 14 Phagenstämme wurden durch Anpassung an Staphylokokkenkulturen, ausgehend von den WILSONschen Phagen oder von anderen durch die Kreuzkulturmethode erhaltenen Phagen, gewonnen. Es handelt sich also bei den zum Typisieren verwendeten Phagen durchweg um Lysogenitätsphagen. In Tabelle 20 sind die Daten über den 1952 vorgeschlagenen Standardsatz der Staphylokokkenphagen zusammengestellt (nach WILLIAMS und RIPPON).

Die Nomenklatur der Staphylokokkenphagen wird verständlich, wenn man die Entwicklung der Methode betrachtet. Wie WILLIAMS und RIPPON ausführen, bekamen die Phagen, welche durch Kreuzkulturen gewonnen wurden, eine bestimmte fortlaufende Nummer, und die angepaßten Phagen erhielten die Nummer der Ausgangsphagen mit einem Buchstaben. So stammt beispielsweise der Phage 52 A vom Phagen 52 ab. Eine gewisse Schwierigkeit ergab sich allerdings in den Fällen, wo von einem Phagen mehrere Anpassungen vorgenommen wurden, daher erfolgte die Bezeichnung der angepaßten Phagen nicht immer folgerichtig. So ist z. B. der 3 C-Phage ausgehend von dem Phagen 3 B angepaßt und nicht vom Originalphagen 3, der jetzt nicht mehr vorhanden ist. Neuerdings erhalten Phagen, die von bereits adaptierten Phagen ausgehend auf Staphylokokkenstämme angepaßt werden, auch fortlaufende Nummern. So bildete beispielsweise der Phage 52 A den Ausgang für den Phagen 80 (ROUNTREE) und der Phage 81 ist vom Phagen 42 B abgeleitet (BYNOE und Mitarbeiter 1956). In diesem Zusammenhang betonen WILLIAMS und RIPPON, daß es vielfach nicht leicht zu entscheiden ist, ob tatsächlich ein angepaßter Phage, d. h. also eine Variante des Ausgangsphagen vorliegt. Da vermutlich alle Staphylokokkenstämme Phagen beherbergen und während der Züchtung abgeben können, kann es sein, daß ein Phagenfiltrat auch Phagen vom Züchtungsstamm enthält, welche dann den eigentlichen Ausgangspunkt der „Anpassung" bilden. ROUNTREE (1949) zeigte, daß die Herkunft des Phagen 42 C wahrscheinlich auf diese Weise zu erklären ist. Ähnlich wie der Vi-II-Phage auf Typhusstämmen können aber auch Staphylokokkenphagen bei Züchtung auf verschiedenen Staphylokokkenstämmen Änderungen ihrer Wirtsspezifität erwerben. Diese Erscheinung muß nach ROUNTREE (1956) auf den Einfluß des Züchtungsstammes im Sinne einer phänotypischen Modifikation zurückgeführt werden.

Die praktische Verwendung der Phagen setzt zunächst ihre Anreicherung auf den jeweiligen zur Phagenvermehrung geeigneten Staphylokokkenstämmen voraus. Hierfür kommen nach WILLIAMS und RIPPON vor allem 2 Methoden in Frage: a) Die Agarplattenmethode und b) die Bouillonmethode. Die Plattenmethode wird so ausgeführt, daß Nähragar nach HARTLEY (1922) mit einer 5—6stündigen Bouillonkultur des Züchtungsstammes gleichmäßig beimpft wird. Danach verteilt man die Suspension des Phagen, der angereichert werden soll, in der benötigten Menge und in einer etwas weniger starken Verdünnung als die der Routinetestverdünnung (RTD) ebenfalls gleichmäßig über die Platte, nur ein kleines Feld zur Kontrolle des Bakterienwachstums bleibt frei. Die Bebrütung erfolgt über Nacht bei 30°. Nach der Bebrütung soll das Kontrollfeld keinerlei Zeichen einer spontanen Lysis zeigen. Ist das Bakterienwachstum typisch, wird dieser Bezirk herausgeschnitten und der übrige Agar bei -10 bis -20^{0} C für 24 Std eingefroren. Während des anschließend vorgenommenen Auftauens bei Zimmertemperatur tritt Flüssigkeit aus dem Agar aus, welche abpipettiert, zentrifugiert und auf ihre Phagenwirksamkeit austitriert wird. Falls ein genügend hoher Titer erreicht ist, erfolgt die Filtration durch ein Seitzfilter. Bei der Plattenmethode kann man nach WILLIAMS SMITH u. a. auch so vorgehen, daß die mit Phagen und dem Staphylokokkenzüchtungsstamm beimpfte Platte nach der Bebrütung von 5—6 Std bei 37° über Nacht im Kühlschrank aufgehoben und dann mit 5 cm^3 steriler Bouillon abgeschwemmt wird. Nach Zentrifugieren dieser Flüssigkeit läßt sich diese dann zum Abschwemmen einer zweiten und eventuell einer dritten Platte benutzen. Hierdurch wird die Phagenkonzentration gesteigert.

Bei geeigneter Bakterieneinsaat und Zusatz einer entsprechenden Phagenverdünnung ist nach den Erfahrungen von Williams Smith, Williams und Rippon u. a. auch Nährbouillon für Anreicherungszwecke brauchbar (als Endkonzentration in der Anreicherungsbouillon sind nach Anderson und Williams (1956) geeignet $^1/_{100}$ Verdünnung einer 2stündigen Staphylokokkenbouillonkultur und Phagen zwischen 10facher und $^1/_{10}$ RTD). Der Vorteil dieses Verfahrens liegt in der Einfachheit. Pöhn erzielte durch Belüftung der Bouillon bei einer Reihe von Phagen hohe Titer. Für manche Phagen ist ein Zusatz von $CaCl_2$ (0,1 bis 1 mg/cm^3) notwendig. Da die Staphylokokkenphagen wärmeempfindlich sind, müssen die Kokken durch Filtration abgetrennt werden. Gelegentlich enthalten die Phagenfiltrate, worauf Williams und Rippon hinweisen, einen Stoff, der das Wachstum von Staphylokokken hemmt (s. Wahl und Fouace 1954). Dieser Effekt muß von einer echten Phagenwirkung unterschieden werden.

Für die Routinetypisierung erfolgt die Anwendung der Phagen in der kritischen Testverdünnung, das ist diejenige Phagenverdünnung, welche mit einem Tropfen auf dem Züchtungsstamm beim Agarplattentest noch eine konfluierende Lysis ergibt. Filtrate mit einer RTD unter $^1/_{1000}$ sollen nicht benutzt werden.

Von ganz besonderer Bedeutung für die Konstanz und Vergleichbarkeit der Typisierungsergebnisse ist es, das lytische Spektrum der neuangereicherten Phagen auf dem Satz der Teststämme zu prüfen, wobei die einzelnen Phagen stets ihr charakteristisches Verhalten auf allen Teststämmen aufweisen müssen. Zunächst werden nach Williams und Rippon die Phagen unverdünnt auf alle Teststämme gebracht und danach auf denjenigen Stämmen, bei denen sie eine Lysis geben, austitriert. Die hierbei erhaltenen Titer müssen in einem bestimmten konstanten Verhältnis zu dem Titer des Phagen auf seinem Züchtungsstamm stehen. Dieses Verhältnis soll mit dem von dem „Subcommittee on Bacteriophage Typing of Staphylococci“ des „International Committee on Bacteriological Nomenclature“ herausgegebenen Schema übereinstimmen. Es läßt sich so eine Standardisierung des Verfahrens und eine Vergleichbarkeit der Resultate erreichen. In Tabelle 21 sind als Beispiel für das Lysespektrum die Phagen 29, 52 und 52 A angegeben.

Aus Tabelle 21 geht hervor, daß z. B. der Phage 29 außer dem homologen Stamm 29 auch die Teststämme 29 A, 31/44, 47 und 44 A mit jeweils einem bestimmten Titer im Verhältnis zum homologen Stamm angreift. Dieses Verhältnis muß bei Neuanreicherungen, wie zuvor erwähnt, konstant bleiben.

Was die praktische Ausführung der Routinetypisierung von Staphylokokkenstämmen betrifft, so wird jeweils eine Nutrient Broth-(„Difco“-)Agarplatte (pH 7,4) oder ein entsprechender Nährboden mit einer 4—6stündigen Bouillonkultur des zu prüfenden Stammes gleichmäßig durch Übergießen und Absaugen des Flüssigkeitsüberschusses beimpft. Nach Trocknen tropft man dann die einzelnen Phagen in konstanter Reihenfolge (in der RTD) auf. Die Bebrütung erfolgt über Nacht bei 30°. Williams und Rippon benutzen folgende Ablesungssymbole: $++$ = alle Grade von konfluierender Lysis bis mehr als 50 Löcher (starke Lysis), $+$ = 20 bis 50 Löcher (mäßige Lysis), $\pm$ = weniger als 20 Löcher (schwache Lysis), $-$ = keine Lysis. Die Größe der Phagenlöcher bleibt unberücksichtigt. Die Reaktionen der Staphylokokkenteststämme mit den hauptsächlich verwendeten Staphylokokkenphagen sind in Tabelle 22 verzeichnet.

Tabelle 21. *Lytisches Spektrum der Phagen 29, 52, 52 A, austitriert auf den Staphylokokkenstämmen, die von ihnen unverdünnt gelöst werden.* (Nach WILLIAMS)[1]

Teststämme	Phagen			Teststämme	Phagen		
	29	52	52 A		29	52	52 A
3 A	.	.	.	42 D	.	.	.
3 B	.	.	.	42 E	.	.	.
3 C	.	.	.	44 A	1	4	4
6	.	(3)	.	47	3	3	.
7	.	3	.	47 A	.	.	.
29	5	.	.	47 B	.	.	.
29 A	4	3	3	51	.	.	.
31/44	4	4	4	52	.	5	4
31 A	.	.	.	52 A	.	4	5
42 B/47 C	.	3	.	53	.	.	.
42 C	.	.	.	54	.	.	.

Zeichenerklärung: 5 = Maximaltiter auf dem homologen Züchtungsstamm; 4 = 1/10 bis 1/100 des Maximaltiters; 3 = 1/1000 bis 1/10000 des Maximaltiters; 2 = 1/100000 bis 1/1000000 des Maximaltiters; 1 = sehr geringe Lysis; ( ) = starkes sekundäres Bakterienwachstum; . = keine Lysis.

[1] Vergleiche auch die graphische Darstellung bei PÖHN (1955).

Aus der Tabelle 22 ist zu entnehmen, daß die Staphylokokkenstämme in der Mehrzahl nicht nur mit einem einzelnen Phagen reagieren, sondern meistens mit mehreren verschiedenen Phagen. Da aber diese Reaktionen gewissen Variationen unterworfen sind und außerdem bei den Staphylokokkenstämmen eine große Zahl verschiedenster Lysebilder vorkommen kann, ist es nicht möglich, etwa wie bei den Typhusbakterien oder Paratyphusbakterien, gut charakterisierte Typen zu unterscheiden. Vielmehr gibt man für die einzelnen Staphylokokkenstämme nach dem jeweiligen Befund sog. Lysebilder (englisch: phage pattern) an. Um einen Stamm zu kennzeichnen, werden alle Phagen aufgezählt, mit denen er eine starke Lysis gibt. Schwache Lysisreaktionen führt man nicht im einzelnen auf, sondern ersetzt sie durch ein „+", z. B. 7/47 B + (s. Tabelle 25).

WILLIAMS und RIPPON haben die Staphylokokkenphagen nach ihrer Wirkungsweise in Gruppen eingeordnet, wobei diejenigen Phagen zusammengefaßt wurden, die mit den übrigen Phagen möglichst wenige Überschneidungen ergeben (s. Tabelle 22). Sehr deutlich ist die Phagengruppe 3 A, 3 B, 3 C, 55 und 71 (Gruppe II = 3 A-Gruppe) von den anderen Phagen abgegrenzt, d. h. die erstgenannten Phagen wirken im allgemeinen nicht auf Stämme, die von Phagen der anderen Gruppen gelöst werden. Ferner unterscheiden WILLIAMS und RIPPON eine zweite Gruppe mit den Phagen 29, (29 A), 52, 52 A, 79 und 80 (Gruppe I = 52-Gruppe), eine dritte Gruppe mit den Phagen 6, 7, 42 B, 42 E, 47, 47 C, 53, 54 (Gruppe III = 6/47-Gruppe) und eine vierte Gruppe mit dem Phagen 42 D (Gruppe IV = 42 D-Gruppe). Zwischen diesen drei zuletzt genannten Phagengruppen können, wie auch WAHL und FOUACE festgestellt haben, Überschneidungen vorkommen, d. h. ein Staphylokokkenstamm kann beispielsweise von Phagen der Gruppe I wie von solchen der Gruppe III angegriffen werden. Die Gruppeneinteilung der Staphylokokkenphagen und die damit ermöglichte entsprechende Einteilung der Staphylokokkenstämme ist in Tabelle 23 (nach WILLIAMS, RIPPON und DOWSETT 1953) zusammengestellt. In dieser Tabelle sind auch noch neu hinzugekommene Phagen angegeben.

Tabelle 22. *Reaktionsbilder der Züchtungs- und Teststämme mit den Typisierungsphagen des Basissatzes.*
[Nach WILLIAMS und RIPPON (1952) modifiziert]

Gruppe	Staphylokokkenstamm	Testphagen																			
		I				II					III										IV
		29	52	52 A	79	3 A	3 B	3 C	55	71	6	7	42 E	47	53	54	70	73	75	77	42 D
I	29	++	—	—	—	—	—	—	—	—	—	—	—	—	—	—	—	—	—	—	—
	52	—	++	+	—	—	—	—	—	—	—	—	—	—	—	—	—	—	—	—	—
	52 A/79	—	±	++	++	—	—	—	—	—	—	—	—	—	—	—	—	—	—	—	—
II	3 A	—	—	—	—	++	+	+	+	—	—	—	—	—	—	—	—	—	—	—	—
	3 B	—	—	—	—	—	++	++	++	++	—	—	—	—	—	—	—	—	—	—	—
	3 C	—	—	—	—	±	++	++	++	++	—	—	—	—	—	—	—	—	—	—	—
	51*	—	—	—	—	—	—	++	++	—	—	—	—	—	—	—	—	—	—	—	—
	55	—	—	—	—	—	++	++	++	++	—	—	—	—	—	—	—	—	—	—	—
	71	—	—	—	—	—	—	++	++	++	—	—	—	—	—	—	—	—	—	—	—
III	6	—	—	—	—	—	—	—	—	—	++	+	±	++	++	++	+	+	++	+	—
	7	—	—	—	—	—	—	—	—	—	+	++	±	++	++	++	+	++	++	+	—
	42 B/47 C*	—	—	—	—	—	—	—	—	—	—	—	—	—	—	—	—	—	—	—	—
	42 E	—	—	—	—	—	—	—	—	—	—	—	++	—	—	—	—	—	—	—	—
	47	—	—	—	—	—	—	—	—	—	—	—	—	++	++	+	—	—	++	++	—
	47 B*	—	—	—	—	—	—	—	—	—	—	—	—	—	—	—	—	—	—	—	—
	53	—	—	—	—	—	—	—	—	—	—	—	—	—	++	++	—	—	++	++	—
	54	—	—	—	—	—	—	—	—	—	—	++	—	++	++	++	+	+	++	++	—
	70	—	—	—	—	—	—	—	—	—	—	—	—	—	—	—	++	—	—	—	—
	73	—	++	—	—	—	—	++	++	—	++	++	++	++	++	++	++	++	++	++	—
	75	—	—	—	—	—	—	—	—	—	—	—	—	—	—	—	—	—	++	++	—
	77	—	—	—	—	—	—	—	—	—	—	—	—	—	—	—	—	—	—	++	—
IV	42 D	—	—	—	—	—	—	—	—	—	—	—	—	—	—	—	—	—	—	—	++
M	29 A*	+	—	—	—	—	—	—	—	—	—	—	±	—	—	—	—	+	—	±	—
	31/44*	+	+	+	++	—	—	—	—	—	—	±	+	+	—	±	++	++	++	++	—
	31 A*	—	—	—	—	—	—	—	—	—	—	—	—	—	—	—	—	—	—	—	—
	42 C*	—	—	—	—	—	—	—	—	—	—	—	—	—	—	—	+	—	—	—	—
	44 A*	—	+	±	—	—	—	—	—	—	—	++	—	—	—	—	—	—	—	—	—
	47 A*	—	—	—	—	—	—	—	—	—	—	—	—	—	—	—	—	—	—	—	—

++ = starke Lysis; + = mäßige Lysis; ± = schwache Lysis.
Bei den mit * bezeichneten Stämmen handelt es sich um Teststämme, deren homologe Phagen nicht zum Basissatz gehören.

Es hat sich als zweckmäßig herausgestellt, die zu typisierenden Staphylokokkenstämme zunächst mit einem „Basissatz“ von 20 Phagen zu testen[1]. Falls mit diesen Phagen keine Reaktionen auftreten, werden die Stämme mit der 1000fachen RTD geprüft. Vielfach ist es hierdurch möglich, noch zu einem

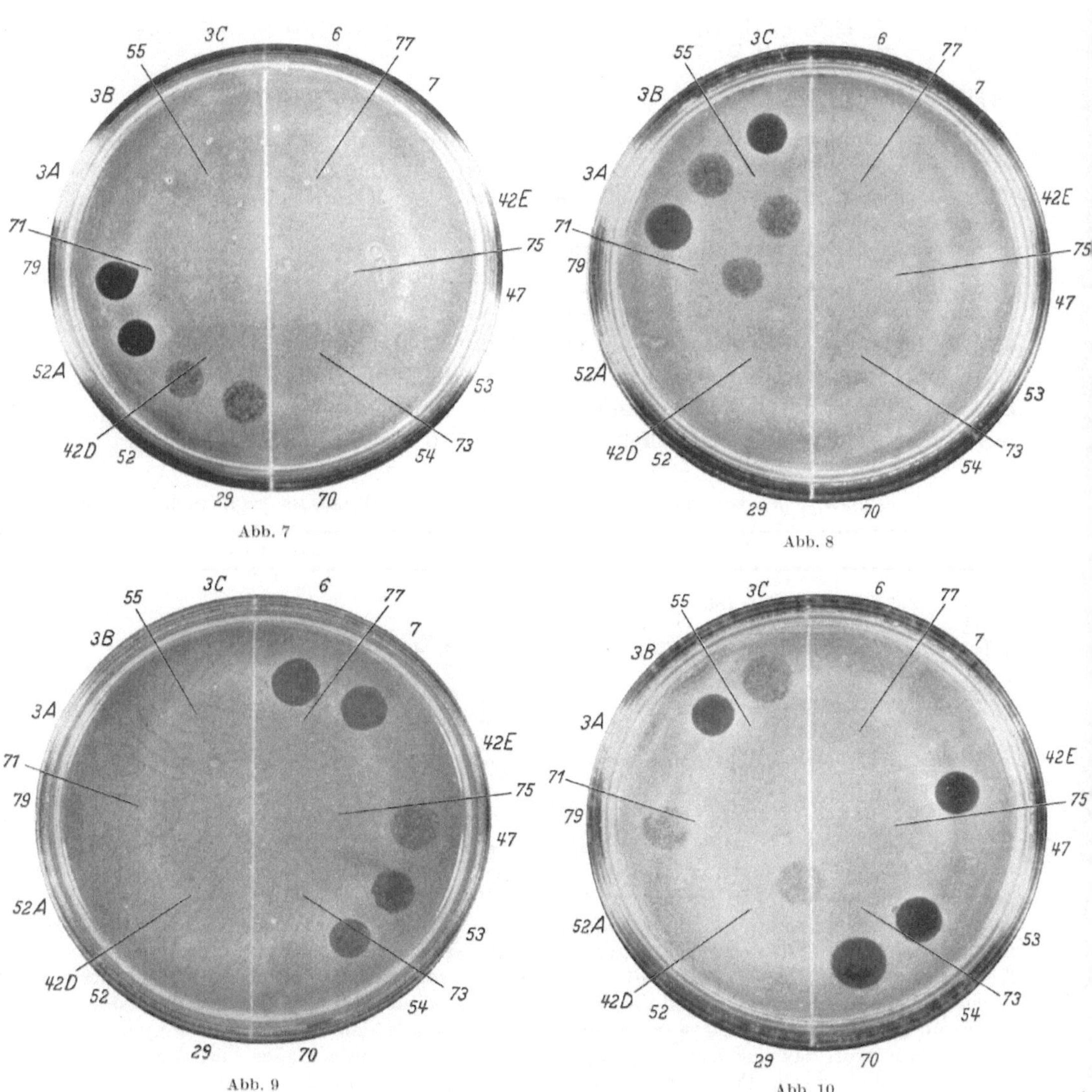

Abb. 7. Staphylokokkenstamm der Lysogruppe I. Lysebild: 29/52/52 A/79. [Die Feldereinteilung für die einzelnen Phagen ist für die photographische Aufnahme fortgelassen]

Abb. 8. Staphylokokkenstamm der Lysogruppe II. Lysebild: 3 A/3 B/3 C/55/71 [s. [ ] Abb. 7)

Abb. 9. Staphylokokkenstamm der Lysogruppe III. Lysebild: 6/7/47/53/54/70 (s. [ ] Abb. 7)

Abb. 10. Nicht für eine Gruppe typisch reagierender Staphylokokkenstamm. Lysebild: 79/3 B/3 C/42 E/54/70 (s. [ ] Abb. 7)

[1] Neuerdings wird der Phage 80 auch noch im Basissatz verwendet.

Tabelle 23. *Einteilung der Staphylokokkenphagen nach Gruppen.* (Nach WILLIAMS, RIPPON und DOWSETT[1])

Gruppe	Phagen, die charakteristisch für die Gruppe sind	Häufig vorkommende Phagenbilder
I	29, 52, 52 A, 79, 80	52 A, 52/52 A, 29/52, 29/31/44/52 +
II	3 A, 3 B, 3 C, 51, 55, 71	3 A, 3 C, 3 B/3 C, 3 C/51, 3 A/3 B/3 C/51, 3 C/55/71
III	6, 7, 42 B, 42 E, 47, 47 B, 47 C, 53, 54, 31 B, 52 B, 70, 73, 75, 75A, 75 B, 76, 77	6/47 +, 6/7/47/53/54 +, 7/47/53/54 +, 53/54, 53/54/75
IV	42 D, 42 F	

\+ Bedeutet, daß noch andere schwache, nicht im einzelnen aufgezeichnete Reaktionen mit weiteren Phagen vorhanden sind.

Die Phagen 7, 42 E, 47 C, 70, 73 und 75 können Lysebilder zusammen mit Phagen der Gruppe I geben.

[1] Nach dem gegenwärtigen Stand berichtigt.

Ergebnis zu kommen. Es werden aber häufiger übergreifende Reaktionen beobachtet, die nach den Beobachtungen von PÖHN (1956) allerdings in geringer Zahl auftreten, wenn statt der 1000fachen RTD nur die 100fache RTD benutzt wird. Erst wenn ein Staphylokokkenstamm auch mit der 1000fachen RTD keine Lysis zeigt, wird er noch mit zusätzlichen Phagen untersucht (Tabelle 24).

Tabelle 24. *Typenphagen von Staphylococcus aureus.* (Nach WILLIAMS [vgl. auch PÖHN])

Gruppe	Basissatz	Zusätzliche Phagen
I	29, 52, 52 A, 79, (80)	
II	3 A, 3 B, 3 C, 55, 71	51
III	6, 7, 42 E, 47, 53, 54, 70, 73, 75, 77	31 B, 42 B, 47 B, 47 C, 52 B, 75 A, 75 B, 76
IV	42 D	42 F
M	—	31, 42 C, 44, 44 A, 57, 58, 69, 78, (81)

( ) = in der letzten Zeit neu angegebene Phagen (ROUNTREE, COMTOIS).
M = „miscellanous".

WAHL und FOUACE (1954) gaben eine Reihe von wichtigen Hinweisen für die Züchtung der Typisierungsphagen und für methodische Fragen, darüber hinaus schlugen die Autoren zwei neue Phagenserien (1. Serie mit den Phagen 735 A, 735 G und 735 E, 2. Serie mit den Phagen 838 A, 838 C, 838 D, 847 B und 841) vor, welche sich zum Klassifizieren von Stämmen, die mit den übrigen Phagen nicht typisierbar sind, eignen sollen. Zur Vereinfachung der Technik empfahlen HOOD (1953) sowie WILLIAMS, RIPPON und DOWSETT, die Phagen der einzelnen Gruppen zu mischen und die Stämme zunächst mit diesen Gemischen zu testen. In einem zweiten Arbeitsgang wird dann das Lysebild mit den Phagen aus demjenigen Gemisch bestimmt, das eine Lysis gegeben hat. Allerdings bedeutet dieses Verfahren einen Zeitverlust. Methodische Fragen haben auch OSWALD und REEDY (1954) sowie GOLDBERG (1954) behandelt.

Da also die bisher verfügbaren Staphylokokkenphagen keine Typenspezifität besitzen, muß man sich mit der Feststellung und der Angabe von individuellen Lysebildern begnügen. Sie machen jedoch gelegentlich die Frage nicht leicht, ob 2 Stämme als identisch oder als verschieden anzusehen sind. Diese Schwierigkeit wurde — vielleicht etwas zu scharf — besonders von VOGELSANG (1953) hervorgehoben, der schrieb: "One and the same strain tested on different days and strains which with great probability derived from a common source have shown so great differences in the lytic effect of the phages that it has been questionable if they should be regarded as belonging to the same phage pattern." Sicher unterscheiden lassen sich im allgemeinen nur solche Kulturen, die zu verschiedenen Gruppen gehören. Stämme aus einer Gruppe dürften nach WILLIAMS und RIPPON dann einen getrennten Ursprung haben, wenn sie sich bei gleichzeitiger Austestung deutlich gegenüber mehr als 2 Phagen verschieden verhalten. Da es nach dem Schema von WILLIAMS und RIPPON keine fest umrissenen Typen gibt, können demnach die Kulturen aus einem einheitlichen Infektionsgeschehen (z. B. bei einer Nahrungsmittelvergiftung) sich nach der Anzahl der Phagen, von denen sie gelöst werden, wie auch nach dem Grade der Lysis unterscheiden (Tabelle 25). Aus der Tabelle 25 geht hervor, daß die Stämme 1—8 einige Reaktionsunterschiede aufweisen, trotzdem aber als einheitlich angesehen werden müssen. Diese Stämme 1—8 kann man von Stamm 9 sicher und mit gewisser Wahrscheinlichkeit von Stamm 10 abgrenzen.

Zusammenfassend läßt sich aus den Untersuchungen von WILLIAMS und RIPPON sowie WAHL und Mitarbeitern sagen, daß bestimmte Staphylokokkenphagen mit einer gewissen Gesetzmäßigkeit gemeinsam auf entsprechende Staphylokokkenstämme wirken, und daß es infolgedessen möglich ist, nach diesem Verhalten einerseits die Phagen, andererseits die Bakterienstämme in Gruppen zu ordnen. Die Gruppen dienen dann als Unterscheidungsmerkmal für die zu prüfenden Kulturen. Innerhalb der Gruppen werden die Stämme durch ihr Lysebild gekennzeichnet, das jedoch, wie schon ausgeführt, nicht mit einem wohlumrissenen Typ gleichgesetzt werden darf. Allerdings kommt es mit den gegenwärtig zur Verfügung stehenden Phagen bei einer Reihe von Stämmen zwischen den Gruppen zu Überschneidungen, so daß nicht in jedem Fall eine Gruppen-

Tabelle 25. *Typisierungsergebnisse von Staphylokokkenstämmen eines Ausbruchs von Lebensmittelvergiftung.* (Nach WILLIAMS und RIPPON)

Nr.		Lysisgrad mit Phagen																Aufgezeichnet als
		3 C	6	7	29	31	42 B	42 C	42 D	42 E	47	47 B	47 C	52	52 A	53	54	
1	Lebensmittel 1	±	±	++	.	.	+	.	.	+	+	++	+	.	.	+	+	7/47 B+
2	Lebensmittel 2	.	±	++	.	.	±	.	.	±	±	+	±	.	.	+	+	7+
3	Lebensmittel 3	.	+	++	.	.	+	.	±	+	+	+	±	.	.	+	+	7+
4	Patient 1, Erbrochenes	.	±	++	.	.	±	.	.	±	++	+	±	.	.	+	+	7/47+
5	Patient 2, Stuhl	.	±	++	.	.	±	.	.	.	.	±	.	.	.	±	±	7+
6	Patient 3, Stuhl	+	+	++	.	±	±	.	.	+	±	++	+	.	.	+	+	7/47 B+
7	Lebensmittelhändler 1	+	+	++	.	.	±	.	.	+	+	++	+	.	.	+	++	7/47 B/54+
8	Lebensmittelhändler 2	±	±	++	±	.	±	.	±	±	±	+	±	.	.	+	+	7+
9	Lebensmittelhändler 3	++	.	.	.	.	.	.	.	.	.	.	.	.	.	.	.	3 C
10	Lebensmittelhändler 4	.	.	++	.	.	.	+	++	++	.	.	.	++	+	.	++	7/42 D/42 E/52/54+

Zeichenerklärung s. S. 136

zuordnung möglich ist. Trotz dieser Einschränkungen hat sich die bisherige Methode der Staphylokokkentypisierung bei der Bearbeitung epidemiologischer Fragen gut bewährt. Hierauf wird noch zurückzukommen sein.

Hinsichtlich der Vergleichbarkeit der bei den Typisierungen erhaltenen Befunde ist zu berücksichtigen, daß von den einzelnen Autoren nicht immer das gleiche Verfahren angewendet wird, z. T. werden auch andere Phagen benützt. So arbeitet G. WALLMARK (1949, 1954) in Schweden mit einem Satz von 14 Phagen. Neun von diesen stammen von WILLIAMS und RIPPON, während die restlichen (KS 6, 819, 1034, 166 und 155) durch WALLMARK von lysogenen

Tabelle 26. *Lysebilder der verschiedenen Staphylokokkentypen.* (Nach WALLMARK)

Lysotyp	Bakteriophagen													
	3 B	3 C	51	6	7	42 B	47	47 A	47 B	KS 6	819	1034	166	155
1 A	+++	+++											+	
1 B	+++	+++	+++											
6		+++												
9			+++											
2 A				+++	+	+	+++		+		+++	+++	++	++
2 C				+++	+	+	+++		+				+	+
3				+	+++		+				+	+	+	+
4						+++				+++				
5 A							+++		+				+++	+++
5 B					+	+	+++		+		+++	+++		
7								+++						
8									+++					
11 A				++	+	+	++		+			+++	++	++
11 B							++		+			+++	+	+
10										+++			+	+
12													+++	+++
13														+++

Die Typen 1 A, 1 B, 6 und 9 bilden die Lysogruppe II (3 A-Gruppe), die Typen 2 A, 2 C, 3, 4, 5 A, 5 B, 7, 8, 11 A und 11 B die Gruppe III (6/47 Gruppe).

+++ = konfluierende Lysis; ++ = halbkonfluierende Lysis oder sehr zahlreiche Einzellöcher; + = bis zu 50 Einzellöcher.

Stämmen isoliert wurden. Mit diesen 14 Phagen in der RTD ergibt sich ein Schema, das 17 „Staphylokokkenlysotypen“ umfaßt. Es sind dies die Lysotypen 1—13 mit den Subtypen 1 A, 1 B, 2 A, 2 C, 5 A, 5 B, 11 A und 11 B. Wie Tabelle 26 zeigt, werden die einzelnen Lysotypen im Schema von WALLMARK ebenfalls wie in demjenigen von WILLIAMS und RIPPON durch Lysebilder gekennzeichnet, d. h. die meisten Typen geben mit mehreren Phagen eine Reaktion. Der Unterschied zu dem Schema von WILLIAMS und RIPPON besteht aber darin, daß die einzelnen Lysebilder jeweils für einen ganz bestimmten Lysotyp charakteristisch sind, und es muß das Typ-Lysisbild ohne Abweichungen vorliegen, wenn man eine Staphylokokkenkultur einem der 17 Typen zuordnen will. Stämme, die auf Grund ihrer Phagenreaktionen keinem der 17 Typen entsprechen, werden als „nichtspezifischer Typ“ bezeichnet. Hinsichtlich der Vergleichbarkeit mit der Gruppeneinteilung von WILLIAMS und RIPPON sei noch angefügt, daß die Lysotypen 1 A, 1 B, 6 und 9 in die Gruppe II von WILLIAMS und RIPPON fallen und die Lysotypen 2 A, 2 C, 3, 4, 5 A, 5 B, 7, 8, 11 A und 11 B in die Gruppe III. Im Phagensatz von WALLMARK sind Phagen der Gruppe I nicht vertreten. Ob vergleichende Untersuchungen zwischen der Methode von Williams und Rippon und derjenigen von Wallmark an einer größeren Zahl von Stämmen vorgenommen worden sind, ist nicht bekannt. Zweifellos dürfte dies aber eine lohnende Aufgabe sein.

Blair und Carr (1953) in den USA benutzen für ihr Verfahren die 21 Phagen von Wilson und Atkinson sowie 4 selbst isolierte Phagen (39, 142, 523 und VA 4). Die Autoren unterscheiden 6 Staphylokokkengruppen („phage types"), die durch entsprechende 6 Serien von Staphylokokkenphagen erkennbar sind:

Staphylokokkengruppe 1	Phagen: 3 A, 3 B, 3 C, 51, 39, 523
Staphylokokkengruppe 2	Phagen: 6, 7, 42 B, 42 C, 42 D, 42 E, 47, 47 A, 47 B, 47 C, VA 4
Staphylokokkengruppe 3	Phagen: 29, 29 A, 31, 52, 52 A
Staphylokokkengruppe 4	nimmt Zwischenstellung unter den Gruppen 2 und 3 ein
Staphylokokkengruppe 5	Phagen: 44, 44 A
Staphylokokkengruppe 6	Phage: 142

Staphylokokkenstämme, die in die Gruppen 1, 2 und 3 fallen, sind mit den „Typen" 1—3 von Wilson und Atkinson vergleichbar. Stämme, die sowohl von den Phagen der 2. Serie als auch der 3. Serie angegriffen werden, ordnen Blair und Carr versuchsweise in eine gesonderte Gruppe (Gruppe 4) ein. Die Kennzeichnung der einzelnen Stämme innerhalb der Gruppen erfolgt wie bei Williams und Rippon nach dem jeweiligen Lysebild.

Die Feststellung, daß mit Hilfe der verschiedenen Testphagenserien eine Gruppeneinteilung der Staphylokokken möglich ist, legte den Gedanken nahe zu prüfen, inwieweit hier Beziehungen zu dem serologischen Verhalten der Staphylokokken aufzudecken sind. Hobbs (1948) zeigte, daß in der Tat gewisse, jedoch keine vollständigen Übereinstimmungen bestehen. So wurden Stämme des serologischen Typs I vor allem durch die Phagen 52, 29, aber auch durch den Phagen 47 C angegriffen. Darüber hinaus waren 21 (= 40,4%) der geprüften Kulturen des serologischen Typs I mit Staphylokokkenphagen nicht klassifizierbar. Stämme des serologischen Typs II gaben mit den Phagen 3 B und 3 C und nur gelegentlich mit dem Phagen 3 A eine Lysis. Der 3 A-Phage löste auch Kulturen, die zu dem von Hobbs aufgestellten serologischen Typ 7163 gehörten, obwohl dieser serologisch von dem Typ II verschieden ist. Auf die Stämme des serologischen Typs III wirkten vor allem nur die Phagen 6 und 47. Für die anderen serologischen Typen konnten keine Beziehungen zu Phagenreaktionen gefunden werden. Wahl und Fouace sowie Blair und Carr stellten ebenfalls gewisse Übereinstimmungen zwischen den ersten 3 Gruppen der Staphylokokkenstämme und den serologischen Typen I—III von Cowan fest. Die zwei erstgenannten Autoren schlugen daraufhin vor, bei den durch die Typenphagen nachweisbaren Staphylokokkengruppen eine mit den drei serologischen Typen Cowans übereinstimmende Numerierung vorzunehmen.

Serologische Gruppe	Staphylokokken-lysogruppe	Phagengruppe
I	I	29, 52 usw.
II	II	3 A, 3 B usw.
III	III	6, 7 usw.

Die Frage der Beziehungen zwischen Phagenreaktionen und serologischen Typen ist eingehend von Oeding (1953) sowie Oeding und Vogelsang (1954) geprüft worden, welche auch die hitzelabilen Staphylokokkenantigene berücksichtigten. Von 846 Stämmen pyogener Staphylokokken waren mit Phagen 63%

typisierbar gegenüber 94,6% mit der serologischen Methode. Es ergab sich, daß klare Beziehungen zwischen einem serologischen Typ und einer Lysogruppe nicht bestehen. Die Autoren schreiben: "One serological group might thus have one phage reaction belonging to phage group I another belonging to group II and a third belonging to group III, although one of these reactions might be in majority among the strains. This shows that a strict correlation between the three serological groups of COWAN and the three phage groups 52, 3 A and 6/47 is untenable. A certain correlation may exist, but this does certainly not justify an identification of the groups of the two systems." Kurz zusammengefaßt fanden OEDING und VOGELsang, daß in COWANS Gruppe I (nach OEDING und VOGELSANG a b e) viele untypisierbare Staphylokokkenstämme (vgl. HOBBS), ferner solche, die zur Lysogruppe III oder gelegentlich zur Lysogruppe II und nur relativ wenige Stämme zur Lysogruppe I gehörten. Die Stämme der serologischen Gruppe II (a b) waren vielfach nicht typisierbar oder fielen in die Lysogruppe III, nur ein kleiner Teil zählte zur Lysogruppe II. Eine gute Übereinstimmung fand sich dagegen zwischen den Stämmen der Lysogruppe III und COWANS serologischer Gruppe III (a b c).

Bei der Beurteilung der von den verschiedenen Autoren erhaltenen, zum Teil widersprechenden Resultate über die Beziehungen zwischen serologischem Verhalten und Phagenempfindlichkeit von Staphylokokkenstämmen ist zu berücksichtigen, daß die Autoren nicht immer die gleiche Untersuchungsmethode angewandt haben. Ferner darf man auch den einzelnen Staphylokokkenphagen, worauf WAHL und FOUACE hinweisen, nicht den gleichen spezifischen Wert für eine Gruppeneinteilung zusprechen. Nach PÖHN (1956) zeigen von den Phagen des Basissatzes insbesondere die Phagen 7, 42 E, 73 und 75 eine geringere Gruppenspezifität als die übrigen Testphagen.

Was die Serologie der zum Typisieren verwendeten Bakteriophagen betrifft, so zeigte ROUNTREE (1949), daß die Phagen 3 A, 3 B, 3 C, 14, 51, 6, 7, 42 B, 47 C, 42 E, 47, 47 A sowie 47 B in die von ihr aufgestellte serologische Gruppe A und die Phagen 29, 29 A, 31, 31 A, 44, 44 A, 52, 52 A, 42 C, 42 D in die serologische Gruppe B einzuordnen waren. Mit anderen Phagen stellte ROUNTREE ferner die serologischen Phagengruppen C, D, E und F auf, von denen die Gruppe F heute die Phagen 76, 77, 42 D und 58 einschließt. RIPPON (1952) fügte dann noch die Gruppe G, welche die Typenphagen 65 und 66 WALLMARKS sowie den Phagen 68 von WAHL umfaßt, hinzu. Die Autorin gab ferner (1956) einen eingehenden Überblick über die Einteilung aller bisher bekannten Staphylokokkenphagen und ihre serologische Klassifizierung, nach der z. Z. die serologischen Gruppen A, B, F, C, D, G, H, E, J und K zu unterscheiden sind.

Die Phagen der serologischen Gruppen A, B, F und G lösen ausschließlich coagulasepositive, dagegen die Phagen der serologischen Gruppen E, J und K nur coagulasenegative Staphylokokkenstämme. Elektronenmikroskopische Untersuchungen über den Phagen 3 A haben FARRANT und ROUNTREE (1953) angestellt.

Nach Lysogenisation von Staphylokokkenstämmen mit den zum Typisieren benutzten Phagen verändert sich diesen gegenüber die Empfindlichkeit der Stämme, d. h. es kommt zu andersartigen Lysebildern (WILLIAMS, SMITH 1948, LOWBURY und HOOD 1953, ROUNTREE 1956).

Über die Bedeutung der Staphylokokkentypisierung für die Aufklärung epidemiologischer Fragen, wie beispielsweise die Verteilung von Staphylokokken bei Gesunden, Kranken und Krankenhauspersonal einschließlich der Rolle von Staphylokokkenausscheidern, der Zusammenhang von Antibioticaresistenz und bestimmten Staphylokokkenlysebildern, die Übertragungsweise der Staphylokokken und vieles andere, liegen bereits zahlreiche Mitteilungen vor, von denen in diesem Überblick nur ein kleiner Teil angeführt werden kann.

Im allgemeinen läßt sich sagen, daß etwa 60% aller Staphylokokkenstämme typisierbar sind. Die untypisierbaren Stämme sind zum Teil bei Verwendung der 1000- oder 100fachen RTD (Wahl) noch klassifizierbar, jedoch treten bei Verwendung von konzentrierten Phagenlösungen vielfach auch stärker übergreifende Reaktionen auf. In Tabelle 27 ist die Verteilung der Staphylokokken auf die Lysogruppen I—III nach den Angaben einer Reihe von Autoren

Tabelle 27. *Verteilung der Staphylokokkenstämme auf die Lysogruppen I—III*

Autor	Zahl der geprüften Stämme	Prozentsatz der typisierbaren Stämme	Lysogruppe			Bemerkungen
			I	II	III	
Vogelsang	858	62,9	4,0	9,5	49,4	Krankenhauspersonal
Blair und Carr	539	53,6	20	13	14	
Williams, Rippon und Dowsett	1349	66,5	21,7	11,8	33	
Fusillo und Mitarbeiter	485	73	7,6	2,5	62,9	Stämme aus klinischem Untersuchungsmaterial
Jackson und Mitarbeiter	541	35,2	5,0	1,5	16,9	
Pöhn	2361	82,3	50,1	7,2	24,8	Einschließlich Verwendung der Phagen in der 1000fachen RTD
Gould und McKillop	460	62	25	15	23	Medizinstudenten
McLean (1956)	262	53 (90)[1]	5	6,5	55,7	Krankenhauspatienten
Rountree und Rheuben (1956)	96	94,8	47,9[2]	26,0	14,6	Krankenhauspatienten

In der Tabelle sind z. T. nicht die Originalzahlen der Autoren angegeben, sondern die Werte wurden umgerechnet auf den prozentualen Anteil innerhalb der Gesamtzahl an untersuchten Stämmen.

[1] Bei zusätzlicher Verwendung von unverdünnten Phagen.

[2] Vorwiegend Typ 80.

zusammengestellt. Allerdings lassen die angegebenen Werte nur einen relativen Vergleich zu, da Untersuchungstechnik, Auswahl der Stämme u. a. nicht völlig übereinstimmen.

Wie Wallmark betont, hängt die Verteilung der „Staphylokokkentypen" sehr von der Herkunft der Stämme ab, d. h. ob diese von Personen der Durchschnittsbevölkerung oder aber von Patienten aus Krankenhäusern stammen. Im erstgenannten Fall herrschen keine besonderen Reaktionsweisen mit den Staphylokokkenphagen vor (s. auch Gould und McKillop). Bemerkenswert ist die große Zahl der Stämme, die zur Gruppe III zählen, im Material von Vogelsang, der Kulturen von Pflegepersonal aus 3 Krankenhäusern in Norwegen untersuchte. Vogelsang fand, daß von seinen Stämmen der Gruppe III 68% penicillinresistent waren, in Gruppe I blieben dagegen nur 35% und in Gruppe II 15% von Penicillin unbeeinflußt. Die Beobachtung der Häufung von penicillinresistenten Stämmen in der Lysogruppe III ist ebenfalls von einer Reihe anderer Autoren gemacht worden (Barber und Mitarbeiter 1949, Rountree und Thomson 1949, Elwood 1951, Rountree 1953, Williams und Mitarbeiter 1953, Wallmark 1954, Fusillo und Mitarbeiter 1954, Jackson und Mitarbeiter

1954 u. a.). FUSILLO und Mitarbeiter wiesen z. B. nach, daß unter 362 penicillinresistenten Stämmen 3,3% zur Gruppe I, 0,3% zur Gruppe II, dagegen 74,5% zur Gruppe III (6/47) gehörten. WILLIAMS und Mitarbeiter (1953) fanden, daß penicillinresistente Stämme in 17% zur Gruppe I, in 13% zur Gruppe II und in 35% zur Gruppe III gehörten, 35% waren unbestimmbar (in der RTD). WALLMARK (1954) wies unter 1775 Staphylokokkenstämmen aus der Phagengruppe III in 94,3% penicillinresistente Kulturen nach. Aufschlußreich ist auch die Tabelle von BARBER und WHITEHEAD über das Phagenverhalten penicillinresistenter Staphylokokkenstämme aus verschiedenen Krankenhäusern Englands (Tabelle 28).

Tabelle 28. *Phagenreaktionen penicillinresistenter Staphylokokkenstämme von Krankenhauspatienten.* (Nach BARBER und WHITEHEAD)

Lysebilder	Hammer-smith-Hospital	St. Thomas-Hospital	General Lying-in Hospital S.E. 1	West London-Hospital	Kent- und Canterbury-Hospital	St. Mary-Hospital Manchester	Gesamt-zahl
6/7/47	25	27	0	1	27	5	85
52 A	1	4	15	0	1	0	21
42	3	0	0	0	0	0	3
3	1	0	0	0	0	0	1
29/31/52	1	0	0	0	0	0	1
Nicht typisierbar	3	8	13	0	4	1	29
Gesamtzahl	34	39	28	1	32	6	140

Von den 140 untersuchten Stämmen fielen, wie Tabelle 28 zeigt, 85 (= 61%) in die 6/47-Gruppe. Es können aber auch penicillinresistente Stämme der Gruppen I und II oder untypisierbare Kulturen in einem Krankenhaus überwiegen (RUYS und WILLEMS 1955, s. auch WILLIAMS und Mitarbeiter 1953, BARBER, WILSON, RIPPON und WILLIAMS 1953). Es sei in diesem Zusammenhang auf die Übersicht von ANDERSON und WILLIAMS (1956) verwiesen, welche die Ergebnisse verschiedener Autoren über die Verteilung penicillin- oder gegen andere Antibiotica resistenter Staphylokokkenstämme auf die einzelnen Lysogruppen in einer Tabelle zusammengestellt haben.

Die Frage, ob bestimmte Staphylokokken bei typischen Krankheitsbildern vorherrschen (Furunkulose, Mastitis, Pemphigus neonatorum u. a.), ist noch nicht sicher zu beantworten, denn zweifellos können die coagulasepositiven Staphylokokken aller Gruppen die für diese Keime charakteristischen Krankheitserscheinungen hervorrufen. Allerdings wiesen PARKER und Mitarbeiter (1955) sowie BARROW (1955) nach, daß beim Impetigo contagiosa der „Staphylokokkentyp" 71 (Gruppe II) eindeutig dominiert. ROUNTREE (1953) und WALLMARK (1954) fanden ein Vorherrschen der Stämme aus der Phagengruppe II bei Furunkulose. Dies konnte jedoch von WILLIAMS, RIPPON und DOWSETT nicht bestätigt werden. Beim Pemphigus neonatorum wurden ebenfalls Staphylokokken der verschiedenen Gruppen als Erreger gefunden (ALLISON: Gruppe II, PARKER und KENNEDY: Gruppe II, WILLIAMS und Mitarbeiter: vor allem Gruppe III, ALLISON und HOBBS: Gruppe I). Innerhalb eines Krankenhauses kann es natürlich auch einmal zur Ausbreitung hauptsächlich nur eines „Typs" kommen (DENTON und Mitarbeiter 1950 u. a.). Bemerkenswert ist die Beobachtung von ROUNTREE (1955), daß im Jahre 1954 bei Patienten im Royal Prince Alfred-Hospital in

Sydney ein neuer Staphylokokken-„Typ" (80) in zunehmendem Maß auftrat, welcher offenbar eine besonders große Virulenz besaß und Epidemien von Furunkulose unter den Krankenhausangehörigen hervorgerufen hat, sowie darüber hinaus bei 19 von 24 Ausbrüchen von Hauteiterungen bei Säuglingen in verschiedenen Krankenhäusern Australiens angetroffen wurde. 94% der Kulturen dieses „Typs" waren penicillinresistent. Ähnliche Feststellungen machten BYNOE und Mitarbeiter (1956) in Kanada. Die Autoren gewannen durch Anpassung des Phagen 42 B an einen mit den Routinetestphagen unbestimmbaren Staphylokokkenstamm einen neuen zusätzlichen Typisierungsphagen „81" (s. Tabelle 24). Sie züchteten in rund 50% aller Staphylokokkenstämme von Patienten mit Furunkulose oder Abscessen in dem von ihnen untersuchten Krankenhaus Kulturen, die von diesem Phagen gelöst wurden. Ob den untypisierbaren Staphylokokkenstämmen eine geringere Pathogenität zukommt als den typisierbaren, wie GOULD und MCKILLOP vermuten, so daß die Phagentypisierbarkeit ein Kriterium der Pathogenität sein würde, bedarf noch weiterer Untersuchungen (s. auch ROUNTREE und RHEUBEN 1956).

Bei einer Staphylokokkenerkrankung besteht ein weitgehender Zusammenhang zwischen den gefundenen Staphylokokkenstämmen und einer bestimmten Lysogruppe: Es ist dies die Nahrungsmittelvergiftung, ausgelöst durch enterotoxinbildende Staphylokokkenstämme. Die hierbei isolierten Kulturen gehören fast ausschließlich zur Gruppe III (ALLISON 1949, WILLIAMS, RIPPON und DOWSETT 1953, WALLMARK 1954, WILSON und ATKINSON 1945, ODDY und CLEGG 1947, MILLAR und POWNALL 1949, SAINT-MARTIN und Mitarbeiter 1951, CROWE 1954). Gerade bei den Nahrungsmittelvergiftungen erwies sich die Phagenmethode zur Aufklärung epidemiologischer Gegebenheiten als sehr brauchbar. So konnten beispielsweise bei dem von ODDY und CLEGG beschriebenen Ausbruch durch Büchsenfleisch Staphylokokken des Reaktionsbildes 47/47 C an der Hand eines Arbeiters, der das Fleisch hergestellt hat, nachgewiesen werden. Bei der Nahrungsmittelvergiftung, die MILLAR und POWNALL beschrieben, litt ein Koch an einer leichten Dermatitis der Hand. Der hiervon isolierte Staphylokokkenstamm hatte das Lysebild 6/47/53/54. Von der Wurst, die dieser Koch zubereitet hatte und welche als die Ursache der Nahrungsmittelvergiftung angesehen werden mußte, sowie aus dem Erbrochenen von Patienten konnten Stämme mit genau der gleichen Phagenreaktion isoliert werden. ALLISON (1949) fand, daß 30 Stämme (= 64%) von insgesamt 47 verschiedenen Ausbrüchen zu dem Reaktionsbild 6/47 gehörten, 8 andere Stämme ließen sich als Typ 42 D, der tierischen Ursprungs[1] ist und bei boviner Mastitis und in roher Kuhmilch häufig vorkommt (WILLIAMS SMITH), identifizieren. Es fielen demnach 81% der Stämme von 47 Ausbrüchen in die zuvor genannten Lysogruppen. Nach WILLIAMS (s. ANDERSON und WILLIAMS 1956) waren von 93 Nahrungsmittelvergiftungen 88mal die ursächlichen Staphylokokkenstämme typisierbar; 77 davon gehörten zur Gruppe III und 6 zur Gruppe IV (42 D), 2 Stämme wurden von den Phagen der Gruppen I und III gelöst, 2 Stämme zählten zur Gruppe II und einer zur Gruppe I. Demnach gehört weitaus die Mehrzahl der nahrungsmittelvergif-

[1] Hinsichtlich der Typisierungsergebnisse bei Staphylokokken tierischen Ursprungs siehe WILLIAMS SMITH (1948), MACDONALD (1946), LUDLAM (1952), PRICE und Mitarbeiter (1954), LEVY, WILLIAMS und RIPPON (1953) u. a.

tenden Staphylokokken zur Gruppe III. Wie WILLIAMS hervorhebt, ist damit aber nicht ausgesagt, daß jeder Staphylokokkenstamm, der zur Gruppe III gehört, ein Enterotoxin bildet. Auf die Bedeutung der Staphylokokken aus der Gruppe III für Nahrungsmittelvergiftungen wiesen erst kürzlich wieder ANDERSON und STONE (1955) hin, die 4 Ausbrüche, hervorgerufen durch getrocknetes Rahmmilchpulver, welches Staphylokokken mit dem Lysebild 42 E/53 w („weak reaction“) enthielt, beschrieben.

Epidemiologisch wertvolle Ergebnisse brachte die Phagenmethode besonders bei Untersuchungen in Krankenhäusern und geburtshilflichen Stationen, in denen die Frage der Staphylokokkenträger und ihre Bedeutung für das Zustandekommen von Infektionen eine wichtige Rolle spielt. BARBER und Mitarbeiter fanden, daß über die Hälfte des Pflegepersonals Träger von meist penicillinresistenten Staphylokokken auf den Schleimhäuten der Nase ist. Auf geburtshilflichen Stationen lassen sich derartige „Krankenhausstämme“ relativ schnell nach der Krankenhausaufnahme bei den Müttern nachweisen. Auch in einem hohen Prozentsatz siedeln sich die Staphylokokken auf den Nasenschleimhäuten der Säuglinge an, die nach den Feststellungen von BARBER die Staphylokokken von dem Pflegepersonal leichter aufnehmen als von ihren Müttern (vgl. BARBER und BURSTON 1955). LUDLAM fand auf geburtshilflichen Stationen oft eine Übereinstimmung im Lysisverhalten bei Staphylokokken, die aus Zimmerstaub isoliert waren und Kulturen, die von Nasenabstrichen der Säuglinge stammten. Die große Schnelligkeit der Besiedlung der Nasenschleimhäute von Säuglingen mit Staphylokokken stellte auch WALLMARK (1954) fest: 34 von 39 Säuglingen in einer geburtshilflichen Station beherbergten Staphylokokken auf den Schleimhäuten der Nase oder des Rachens; die Säuglinge hatten sich die Infektion innerhalb der ersten 4 Lebenstage zugezogen. Bei Erwachsenen finden sich auf der Nasenschleimhaut Staphylokokken bei 30—60%, ohne daß klinische Erscheinungen vorliegen. Staphylokokken, die von der Haut isoliert werden, haben vielfach die gleiche Phagenreaktion wie diejenigen aus der Nase (WILLIAMS 1946). So isolierte WILLIAMS bei 65 Personen Staphylokokkenstämme von der Hand und der Nase. Bei 36 Personen waren die Kulturen in beiden Fällen typisierbar, und sie gaben in 31 Fällen die gleiche Phagenreaktion. WILLIAMS zeigte ferner, daß die Gegenwart von Staphylokokken in der Nase bei einzelnen Menschen nur sporadisch ist, während andere Menschen Ausscheider über lange Zeit sind. Bei diesen chronischen Ausscheidern stellten GOULD und MCKILLOP (1954) bei mehreren Nasenabstrichen im Verlauf eines Jahres bei den einzelnen untersuchten Personen Stämme mit jeweils immer dem gleichen Lysebild fest, nur bei 2% waren im Lauf der Beobachtungszeit verschiedenartige Staphylokokkenstämme nachzuweisen. Schließlich sei noch erwähnt, daß meistenteils bei rezidivierender Staphylokokkenerkrankung (z. B. Furunkulose) alle von den einzelnen Entzündungsherden zu verschiedenen Zeiten gezüchteten Kulturen dieselben Phagenreaktionen geben und somit als übereinstimmend zu betrachten sind (ROODYN 1954).

V. Die Lysotypie bei anderen Bakterienarten

SONNENSCHEIN (1929), BURNET und MCKIE (1930), CLAUBERG und MARCUSE (1932), SARTORIUS und REPLOH (1932), RADOJČIĆ (1936), MILLER (1937), DUNLOP (1943), THOMEN und FROBISHER (1945) u. a. wiesen nach, daß sich mit geeigneten

Phagen eine Einteilung der Dysenteriebakterienstämme ergibt, welche Beziehungen zu dem serologischen Verhalten der Kulturen aufweist. HAMMARSTRÖM (1947, 1949) zeigte ferner die Möglichkeit, bei Shigella sonnei mit 12 nichtadaptierten, auf die kritische Testverdünnung eingestellten Phagen 68 Lysotypen zu unterscheiden, von denen jedoch 37 nur einmal festgestellt wurden. Für die Typisierung erwiesen sich vor allem Kulturen in der R-Form als geeignet. Die Typenstabilität war befriedigend, aber nicht absolut. Diese Beobachtung machte ebenfalls MAYR-HARTING (1952), der auch experimentelle Typenumwandlungen gelangen. Über Ergebnisse der Lysotypie von Shigella sonnei mit den Phagen von HAMMARSTRÖM berichteten ferner COOPER und MAYR-HARTING (1951), RASKA und Mitarbeiter (1950) sowie LUDFORD (1953). TEE (1955) bestätigte mit selbstgewonnenen Phagen die Möglichkeit der Typisierung von Shigella sonnei. Der Autor mißt dieser Methode jedoch nur einen geringen praktischen Wert zu, da die Mehrzahl aller Stämme zum gleichen Lysotyp gehört und die Typen in vivo nicht immer stabil sind.

Zur Typeneinteilung von Ps. pyocyanea veröffentlichte WARNER (1950) ein Typisierungsschema. Dieser Autor teilte 68 Stämme von Ps. pyocyanea mit 19 Phagen in 19 Gruppen bzw. Typen ein. MAYR-HARTING (1949) wies nach, daß deutliche Beziehungen zwischen Phagenreaktionen und serologischem Verhalten der geprüften Pyocyaneusstämme bestehen. Auch VAN DEN ENDE (1952) kam zu ähnlichen Feststellungen.

Zu recht interessanten und auch epidemiologisch verwertbaren Resultaten haben die Untersuchungen von NICOLLE, LE MINOR, BUTTIEAX und DUCREST (1952) über die Typendifferenzierung der Dyspepsiecolibakterien 0111:B 4, 055:B 5 und 026:B 6 geführt, welche von den Autoren mit 14 ausgewählten, zum Teil aus Kinderstühlen, Abwasser oder anderen Quellen stammenden unverdünnten oder nur gering verdünnten Bakteriophagen, deren Wirkung nicht auf Dyspepsiecolibakterien beschränkt ist, vorgenommen wird. Durch die Phagen 4, 10, 28, 22, 42, 32 und 36 sind bei den Colistämmen 0111:B4 7 Lysotypen (Montparnasse, Sèvres, Lille, Dorf, Tourcoing, Vienne, Bretonneau) erkennbar. Ebenfalls 7 Lysotypen (Weiler, Flandre, Graz, Londres, Lomme, Béthune, Vaugirard) werden bei den Colibakterien 055:B 5 mit den Phagen 46, 20, 48, 28, 16, 38 und 14 unterschieden. Bei den Colibakterien 026:B 6 haben die Autoren 5 Lysotypen (Birmingham, La Butte, Styrie, Morin, Warwick) aufgefunden. Sie benutzen für deren Differenzierung die Phagen 44, 10, 42, 20 und 36. NICOLLE und Mitarbeiter sehen das von ihnen aufgestellte Lysotypieschema der Dyspepsiecolibakterien noch nicht als endgültig an. Es ist auch inzwischen bereits modifiziert worden (NICOLLE und Mitarbeiter 1954[c], 1956[c]: 10 Lysotypen bei E. coli 0111:B 4, 10 Lysotypen bei E. coli 055:B 5 und 7 Typen bei E. coli 026:B 6)[1]. Die Autoren führen eine gelegentlich beobachtete Inkonstanz im Lysotyp bei mehreren Stämmen von ein und demselben Patienten auf die noch bestehende Unvollkommenheit der Technik, Degradierung der Kulturen infolge zu langer Aufbewahrung vor der Typisierung oder auf den Einfluß von exogenen Bakteriophagen zurück. Zwischen dem serologischen und biochemischen Verhalten einerseits und den Ergebnissen der Lysotypie andererseits bestehen gewisse Übereinstimmungen (NICOLLE und Mitarbeiter

[1] Die neu hinzugekommenen Typen heißen bei E. coli 0111: Clichy, Paris, Palerme; bei E. coli 055: St. Christopher, Finlande, Jerusalem, Rostock; bei E. coli 026: Zürich, Liège.

1954, 1956[c]). So zeigen die Lysotypen Montparnasse, Sèvres, Tourcoing und Vienne von E. coli 0111 : B 4 immer einen positiven Ausfall des β-phenyl-propionsäure-Testsnach d'Alessandro und Comes und haben das Geissel-Antigen H 2; die anderen Lysotypen dieser Serogruppe geben dagegen eine negative Reaktion nach d'Alessandro und weisen andere H-Antigene auf oder sind unbeweglich. Ähnliche Verhältnisse finden sich auch bei E. coli 055:B 5. Hier reagiert nur der Lysotyp St. Christopher, der bisher ausschließlich bei Kulturen aus England festgestellt wurde, nach d'Alessandro positiv und besitzt das Geissel-Antigen H 2. Zweifellos lassen sich mit der Methode von Nicolle tiefere Einblicke in die Epidemiologie der Dyspepsiecolibakterien gewinnen. In den einzelnen Kinderkrankenhäusern herrschen zum Teil untereinander völlig verschiedene Lysotypen vor. Während wir z. B. in der Kinderklinik in Mainz die Typen Tourcoing (0111 : B 4) und Lomme (055:B 5) beobachtet haben, treten in der Kinderklinik in Frankfurt a. M. die Typen Sèvres (0111:B 4) und Béthune (055:B 5) auf[1]. Auch Eörsi und Mitarbeiter (1954) haben auf die Möglichkeit der Typisierung von Dyspepsiecolibakterien mit Phagen hingewiesen.

Unter den von Nicolle verwendeten Phagen gibt es einige, die nur jeweils einen serologischen Colityp (z. B. nur Stämme der Antigenstruktur 0111:B 4) angreifen. Es erscheint daher möglich, daß diese Phagen spezifisch auf das B-Antigen eingestellt sind. Toft (1947) zeigte, daß bei den Colibakterien solche K-Phagen vorkommen, deren Wirkung von dem Vorhandensein eines bestimmten K-Antigens (A oder L) abhängt; eine Feststellung, die der Auffindung von typenspezifischen Phagen bei Klebsiellen durch Rakieten und Mitarbeitern (1940) entspricht.

Bei einer Reihe von anderen Bakterienarten sind ebenfalls Versuche einer Typendifferenzierung mit Phagen unternommen worden, die jedoch bisher nur vorläufiger Natur sind und noch keine praktische Bedeutung erlangt haben. Es sollen abschließend hier nur kurz die Untersuchungen von Keogh, Simmons, und Anderson (1938), Hewitt (1952), Toshach (1950) sowie Fahey (1952) bei Diphtheriebakterien erwähnt werden, wobei der letztgenannte Autor nach dem Lysisbild mit 5 ausgewählten Phagen 9 Lysotypen feststellte. Mit der Differenzierung der Mycobakterien beschäftigten sich vor allem Penso und Mitarbeiter (1949), Hnatko (1953, 1954, 1956), Froman und Bogen (1953) sowie Froman, Drake und Bogen (1954). Froman und Mitarbeiter beschrieben Phagen, die neben einer Wirkung auf säurefeste Saprophyten auch virulente Tuberkelbakterien angreifen, und unterschieden bei diesen nach der Phagenempfindlichkeit 4 Gruppen. Rifkind und Pickett (1954) zeigten, daß bei Past. multocida mit Phagen aus lysogenen Kulturen eine Einteilung in Lysotypen möglich ist, während Gunnison und Mitarbeiter (1951) nachwiesen, daß ein Pestphage („P"-Phage) bei Anwendung in der kritischen Testverdünnung und Bebrütung der Kulturen bei 20° eine Unterscheidung zwischen Past. pestis und Past. pseudotuberculosis gestattet. Wenig ist bisher über eine Differenzierung der Streptokokken durch Phagen bekannt. Evans (1934—1943) arbeitete mit Phagen (Bouillonmethode), deren Wirkung der serologischen Gruppeneinteilung entsprach. Kjems (1955) fand, daß mit geeigneten Streptokokkenphagen innerhalb der serologischen

[1] Die Typisierung unserer Stämme wurde liebenswürdigerweise von Herrn Dr. P. Nicolle, Paris, vorgenommen.

Gruppe A der hämolysierenden Streptokokken eine Beziehung zwischen serologischen Typen und Empfindlichkeit für Phagen festzustellen ist. Zur Trennung serologisch verwandter und daher schwer differenzierbarer Streptokokkentypen (z. B. 11 und 12) können solche Streptokokkenphagen von Nutzen sein.

Schlußbetrachtung

Es dürfte kein Zweifel bestehen, daß die Bakteriophagen, die ein überaus feines Reagens für strukturelle Unterschiede an der Bakterienoberfläche darstellen, ein brauchbares Hilfsmittel sowohl zur Differenzierung zwischen verschiedenen Bakterienarten als auch zur Typeneinteilung innerhalb einer Bakterienart sind. Die praktischen Erfolge vor allem bei der Lysotypie der Typhus- und Paratyphus B-Bakterien beweisen dies. Freilich hat die Phagenmethode auch Grenzen und Fehlermöglichkeiten, die man bei ihrer Anwendung kennen muß. Ebenso kann, wie besonders in vitro-Versuche ergeben haben, nicht erwartet werden, daß die Konstanz der Lysotypen absolut ist. Entscheidend für den Wert der Lysotypie sind die praktischen Erfahrungen. Die bei einer Bakterienart gemachten Beobachtungen dürfen aber nicht ohne weiteres auf andere Arten übertragen werden. Der Gebrauch von Phagen für diagnostische Zwecke würde, wie STOCKER betont, wahrscheinlich ausgedehnter sein und verläßliche Resultate erwarten lassen, wenn es Laboratorien gäbe, die geprüfte und spezifisch wirkende Phagenpräparationen herstellen und abgeben würden, wie es schon z. B. von dem International Reference Laboratory for Enteric Phage Typing für die Typhus-Vi-Phagen gehandhabt wird.

Literatur

ALLISON, V. D.: Food poisoning. Proc. Roy. Soc. Med. **42**, 216 (1949).

ANDERSON, E. S.: Consideration of the Vi-phage types of Salmonella typhi on a structural basis. Nature (Lond.) **175**, 171 (1955).

— The significance of Vi-phage types F_1 and F_2 of Salmonella typhi. J. of Hyg. **49**, 458 (1951).

— Diskussionsbeitrag. J. Gen. Microbiol. **12**, 379 (1955).

— A new Vi-phage type of Salmonella typhi; with a discussion of methods of preparation of typing phages for new Vi-types. J. Gen. Microbiol. **14**, 676 (1956).

—, and A. FELIX: Variation in Vi-phage II of Salmonella typhi. Nature (Lond.) **170**, 492 (1952).

— — „Degraded Vi-strains" and variation in Vi-phage II of Salmonella typhi. J. Gen. Microbiol. **8**, 408 (1953).

— — The Vi-type determining phages carried by Salmonella typhi. J. Gen. Microbiol. **9**, 65 (1953).

— — Vi-phage type specificity and degradation. Atti del VI. congresso internazionale di Microbiologia. Rom 6.—12. Nov. 1953. Vol. 3, Sez. X, 462.

—, and A. FRASER: The influence of the factors determining Vi-type specificity in Salmonella typhi on the adaptation of Vi-phage. II. J. Gen. Microbiol. **13**, 519 (1955).

— — The statistical distribution of phenotypically modifiable particles and host-range mutants in populations of Vi-phage II. J. Gen. Microbiol. **15**, 225 (1956).

—, and R. E. O. WILLIAMS: Bacteriophage typing of enteric pathogens and staphylococci and its use in epidemiology. J. Clin. Path. **9**, 94 (1956).

ANDERSON, P. H. R., and D. M. STONE: Staphylococcal food poisoning associated with spray-dried milk. J. of Hyg. **53**, 387 (1955).

ATKINSON, N., and H. G. GEYTENBEEK: Salmonella bacteriophages. I. Bacteriophages of S. adelaide. Austral. J. Exper. Biol. a. Med. Sci. **31**, 441 (1953).

— — M. C. SWANN and J. M. WALLASTON: Lysogenicity and lysis patterns in the Salmonellas. Austral. J. Exper. Biol. a. Med. Sci. **30**, 333 (1952).

BAIL, O.: Bakteriophage Wirkungen gegen Flexner- und Colibakterien. Wien. klin. Wschr. **1921**, 447.

BANKER, D. D.: Paratyphoid A phage-typing. Nature (Lond.) **175**, 309 (1955).

BARBER, M.: Staphylococcal infection due to penicillin-resistent strains. Brit. Med. J. **1947 II**, 863.

—, and J. BURSTON: Antibiotic resistance in staphylococcal infection. A study of antibiotic sensitivity in relation to bacteriophage type. Lancet **1955 II**, 578.

— F. G. J. HAYHOE and J. E. M. WHITEHEAD: Penicillin-resistant staphylococcal infection in a maternity hospital. Lancet **1949 II**, 1120.

—, and M. ROZWADOWSKA: Infection by penicillin-resistant staphylococci. Lancet **1948 II**, 641.

—, and J. E. M. WHITEHEAD: Bacteriophage types in penicillin-resistant staphylococcal infections. Brit. Med. J. **1949 II**, 565.

— B. D. WILSON, J. E. RIPPON and R. E. O. WILLIAMS: Spread of staphylococcus aureus in a maternity department in the absence of severe sepsis. J. Obstetr. **60**, 476 (1953).

BARROW, G. I.: Clinical and bacteriological aspects of impetigo contagiosa. J. of Hyg. **53**, 495 (1955).

BERTANI, G., and J. J. WEIGLE: Host-controlled variation in bacterial viruses. J. Bacter. **65**, 113 (1953).

BEUMER, J., et A. VANDEWEYER-DEVEEN: Premiers résultats de la lysotypie de S. paratyphi B en Belgique. Antonie van Leeuwenhoek **20**, 263 (1954).

BLAIR, J. E., and M. CARR: Bacteriophage typing of staphylococci. J. Inf. Dis. **93**, 1 (1953).

BLASI, R. DE, e A. BUOGO: La tipizzazione fagica delle SS. typhi e paratyphi B. Riv. ital. d'Igiene **12**, 16 (1952).

BOYD, J. K. S.: The symbiotic bacteriophages of Salmonella typhi-murium. J. of Path. **62**, 501 (1950).

— M. T. PARKER and N. S. MAIR: Symbiotic bacteriophage as a „marker" in the identification of strains of S. typhi-murium. J. of Hyg. **49**, 442 (1951).

BRANDIS, H.: Ergebnisse der Vi-Phagentypisierung. Klin. Wschr. **1953**, 869.

— Über einen Paratyphus B-Stamm mit atypischem Verhalten bei der Phagentypisierung. Z. Hyg. **138**, 296 (1953).

— Zur Unterteilung des Typhusbakterientyps E_1. Zbl. Bakter. I Orig. **162**, 223 (1955).

— Die Lysotypie von Typhus- und Paratyphus B-Bakterien. Zbl. Bakter. I Orig. **164**, 149 (1955).

— Die Lysotypie mit Typhus- und Paratyphus B-Stämmen von Dauerausscheidern. Zbl. Bakter. I Orig. **162**, 437 (1955).

— Über die Beeinflussung des Reaktionsbildes bei der Lysotypie von Paratyphus B-Stämmen durch einen Paratyphus B-Phagen. Z. Hyg. **142**, 197 (1955).

—, u. P. S. IMAMURA: Vergleichende Untersuchungen über Vi-Bakteriophagen. Z. Hyg. **143**, 50 (1956).

—, u. H. MAURER: Über die Beziehungen zwischen Phagentyp und Xyloseverhalten bei Typhusstämmen. Z. Hyg. **140**, 138 (1954).

—, u. I. STORCH: Unveröffentlicht.

—, u. M. THOMSEN: Zur Kenntnis von Salm. java. Z. Hyg. **141**, 551 (1955).

— — Beziehungen zwischen Lysotyp und kulturellem Verhalten bei Paratyphus B-Stämmen. Z. Hyg. **142**, 227 (1956).

BUCKLE, G.: The typing of Bact. typhosum. Med. J. Austral. **1946 II**, 325.

BUONOMINI, G., e V. D'AMELIO: Rilievi sulla tipizzazione fagica di S. typhi e S. paratyphi nell'Italia centrale. Riv. ital. d'Igiene **14**, 202 (1954).

BURNET, F. M., and D. LUSH: The staphylococcal bacteriophages. J. of Path. **40**, 455 (1935).

—, and M. MCKIE: Bacteriophage reactions of flexner dysentery strains. J. of Path. **33**, 637 (1930).

BYNOE, E. T., R. H. ELDER and R. D. COMTOIS: Phage-typing and antibiotic-resistance of staphylococci isolated in a general hospital. Canad. J. Microbiol. **2**, 346 (1956).

CHERRY, W. B., B. R. DAVIS and PH. R. EDWARDS: Observations on the types and typing of Salmonella paratyphi B cultures in the United States. Amer. J. Publ. Health **43**, 1280 (1953).

— — — Lysis of cultures devoid of Vi-antigen by Vi I bacteriophage of Salmonella typhosa. Science (Lancaster, Pa.) **120**, 309 (1954).

Cherry, W. B., B. R. Davis, Ph. R. Edwards and R. B. Hogan: A simple procedure for the identification of the genus Salmonella by means of a specific bacteriophage. J. Labor. a. Clin. Med. **44**, 51 (1954[a]).

Ciuca, M., u. C. Combiescu: Die Verbreitung der Lysotypen von S. typhi und S. paratyphi B in der Volksrepublik Rumänien, die Lysotypie in der Praxis der epidemiologischen Forschungen. 2. Colloquium über Fragen der Lysotypie. Wernigerode, 8.—10. 10. 1956.

Clauberg, K. W., and K. Marcuse: Über die Bacteriophagendiagnostik als Hilfsmittel für die Typendifferenzierung innerhalb der Ruhrbazillengruppe. Zbl. Bakter. I Orig. **124**, 29 (1932).

Combiesco, G., et M. Popvici: Détermination du type bactériophagique des bacilles typhiques isolés en Roumanie. Arch. roum. Path. expér. **14**, 69 (1948).

Cooper, K. E., and A. Mayr-Harting: Phage-typing and epidemiology of Shigella dysenteriae Sonne. Brit. Med. J. **1951 II**, 271.

Cowan, S. T.: Classification of staphylococci by slide agglutination. J. of Path. **48**, 169 (1939).

Craigie, J.: Notes on the typing of B. typhosus with special reference to types E_2 and F_2. Canad. Publ. Health J. **30**, 37 (1939).

— The present status of phage typing of Bact. typhosum. Canad. Publ. Health J. **33**, 41 (1942).

— The significance and applications of bacteriophage in bacteriological and virus research. Bacter. Rev. **10**, 73 (1946).

—, and K. F. Brandon: Bacteriophage specific for the O-resistant V form of B. typhosus. J. of Path. **43**, 233 (1936).

—, and A. Felix: Typing of typhoid bacilli with Vi-Bacteriophage. Lancet **1947 I**, 823.

—, and Ch. H. Yen: The demonstration of types of B. typhosus by means of preparations of type II Vi bacteriophage. Canad. Publ. Health J. **29**, 448 (1938).

Crowe, E. J.: Acute staphylococcal enterotoxin food poisoning in a Victorian town. Med. J. Austral. **1954**, 468.

Denton, G. D., G. Kalz and A. R. Foley: An investigation of an outbreak of staphylococcus folliculitis (pemphigus neonatorum) by the use of bacteriophage typing of staphylococcus pyogenes. Canad. Med. Assoc. J. **62**, 219 (1950).

Desranleau, J. M.: Typing of B. typhosus with bacteriophage in the province of Quebec. Canad. J. Publ. Health **33**, 122 (1942).

— Typing of Bact. typhosum and Bact. paratyphosum B by means of bacteriophage. Canad. J. Publ. Health **38**, 343 (1947).

—, and I. Martin: Bacteriophage typing in the province of Quebec. Canad. J. Publ. Health **41**, 128 (1950).

— — and M. Saint-Martin: Studies on staphylococcal infections III. Some epidemiological aspects and bacteriophage typing. Canad. J. Publ. Health **46**, 67 (1955).

Dhayagude, R. G., and D. D. Banker: Typing of locally isolated cultures of Salmonella typhi by means of Vi-bacteriophage. Ind. J. Med. Res. **39**, 1 (1951).

Dunbar, J. M.: Bacteriophage typing of untypable Salmonella typhi organisms. Nature (Lond.) **162**, 851 (1948).

Dunlop, S. J. C.: On bacteriophage anti-flexner. Antonie van Leeuwenhoek **9**, 41 (1943).

Edlinger, E.: Die Lysotypie der Salmonella typhi und paratyphi B. Wien. klin. Wschr. **1952**, 343.

— P. Nicolle u. Y. Hamon: Die Methoden der Lysotypie von S. typhi und S. paratyphi B und die Ergebnisse mit Stämmen aus Österreich. Arch. f. Hyg. **138**, 157 (1954).

—, et J. Vieuchange: Fixation des bactériophages spécifiques de l'antigène Vi sur les hématies sensibilisées avec cet antigène. Ann. Inst. Pasteur **84**, 386 (1953).

Elwood, J. S.: The penicillin sensitivity and phage types of staphylococci isolated from hospital patients. J. of Hyg. **49**, 263 (1951).

Ende, M. van den: Observations on the antigenic structure of Ps. aeruginosa. J. of Hyg. **50**, 405 (1952).

Eörsi, M.: Phage types of S. typhi strains isolated in Hungary and relevant investigations made from 1950 to 1954. Acta microbiol. **3**, 285 (1956).

— Jablonszky u. H. Milch: Significance of bacteriophage in infantile enterale infections. I. Enteritis due to E. coli 0111 and 055. Acta microbiol. **1**, 1 (1954).

Erber, M., en C. E. de Moor: Buiktyphus en phaagtypering ter Sumatra's oostkust. Nederl. Tijdschr. Geneesk. **98**, 749 (1954).
— — Typhoid fever and phage typing on the East Cost of Sumatra. Documenta med. geogr. trop. **6**, 144 (1954).
Evans, A. C.: Streptococcus bacteriophage: A study of four serological types. Publ. Health Rep. **49**, 1386 (1934).
— Studies on hemolytic streptococci. I. Methods of classification. J. Bacter. **31**, 423 (1936).
— The potency of nascent streptococcus bacteriophage B. J. Bacter. **39**, 597 (1940).
— Technique for the determination of the sensitivity of a strain of streptococcus to bacteriophage of types A, B, C or D. J. Bacter. **44**, 207 (1942).
—, and E. M. Sockrider: Another serologic type of streptococcic bacteriophage. J. Bacter. **44**, 211 (1942).
Fahey, J. E.: Preliminary observations on phage typing of Corynebacterium diphtheriae. Canad. J. Publ. Health **43**, 167 (1952).
Farrant, J. L., and Ph. M. Rountree: Electron microscopy of a staphylococcal bacteriophage. J. Gen. Microbiol. **9**, 288 (1953).
Felix, A.: Experiences with typing of typhoid bacilli by means of Vi-bacteriophage. Brit. Med. J. **1943 I**, 435.
— Modern laboratory methods in the control of typhoid and paratyphoid B fever. Brit. Med. Bull. **2**, 269 (1944).
— Laboratory control of the enteric fevers. Brit. Med. Bull. **7**, 153 (1951).
— World survey of typhoid and paratyphoid-B phage types. Bull. Organ mond. Santé **13**, 109 (1955).
— Phage typing of Salmonella typhi murium: its place in epidemiological and epizootiological investigations. J. Gen. Microbiol. **14**, 208 (1956).
—, and E. S. Anderson: Bacteriophage, virulence and agglutination tests with a strain of Salmonella typhi of low virulence. J. of Hyg. **49**, 349 (1951).
— — Bacteriophages carried by the Vi-phage types of Salmonella typhi. Nature (Lond.) **167**, 603 (1951).
—, and B. R. Callow: Typing of paratyphoid B bacilli by means of Vi-bacteriophage. Brit. Med. J. **1943 II**, 127.
— — Paratyphoid B Vi-phage typing. Lancet **1951 II**, 10.
Ferguson, W. W., A. Juenker and R. A. Ferguson: Characterization of latent phages from strains of Salmonella typhi typable and untypable with Vi-phage. Amer. J. Hyg. **62**, 306 (1955).
Fisk, R. T.: Studies on staphylococci. I. Occurence of bacteriophage carriers among strains of Staphylococcus aureus. J. Inf. Dis. **71**, 153 (1942).
— Studies on staphylococci. II. Identification of Staphylococcus aureus strains by means of bacteriophage. J. Inf. Dis. **71**, 161 (1942[2]).
—, and O. E. Mordvin: Studies on Staphylococci. III. Further observations on bacteriophage typing of Staphylococcus aureus. Amer. J. Hyg. **40**, 232 (1944).
Freytag, Bl., u. E. Plochmann: Erfahrung mit der Lysotypie bei der Typhusdiagnose. Z. Hyg. **142**, 188 (1955).
Froman, S., and E. Bogen: Mycobacteriophage. Transact. 49. annual Meeting Nat. Tuberculosis Assoc. 1953.
— D. W. Will and E. Bogen: Bacteriophage active against virulent Mycobacterium tuberculosis. I. Isolation and activity. Amer. J. Publ. Health **44**, 1326 (1954).
Fuace, J., et A. Lutz: Type bactériophagique des staphylocoques pathogènes sécréteurs de pénicillinase. Ann. Inst. Pasteur **85**, 387 (1953).
Fusillo, M. H., R. N. Roerig and K. F. Ernst: Phage typing the antibiotic-resistant staphylococci. IV. Incidence and phage type relationship of antibiotic-resistant staphylococci among hospital and nonhospital groups. Antibiotics a. Chemother. **4**, 1202 (1954).
— — J. F. Metzger and K. F. Ernst: Phage typing antibiotic resistant staphylococci. Amer. J. Publ. Health **44**, 317 (1954).
Gärtner, H.: Beitrag zur regionären Verteilung der Paratyphus B-Phagentypen. Z. Hyg. **142**, 432 (1956).

GIUNTINI, J., E. EDLINGER et P. NICOLLE: Étude de quelques bactériophages typhiques Vi. Morphologie des corpuscules au microscope électronique, aspect des plages et thermosensibilité. Ann. Inst. Pasteur **84**, 787 (1953).

GOLDBERG, S.: Slide technique for bacteriophage typing of staphylococcus aureus. Science (Lancaster, Pa.) **120**, No 3129 (1954).

GOULD, J. C., and E. J. MCKILLOP: The carriage of Staphylococcus pyogenes var. aureus in the human nose. J. of Hyg. **52**, 304 (1954).

GUNNISON, J. B., A. LARSON and A. S. LAZARUS: Rapid differentiation between Pasteurella pestis and Pasteurella pseudotuberculosis by action of bacteriophage. J. Inf. Dis. **88**, 254 (1951).

HAMMARSTRÖM, E.: Bacteriophage classification of Shigella sonnei. Lancet **1947 I**, 102.
— Phage typing of Shigella sonnei. Acta med. scand. (Stockh.) Suppl. **223** (1949).

HAMON, Y., et P. NICOLLE: Recherches sur les facteurs qui conditionnent l'appartenance des bacille paratyphiques B aux différents types de FELIX et CALLOW. II. Obtention expérimentale, a partir des cultures non lysogènes, des types de S. paratyphi B par contamination avec les bactériophages extraits des types lysogènes. Ann. Inst. Pasteur **80**, 496 (1951).

HARTLEY, P.: The value of Douglas's medium for the preparation of diphtheria toxin. J. of Path. **25**, 479 (1922).

HELMER, D. E., D. E. KERR, C. E. DOLMAN and L. E. RANTA: Two phage-susceptible types of B. typhosus isolated from a typhoid fever case. Canad. J. Publ. Health J. **31**, 433 (1940).

HENDERSON, N. D., and W. W. FERGUSON: Bacteriophage typing of Salm. typhi. A report of typing in Michigan. J. Labor. a. Clin. Med. **34**, 739 (1949).

HENRIKSEN, S. D.: Bakteriofagtyping av Salmonella typhi og Salmonella paratyphi B. Nord. hyg. Tidskr. **1952**, 93.

HEWITT, L. F.: Diphtheria bacteriophages and their relation to the development of bacterial variants. J. Gen. Microbiol. **7**, 362 (1952).

HIRSZFELD, L., A. GALIS-MALEJCZYK, Z. SEMBRAT-NIEWIADOMSKA u. C. ZWIERZ: Bakteriofagi i ich rola w rozpoznawanui duru brzusznego. Polski Tygodnik Lek. **3**, 417 (1948).

HNATKO, ST. I.: The isolation of bacteriophages for mycobacteria with reference to phage typing of the genus. Canad. J. Med. Sci. **31**, 462 (1953).
— The investigation of soil for bacteriophages against pathogenic and saprophytic acid-fast micro-organisms. Canad. J. Publ. Health **45**, 70 (1954).
— The application of bacteriophages to the study of acid-fast microorganisms from tuberculous patients. Canad. J. Microbiol. **2**, 39 (1956).

HOBBS, B. C.: A study of the serological type differentiation of staphylococcus pyogenes. J. of Hyg. **46**, 222 (1948).
—, and M. E. SMITH: Outbreaks of paratyphoid B fever associated with imported frozen egg. II. Bacteriology. J. Appl. Bacter. **18**, 471 (1955).

HOFMANN, S.: Über die Spezifität des O-Phagentestes nach CHERRY, DAVIS, EDWARDS und HOGAN. 2. Colloquium über Fragen der Lysotypie. Wernigerode, 8.—10. 10. 1956.

HOOD, A. M.: Phage typing of Staphylococcus aureus. J. Med. Labor. Technol. **11**, 45 (1953).
— Phage typing of staphylococcus aureus. J. of Hyg. **51**, 1 (1953).

JACKSON, G. G., H. F. DOWLING and M. H. LEPPER: Bacteriophage typing of staphylococci. I—III. J. Labor. a. Clin. Med. **44**, 14 (1954).

JUDE, A., P. NICOLLE et P. DUCREST: Sur la présence simultanée de deux types bactériophagiques (D_1 et D_6) dans une culture de Salmonella typhi. Ann. Inst. Pasteur **81**, 245 (1951).

KAUFFMANN, F.: On the serology of the Salmonella V Antigen. Acta path. scand. (Københ.) **24**, 591 (1947).

KEOGH, E. V., R. T. SIMMONS and G. ANDERSON: Type-specific bacteriophages for Corynebacterium diphtheriae. J. of Path. **46**, 565 (1938).

KJEMS, E.: Studies on streptococcal bacteriophages. I. Technique of isolating phage-producing strains. Acta path. scand. (Københ.) **36**, 433 (1955).

LACHOWICZ, K., i Z. BUCZOWSKI: Untersuchungen über die biochemische und bakteriophagische Typisierung von S. typhi und ihre epidemiologische Ausnutzung. [Polnisch.] Med. dóswiadcz. i mikrobiol. **2**, 370 (1950).

LAURELL, G., u. G. WALLMARK: Studies on Staphylococcus aureus pyogenes in a children hospital. I., III., IV. Acta path. scand. (København.) **32**, 424, 438, 554 (1953).

LEVY, E., J. E. RIPPON and R. E. O. WILLIAMS: The relation of bacteriophage pattern to some biological properties of staphylococci. J. Gen. Microbiol. **9**, 97 (1953).

LIBIKOVA, E.: Typizacia Salmonella typhi abdominalis pomocon Vi-bacteriofaga na Slovensku. Slovensky Lek. **12**, 390 (1950).

LIE KIAN JOE: Typing of typhoid bacilli from Indonesia by means of Vi-bacteriophage II. Documenta neerl. et indones. morbis trop. **1**, 145 (1949).

LILLEENGEN, K.: Typing of Salmonella typhi-murium by means of bacteriophage. Acta path. scand. (København.) Suppl. **77** (1948).

— Typing of Salmonella dublin and Salmonella enteritidis by means of Bacteriophage. Acta path. scand. (København.) **27**, 625 (1950).

— Typing of Salmonella gallinarum and Salmonella pullorum by means of bacteriophages. Acta path. scand. (København.) **30**, 194 (1952).

LOWBURY, E. J. L., and A. M. HOOD: The acquired resistance of Staphylococcus aureus to bacteriophage. J. Gen. Microbiol. **9**, 524 (1953).

LUDFORD, C. G.: Bacteriophage types of Shigella sonnei in Queensland. Austral. J. Exper. Biol. a. Med. Sci. **31**, 545 (1953).

LUDLAM, G. B.: Staphylococcus aureus in a slaughterhouse. Monthly Bull. Min. Health Publ. Health Labor. Serv. **11**, 138 (1952).

— Incidence and penicillin sensitivity of staphylococcus aureus in the nose in infants and their mothers. J. of Hyg. **51**, 64 (1953).

MACCORMACK, J. D.: Typhoid fever control. Publ. Health **67**, 128 (1954).

MACIEREWICZ, M.: Die Bestimmung der Typen der Typhusbakterien mit Hilfe des Anti-Vi-Bakteriophagen im Bezirk Warschau. [Polnisch.] Med. doswiadcz. i mikrobiol. **2**, 388 (1950).

MARCUSE, K.: Über Typhus-Diagnostik mit spezifischen Bakteriophagen (nach SONNENSCHEIN). Zbl. Bakter. I Orig. **131**, 206 (1934).

MAURER, H.: Über die Typendifferenzierung von Typhus- und Paratyphus B-Bakterien mit Bakteriophagen. Inaug.-Diss. Frankfurt a. M. 1954.

MAYR-HARTING, A.: A study of Ps. pyocyanea. Thesis Univ. of Bristol 1949.

— The phage-typing of Shigella sonnei, and the limits of type stability. J. Gen. Microbiol. **7**, 382 (1952).

MCCLOY, E. W.: Studies on a lysogenic bacillus strain. I. A bacteriophage specific for Bacillus anthracis. J. of Hyg. **49**, 114 (1951).

MCLEAN, S. J.: Bacteriophage typing of strains of staphylococcus aureus in South Australia. Med. J. Austral. **1956 I**, 53.

MILLAR, E. L. M., and M. POWNALL: Food poisoning in Sheffield in 1949. Brit. Med. J. **1950 II**, 551.

MILLER, A. A.: Identification des Bacilles dysentériques au moyen du bactériophage spécifique. Ann. Inst. Pasteur **58**, 709 (1937).

MINOR, LE L.: Lysotypie des Escherichia coli isolés dans les gastro-entérites infantiles. Bull. Acad. Méd. Paris **24**, 480 (1952).

MORRIS, J. F., A. BRIM and T. F. SELLERS: Types of Eberthella typhosa found in Georgia during the period of four years 1941—44. J. Inf. Dis. **77**, 25 (1945).

NAIDU, P. M. N.: Zit. nach LILLEENGEN 1948. Bull. Acad. vet. France **8**, 306 (1935).

NEWELL, K. W.: Paratyphoid B fever possibly associated with Chinese frozen egg. Monthly Bull. Min. Health **14**, 146 (1955). — J. Appl. Bacter. **18**, 462 (1955).

NICOLLE, P., et G. DIVERNEAU: Distribution des lysotypes Vi du bacille typhique en Afrique du Nord. J. Nord. Africaines de Pédratr., Tunis Mai 1954.

— — Sur les différents états d'un même lysotype (Type A) de Salmonella typhi. C. r. Acad. Sci. Paris **240**, 126 (1955).

— J. FOURNIER, P. KIRSCHE, P. LAJUDIE, E. BRYGOO, G. SOUVEINE et J. VOELCKEL: Distribution des lysotypes du bacille typhique au Vietnam. Bull. Soc. Path. exot. Paris **46**, 665 (1953[a]).

—, et Y. HAMON: Recherches sur les facteurs qui conditionnent l'appartenance des bacilles paratyphiques B aux différents types bactériophagiques de FELIX et CALLOW. III. Nouvelle étude sur les transformations de types par l'action des bactériophages extraits des bacilles lysogènes. Ann. Inst. Pasteur **81**, 614 (1951[a]).

Nicolle, P., et Y. Hamon: La lysogénéité, facteur qui conditionne l'appartenance des bacilles paratyphiques B aux différents types de Felix et Callow. C. r. Acad. Sci. Paris **232**, 898 (1951[a]).

— — Distribution des lysotypes du bacille typhique et du bacille paratyphique B en France, dans les territoires d'outre-mer et dans quelques autres pays. Rev. d'Hyg. **2**, 424 (1954).

— — et E. Edlinger: Recherches sur les facteur qui conditionnent l'appartenance des bacilles paratyphiques B aux différents types bactériophagiques de Felix et Callow. I. La lysogénéité des différents types de Salmonella paratyphi B. Ann. Inst. Pasteur **80**, 479 (1951[a]).

— — — Essai de subdivision de quelques lysotypes B Vi de Salmonella paratyphi B. Ann. Inst. Pasteur **85**, 706 (1953[b]).

— — — Aspects théoretiques et pratiques de la lysotypie des bacilles typhiques et paratyphiques B. Biol. méd. **42**, Nr. 5 (1953).

— A. Jude et R. Buttiaux: Stabilité des types bactériophagiques de S. paratyphi B et valeur épidemiologique de la lysotypie par la méthode de Felix et Callow. Ann. Inst. Pasteur **79**, 246 (1950).

— — — Antigènes entravant l'action de certains bactériophages. Ann. Inst. Pasteur **84**, 27 (1953).

— L. le Minor, R. Buttiaux et P. Ducrest: Lysotypie des „Escherichia coli" isolés dans les gastro-entérides infantiles. I. Schémas des types actuellement individualisés. Bull. Acad. Nat. Méd. Paris **1952**, 480.

— — — — Lysotypie des „Escherichia coli" isolés dans les gastro-entérides infantiles. II. Fréquence relative des types dans différents foyers et valeur épidémiologique de la méthode. Bull. Acad. Nat. Méd. **1952**, 483.

— — S. le Minor et R. Buttiaux: Relation entre la sensibilité aux bacteriophages des Escherichia coli des gastro-entérites infantiles et leurs caractères antigéniques et biochimiques. C. r. Acad. Sci. Paris **239**, 462 (1954[c]).

— — — — Lysotypie von Escherichia coli isoliert aus dyspeptischen Stühlen von Neugeborenen. 2. Colloquium über Fragen der Lysotypie. Wernigerode, 8.—10. 10. **1956**[c].

— L. et S. le Minor et G. Diverneau: Schémas lysotypiques anormaux. Leur intérêt pour la création éventuelle de nouveaux types bactériophagiques de S. typhi et de S. paratyphi B et la controle de l'identité de ces salmonella. Ann. Inst. Pasteur **82**, 19 (1952).

— E. van Oye, C. G. Crocker et J. Brault: Sur une variété du lysotype C de Salmonella typhi rencontrée en Afrique équatoriale et à Madagascar. Bull. Soc. Path. exot. Paris **48**, 492 (1955).

— M. Pavlatou et G. Diverneau: Les lysotypies auxiliaires de Salmonella typhi. I. Subdivision du type A et du groupe I + IV par une nouvelle série de phages. Ann. Inst. Pasteur **87**, 493 (1954).

— — — Subdivision de quelques types Vi fréquents de Salmonella typhi par des lysotypies auxiliaires. C. r. Acad. Sci. Paris **236**, 2453 (1953).

— J.-F. Vieu et G. Diverneau: Lysotypes ubiquitaires, exotiques et rares du bacille typhique. C. r. Acad. Sci. Paris **243**, 454 (1956).

— — R. Skalova et J. Brault: Uniformité et diversité biochimiques de lysotypes du bacille typhique. C. r. Acad. Sci. Paris **243**, 994 (1956).

Nobrega, P.: Differenciação entre S. pullorum e S. gallinarum. Papel importante do bacteriophago. Arch. Inst. biol. São Paulo **6**, 71 (1935).

— Amostras atypicas de „Salmonella pullorum". Arch. Inst. biol. São Paulo **7**, 61 (1936).

Oddy, J. G., and H. W. Clegg: Outbreak of staphylococcal foodpoisoning. Brit. Med. J. **1947**, **I**, 442.

Oeding, P.: Serological typing of staphylococci. III. Further investigations and comparison to phage typing. Acta path. scand. (Kobenh.) **33**, 324 (1953).

—, u. Th. Vogelsang: Staphylococcal studies in hospital staffs. V. Comparison between serological typing and phage typing. Acta path. scand. (København.) **34**, 47 (1954).

Olitzki, L., Z. Olitzki and M. Shelubsky: Types of Eberthella typhosa in Palestine. Trans. Roy. Soc. Trop. Med. Lond. **39**, 167 (1945).

OYE, E. VAN, et P. NICOLLE: La lysotypie des bacilles typhiques isolés au Congo belge. Bull. Soc. Path. exot. Paris **46**, 48 (1953).

OSWALD, E. J., and R. J. REEDY: A technic for the bacteriophage typing of staphylococci. Antibiotics a. Chemother. **4**, 574 (1954).

PARKER, M. T., A. J. H. TOMLINSON and R. E. O. WILLIAMS: Impetigo contagiosa. The association of certain types of staphylococcus aureus and of streptococcus pyogenes with superficial skin infections. J. of Hyg. **53**, 458 (1955).

PAVLATOU, M., et P. NICOLLE: Incidence des types biochimiques parmi les types bactériophagiques de Salmonella typhi. Ann. Inst. Pasteur **85**, 185 (1953).

PENSO, G., e V. ORTALI: Studi e ricerche sui micobatteri. II. I fagi dei micobatteri. Rend. Ist. sup. Sanità **12**, 903 (1949).

PILLET, J., J. CALMELS, B. ORTA et G. CHABANIER: Étude comparée du type sérologique, de la sensibilité aux bactériophages et de l'antibiogramme de 201 souches de staphylocoques isolés par prélèvements systematiques chez le nourrisson. Ann. Inst. Pasteur. **86**, 309 (1954).

PÖHN, H.-PH.: Die Differenzierung von Staphylococcus aureus mittels Bakteriophagen. Zbl. Bakter. I Orig. **164**, 164 (1955).

— Die Lysotypie bei Staphylococcus aureus. Colloquium über Lysotypie. Wernigerode 29. 9.—1. 10. 1955.

— Die Spezifität der Typenphagen für die einzelnen Staphylokokken-Gruppen. 2. Colloquium über Fragen der Lysotypie. Wernigerode, 8.—10. 10. 1956.

PRAZMOWSKI i R. STEMPIEŃ: Rozmieszczenie typow pałeczki durowej klasyfikowanej bakteriofagiem na terenie wojewodztwa Lodzkiego i Lodzi. Med. dóswiadcz. i mikrobiol. **2**, Nr 1 (1950).

PRICE, P., F. K. NEAVE, J. E. RIPPON and R. E. O. WILLIAMS: The use of phage typing and penicillin sensitivity tests in studies of staphylococci from bovine mastitis. J. Dairy Res. **21**, 342 (1954).

RADOJČIĆ, M. M.: Biochemische, bakteriophagische und serologische Studien über Dysenteriebazillen der Flexnergruppe. Zbl. Bakter. I Orig. **136**, 326 (1936).

RAKIETEN, M. L., A. H. EGGERTH, T. L. RAKIETEN: Studies with bacteriophages active against mucoid strains of bacteria. J. Bacter. **40**, 529 (1940).

RAŠKA, K., V. MALIŠOVÁ u. M. MAZÁČEK: Praktický význam fagotypisace v epidemiologii střevnich nákaz. Čas. lék. ćesk. **89**, 835 (1950).

REUL, R.: Caractérisation, au moyen des Vi-bactériophages, des souches de Salmonella typhi (Bacille d'Eberth) isolées dans la region de Coquilhatville, au Congo belge. Ann. Soc. belge Méd. trop. **29**, 339 (1949).

RIFKIND, D., and M. PICKETT: Bacteriophage studies on the hemorrhagic septicemia pasteurellae. J. Bacter. **67**, 243 (1954).

RIPPON, J. E.: The classification of bacteriophages lysing staphylococci. J. of Hyg. **54**, 213 (1956).

— A new serological division of staphylococcus aureus bacteriophages: Group G. Nature (Lond.) **170**, 287 (1952).

RISCHE, H.: Verteilung der Lysotypen von Salm. typhi und Salm. paratyphi B in der DDR. Colloquium über Lysotypie. Wernigerode 28. 9.—1. 10. 1955. Diskussionsbemerkung.

— Die epidemiologische Bedeutung der Typendifferenzierung von Typhus- und Paratyphus-B-Bakterien. Z. ges. Hyg. u. Grenzgeb. **1**, 110 (1955).

— Typenbestimmende Phagen bei degradierten Typhusstämmen. Tagg Österr. Ges. für Mikrobiologie und Hygiene, Gmunden, 2.—6. 9. 1956.

— K. ROHNE u. H. SCHNEIDER: Ergebnisse der Vi-Phagen-Typisierung von Salmonella typhi und paratyphi B. Z. inn. Med. **9**, 896 (1954).

ROHNE, K.: Die Anwendung der Lysotopie für S. typhi-murium nach FELIX und CALLOW anläßlich einer Lebensmittelvergiftung. Tagg. Österr. Ges. für Mikrobiologie und Hygiene, Gmunden, 2.—6. 9. 1956.

ROODYN, L.: Staphylococcal infections in general practice. Brit. Med. J. **1954 II**, 1322.

ROSCHKA, R.: Über das Vorkommen serologischer Varianten von S. typhi und paratyphi B in Österreich. Z. Immun.forsch. **110**, 106 (1953).

ROUNTREE, PH. M.: Bacteriophage typing of strains of staphylococci isolated in Australia. Lancet **1953 I**, 514.

— The serological differentiation of staphylococcal bacteriophages. J. Gen. Microbiol. **3**, 164 (1949).

ROUNTREE, PH. M.: Variations in a related series of staphylococcal bacteriophages. J. Gen. Microbiol. **15**, 266 (1956).
—, and B. M. FREEMAN: Infections caused by a particular phage type of staphylococcus aureus. Med. J. Austral. **1955 II**, 157.
— M. HESELTINE, J. RHEUBEN and R. P. SHEARMAN: Control of staphylococcal infection of the newborn by the treatment of nasal carriers in the staff. Med. J. Austral. **1956 I**, 528.
—, and J. RHEUBEN: Penicillin-resistant staphylococci in the general population. Med. J. Austral. **1956 I**, 399.
—, and E. F. THOMSON: Incidence of penicillinresistant and streptomycinresistant staphylococci in a hospital. Lancet **1949 II**, 501.
RUYS, A. CH., en M. J. WILLEMS: Onderzoekingen over de verspreiding van staphylococcus pyogenes bij gezonden en zieken. Nederl. Tijdschr. Geneesk. **99**, 406 (1955).
SAINT-MARTIN, M., G. CHAREST and J. M. DESRANLEAU: Bacteriophage typing in investigations of staphylococcal food poisoning outbreaks. Canad. J. Publ. Health **42**, 351 (1951).
SARTORIUS, F., u. H. REPLOH: Über die weitere Entwicklung der Ruhrrassenforschung, insbesondere in der Pseudodysenterie- resp. Flexnergruppe. Zbl. Bakter. I Orig. **126**, 10 (1932).
SAVIC, A.: O znacaju fagotipizacije Salmonella typhi. Srpski Ark. Lekarst. (Beograd) **81**, 164 (1953).
SCHMIDT, A.: Gewinnung spezifisch eingestellter Diagnostikphagen durch Umzüchtung. Zbl. Bakter. I Orig. **123**, 202 (1931).
SCHOLTENS, R. TH.: The resistance developed against bacteriophage. J. of Hyg. **36**, 452 (1936).
— Phage typing of Salmonella typhi in the Netherlands. Antonie van Leeuwenhoek **16**, 245 (1950).
— Lysogenicity as a tool for bacteriophage typing of Salmonella paratyphi B. Antonie van Leeuwenhoek **16**, 256 (1950).
— Bacteriologische resultaten der phagotypering van in Nederland geïsoleerde stammen van Salmonella typhi. Geneesk. Gids **29**, 34 (1951).
— La formation de bactériophages dans les cultures mixtes de souches de Salm. paratyphi B. C. r. Soc. Biol. Paris **146**, 504 (1952).
— The formation of bacteriophages in mixed cultures by recombination of genetic elements; characterisation of phage types of Salmonella paratyphi B by phage reactions and lysogenic properties. Antonie van Leeuwenhoek **18**, 257 (1952).
— Characterization and grouping of phage types of Salmonella paratyphi B, especially of a new type „Sittard" by sensitivity to type phages and by lysogenic properties. J. of Hyp. **53**, 1 (1955).
— Vi-phage typing of Salmonella typhi. Extension of Vi-type group E and observations on group „29". Ann. Inst. Pasteur **89**, 216 (1955).
— Phage typing of Salmonella paratyphi B; an arrangement of phage types based on the study of lysogenicity. Antonie van Leeuwenhoek **22**, 65 (1956).
— The relationship of Vi types E_3 and E_4 of Salmonella typhi to Vi types E_1 and D_1. Antonie van Leeuwenhoek **22**, 281 (1956).
ŠEFČOVICOVÁ, L.: Výsledky fagotypizácie Salmonella typhi abdominalis na Slovensku v r. 1950—52. Lek. obzor (Bratislava) **2**, 773 (1953).
SERTIC, V., et N.-A. BOULGAKOV: Sur la sensibilité des souches d'Eberthella typhi au bacteriophage, en relation avec les caractères antigéniques. C. r. Soc. Biol. Paris **122**, 35 (1936).
SICCA, G. T., e V. D'AMELIO: Rapporti tra batteriofagi cosidetti anti-Vi e struttura antigene della S. paratyphi B. Riv. Ist. sieroter. ital. **28**, Nr 6 (1953).
SONNENSCHEIN, C.: Die Verwendbarkeit der Bakteriophagen für die bakteriologische Diagnose. Münch. med. Wschr. **1925**, **1443**.
— Bakteriendiagnose mit Bakteriophagen. Dtsch. med. Wschr. **1928**, 1034.
— Wirkung von Bakteriophagen auf Typhus-Bakterien. Münch. med. Wschr. **1929**, 355.
SOTO, B., and P. M. OTERO: Types of S. typhosa isolated in Puerto Rico. Puerto Rico J. Publ. Health **24**, 31 (1948).
STOCKER, B. A. D.: Bacteriophage and bacterial classification. J. Gen. Microbiol. **12**, 375 (1955).

Tee, G. H.: Bacteriophage typing of Shigella sonnei and its limitations in epidemiologica investigation. J. of Hyg. 53, 54 (1955).

Thal, E., u. L. O. Kallings: Zur Bestimmung des Genus Salmonella mit Hilfe eines Bakteriophagen. Nord. Veterinaermed. 7, 1063 (1955).

Thomen, L. F., and M. Frobisher jr.: A study of Shigella by means of bacteriophage. Amer. J. Hyg. 42, 225 (1945).

Thumes, M.: Vergleichende Untersuchungen über die Einwirkung von Bakteriophagen des Hühnerdarmes auf Bact. gallinarum Klein und Bact. pullorum Rettger. Inaug.-Diss. Leipzig 1933.

Toft, G.: Studies on the specificity of some colon bacteriophages, with a special view to the capsulophages (K-phages). Acta path. scand. (Kobenh.) 24, 260 (1947).

Tomašić, P., i M. Bajzer: Fagotipizacija S. typhi na području NR Hrvatske. Hygijena 6, 75 (1954).

Toshach, Sh.: Bacteriophages for C. diphtheriae. Canad. J. Publ. Health 41, 332 (1950).

Trüb, C. L., u. W. Sauer: Über die Ergebnisse der Bakteriophagen-Lysotypie von S. typhi und paratyphi B im Reg. Bez. Düsseldorf in den Jahren 1952—1954. Jahrbuch 1955 Akademie für Staatsmed. Düsseldorf.

Tulloch, L. G.: Nasal carriage in staphylococcal skin infections. Brit. Med. J. 1954 II, 912.

Vogelsang, Th. M.: Staphylococcal studies in hospital staffs. Acta path. scand. (København.) 33, 435 (1953).

Wahl, R., et J. Fouace: Sur les principes de classification des staphylocoques pathogènes par la méthode des phages. Ann. Inst. Pasteur 82, 542 (1952).

— — Isolement et emploi de phages nouveaux pour identifier les souches de staphylocoques pathogènes insensibles aux phages classiques. Ann. Inst. Pasteur. 86, 161 (1954).

— — I. A propos des techniques d'identification des staphylocoques pathogènes par les phages. Ann. Inst. Pasteur 87, 159 (1954).

— — A propos des techniques d'identification des staphylocoques pathogènes par les phages. II. Choix d'une méthode de dilution des Phages. Etude des fausses réactions produites par les lysats. Ann. Inst. Pasteur 87, 279 (1954).

—, et P. Lapeyre-Mensignac: L'identification des staphylocoques par les bactériophages. 1. Application aux staphylococcies cutanées récidivantes. Ann. Inst. Pasteur 78, 353 (1950).

— — L'identification des staphylocoques par les bactériophages. II. Essai de classification des staphylocoques par la méthode des phages. Ann. Inst. Pasteur 78, 765 (1950).

Wallmark, G.: Fagtypning av patogena stafylokocker. Nord. Med. 41, 806 (1949).

— Bacteriophage types, sensitivity to antibiotics, and penicillinase production of Staphylococcus aureus (pyogenes). Acta Soc. Med. upsal. 1954, 209.

— Bacteriophage typing of Staphylococcus aureus pyogenes. I. A report of the method and some experimental results. Acta path. scand. (København.) 34, 57 (1954).

— Bacteriophage-typing of Staphylococcus aureus pyogenes. II., III., IV. Acta. path. scand. (København.) 34, 497, 577, 585 (1954).

Warner, P. T. J. C. P.: The isolation of bacteriophages of Ps. pyocyanea. Brit J. Exper. Path. 31, 112 (1950).

Wassermann, M. M., and I. Saphra: The use of bacteriophages in typing Salmonella cultures. J. Bacter. 69, 97 (1955).

—, and E. Seligmann: Serratia marcescens bacteriophages. J. Bacter. 66, 119 (1953).

Weigle, J. J., et G. Bertani: Variations des bactériophages conditionnées par les bactéries hôtes. Ann. Inst. Pasteur 84, 175 (1953).

Williams, R. E. O.: Skin and nose carriage of bacteriophage types of Staph. aureus. J. of Path. 58, 259 (1946).

—, and J. E. Rippon: Bacteriophage typing of Staphylococcus aureus. J. of Hyg. 50, 320 (1952).

— — and L. M. Dowsett: Bacteriophage typing of strains of staphylococcus aureus from various sources. Lancet 1953 I, 510.

Williams, St., and C. Timmins: An investigation of the source of staphylococcal infection in acute osteomyelitis. Med. J. Austral. 1938 II, 687.

WILLIAMS SMITH, H.: Investigations on the typing of staphylococci by means of bacteriophage. I. Origin and nature of lysogenic strains. J. of Hyg. **46**, 74 (1948).

— Investigations on the typing staphylococci by means of bacteriophage. II. The significance of lysogenic strains in staphylococcal type designation. J. of Hyg. **46**, 82 (1948).

— The typing of staphylococci of animal origin by the bacteriophage method. J. Comp. Path. a. Ther. **58**, 179 (1948).

— The typing of Salmonella thompson by means of bacteriophage. J. Gen. Microbiol. **5**, 472 (1951).

— The typing of Salmonella dublin by means of bacteriophage. J. Gen. Microbiol. **5**, 919 (1951).

WILSON, G. S., and J. D. ATKINSON: Typing of Staphylococci by the bacteriophage method. Lancet **1945 I**, 647.

WILSON, V. R., and P. R. EDWARDS: Unveröffentlicht.

WIZA, J.: Typing of typhoid bacilli isolated in the province Poznan in 1949—1950 by means of standard bacteriophage of CRAIGIE and FELIX. [Polnisch.] Med. dóswiadcz. i mikrobiol. **4**, 115 (1952).

WÜSTENBERG, J., u. O. KORELL: Die Vi-Phagentypisierung in ihrer Bedeutung für die Epidemiologie des Typhus und Paratyphus B. Arch. f. Hyg. **139**, 85 (1955).

YEN, C. H.: Bacteriophage typing of B. typhosus isolated in Peiping. Proc. Soc. Exper. Biol. a. Med. **41**, 162 (1939).

IV. Das Kauffmann-White-Schema[1, 2]

(Diagnostisches Salmonella-Antigenschema)

Von

F. Kauffmann[3]

Vorwort

Es sind bereits über 30 Jahre vergangen, seit P. Bruce White das erste *Salmonella*-Antigenschema „Table of the relative antigenic constitutions of known Salmonella types" veröffentlichte und über 22 Jahre, seit das Salmonella Subcommittee des Internationalen Nomenklatur-Komitees unter dem Vorsitze von H. Schütze das erste „*Kauffmann-White-Schema*" (1934) publizierte.

In der folgenden Review werden nach einer historischen Übersicht die wichtigsten serologischen Daten zusammengestellt und auf den neuesten Stand der Dinge gebracht, um die serologische Typendiagnose in der *Salmonella*-Gruppe zu erleichtern.

Die Bestimmung aller bekannten *Salmonella*-Typen soll prinzipiell nur in den hierzu eingerichteten *Salmonella*-Zentralen ausgeführt werden, also nicht in den gewöhnlichen bakteriologischen Laboratorien, die sich mit der Diagnose pathogener Darmbakterien befassen. Diese Laboratorien sollen nur die *Salmonella*-Gruppendiagnose stellen und die wichtigsten Typen diagnostizieren, wie z. B. *S. paratyphi A, B* und *C, S. typhi, S. typhi murium, S. cholerae suis* und *S. enteritidis*.

Zu diesem Zwecke genügen 15 nichtabsorbierte Immunseren: *O-Seren*: A = 1, 2, 12; B = 4, 5, 12; C = 6, 7, 8; D = 9, 12; E = 3, 10, 15 und ein polyvalentes O-Serum A + B + C + D + E. Ferner ein *Vi-Serum* und die *H-Seren*: a; b; c; d; e, n, x; g, p; i und 1, 2, 5.

Absorbierte Seren, sog. Faktor-Seren, sollen nur in *Salmonella*-Zentralen für eigenen Gebrauch hergestellt werden. Die fabrikmäßige Herstellung derartiger Faktor-Seren und ihre Lieferung an bakteriologische Laboratorien werden dringend widerraten.

Die vorliegende Review ist in praktischer Hinsicht für den Gebrauch in *Salmonella*-Zentralen bestimmt, in wissenschaftlicher Hinsicht darf sie aber allgemeineres Interesse beanspruchen, da sie die Resultate der *Salmonella*-Antigenanalyse während der letzten 30 Jahre wiedergibt.

Kopenhagen, Ende Dezember 1956. F. Kauffmann.

---

[1] In memoriam P. Bruce White.

[2] Aus Statens Seruminstitut, Kopenhagen (Direktor: Dr. J. Ørskov).

[3] Leiter der Internationalen Salmonella- und Escherichia-Centrale.

Definition des Kauffmann-White-Schemas und der Salmonella-Gruppe

Das Kauffmann-White Schema ist eine diagnostische Antigentabelle, in der die verschiedenen Körper- und Geißel-Antigene der *Salmonella*-Typen gekürzt und formelmäßig wiedergegeben sind. Es handelt sich also um eine serologische Bestimmungstabelle der Typen und nicht um ein Verzeichnis aller nachweisbaren Antigene.

Die *Salmonella*-Gruppe besteht aus *Gram*-negativen, aeroben, nicht sporenbildenden Stäbchen, die auf Grund peritricher Geißeln beweglich sind, doch kommen auch unbewegliche Formen vor. Typische *Salmonella*-Kulturen spalten nicht Adonit, Lactose und Saccharose, verflüssigen nicht Gelatine, bilden kein Indol und greifen Urea nicht an. Sie vergären regelmäßig Glucose, meist unter Gasbildung, und wachsen nicht im KCN-Substrat. Alle bekannten Typen sind pathogen für Menschen, Tiere oder beide und können auf Grund ihrer Antigenstruktur in das Kauffmann-White-Schema eingefügt werden. Es sei aber beachtet, daß der Ton auf dem Worte „*können*" liegt, d. h., daß bestimmte Typen in das Antigenschema eingefügt werden *können*, daß sie aber nicht zwangsläufig eingefügt werden *müssen*. Es gibt nämlich eine Menge anderer Bakterien, die zwar *Salmonella*-Antigene besitzen, aber infolge ihres morphologischen, färberischen oder biochemischen Verhaltens nicht zur *Salmonella*-Gruppe gerechnet werden. Das *Salmonella*-Antigenschema ist daher nur ein willkürlicher Ausschnitt aus einem viel umfassenderen Antigenschema, das alle Bakterien umschließt.

Hinsichtlich Nomenklatur sei nur erwähnt, daß wir die Ausdrücke „*genus*" und „*species*" vermeiden und von einer „*Salmonella*-Gruppe", die in verschiedene *Sero-*, *Bio-* und *Phag-Typen* eingeteilt wird, sprechen. Die letzte Einheit ist der *serofermentative Phag-Typ*, d. h. ein Typ, der durch seine serologischen und biochemischen Eigenschaften sowie sein Verhalten gegen Phagen gekennzeichnet ist.

Historischer Überblick

Die Serologie der *Salmonella*-Gruppe beruht auf den Untersuchungen über die *O- und H-Formen* (SMITH und REAGH, BEYER und REAGH, WEIL und FELIX), die *O- und R-Formen* [ARKWRIGHT, SCHÜTZE (1)], die *Phasen* (ANDREWES) und die Typendifferenzierung mit Hilfe des Castellanischen Absorptionsversuches [BOYCOTT, BAINBRIDGE und O'BRIEN, SCHÜTZE (2) u. a.]. Durch einen systematischen Ausbau dieser grundlegenden Untersuchungen gelang es P. BRUCE WHITE, das erste Antigenschema, das in Tabelle 1 wiedergegeben ist, zu schaffen.

Das erste Antigenschema wurde von P. BRUCE WHITE im Jahre 1926 publiziert und in etwas veränderter Form im „*System of Bacteriology*" (Band 4) 1929 abgedruckt. In der Tabelle 1 ist ein Schreibfehler in der Tabelle von BRUCE WHITE verbessert worden, da das O-Antigen von *Salmonella sendai* VI, nicht XI heißen soll. Während BRUCE WHITE in seinem ersten Schema (1926) das O-Antigen von *S. paratyphi A* mit XI und das von *S. sendai* mit XI $\pm$ angab, hat er im zweiten Schema (1929) das O-Antigen von *S. paratyphi A* mit VI bezeichnet und dann versehentlich bei *S. sendai* die Zahl XI des ersten Schemas übernommen.

Nach den Befunden von BRUCE WHITE sollten nämlich die O-Antigene von *S. paratyphi A* und *S. sendai* identisch oder fast identisch sein, so daß es im zweiten Schema bei *S. sendai* VI und nicht XI heißen müßte, zumal das letzte O-Antigen von *S. bombay* die Nummer VIII hat.

Tabelle 1. „*Relative Antigenic Constitutions of Salmonella Types*“
(according to P. Bruce White)

	Meta-Salmonella Types		Monophasic Series									
			„Specific“								„Non-specific“	
Type	Pullorum	Sanguinarium	Paratyphosus A	Typhosus	Enteritidis	Dublin	Derby	Moscow	Abortus equi	Dar-es-Salaam	Binns	European Suipestifer
Stable „O“ Somatic Factors	III	III	VI	III (? +)	III	III	II	III	II (+)	III (±)	I II	V
Labile, „H“, flagellar factors	Nil.	Nil.	T	S	R P_1 P_2	R P_1 P_3	R Q	R (+)	D_1 N_1 N_2	D_1 N_1 M_1 Y	G A B C	G E_1 E_2

Explanation: Somatic antigenic factors are expressed by Roman numerals; flagellar factors by letter, by name or by both. (+) or (—) indicates that there is some qualitative excess or deficiency in the factors stated; (±) that both conditions are present.

Die Identität der O-Antigene von *S. paratyphi A* und *S. sendai* konnte von Kauffmann (1) nicht bestätigt werden, da *S. sendai* zur O-Gruppe D gehört. Dieser Irrtum von Bruce White ist durchaus verständlich, weil damals das O 1-Antigen, das bei beiden Typen vorkommt, noch nicht bekannt war. Arbeitet man nämlich mit O-Seren, die starke O 1-Agglutinine enthalten, so kommt es zu stark übergreifenden Reaktionen zwischen *S. paratyphi A* und *S. sendai*. Ferner können stark entwickelte O 12-Antigene die Differentialdiagnose zwischen 1, 2, 12 = *S. paratyphi A* und 1, 9, 12 = *S. sendai* erschweren.

Im Antigenschema von Bruce White wurde der *S. dar-es-salaam*-Typ als monophasisch betrachtet, doch handelt es sich hierbei auf Grund der Untersuchungen von Kauffmann und Mitsui (2) um einen diphasischen Typ mit α—β Phasenwechsel = 1, 9, 12 : 1, w : e, n . . .

Tabelle 1 (Fortsetzung). „*Modified and Simplified from the Scheme presented in Medical Research Council, Special Report Series, No. 103*".

Diphasic Series

	Tokyo	Bombay	Stanley	Paratyphosus B	Aertrycke	Newport	Reading	Thompson	Morbificans Bovis	American Suipestifer	Hirschfeld	Sendai	L
	III	VIII	I II	I II	I II	IV	II	V	IV	V	V	VI	VII
Characterizing nonspecific phase	(Insufficiently examined)	(Insufficiently examined)	G A	G A	G A B C	G A B C	G B (—) E_1 E_2	G E_1 E_2	G E_1 E_2	G E_1 E_2	G E_1 (? +)	G E_1	G A (±)
Characterizing specific phase	Tokyo specific = R P_1 P_2	Bombay specific = S ±	Stanley specific = S	Paratyphosus B specific	Aertrycke specific	Newport specific = D_1 D_2	Reading specific = D_1 D_2	Thompson specific	Morbificans Bovis specific	Suipestifer specific = W	Hirschfeld specific = W (±)	Sendai specific = T	L specific = M_1 M_2

To avoid undue complication a number of minor factors, somatic and flagellar, have been omitted from the table.

Die beiden Typen *S. tokyo* und *S. bombay* konnten nicht bestätigt werden, da es sich bei *S. tokyo* um eine Kultur von *S. enteritidis* handelt, während *S. bombay* völlig rauh ist und daher nicht als besonderer Typ anerkannt werden konnte. Wie Kauffmann und Petersen kürzlich erwähnten, hat der „*Bombay*"-Stamm die H-Antigene d : 1, 7.

Der „*Binns*"-Typ von Bruce White ist die 2. Phase von *S. typhi murium*, während der „*European Suipestifer*" die 2. Phase von *S. cholerae suis* ist.

S. pullorum und *S. sanguinarium* von Bruce White werden heute mit *S. gallinarum-pullorum* bezeichnet und als ein einziger Serotyp aufgefaßt.

Auf Grund unserer heutigen Klassifikation und Nomenklatur enthält das Antigenschema von Bruce White 20 verschiedene Serotypen, die in der Tabelle 2, dem Kauffmann-White-Schema entsprechend, angegeben sind.

Tabelle 2. *Auszug aus dem Kauffmann-White-Schema*

Group	Type	Somatic Antigens	Flagellar Antigens	
			Phase 1	Phase 2
A	S. paratyphi A	1, 2, 12	a	—
B	S. paratyphi B	1, 4, 5, 12	b	1, 2
	S. stanley	4, 5, 12	d	1, 2
	S. reading	4, 12	e, h	1, 5
	S. derby	1, 4, 12	f, g	—
	S. typhi murium	1, 4, 5, 12	i	1, 2
	S. abortus equi	4, 12	—	e, n, x
C	S. paratyphi C	6, 7, Vi	c	1, 5
	S. cholerae suis	6, 7	c	1, 5
	S. thompson	6, 7	k	1, 5
	S. newport	6, 8	e, h	1, 2
	S. morbificans bovis	6, 8	r	1, 5
D	S. sendai	1, 9, 12	a	1, 5
	S. typhi	9, 12, Vi	d	—
	S. enteritidis	1, 9, 12	g, m	—
	S. dublin	1, 9, 12	g, p	—
	S. moscow	9, 12	g, q	—
	S. dar-es-salaam	1, 9, 12	l, w	e, n...
	S. gallinarum-pullorum	1, 9, 12	—	—
E	S. london	3, 10	l, v	1, 6

Die Tabelle 2 enthält die 20 Serotypen aus dem Antigenschema von BRUCE WHITE, dem *Kauffmann-White-Schema* entsprechend angeordnet.

Die erste Antigentabelle wurde von KAUFFMANN (2), (3) am 16. Dezember 1929 in einer Sitzung der Berliner Mikrobiologischen Gesellschaft demonstriert und im Jahre 1930 in einer Arbeit „Der Antigenaufbau der Typhus-Paratyphus-Gruppe" publiziert. Erweiterungen dieser Antigentabelle erfolgten im gleichen Jahre in einer Mitteilung von KAUFFMANN und MITSUI (1) und im Jahre 1931 in einem Übersichtsreferat von KAUFFMANN (4) „Der heutige Stand der Paratyphusforschung".

Als auf Veranlassung der Internationalen Vereinigung für Mikrobiologie ein *Salmonella*-Subkomitee unter dem Vorsitze von H. SCHÜTZE gegründet wurde, kam es innerhalb dieses Komitees zu einer Verständigung zwischen P. BRUCE WHITE und F. KAUFFMANN, daß die von KAUFFMANN angegebene Nomenklatur und Einteilung international benutzt werden solle. Das *Salmonella*-Subkomitee billigte diese Vereinbarung und hat den Namen „*Kauffmann-White Schema*" zum ersten Male im J. of Hyg. **34**, 333 (1934) veröffentlicht:

"The classification employed (The Kauffmann-White-Schema) is based upon the taxonomic schema finally presented by KAUFFMANN, 1931, which is in itself a re-examination and amplification of the pioneer work of SCHÜTZE, BRUCE WHITE, SCOTT and others. The species recognized are taken in the order followed by KAUFFMANN, with a few minor alterations and additions. The names employed are those finally approved by the subcommittee. They are followed, in each case, by brief particulars of first isolation and by a list of the principal synonyms employed in the literature. The antigenic structure is indicated according to KAUFFMANN's method . . ."

Dieses Zitat ist hier abgebrochen, da die Antigenbezeichnungen inzwischen auf Grund internationaler Vereinbarungen hinsichtlich der O-Antigene abgeändert wurden. An Stelle der früher gebrauchten römischen Zahlen werden die O-Antigene jetzt mit arabischen Zahlen bezeichnet. Die Bezeichnungen der H-Antigene sind dagegen unverändert geblieben.

Zum Abschlusse dieser kurzen, historischen Übersicht sollen die Fehler des Verfassers, die bereits bei früheren Gelegenheiten berichtigt wurden, zusammengefaßt werden:

1. In der Arbeit „Über grob- und feinflockige Typhusagglutination" sind die beiden Agglutinationstypen O und H falsch bewertet, da sie nicht qualitativ, sondern quantitativ aufgefaßt wurden [Z. Hyg. **106**, 241 (1926)].

2. In der Arbeit „Der Typus ‚Berlin' der Paratyphus C-Gruppe" entspricht die „relativ spezifische" Phase der „spezifischen" Phase von *Salmonella thompson*, so daß der Typ „Berlin" mit *S. thompson* identisch ist [Z. Hyg. **110**, 537 (1929)].

3. In der Arbeit von F. KAUFFMANN und CH. MITSUI „Zwei neue Paratyphustypen mit bisher unbekanntem Phasenwechsel" sind die Phasen mit l, v : e, n, v bzw. l, w : e, n, w angegeben, während sie tatsächlich die Formel l, v : e, n bzw. l, w : e, n haben [Z. Hyg. **111**, 740 (1930)].

4. In der Arbeit „Über die diphasische Natur der Typhusbazillen" ist die künstlich erzeugte, d. h. induzierte Phase als „zweite" Phase bezeichnet worden, während sie als „dritte" Phase aufzufassen ist [Z. Hyg. **119**, 103 (1936)].

5. In der Arbeit „Über weitere neue Salmonella-Typen" wurde auf Grund einer unvollständigen Untersuchung ein neuer Typ „*S. new york*", der von P. R. EDWARDS als *S. javiana* diagnostiziert wurde, aufgestellt [Acta path. scand. (Copenh.) Suppl. **54**, 33 (1944)].

6. Die in der Arbeit „Über das Vorkommen von Salmonella-Antigenen in Colikulturen" benutzte Bezeichnung „*Salmonella coli*" wurde aufgegeben, so daß diese Coli-ähnlichen Kulturen mit Salmonella-Antigenen aus dem Kauffmann-White-Schema ausgeschlossen wurden [Acta path. scand. (Copenh.) **18**, 225 (1941).]

Antigen-Bestimmung

Die *Salmonella*-Antigene werden in zwei große Gruppen von Antigenen eingeteilt:

I. Die Körper-Antigene

II. Die Geißel-Antigene (H-Antigene)

Die Körper-Antigene können in 2 Gruppen unterteilt werden:

1. Die O-Antigene

2. Die K-Antigene

Typische K-Antigene, die nach KAUFFMANN und VAHLNE Hüllen- oder Kapsel-Antigene darstellen, sind das *Vi-Antigen* und das *M-Antigen* (Schleim-Antigen). Das somatische 5-Antigen, das in der 2. Auflage der „*Enterobacteriaceae*" von KAUFFMANN (5) ebenfalls als K-Antigen bezeichnet wurde, hat eine besondere Stellung ("has a special position and differs from other somatic antigens"), es ist weder ein typisches O-Antigen, noch ein typisches K-Antigen. Es wird daher vorgeschlagen, dieses Antigen nicht mehr als K-Antigen zu bezeichnen, sondern ihm eine

Sonderstellung einzuräumen und es „O“ 5 zu nennen. Wahrscheinlich kommt auch verschiedenen anderen O-Antigenen eine ähnliche Sonderstellung zu, doch liegen hierüber noch keine näheren Untersuchungen vor. Aus praktisch-diagnostischen Gründen ist es nicht notwendig, das 5-Antigen von anderen O-Antigenen zu unterscheiden[1].

Die serologische Variation

Zum Verständnis der *Salmonella*-Serologie ist eine Kenntnis der serologischen Variation erforderlich, doch muß betreffs Einzelheiten auf das Buch „*Enterobacteriaceae*“ (KAUFFMANN) verwiesen werden.

1. *H-O-Variation* = Verlust des H-Antigens, Übergang der O-H-Form zur O-Form.

2 *S-T-R-Variation* = Verlust des O-Antigens, Übergang der S-Form zur T-Form oder R-Form [s. bei KAUFFMANN (6)]. Die Ausdrücke S, T und R werden rein serologisch gebraucht, während die morphologisch glatt oder rauh wachsenden Kulturen „*Glatt-Formen*“ oder „*Rauh-Formen*“ genannt werden.

3. *Form-Variation* = Quantitative Veränderungen der Körper-Antigene:

a) *O-Variation*, innerhalb bestimmter O-Antigene.

b) *V-W-Variation*, innerhalb des Vi-Antigens.

c) *M-N-Variation*, Übergang der M-Form (Schleim-Form) zur N-Form (Normal-Form).

4. *Phasen-Variation* (Phasenwechsel) = qualitative Veränderungen innerhalb der H-Antigene.

a) „Spezifisch-unspezifischer Phasenwechsel“ (ANDREWES), z. B. b : 1, 2.

b) „α-β-Phasenwechsel“ [KAUFFMANN und MITSUI (2)], z. B. b : e, n, x.

c) Phasenwechsel nach EDWARDS und BRUNER (1), z. B. b : l, w.

Um einen gemeinsamen, neutralen Ausdruck zu haben, werden die natürlich auftretenden Phasen mit 1. und 2. Phase bezeichnet. Die künstlichen oder „induzierten“ Phasen sind im *Kauffmann-White-Schema* nicht enthalten.

Die Gruppen- und Typen-Einteilung

Auf Grund verschiedener O-Antigene wird die *Salmonella*-Gruppe in O-Gruppen und Untergruppen, die folgendermaßen bezeichnet sind, eingeteilt:

[1], 2, 12	= A	11	35
[1], 4, [5], 12	= B	13, 22	38
[1], 4, 12, 27		[1], 13, 23	39
6, 7	= C	6, 14, 24	40
6, 8		(1), 6, 14, 25	41
(8), 20		16	42
[1], 9, 12	= D	17	43
(9), 46		18	44
3, 10	= E	21	45
3, 15		28	47
(3), (15), 34		30	48
1, 3, 19		31	49
			50

[1] Die Behauptung von G. S. WILSON im J. of Path. **73**, 281 (1957), ich hätte früher das 5-Antigen als ein echtes Vi-Antigen betrachtet, ist unzutreffend. Das sog. „Vi“-Antigen von *S. paratyphi B* und *S. typhi murium* ist mit dem „O“ 5-Antigen identisch, während das sog. „Vi“-Antigen von *S. paratyphi A* mit dem „O“ 2-Antigen identisch ist.

Die eckigen Klammern bedeuten, daß die betreffenden Antigene fehlen können, während die runden Klammern anzeigen, daß die betreffenden Antigene nicht vollständig vorhanden sind. Im Kauffmann-White-Schema sind die eckigen Klammern zwecks Vereinfachung fortgelassen.

Abgesehen von der ersten O-Gruppe (1, 2, 12), die nur einen einzigen Serotyp (*Salmonella paratyphi A*) enthält, kommen bei allen anderen O-Gruppen zahlreiche Serotypen vor. Wir sehen hier von einigen der letzten O-Gruppen, die noch nicht lange genug bekannt sind, ab. Da aber auch innerhalb dieser Gruppen bereits verschiedene Typen, die zum Teil noch nicht publiziert sind, gefunden wurden, so können wir mit mindestens 30 verschiedenen größeren O-Gruppen, die zahlreiche Typen enthalten, rechnen.

Bei der Einteilung dieser O-Gruppen in Typen, auf Grund verschiedener H-Antigene, müssen wir 2 Arten von Typen unterscheiden, diphasische und monophasische Typen. Bei den diphasischen Typen kommen mindestens 20 verschiedene H-Antigenkomplexe der 1. Phase mit mindestens 10 verschiedenen H-Antigenkomplexen der 2. Phase in allen möglichen Kombinationen vor. Die H-Antigenkomplexe der 1. Phase sind wie folgt bezeichnet: Bei meist diphasischen Typen = a—b—c—d—e, h—i—k—l, v—l, w—l, z_{13}—l, z_{28}—l, z_{40}—r—y—z—z_{29}—z_{35}—z_{41} usw. Bei meist monophasischen Typen = f, g—f, g, t—g, m—g, m, q—g, m, s—g, m, t—g, p—g, p, u—g, q—g, s, t—m, t—z_4, z_{23}—z_4, z_{32}—z_{36}—z_{38}. usw.

Als 2. Phasen treten folgende Antigenkomplexe auf:

1, 2—1, 5—1, 6—1, 7—z_6—z_{27}—z_{37}—z_{39}—e, n, x—e, n, z_{15}—e, n, z_{15}, z_{16}—e, n, z_{16}, z_{18}—l, w usw.

Betrachtet man das Kauffmann-White-Schema, so kann man feststellen, daß fast alle möglichen Kombinationen zwischen den Antigenen der 1. und 2. Phase verwirklicht sind, wenn wir alle O-Gruppen zusammenrechnen. Betrachtet man aber nur eine einzige Gruppe, wie z. B. die O-Gruppe 6, 7, so sieht man, daß bisher weniger als die Hälfte aller Kombinationen nachgewiesen sind.

Da es sich hierbei um etwa 20 verschiedene H-Antigenkomplexe der 1. Phase, die mit etwa 10 verschiedenen H-Antigenkomplexen der 2. Phase kombiniert sind, handelt, so sind im ganzen etwa 200 verschiedene Typen innerhalb einer jeden O-Gruppe möglich. Bei einer Zahl von 30 O-Gruppen kommen wir also auf etwa 6000 mögliche *Salmonella*-Typen, die mit Hilfe des Kauffmann-White-Schemas diagnostiziert werden können. Von diesen möglichen 6000 Typen sind bisher nur etwa 500 Typen festgestellt worden, so daß es kein Wunder ist, wenn im Laufe der letzten Jahre jährlich etwa 50 neue Typen gefunden wurden. Während der letzten 30 Jahre ist die Zahl der Serotypen von 20, die Bruce White analysiert hatte, auf etwa 500 gestiegen.

Es ist zu beachten, daß die Zahl von 6000 möglichen *Salmonella*-Typen als Minimalzahl betrachtet werden muß, weil das Kauffmann-White Schema aus praktischen Gründen vereinfacht ist. Wenn wir alle nachweisbaren H-Partialantigene berücksichtigen, so können wir etwa 40 H-Antigenkomplexe der 1. Phase und etwa 20 der 2. Phase unterscheiden. Wir gelangen so zu 800 verschiedenen H-Antigen-Kombinationen, die innerhalb der 30 O-Gruppen auftreten können, so daß wir zu einer Zahl von über 20000 Serotypen gelangen.

Um eine Vereinfachung zu erreichen, haben EDWARDS und KAUFFMANN ein gekürztes Antigenschema, mit dessen Hilfe die O-Gruppen- und Typen-Diagnose durch unabsorbierte Seren gestellt werden kann, vorgeschlagen. Man kann auf diese Weise die Zahl der H-Antigen-Komplexe der 1. Phase auf etwa 15 und diejenige der 2. Phase auf 2 H-Antigen-Komplexe beschränken. Mit Hilfe dieses vereinfachten Schemas kann man über 600 verschiedene *Salmonella*-Typen diagnostizieren, während mit Hilfe des originalen Kauffmann-White-Schemas etwa 6000 verschiedene Serotypen bestimmt werden können.

Man kann hieraus ersehen, daß man das diagnostische Antigenschema vereinfachen oder komplizieren kann, je nach den speziellen Bedürfnissen des Untersuchers. Das Kauffmann-White-Schema steht zwischen dem vereinfachten und dem kompletten Schema und ist mit Rücksicht auf die diagnostische Praxis ein Kompromiß.

Liste und Nachweis der Körper-Antigene

O 1

S. paratyphi A = 1, 2, 12.

Zum Nachweis des O 1-Antigens kann ein O-Serum von *S. paratyphi A* = 1, 2, 12, absorbiert mit *S. paratyphi A var. durazzo* = 2, 12, benutzt werden. In der Praxis ist dieses aber nicht nötig, weil Kulturen mit O 1-Antigen durch unabsorbierte 1, 2, 12- und 1, 3, 19-Seren agglutiniert werden. Außerhalb der E-Gruppe genügt es, zur Bestimmung des O 1-Antigens ein unabsorbiertes 1, 3, 19-Serum von *S. senftenberg* mit starkem 1 Titer, 1:5 verdünnt, in der Objektglas-Agglutination anzuwenden.

O 2

S. paratyphi A = 1, 2, 12.

Zum Nachweis des O-2-Antigens kann ein O-Serum von *S. paratyphi A* = 1, 2, 12, absorbiert mit *S. senftenberg* = 1, 3, 19 + *S. typhi* = 9, 12, gebraucht werden. In der Praxis ist dieses nicht nötig, da die Diagnose von *S. paratyphi A* durch unabsorbierte Seren (1, 2, 12 und 1, 3, 19) gestellt werden kann. Alle Stämme, die das O 1-Antigen enthalten, werden von beiden Seren agglutiniert, während Stämme ohne das O 1-Antigen vom 1, 2, 12-Serum agglutiniert werden. Bei dem typischen, biochemischen Verhalten von *S. paratyphi A* ist eine genaue Bestimmung der einzelnen O-Partialantigene überflüssig, speziell wenn gleichzeitig das H-Antigen nachweisbar ist.

O 3

S. london = 3, 10, 26.

Nachweis durch Agglutination in unabsorbierten 3, 10- und 3, 15-Seren.

O 4

S. essen = 4, 12.

Nachweis durch Agglutination in einem unabsorbierten 4, 12 Serum. Ein reines O4-Serum kann durch Absorption des 4, 12-Serums mit einer Kultur von *S. typhi* = 9, 12 (Stamm H 901) hergestellt werden.

„O“ 5

S. paratyphi B = 4, 5, 12.

Nachweis durch Agglutination in einem reinen „O“ 5-Serum, das durch Absorption eines 4, 5, 12-Serums mit einer 4, 12-Kultur hergestellt ist. Zur Erzielung hoher „O“ 5-Titer ist es ratsam, zur Immunisierung eine formalinisierte 4, 5, 12 O-Form (z. B. von *S. paratyphi B*) zu benutzen. In der Praxis kann auch ein unabsorbiertes Coli-Serum, hergestellt mit einer Coli-Kultur, die einen Teil des „O“5 Antigens enthält, angewandt werden.

O 6

S. thompson = 6, 7.

Nachweis durch Agglutination in unabsorbierten 6, 7-und 6,8-Seren. Das 6_1-Partialantigen kann durch ein *S. thompson*-Serum = 6_1, 6_2, 7, absorbiert mit *S. cholerae suis var. kunzendorf* Nr. 5210 = 6_2, 7, bestimmt werden. Das 6_2-Partialantigen kann durch ein *S. thompson*-Serum, absorbiert mit *S. cholerae suis var. kunzendorf* Nr. 1350 = 6_1, 7, bestimmt werden. Bei der näheren Analyse des O 6-Antigens muß der von EDWARDS (1) festgestellte 6_1-Formenwechsel berücksichtigt werden.

O 7

S. thompson = 6, 7.

Nachweis durch Agglutination in einem 6, 7-Serum, das mit *S. newport* = 6, 8 absorbiert ist. In der Praxis kann auch ein unabsorbiertes Serum von *S. lomita* = 7 benutzt werden, da ein bestimmter Stamm von *S. lomita* das O 6-Antigen nicht enthält.

O 8

S. newport = 6, 8.

Nachweis durch Agglutination in einem 6, 8-Serum, das mit einer 6,7-Kultur absorbiert ist. In der Praxis kann ein unabsorbiertes Serum von *S. amherstiana* oder *S. virginia* = (8) benutzt werden. Diese Typen enthalten nicht das O 6-Antigen und nur einen Teil des O 8-Antigens.

O 9

S. enteritidis = 9, 12.

Nachweis durch Agglutination in einem unabsorbierten 9, 12-Serum. Ein reines O 9-Serum kann durch Absorption des 9, 12-Serums mit *S. paratyphi A* + *S. reading* hergestellt werden.

O 10

S. london = 3, 10, 26.

Nachweis durch Agglutination in einem O-Serum von *S. london*, absorbiert mit *S. newington* = 3, 15 + *S. minnesota* = 21, 26. Das O 10-Antigen ist komplex gebaut und kann in mehrere Partialantigene eingeteilt werden (s. bei KAUFFMANN: „Die Bakteriologie der Salmonella-Gruppe“, S. 138).

O 11

S. aberdeen = 11.

Nachweis durch Agglutination in einem unabsorbierten O-Serum von *S. aberdeen*.

O 12

Komplex gebautes, übergreifendes Antigen = $12_1, 12_2, 12_3$.

S. paratyphi A und *S. typhi* (T 2) = $12_1, 12_3$.

S. reading = $12_1, 12_2$.

S. typhi (T 4) = $12_1, 12_2, 12_3$.

Ein 12_2 Serum kann durch Absorption des O-Serums von *S. typhi* (T 4) = 9, 12_1, 12_2, 12_3 mit dem Stamme von *S. typhi* (T 2) = 9, 12_1, 12_3 hergestellt werden [s. bei KAUFFMANN (7)].

O 13

S. poona = 13, 22.

Nachweis durch Agglutination in unabsorbierten 13, 22- und 13, 23-Seren.

O 14

S. boecker = 6, 14.

Nachweis durch Agglutination in einem O-Serum von *S. boecker* = 6, 14, absorbiert mit einem 6, 7-Stamme.

O 15

S. newington = 3, 15.

Nachweis in einem O-Serum von *S. newington*, absorbiert mit einem 3, 10-Stamme.

O 16

S. hvittingfoss = 16.

Nachweis durch Agglutination in einem unabsorbierten O-Serum von *S. hvittingfoss*.

O 17

S. kirkee = 17.

Nachweis durch Agglutination in einem unabsorbierten O-Serum von *S. kirkee*.

O 18

S. cerro = 18.

Nachweis durch Agglutination in einem unabsorbierten O-Serum von *S. cerro*.

O 19

S. senftenberg = 1, 3, 19.

Nachweis durch Agglutination in einem O-Serum von *S. senftenberg*, absorbiert mit *S. paratyphi A* + *S. london*.

O 20

S. kentucky = (8), 20.

Nachweis durch Agglutination in einem O-Serum von *S. kentucky*, absorbiert mit *S. newport*.

O 21

S. minnesota = 21, 26.

Nachweis durch Agglutination in einem O-Serum von *S. minnesota*, absorbiert mit *S. london*. In der Praxis kann ein unabsorbiertes O-Serum von *S. minnesota* benutzt werden.

0 22

S. poona = 13, 22.

Nachweis durch Agglutination in einem O-Serum von *S. poona*, absorbiert mit *S. grumpensis* = 13, 23.

0 23

S. worthington = 1, 13, 23.

Nachweis durch Agglutination in einem O-Serum von *S. worthington*, absorbiert mit *S. paratyphi A* + *S. poona*.

0 24

S. carrau = 6, 14, 24.

Nachweis durch Agglutination in einem O-Serum von *S. carrau*, absorbiert mit *S. onderstepoort*.

0 25

S. onderstepoort = (1), 6, 14, 25.

Nachweis durch Agglutination in einem O-Serum von *S. onderstepoort*, absorbiert mit *S. paratyphi A* + *S. carrau*.

0 26

S. minnesota = 21, 26.

Nachweis durch Agglutination in unabsorbierten 3, 10, 26- und 21, 26-Seren.

0 27

S. schleissheim = 4, 12, 27.

Nachweis durch Agglutination in einem O-Serum von *S. schleissheim*, absorbiert mit *S. reading* = 4, 12.

0 28

S. tel-aviv = 28.

Nachweis durch Agglutination in einem unabsorbierten O-Serum von *S. tel-aviv*.

0 29

Ballerup-Gruppe = 29.

Aus der *Salmonella*-Gruppe ausgeschlossen.

0 30

S. urbana = 30.

Nachweis durch Agglutination in einem O-Serum von *S. urbana*, absorbiert mit *S. riogrande* = 40.

0 31

S. djakarta = 31.

Nachweis durch Agglutination in einem unabsorbierten O-Serum 31.

0 32

Aus der *Salmonella*-Gruppe ausgeschlossen [s. KAUFFMANN (8)].

0 33

Arizona-Gruppe = 33.

Aus der *Salmonella*-Gruppe ausgeschlossen.

O 34

S. illinois = (3), (15), 34.

Nachweis durch Agglutination in einem O-Serum von *S. illinois*, absorbiert mit *S. newington* = 3, 15.

O 35

S. adelaide = 35.

Nachweis durch Agglutination in einem unabsorbierten O-Serum von *S. adelaide.*

O 36

S. poona und andere Typen.
S. poona = 13, 22, 36.
S. grumpensis = 13, 23, 36.
S. worthington = 1, 13, 23, 37.
S. mississippi = 1, 13, 23, 36, 37.

Nachweis durch Agglutination in einem O-Serum von *S. mississippi*, absorbiert mit *S. worthington.*

O 37

S. worthington und andere Typen.

Nachweis durch Agglutination in einem O-Serum von *S. worthington*, absorbiert mit *S. paratyphi A* + *S. grumpensis* [s. bei Kauffmann (9) und bei Edwards, McWhorter und Fife].

Auf Grund der Angaben von Edwards, McWhorter und Fife haben die O-Antigene weiterer Typen dieser O-Gruppe folgende Formeln:

S. atlanta = 13, 23, 36.
S. cubana = 1, 13, 23, 37.
S. wichita = 1, 13, 23, 37.

Ferner wurden folgende Kulturen mit ungewöhnlicher Antigenkombination gefunden:

13, 23, 36 : b : 1, 5
1, 13, 23, 37 : b : 1, 5
1, 13, 23, 36, 37 :—: 1, 5
1, 13, 22, 36, 37 : z : 1, 6

O 38

S. inverness = 38.

Nachweis durch Agglutination in einem unabsorbierten O-Serum von *S. inverness.*

O 39

S. champaign = 39.

Nachweis durch Agglutination in einem unabsorbierten O-Serum von *S. champaign.*

O 40

S. riogrande = 40.

Nachweis durch Agglutination in einem O-Serum von *S. riogrande*, absorbiert mit *S. urbana* = 30.

O 41

S. waycross = 41.

Nachweis durch Agglutination in einem unabsorbierten O-Serum von *S. waycross*.

O 42

S. weslaco = 42.

Nachweis durch Agglutination in einem unabsorbierten O-Serum von *S. weslaco*.

O 43

S. milwaukee = 43.

Nachweis durch Agglutination in einem unabsorbierten O-Serum von *S. milwaukee*.

O 44

S. niarembe = 44.

Nachweis durch Agglutination in einem unabsorbierten O-Serum von *S. niarembe*.

O 45

S. deversoir = 45.

Nachweis durch Agglutination in einem polyvalenten O-Serum von *S. deversoir* + *S. dugbe*, da die O-Antigene dieser beiden Typen verschieden sind.

O 46

S. strasbourg = (9), 46.

Nachweis durch Agglutination in einem O-Serum von *S. strasbourg*, absorbiert mit *S. enteritidis* = 9, 12.

O 47

S. bergen = 47.

Nachweis durch Agglutination in einem polyvalenten O-Serum von *S. bergen* + *S. kaolack*, da die O-Antigene dieser beiden Typen verschieden sind.

O 48

S. dahlem = 48.

Nachweis durch Agglutination in einem unabsorbierten O-Serum von *S. dahlem*.

O 49

S. bulawayo = (1), 49.

Nachweis durch Agglutination in einem O-Serum von *S. bulawayo*, absorbiert mit *S. allandale*.

O 50

S. greenside = 50.

Nachweis durch Agglutination in einem unabsorbierten O-Serum von *S. greenside*.

Liste und Nachweis der Geißel-Antigene

H a

S. paratyphi A = a.

Nachweis durch Agglutination in einem unabsorbierten H-Serum von *S. paratyphi A*. Das H a-Antigen ist komplex gebaut, doch sind die Partialantigene nicht im Kauffmann-White Schema angegeben. Auf Grund der Untersuchungen von L. le Minor (1, 2) lauten die Formeln für *S. sendai* und verwandte Typen wie folgt:

S. sendai und *S. bambesa* = a_1, a_2, a_4
S. miami = a_1, a_2, a_3
S. wuerzburg = a_1, a_3

Die H-Antigene von *S. paratyphi A* und *S. sendai* sind identisch [s. bei Kauffmann (1)].

Zwischen H a und H z_6 sowie H z_{10} können Antigenbeziehungen bestehen.

H b

S. paratyphi B (1. Phase) = b.

Nachweis durch Agglutination in einem unabsorbierten H-Serum von *S. hvittingfoss* (1. Phase). Das H b-Antigen ist komplex gebaut, so enthält z. B. *S. schleissheim* einen Sonderfaktor, der mit z_{12} bezeichnet wurde.

H c

S. paratyphi C (1. Phase) = c.

Nachweis durch Agglutination in einem unabsorbierten H-Serum von *S. cholerae suis* (1. Phase). Das H c-Antigen ist komplex gebaut und wurde von Edwards, Kauffmann und van Oye sowie von Kauffmann, Edwards und McWhorter wie folgt angegeben:

S. paratyphi C = c_1, c_2
S. cholerae suis = c_1
S. quiniela = c_1, c_2, c_3
S. decatur = c_1, c_2, c_3

H d

S. typhi = d.

Nachweis durch Agglutination in einem unabsorbierten H-Serum von *S. typhi*. Das H d-Antigen ist komplex gebaut und wurde von Hormaeche und Peluffo wie folgt angegeben:

S. typhi = d, d_1
S. stanley = d, d_3
S. amersfoort = d, d_2, d_3
S. muenchen = d, d_3, d_4

H e

S. eastbourne (1. Phase) = e, h.

S. abortus equi = e, n, x.

Nachweis durch Agglutination in unabsorbierten H-Seren von *S. eastbourne* (1. Phase) und *S. abortus equi*.

H f

S. derby = f, g.

Nachweis durch Agglutination in einem H-Serum von *S. derby*, absorbiert mit *S. essen* + *S. budapest*.

H g

S. enteritidis = g, m.

Nachweis durch Agglutination in unabsorbierten H-Seren von *S. enteritidis* und *S. dublin* = g, p.

Der g . . .-Antigenkomplex wurde von KAUFFMANN (10) wie folgt angegeben:

S. enteritidis = g, o, m, z_1, z_2
S. blegdam = g, o, m, q, z_1
S. moscow = g, o, q, z_3.

H h

S. eastbourne (1. Phase) = e, h.

Nachweis durch Agglutination in einem H-Serum von *S. eastbourne* (1. Phase), absorbiert mit *S. dar-es-salaam* (2. Phase).

Das h-Antigen ist komplex gebaut, doch sind die Partial-Antigene nicht im Kauffmann-White-Schema angegeben. Neuerdings ist der Typ *S. kaposvár* = 4, 5, 12:e, (h):1, 5, der nicht das volle h-Antigen enthält, fortgelassen und mit *S. rea, ding* kombiniert worden, so daß die Formel von *S. reading* nun mit 4, 5, 12:e-h:1, 5 angegeben ist.

H i

S. typhi murium (1. Phase) = i.

Nachweis durch Agglutination in einem unabsorbierten H-Serum von *S. aberdeen* (1. Phase) oder von *S. bonariensis* (1. Phase). Zwischen H i und H r können Antigenbeziehungen bestehen.

H j

Induzierte Phase von *S. typhi* = j.

[Siehe bei KAUFFMANN (11).]

H k

S. thompson (1. Phase) = k.

Nachweis durch Agglutination in einem unabsorbierten H-Serum von *S. thompson* (1. Phase).

H l

S. london (1. Phase) = l, v.

Nachweis durch Agglutination in unabsorbierten H-Seren von *S. london* (1. Phase) und von *S. dar-es-salaam* (1. Phase) = l, w.

H m

S. oranienburg = m, t.

Nachweis durch Agglutination in einem H-Serum von *S. oranienburg*, absorbiert mit *S. senftenberg* + *S. berta*.

H n

S. abortus equi = e, n, x.

Nachweis durch Agglutination in e, n . . .-Seren, die mit *S. eastbourne* (1. Phase) absorbiert sind.

H o

S. enteritidis = g, o, m.

Nachweis in der Praxis nicht nötig.

H p

S. dublin = g, p.

Nachweis durch Agglutination in einem H-Serum von *S. dublin*, absorbiert mit *S. enteritidis*.

H q

S. moscow = g, q.

Nachweis durch Agglutination in einem H-Serum von *S. moscow*, absorbiert mit *S. enteritidis*.

H r

S. virchow (1. Phase) = r.

Nachweis durch Agglutination in einem unabsorbierten H-Serum von *S. virchow*. Zwischen H r und H i können Antigenbeziehungen bestehen.

H s

S. montevideo = g, m, s.

Nachweis durch Agglutination in einem H-Serum von *S. montevideo*, absorbiert mit *S. oranienburg* + *S. enteritidis*.

H t

S. oranienburg = m, t.

Nachweis durch Agglutination in einem H-Serum von *S. oranienburg*, absorbiert mit *S. montevideo*.

Das komplex gebaute t-Antigen wurde von Kauffmann (12) wie folgt eingeteilt:

S. oranienburg = t_1, t_2, t_3
S. senftenberg = t_1, t_2
S. berta = t_1, t_3

H u

S. rostock = g, p, u.

Nachweis durch Agglutination in einem H-Serum von *S. rostock*, absorbiert mit *S. dublin*.

H v

S. london = l, v (1. Phase).

Nachweis durch Agglutination in einem H-Serum von *S. london* (1. Phase), absorbiert mit *S. dar-es-salaam* (1. Phase) + *S. uganda* (1. Phase).

H w

S. dar-es-salaam (1. Phase) = l, w.

Nachweis durch Agglutination in einem H-Serum von *S. dar-es-salaam* (1. Phase), absorbiert mit *S. london* (1. Phase) + *S. uganda* (1. Phase).

H x

S. abortus equi = e, n, x.

Nachweis durch Agglutination in einem H-Serum von *S. abortus equi*, absorbiert mit *S. san diego* (2. Phase). Dieses „x"-Serum enthält in Wirklichkeit die x, z_{16} Antikörper, doch ist es praktischer, ein derartiges Serum zu benutzen, da ein reines x-Serum für die Objektglas-Agglutination oft zu schwach ist.

Während die Mehrzahl der Typen, welche in der 2. Phase den e, n . . .-Antigenkomplex enthalten, entweder vom diagnostischen „x"-Serum (= x, z_{16}) oder vom „z_{15}"-Serum agglutiniert werden, reagieren einige Typen, wie z. B. *Salmonella uphill*, sowohl in „x"-, als auch in „z_{15}"-Seren. Diese Typen haben die Antigenstruktur e, n, z_{15}, z_{16}, sind aber im Kauffmann-White Schema mit e, n, z_{15} angegeben, weil die H-Antigene z_{16}, z_{17}, z_{18} und z_{19} in diesem vereinfachten Schema nicht berücksichtigt sind.

S. abortus equi (2. Phase) = e, n, x, z_{16}
S. san diego (2. Phase) = e, n, z_{15}, z_{17}
S. chester (2. Phase) = e, n, x, z_{17}
S. dar-es-salaam (2. Phase) = e, n, z_{16}, z_{18}.

[Siehe bei KAUFFMANN (13).]

H y

S. bareilly (1. Phase) = y.

Nachweis durch Agglutination in einem unabsorbierten H-Serum von *S. bareilly*.

H z

S. poona (1. Phase) = z.

Nachweis durch Agglutination in einem unabsorbierten H-Serum von *S. poona* (1. Phase). Zwischen H z und H z_{35} bestehen Antigenbeziehungen, so daß es ratsam ist, absorbierte H-Seren zu benutzen.

H z_1

S. blegdam = g, o, m, q, z_1.

Nachweis in der Praxis nicht nötig.

H z_2

S. enteritidis = g, o, m, z_1, z_2.

Nachweis in der Praxis nicht nötig.

H z_3

S. moscow = g, o, q, z_3.

Nachweis in der Praxis nicht nötig.

H z_4

S. cerro = z_4, z_{23}, z_{25}.

Nachweis durch Agglutination in unabsorbierten H-Seren von z_4, z_{23} ... und z_4, z_{24} ...

H z_5

Induzierte Phase von *S. schleissheim* = z_5.

(Siehe bei Kauffmann und Tesdal.)

H z_6

S. kentucky (2. Phase) = z_6.

Nachweis durch Agglutination in einem H-Serum von *S. kentucky* (2. Phase), absorbiert mit *S. thompson* (2. Phase). Die z_6-Phase kann als „unspezifische" Phase, die mit der 1, 5-Phase nahe verwandt ist, aufgefaßt werden. Zwischen H a und H z_6 können Antigenbeziehungen bestehen.

H z_7

S. berta = f, g, t, z_7, z_8.

Nachweis in der Praxis nicht nötig.

(Siehe bei Hormaeche, Peluffo und Salsamendi.)

H z_8

S. senftenberg = g, s, t, z_8.

S. derby = f, g, z_8.

Nachweis in der Praxis nicht nötig.

(Siehe bei Hormaeche, Peluffo und Salsamendi.)

H z_9

S. budapest = g, t, z_8, z_9.

S. senftenberg = g, s, t, z_8, z_9.

Nachweis in der Praxis nicht nötig.

(Siehe bei Rauss.)

H z_{10}

S. glostrup (1. Phase) = z_{10}.

Nachweis durch Agglutination in einem unabsorbierten H-Serum von *S. glostrup* (1. Phase). Zwischen H a und H z_{10} können Antigenbeziehungen bestehen.

H z_{11}

Induzierte Phase von *S. paratyphi A* = z_{11}.

[Siehe bei Bruner und Edwards (1).]

H z_{12}

S. schleissheim = b, z_{12}.

Nachweis in der Praxis nicht nötig.

[Siehe bei Kauffmann (14).]

H z_{13}

S. uganda (1. Phase) = l, z_{13}.

Nachweis durch Agglutination in einem H-Serum von *S. uganda* (1. Phase), absorbiert mit *S. london* (1. Phase) + *S. dar-es-salaam* (1. Phase) + *S. javiana* (1. Phase).

[Siehe bei EDWARDS und BRUNER (2).]

H z_{14}

Ballerup-Gruppe = z_{14}.

Aus der *Salmonella*-Gruppe ausgeschlossen.

H z_{15}

S. san diego (2. Phase) = e, n, z_{15}, z_{17}.

Nachweis durch Agglutination in einem H-Serum von *S. san diego*, absorbiert mit *S. chester* (2. Phase).

[Siehe bei KAUFFMANN (13).]

H z_{16}

S. abortus equi = e, n, x, z_{16}.

Nachweis durch Agglutination in einem H-Serum von *S. abortus equi*, absorbiert mit *S. chester* (2. Phase).

H z_{17}

S. chester (2. Phase) = e, n, x, z_{17}.

Nachweis in der Praxis nicht nötig.

[Siehe bei KAUFFMANN (13).]

H z_{18}

S. dar-es-salaam (2. Phase) = e, n, z_{16}, z_{18}.

Nachweis in der Praxis nicht nötig.

[Siehe bei KAUFFMANN (13).]

H z_{19}

Partialantigen verschiedener e, n . . .-Phasen.

Nachweis in der Praxis nicht nötig.

EDWARDS und BRUNER (3) gaben die Formeln verschiedener Typen mit e, n . . .-Phase wie folgt an:

S. abortus equi = e, n, x, z_{16}, z_{19}
S. urbana = e, n, x, z_{16}
S. chester = e, n, x, z_{17}, z_{19}
S. san diego = e, n, z_{15}, z_{17}, z_{19}
S. dar-es-salaam = e, n, z_{16}, z_{18}

H z_{20}, z_{21} und z_{22}

Aus der *Salmonella*-Gruppe ausgeschlossen.

[Siehe bei KAUFFMANN (8).]

H z_{23}

S. cerro = z_4, z_{23}, z_{25}.

Nachweis durch Agglutination in einem H-Serum von *S. cerro*, absorbiert mit *S. tallahassee* + *S. duesseldorf*.

H z_{24}

S. duesseldorf = z_4, z_{24}.

Nachweis durch Agglutination in einem H-Serum von *S. duesseldorf*, absorbiert mit *S. cerro* + *S. tallahassee*.

H z_{25}

S. cerro = z_4, z_{23}, z_{25}.

Nachweis in der Praxis nicht nötig.

H z_{26}

Arizona-Gruppe = z_4, z_{23}, z_{26}.

H z_{27}

S. senftenberg (2. Phase) = z_{27} oder *S. simsbury*.

Nachweis in einem unabsorbierten H-Serum von *S. simsbury*.

H z_{28}

S. javiana (1. Phase) = l, z_{28}.

Nachweis durch Agglutination in einem H-Serum von *S. javiana* (1. Phase), absorbiert mit *S. uganda* (1. Phase) + *S. potsdam* (1. Phase).

H z_{29}

S. tennessee = z_{29}.

Nachweis durch Agglutination in einem unabsorbierten H-Serum von *S. tennessee*.

H z_{30} und z_{31}

Ballerup-Gruppe.

Aus der *Salmonella*-Gruppe ausgeschlossen.
[Siehe bei Edwards (2).]

H z_{32}

S. tallahassee = z_4, z_{32}.

Nachweis durch Agglutination in einem H-Serum von *S. tallahassee*, absorbiert mit *S. cerro* + *S. duesseldorf* + *S. wangata*.

H z_{33}

Induzierte Phase von *S. minnesota*, kommt aber auch unter natürlichen Verhältnissen vor.

(Siehe bei Edwards und Moran.)

H z_{34}

Induzierte Phase von *S. senftenberg*.
(Siehe bei EDWARDS, MORAN und BRUNER.)

H z_{35}

S. chittagong (2. Phase) = z_{35}.

Nachweis durch Agglutination in einem H-Serum von *S. chittagong* (2. Phase), absorbiert mit *S. paratyphi A* + *S. paratyphi B* (1. Phase) + *S. poona* (z).

Falls man ein H-Serum von *S. cairina* (z_{35}) benutzt, so wird es mit *S. paratyphi A* + *S. poona* (z) absorbiert.

H z_{36}

S. weslaco = z_{36}.

Nachweis durch Agglutination in einem H-Serum von *S. weslaco*, absorbiert mit *S. lille*.

H z_{37}

S. wichita (2. Phase) = z_{37}.

Nachweis durch Agglutination in einem unabsorbierten H-Serum von *S. wichita* (2. Phase).

H z_{38}

S. lille = z_{38}.

Nachweis durch Agglutination in einem H-Serum von *S. lille*, absorbiert mit *S. weslaco*.

H z_{39}

S. springs (2. Phase) = z_{39}.

Nachweis durch Agglutination in einem H-Serum von *S. springs* (2. Phase), absorbiert mit *S. paratyphi A*.

H z_{40}

S. rutgers (1. Phase) = l, z_{40}.

Nachweis durch Agglutination in einem H-Serum von *S. rutgers* (1. Phase), absorbiert mit *S. give* (1. Phase).

H z_{41}

S. karamoja (1. Phase) = z_{41}.

Nachweis durch Agglutination in einem H-Serum von *S. karamoja* (1. Phase).

H 1

Gemeinsames Antigen der ,,unspezifischen“ Phasen. Nachweis durch Agglutination in unabsorbierten 1, 2- und 1, 5-Seren.

H 2

S. paratyphi B (2. Phase) = 1, 2.

Nachweis durch Agglutination in einem H-Serum von *S. paratyphi B* (2.Phase), absorbiert mit *S. bredeney* (2. Phase) + *S. thompson* (2. Phase) + *S. panama* (2. Phase).

H 3

S. newport (2. Phase) = 1, 2, 3.

Nachweis in der Praxis nicht nötig, da das H-3-Antigen aus dem Kauffmann-White-Schema zwecks Vereinfachung fortgelassen wurde. Nachweis durch Agglutination in einem H-Serum von *S. newport* (2. Phase), absorbiert mit *S. muenchen* (2. Phase).

H 4

Gemeinsames Antigen ,,unspezifischer" Phasen, im Kauffmann-White-Schema nicht angegeben.

H 5

S. thompson (2. Phase) = 1, 5.

Nachweis durch Agglutination in einem H-Serum von *S. thompson* (2. Phase), absorbiert mit *S. anatum* (2. Phase) + *S. newport* (2. Phase).

Es ist seit langem bekannt, daß die 1, 5-Phasen komplex gebaut sind und in mindestens 3—4 verschiedene Phasen aufgeteilt werden können. Kürzlich hat L. LE MINOR (2) bei der Untersuchung von *S. sendai* und verwandten Typen diese Unterschiede innerhalb der 1, 5-Phasen wie folgt ausgedrückt:

S. sendai, 2. Phase = A, B
S. bambesa, 2. Phase = A, C
S. miami, 2. Phase = A, B, C, D
S. wuerzburg, 2. Phase = A, B, C, D

H 6

S. anatum (2. Phase) = 1, 6.

Nachweis durch Agglutination in einem H-Serum von *S. anatum* (2. Phase), absorbiert mit *S. muenster* (2. Phase) + *S. thompson* (2. Phase).

H 7

S. bredeney (2. Phase) = 1, 7.

Nachweis durch Agglutination in einem H-Serum von *S. bredeney* (2. Phase), absorbiert mit *S. newport* (2. Phase) + *S. panama* (2. Phase) + *S. muenster* (2. Phase).

Die 1, 7-Phasen sind komplex gebaut, doch sind diese Unterschiede nicht im Kauffmann-White-Schema berücksichtigt. *S. gaminara* hat einen besonderen Faktor in der 2. Phase = 1, 7 +.

H 8 und 9

Partialantigene verschiedener 2. Phasen.
[Siehe bei KAUFFMANN (15).]

H 10 und 11

Induzierte Phasen der 1, 5-Phase.
[Siehe bei BRUNER und EDWARDS (2).]

Polyvalente Immunseren

Zur Vereinfachung der Antigenbestimmung werden polyvalente Immunseren in der Objektglas-Agglutination benutzt, worauf zur näheren Bestimmung monovalente Seren angewandt werden.

Polyvalente O-Seren

OA

S. paratyphi A	= 1, 2, 12
S. paratyphi B	= 4, 5, 12
S. enteritidis	= 9, 12
S. strasbourg	= (9), 46
S. london	= 3, 10, 26
S. senftenberg	= 1, 3, 19
S. minnesota	= 21, 26

OB

S. thompson	= 6, 7
S. newport	= 6, 8
S. aberdeen	= 11
S. poona	= 13, 22
S. grumpensis	= 13, 23
S. carrau	= 6, 14, 24
S. kentucky	= (8), 20

OC

S. hvittingfoss	= 16
S. kirkee	= 17
S. cerro	= 18
S. tel-aviv	= 28
S. urbana	= 30
S. adelaide	= 35

OD

S. inverness	= 38
S. champaign	= 39
S. riogrande	= 40
S. waycross	= 41
S. weslaco	= 42
S. milwaukee	= 43

OE

S. djakarta	= 31
S. niarembe	= 44
S. deversoir + *S. dugbe*	= 45
S. bergen + *S. kaolack*	= 47
S. dahlem	= 48
S. bulawayo	= (1), 49
S. greenside	= 50

Polyvalente H-Seren

HA

S. paratyphi A	= a
S. hvittingfoss	= b
S. cholerae suis	= c
S. typhi	= d
S. typhi murium	= i
S. glostrup	= z_{10}
S. tennessee	= z_{29}

HB

S. eastbourne	= e, h
S. abortus equi	= e, n, x
S. san diego	= e, n, z_{15}
S. derby	= f, g
S. montevideo	= g, m, s
S. dublin	= g, p
S. oranienburg	= m, t

HC

S. thompson	= k
S. london	= l, v
S. dar-es-salaam	= l, w
S. virchow	= r
S. bareilly	= y
S. poona	= z
S. cerro	= z_4, z_{23}

HD

S. cairina	= z_{35}
S. weslaco	= z_{36}
S. wichita	= z_{37}
S. lille	= z_{38}
S. springs	= z_{39}
S. karamoja	= z_{41}

HE

S. paratyphi B	= 1, 2
S. thompson	= 1, 5
S. anatum	= 1, 6
S. bredeney	= 1, 7
S. kentucky	= z_6

Kauffmann-White-Schema (1956)

Diagnostisches Antigenschema

Typ	Körper-Antigene	Geißel-Antigene	
		1. Phase	2. Phase
Gruppe A			
S. paratyphi A	1, 2, 12	a	—
Gruppe B			
S. kisangani	1, 4, 5, 12	a	1, 2
S. hessarek	4, 12	a	1, 5
S. fulica	4, 5, 12	a	1, 5
S. arechavaleta	4, 5, 12	a	1, 7
S. bispebjerg	1, 4, 12	a	e, n, x
S. abortus equi	4, 12	—	e, n, x
S. tinda	1, 4, 12, 27	a	e, n, z_{15}
S. paratyphi B	1, 4, 5, 12	b	1, 2
S. java	1, 4, 5, 12	b	[1, 2]
S. limete	1, 4, 12, 27	b	1, 5
S. uppsala	4, 12, 27	b	1, 7
S. abony	1, 4, 5, 12	b	e, n, x
S. abortus bovis	1, 4, 12, 27	b	e, n, x
S. wagenia	1, 4, 12, 27	b	e, n, z_{15}
S. schleissheim	4, 12, 27	b	—
S. wien	4, 12	b	l, w
S. abortus ovis	4, 12	c	1, 6
S. altendorf	4, 12	c	1, 7
S. womba	4, 12, 27	c	1, 7
S. bury	4, 12, 27	c	z_6
S. stanley	4, 5, 12	d	1, 2
S. cairo	1, 4, 12, 27	d	1, 2
S. schwarzengrund	1, 4, 12, 27	d	1, 7
S. sarajane	4, 12, 27	d	e, n, x
S. duisburg	1, 4, 12	d	e, n, z_{15}
S. salinatis	4, 12	d, e, h	d, e, n, z_{15}
S. saint-paul	1, 4, 5, 12	e, h	1, 2
S. reading[1]	4, 5, 12	e, h	1, 5
S. kaapstad	4, 12	e, h	1, 7
S. chester	4, 5, 12	e, h	e, n, x
S. san diego	4, 5, 12	e, h	e, n, z_{15}
S. derby	1, 4, 12	f, g	—
S. essen	4, 12	g, m	—
S. hato	4, 5, 12	g, m, s	—
S. california	4, 12	g, m, t	—
S. kingston	1, 4, 12, 27	g, s, t	—
S. budapest	1, 4, 12	g, t	—
S. banana	4, 5, 12	m, t	—
S. typhi murium	1, 4, 5, 12	i	1, 2
S. agama	4, 12	i	1, 6
S. gloucester	1, 4, 12, (27)	i	l, w
S. texas	4, 5, 12	k	e, n, z_{15}
S. azteca	4, 5, 12	l, v	1, 5
S. bredeney	1, 4, 12, 27	l, v	1, 7
S. kimuenza	1, 4, 12, 27	l, v	e, n, x
S. brandenburg	4, 12	l, v	e, n, z_{15}
S. kilwa	4, 12	l, w	e, n, x
S. vom	4, 12, 27	l, z_{13}, z_{28}	e, n, z_{15}
S. kunduchi	1, 4, 12, 27	l, z_{28}	1, 2
S. heidelberg	4, 5, 12	r	1, 2
S. bradford	4, 12	r	1, 5
S. africana	4, 12	r (i)	l, w
S. coeln	4, 5, 12	y	1, 2
S. teddington	4, 12, 27	y	1, 7
S. ball	1, 4, 12	y	e, n, x
S. shubra	4, 5, 12	z	1, 2
S. kiambu	4, 12	z	1, 5
S. indiana	1, 4, 12	z	1, 7
S. entebbe	1, 4, 12, 27	z	z_6
S. preston	1, 4, 12	z	l, w
S. stanleyville	4, 5, 12	z_4, z_{23}	[1, 2]
S. kalamu	4, 12	z_4, z_{24}	—
S. haifa	1, 4, 5, 12	z_{10}	1, 2
S. ituri	1, 4, 12	z_{10}	1, 5
S. tudu	4, 12	z_{10}	1, 6
S. fortune	4, 12, 27	z_{10}	z_6
S. brancaster	1, 4, 12	z_{29}	—
Gruppe C 1			
S. san juan	6, 7	a	1, 5
S. umhlali	6, 7	a	1, 6
S. austin	6, 7	a	1, 7
S. oslo	6, 7	a	e, n, x
S. denver	6, 7	a	e, n, z_{15}
S. brazzaville	6, 7	b	1, 2
S. edinburg	6, 7	b	1, 5
S. leopoldville	6, 7	b	z_6
S. georgia	6, 7	b	e, n, z_{15}
S. kotte	6, 7	b	z_{35}
S. paratyphi C	6, 7, Vi	c	1, 5
S. cholerae suis	6, 7	c	1, 5
S. typhi suis	6, 7	c	1, 5
S. decatur	6, 7	c	1, 5
S. birkenhead	6, 7	c	1, 6
S. kaduna	6, 7	c	e, n, z_{15}
S. mission	6, 7	d	1, 5
S. amersfoort	6, 7	d	e, n, x
S. livingstone	6, 7	d	l, w
S. lomita	6, 7	e, h	1, 5
S. norwich	6, 7	e, h	1, 6
S. braenderup	6, 7	e, h	e, n, z_{15}

[1] *S. reading* und *S. kaposvár* sind zu einem Typ vereint.

Typ	Körper-Antigene	Geißel-Antigene 1. Phase	Geißel-Antigene 2. Phase
S. ardwick	6, 7	f, g	—
S. montevideo	6, 7	g, m, s	—
S. menston	6, 7	g, s, t	—
S. oranienburg	6, 7	m, t	—
S. garoli	6, 7	i	1, 6
S. thompson	6, 7	k	1, 5
S. daytona	6, 7	k	1, 6
S. singapore	6, 7	k	e, n, x
S. concord	6, 7	l, v	1, 2
S. irumu	6, 7	l, v	1, 5
S. gelsenkirchen	6, 7	l, v	z_6
S. bonn	6, 7	l, v	e, n, x
S. potsdam	6, 7	l, v	e, n, z_{15}
S. colorado	6, 7	l, w	1, 5
S. ness-ziona	6, 7	l, z_{13}	1, 5
S. kenya	6, 7	l, z_{13}	e, n, x
S. makiso	6, 7	l, z_{28}	z_6
S. neukoelln	6, 7	l, z_{28}	e, n, z_{15}
S. virchow	6, 7	r	1, 2
S. infantis	6, 7	r	1, 5
S. nigeria	6, 7	r	1, 6
S. colindale	6, 7	r	1, 7
S. papuana	6, 7	r	e, n, z_{15}
S. richmond	6, 7	y	1, 2
S. bareilly	6, 7	y	1, 5
S. hartford	6, 7	y	e, n, x
S. mikawasima	6, 7	y	e, n, z_{15}
S. tosamanga	6, 7	z	1, 5
S. aequatoria	6, 7	z_4, z_{23}	e, n, z_{15}
S. eschweiler	6, 7	z_{10}	1, 6
S. djugu	6, 7	z_{10}	e, n, x
S. mbandaka	6, 7	z_{10}	e, n, z_{15}
S. jerusalem	6, 7	z_{10}	l, w
S. tennessee	6, 7	z_{29}	—
S. lille	6, 7	z_{38}	—

Gruppe C 2

Typ	Körper-Antigene	Geißel-Antigene 1. Phase	Geißel-Antigene 2. Phase
S. curacao	6, 8	a	1, 6
S. narashino	6, 8	a	e, n, x
S. nagoya	6, 8	b	1, 5
S. banalia	6, 8	b	z_6
S. gatuni	6, 8	b	e, n, x
S. utah	6, 8	c	1, 5
S. bronx	6, 8	c	1, 6
S. belem	6, 8	c	e, n, x
S. quiniela	6, 8	c	e, n, z_{15}
S. muenchen	6, 8	d	1, 2
S. manhattan	6, 8	d	1, 5
S. labadi	6, 8	d	z_6
S. newport	6, 8	e, h	1, 2
S. kottbus	6, 8	e, h	1, 5
S. tshiongwe	6, 8	e, h	e, n, z_{15}
S. lindenburg	6, 8	i	1, 2
S. takoradi	6, 8	i	1, 5
S. bonariensis	6, 8	i	e, n, x
S. aba	6, 8	i	e, n, z_{15}
S. blockley	6, 8	k	1, 5
S. litchfield	6, 8	l, v	1, 2
S. manchester	6, 8	l, v	1, 7
S. edmonton	6, 8	l, v	e, n, z_{15}
S. fayed	6, 8	l, w	1, 2
S. baragwanath	6, 8	m, t	1, 5
S. germiston	6, 8	m, t	e, n, x
S. bovis morbificans	6, 8	r	1, 5
S. hidalgo	6, 8	r	e, n, z_{15}
S. gold coast	6, 8	r	l, w
S. tananarive	6, 8	y	1, 5
S. alagbon	6, 8	y	1, 7
S. praha	6, 8	y	e, n, z_{15}
S. chailey	6, 8	z_4, z_{23}	—
S. duesseldorf	6, 8	z_4, z_{24}	—
S. tallahassee	6, 8	z_4, z_{32}	—
S. hadar	6, 8	z_{10}	e, n, x
S. glostrup	6, 8	z_{10}	e, n, z_{15}

Gruppe C 3

Typ	Körper-Antigene	Geißel-Antigene 1. Phase	Geißel-Antigene 2. Phase
S. sanga	(8)	b	1, 7
S. shipley	(8), 20	b	e, n, z_{15}
S. virginia	(8)	d	—
S. emek	(8), 20	g, m, s	—
S. kentucky	(8), 20	i	z_6
S. amherstiana	(8)	l, (v)	1, 6
S. hindmarsh	(8)	r	1, 5
S. corvallis	(8), 20	z_4, z_{23}	—
S. albany	(8), 20	z_4, z_{24}	—

Gruppe D 1

Typ	Körper-Antigene	Geißel-Antigene 1. Phase	Geißel-Antigene 2. Phase
S. sendai	1, 9, 12	a	1, 5
S. miami	1, 9, 12	a	1, 5
S. saarbruecken	1, 9, 12	a	1, 7
S. loma-linda	9, 12	a	e, n, x
S. durban	9, 12	a	e, n, z_{15}
S. onarimon	1, 9, 12	b	1, 2
S. frintrop	1, 9, 12	b	1, 5
S. mjimwema	1, 9, 12	b	e, n, x

Typ	Körper-Antigene	Geißel-Antigene 1. Phase	Geißel-Antigene 2. Phase
S. ridge	9, 12	c	z_6
S. alabama	9, 12	c	e, n, z_{15}
S. typhi	9, 12, Vi	d	—
S. ndolo	9, 12	d	1, 5
S. zega	9, 12	d	z_6
S. jaffna	1, 9, 12	d	z_{35}
S. eastbourne	1, 9, 12	e, h	1, 5
S. israel	9, 12	e, h	e, n, z_{15}
S. berta	9, 12	f, g, t	—
S. enteritidis	1, 9, 12	g, m	—
S. blegdam	9, 12	g, m, q	—
S. pensacola	9, 12	g, m, t	—
S. dublin	1, 9, 12	g, p	—
S. rostock	1, 9, 12	g, p, u	—
S. moscow	9, 12	g, q	—
S. neasden	9, 12	g, s, t	e, n, x
S. hamburg	1, 9, 12	g, t	—
S. hannover	9, 12	m, t	—
S. seremban	9, 12	i	1, 5
S. marylebone	9, 12	k	1, 2
S. claibornei	1, 9, 12	k	1, 5
S. mendoza	9, 12	l, v	1, 2
S. panama	1, 9, 12	l, v	1, 5
S. kapemba	9, 12	l, v	1, 7
S. goettingen	9, 12	l, v	e, n, z_{15}
S. victoria	1, 9, 12	l, w	1, 5
S. dar-es-salaam	1, 9, 12	l, w	e, n
S. napoli	1, 9, 12	l, z_{13}	e, n, x
S. javiana	1, 9, 12	l, z_{28}	1, 5
S. shoreditch	9, 12	r	e, n, z_{15}
S. wangata	9, 12	z_4, z_{23}	—
S. portland	9, 12	z_{10}	1, 5
S. canastel	9, 12	z_{29}	1, 5
S. penarth	9, 12	z_{35}	z_6
S. fresno	9, 12	z_{38}	—
S. gallinarum-pullorum	1, 9, 12	—	—
Gruppe D 2			
S. strasbourg	(9), 46	d	1, 7
S. haarlem	(9), 46	z	e, n, x
Gruppe E 1			
S. oxford	3, 10	a	1, 7
S. kalina	3, 10	b	1, 2
S. butantan	3, 10	b	1, 5
S. huvudsta	3, 10	b	1, 7
S. okefoko	3, 10	c	z_6
S. shangani	3, 10	d	1, 5
S. weybridge	3, 10	d	z_6
S. souza	3, 10	d	e, n, x
S. vejle	3, 10	e, h	1, 2
S. muenster	3, 10	e, h	1, 5
S. anatum	3, 10	e, h	1, 6
S. nyborg	3, 10	e, h	1, 7
S. newlands	3, 10	e, h	e, n, x
S. meleagridis	3, 10	e, h	l, w
S. westhampton	3, 10	g, s, t	—
S. falkensee	3, 10	i	e, n, z_{15}
S. zanzibar	3, 10	k	1, 5
S. nchanga	3, 10	l, v	1, 2
S. london	3, 10	l, v	1, 6
S. give	3, 10	l, v	1, 7
S. uganda	3, 10	l, z_{13}	1, 5
S. rutgers	3, 10	l, z_{40}	1, 7
S. seegefeld	3, 10	r (i)	1, 2
S. elisabethville	3, 10	r	1, 7
S. weltevreden	3, 10	r	z_6
S. simi	3, 10	r	e, n, z_{15}
S. amager	3, 10	y	1, 2
S. orion	3, 10	y	1, 5
S. stockholm	3, 10	y	z_6
S. bolton	3, 10	y	e, n, z_{15}
S. alexander	3, 10	z	1, 5
S. clerkenwell	3, 10	z	l, w
S. adabraka	3, 10	z_4, z_{23}	—
S. lexington	3, 10	z_{10}	1, 5
S. coquilhatville	3, 10	z_{10}	1, 7
S. cairina	3, 10	z_{35}	z_6
S. macallen	3, 10	z_{36}	—
Gruppe E 2			
S. goerlitz	3, 15	e, h	1, 2
S. new-haw	3, 15	e, h	1, 5
S. newington	3, 15	e, h	1, 6
S. selandia	3, 15	e, h	1, 7
S. cambridge	3, 15	e, h	l, w
S. new brunswick	3, 15	l, v	1, 7
S. kinshasa	3, 15	l, z_{13}	1, 5
S. lanka	3, 15	r	z_6
S. tuebingen	3, 15	y	1, 2
S. binza	3, 15	y	1, 5
S. manila	3, 15	z_{10}	1, 5

Typ	Körper-Antigene	Geißel-Antigene	
		1. Phase	2. Phase
Gruppe E 3			
S. minneapolis	(3), (15), 34	e, h	1, 6
S. canoga	(3), (15), 34	g, s, t	—
S. thomasville	(3), (15), 34	y	1, 5
S. illinois	(3), (15), 34	z_{10}	1, 5
S. harrisonburg	(3), (15), 34	z_{10}	1, 6
Gruppe E 4			
S. accra	1, 3, 19	b	z_6
S. niloese	1, 3, 19	d	z
S. liverpool	1, 3, 19	d	e, n, z_{15}
S. senftenberg	1, 3, 19	g, s, t	—
S. taksony	1, 3, 19	i	z_6
S. krefeld	1, 3, 19	y	l, w
S. schoeneberg	1, 3, 19	z	e, n, z_{15}
S. simsbury	1, 3, 19	—	z_{27}
S. llandoff	1, 3, 19	z_{29}	—
S. chittagong	(1), 3, 10, (19)	b	z_{35}
Gruppe F			
S. marseille	11	a	1, 5
S. luciana	11	a	e, n, z_{15}
S. pharr	11	b	e, n, z_{15}
S. chandans	11	d	e, n, x
S. chingola	11	e, h	1, 2
S. redhill	11	e, h	l, z_{13}, z_{28}
S. aberdeen	11	i	1, 2
S. veneziana	11	i	e, n, x
S. pretoria	11	k	1, 2
S. abaetetuba	11	k	1, 5
S. kisarawe	11	k	e, n, x
S. maracaibo	11	l, v	1, 5
S. senegal	11	r	1, 5
S. rubislaw	11	r	e, n, x
S. solt	11	y	1, 5
S. wentworth	11	z_{10}	1, 2
S. tel-hashomer	11	z_{10}	e, n, x
Gruppe G 1			
S. ibadan	13, 22	b	1, 5
S. friedenau	13, 22	d	1, 6

Typ	Körper-Antigene	Geißel-Antigene	
		1. Phase	2. Phase
S. borbeck	13, 22	l, v	1, 6
S. poona	13, 22	z	1, 6
S. bristol	13, 22	z	1, 7
S. roodepoort	13, 22	z_{10}	1, 5
S. clifton	13, 22	z_{29}	1, 5
Gruppe G 2			
S. mississippi	1, 13, 23	b	1, 5
S. bracknell	13, 23	b	1, 6
S. atlanta	13, 23	b	—
S. durham	13, 23	b	e, n, z_{15}
S. mishmar-haemek	1, 13, 23	d	1, 5
S. grumpensis	13, 23	d	1, 7
S. tel-el-kebir	13, 23	d	e, n, z_{15}
S. wichita	1, 13, 23	d	[z_{37}]
S. havana	1, 13, 23	f, g	—
S. luanshya	13, 23	g, s, (t)	—
S. okatie	13, 23	g, s, t	—
S. worcester	1, 13, 23	m, t	e, n, x
S. tunis	1, 13, 23	y	z_6
S. nachshonim	1, 13, 23	z	1, 5
S. worthington	1, 13, 23	z	l, w
S. ajiobo	13, 23	z_4, z_{23}	—
S. cubana	1, 13, 23	z_{29}	—
S. fanti	13, 23	z_{38}	—
Gruppe H			
S. heves	6, 14, 24	d	1, 5
S. florida	(1), 6, 14, 25	d	1, 7
S. albuquerque	6, 14, 24	d	z_6
S. charity	6, 14	d	e, n, x
S. teko	(1), 6, 14, 25	d	e, n, z_{15}
S. onderstepoort	(1), 6, 14, 25	e, (h)	1, 5
S. warragul	(1), 6, 14, 25	g, m	—
S. caracas	(1), 6, 14, 25	g, m, s	—
S. buzu	(1), 6, 14, 25	i	1, 7
S. boecker	6, 14	l, v	1, 7
S. horsham	(1), 6, 14, 25	l, v	e, n, x
S. carrau	6, 14, 24	y	1, 7
S. madelia	(1), 6, 14, 25	y	1, 7
S. homosassa	(1), 6, 14, 25	z	1, 5
S. sundsvall	(1), 6, 14, 25	z	e, n, x
S. siegburg	6, 14, 18	z_4, z_{23}	—
S. uzaramo	(1), 6, 14, 25	z_4, z_{24}	—

Typ	Körper-Antigene	Geißel-Antigene	
		1. Phase	2. Phase
Gruppe I			
S. brazil	16	a	1, 5
S. hull	16	b	1, 2
S. hvittingfoss	16	b	e, n, x
S. vancouver	16	c	1, 5
S. shamba	16	c	e, n, x
S. gaminara	16	d	1, 7
S. nottingham	16	d	e, n, z_{15}
S. weston	16	e, h	z_6
S. adeoyo	16	g, m	—
S. mobeni	16	g, s, (t)	—
S. merseyside	16	g, t	1, 5
S. rowbarton	16	m, t	—
S. szentes	16	k	1, 2
S. orientalis	16	k	e, n, z_{15}
S. shanghai	16	l, v	1, 6
S. welikada	16	l, v	1, 7
S. salford	16	l, v	e, n, x
S. kikoma	16	y	e, n, x
S. lisboa	16	z_{10}	1, 6
S. jacksonville	16	z_{29}	—
Gruppe J			
S. jangwani	17	a	1, 5
S. kinondoni	17	a	e, n, x
S. kirkee	17	b	1, 2
S. hillbrow	17	b	e, n, z_{15}
S. victoriaborg	17	c	1, 6
S. bleadon	17	(f), g, t	—
S. matadi	17	k	e, n, x
S. morotai	17	l, v	1, 2
S. michigan	17	l, v	1, 5
S. carmel	17	l, v	e, n, x
S. verity	17	e, n, z_{15}	1, 6
Gruppe K			
S. usumbura	18	d	1, 7
S. memphis	18	k	1, 5
S. cerro	18	z_4, z_{23}	—
S. blukwa	18	z_4, z_{24}	—
Gruppe L			
S. minnesota	21	b	e, n, x
S. magwa	21	d	e, n, x
S. ruiru	21	y	e, n, x
S. gwaai	21	z_4, z_{24}	—
Gruppe M			
S. solna	28	a	1, 5
S. dakar	28	a	1, 6
S. seattle	28	a	e, n, x
S. kaltenhausen	28	b	z_6
S. langford	28	b	e, n, z_{15}
S. mundonobo	28	d	1, 7
S. patience	28	d	e, n, z_{15}
S. ona	28	g, s, t	—
S. vinohrady	28	m, t	—
S. ilala	28	k	1, 5
S. taunton	28	k	e, n, x
S. leoben	28	l, v	1, 5
S. nashua	28	l, v	e, n, z_{15}
S. chicago	28	r	1, 5
S. kibusi	28	r	e, n, x
S. nima	28	y	1, 5
S. pomona	28	y	1, 7
S. tel-aviv	28	y	e, n, z_{15}
S. ezra	28	z	1, 7
S. umbilo	28	z_{10}	e, n, x
S. moroto	28	z_{10}	l, w
Gruppe N			
S. urbana	30	b	e, n, x
S. godesberg	30	g, m	—
S. landau	30	i	1, 2
S. morehead	30	i	1, 5
S. aqua	30	k	1, 6
S. angoda	30	k	e, n, x
S. donna	30	l, v	1, 5
S. matopeni	30	y	1, 2
S. bodjonegoro	30	z_4, z_{24}	—
Gruppe O			
S. djakarta	31	z_4, z_{24}	—
Gruppe P			
S. umhlatazana	35	a	e, n, z_{15}
S. yolo	35	c	—

Typ	Körper-Antigene	Geißel-Antigene 1. Phase	Geißel-Antigene 2. Phase
S. adelaide	35	f, g	—
S. agodi	35	g, t	—
S. monschaui	35	m, t	—
S. gambia	35	i	e, n, z_{15}
S. alachua	35	z_4, z_{23}	—

Gruppe Q

Typ	Körper-Antigene	Geißel-Antigene 1. Phase	Geißel-Antigene 2. Phase
S. sheffield	38	c	1, 5
S. thiaroye	38	e, h	1, 2
S. kasenyi	38	e, h	1, 5
S. korovi	38	g, m, s	—
S. mgulani	38	i	1, 2
S. inverness	38	k	1, 6
S. roan	38	l, v	e, n, x
S. lindi	38	r	1, 5
S. emmastad	38	r	1, 6
S. freetown	38	y	1, 5
S. colombo	38	y	1, 6

Gruppe R

Typ	Körper-Antigene	Geißel-Antigene 1. Phase	Geißel-Antigene 2. Phase
S. wandsworth	39	b	1, 2
S. champaign	39	k	1, 5

Gruppe S

Typ	Körper-Antigene	Geißel-Antigene 1. Phase	Geißel-Antigene 2. Phase
S. springs	40	a	z_{39}
S. rio grande	40	b	1, 5
S. johannesburg	1, 40	b	e, n, x
S. duval	1, 40	b	e, n, z_{15}
S. boksburg	40	g, s	e, n, z_{15}
S. allandale	1, 40	k	1, 6
S. degania	40	z_4, z_{24}	—
S. karamoja	40	z_{41}	1, 2

Gruppe T

Typ	Körper-Antigene	Geißel-Antigene 1. Phase	Geißel-Antigene 2. Phase
S. waycross	41	z_4, z_{23}	—

Gruppe U

Typ	Körper-Antigene	Geißel-Antigene 1. Phase	Geißel-Antigene 2. Phase
S. chinovum	42	b	1, 5
S. uphill	42	b	e, n, z_{15}
S. kampala	1, 42	c	z_6
S. kaneshie	1, 42	i	l, w
S. nairobi	42	r	—
S. rand	42	z	e, n, z_{15}
S. weslaco	42	z_{36}	—

Gruppe V

Typ	Körper-Antigene	Geißel-Antigene 1. Phase	Geißel-Antigene 2. Phase
S. berkeley	43	a	1, 5
S. milwaukee	43	f, g	—
S. ahuza	43	k	1, 5
S. kingabwa	43	y	1, 5

Gruppe W

Typ	Körper-Antigene	Geißel-Antigene 1. Phase	Geißel-Antigene 2. Phase
S. niarembe	44	a	l, w
S. christiansborg	44	z_4, z_{24}	—
S. guinea	44	z_{10}	—

Gruppe X

Typ	Körper-Antigene	Geißel-Antigene 1. Phase	Geißel-Antigene 2. Phase
S. deversoir	45	c	e, n, x
S. dugbe	45	d	1, 6
S. windhoek	45	g, t	1, 5

Gruppe Y

Typ	Körper-Antigene	Geißel-Antigene 1. Phase	Geißel-Antigene 2. Phase
S. bergen	47	i	e, n, z_{15}
S. kaolack	47	z	1, 6

Gruppe Z

Typ	Körper-Antigene	Geißel-Antigene 1. Phase	Geißel-Antigene 2. Phase
S. dahlem	48	k	e, n, z_{15}

Gruppe 49

Typ	Körper-Antigene	Geißel-Antigene 1. Phase	Geißel-Antigene 2. Phase
S. bulawayo	(1), 49	z	1, 5

Gruppe 50

Typ	Körper-Antigene	Geißel-Antigene 1. Phase	Geißel-Antigene 2. Phase
S. greenside	50	z	e, n

Bemerkungen zum Kauffmann-White-Schema

Im Kauffmann-White Schema sind nur diejenigen Typen mit gleicher Antigenformel angegeben, welche klinische Bedeutung haben, wie z. B. *S. paratyphi B* und *S. java*.

Aus diesem Grunde sind auch *S. sendai* und *S. miami* = [1], 9, 12 : a : 1, 5 in der Tabelle enthalten, während *S. wuerzburg* und *S. bambesa* = 9, 12 : a : 1, 5 ausgelassen sind. In der Praxis können diese beiden zuletzt genannten Typen als „*S. miami*" diagnostiziert werden.

S. isangi = 6, 7 : d : 1, 5 ist mit *S. mission* vereint, während *S. maritza* = 16 : 1, v : e, n, x zu *S. salford* gehört, obwohl biochemische und serologische Unterschiede zwischen diesen Typen bestehen.

S. reading = 4, 12 : e, h : 1,5 und *S. kaposvár* = 4, 5, 12 : e, (h) : 1, 5 sind zwecks Vereinfachung der Tabelle zu einem einzigen Typ vereint, so daß die Formel von *S. reading* mit 4, 5, 12 : e, h : 1, 5 angegeben ist.

„*S. kralendyk*" = 6, 7 : z_4, z_{24} :— wurde aus der Tabelle fortgelassen, da dieser Typ als *Arizona*-Kultur betrachtet wird. Dasselbe gilt auch von „*S. delplata*" = 1, 13, 23 : z_4, z_{23} :—.

Der im letzten Antigenschema (*Enterobacteriaceae*, 2. Auflage) angegebene Typ 6, 8 : b : e, n, z_{15} hat die Formel (8), 20 : b : e, n, z_{15} und heißt *S. shipley*.

S. verity = 17 : e, n, z_{15} : 1, 6 ist wahrscheinlich eine transduzierte Form, die aus zwei zweiten Phasen besteht. Dieser Typ wurde in das Antigenschema aufgenommen, weil er in der Natur gefunden wurde.

Vereinfachung des Kauffmann-White-Schemas

Ein vereinfachtes Antigenschema wurde von EDWARDS und KAUFFMANN vorgeschlagen, um die Typendiagnose zu erleichtern und absorbierte Seren möglichst zu vermeiden. Man muß hierbei aber in Kauf nehmen, daß eine größere Anzahl der Typen nicht exakt bestimmt und benannt werden kann. Im folgenden sind die zur vereinfachten Typenbestimmung nötigen Seren angegeben.

Körper-Antigen-Sera

A = S. paratyphi A
B = S. paratyphi B + S. schleissheim
C = S. thompson + S. newport + S. carrau + S. florida
D = S. gallinarum + S. strasbourg
E = S. anatum + S. newington
11 = S. aberdeen
13 = S. poona + S. grumpensis
16 = S. gaminara
17 = S. kirkee
18 = S. cerro
21 = S. minnesota
28 = S. tel-aviv
30 = S. urbana
31 = S. djakarta

35 = S. adelaide
38 = S. inverness
39 = S. champaign
40 = S. riogrande
41 = S. waycross
42 = S. weslaco
43 = S. milwaukee
44 = S. niarembe
45 = S. deversoir + S. dugbe
47 = S. bergen + S. kaolack
48 = S. dahlem
49 = S. bulawayo
50 = S. greenside
Vi = Ballerup original

Geißel-Antigen-Sera

a = S. paratyphi A
b = S. java oder S. hvittingfoss (1. Phase)
c = S. cholerae suis (1. Phase)
d = S. typhi
h = S. 4, 5, 12 : e, h :—
G = S. derby + S. oranienburg + S. dublin + S. simsbury
i = S. typhi murium oder S. bonariensis (1. Phase)
k = S. thompson (1. Phase)
L = S. worthington (1. Phase)
r = S. rubislaw (1. Phase)
y = S. madelia (1. Phase)
z = S. poona (1. Phase)
z_4 = S. cerro + S. duesseldorf + S. tallahassee
z_{10} = S. illinois (1. Phase)
z_{29} = S. tennessee
z_{35} = S. cairina
z_{36} = S. weslaco
z_{37} = S. wichita (2. Phase)
z_{38} = S. lille
z_{39} = S. springs (2. Phase)
z_{41} = S. karamoja (1. Phase)
l = S. newport (2. Phase) + S. thompson (2. Phase) + S. kentucky (2. Phase)
e, n = S. abortus equi

Unter der Bezeichnung „G“ sind alle diejenigen H-Antigene, die zum g . . . Komplex gehören, zusammengefaßt.:

f, g	g, m, s	g, q
f, g, t	g, m, t	g, s, t
g, m	g, p	m, t
g, m, q	g, p, u	z_{27}

In der entsprechenden Weise sind die l, v — l, w — l, z_{13}, l, z_{28} und l, z_{40} Phasen unter der Bezeichnung „L“ zusammengefaßt.

Die miteinander verwandten H-Antigene z_4, z_{23} — z_4, z_{24} und z_4, z_{32} sind mit „z_4“ bezeichnet.

Die H-Antigene der 2. Phase mit dem l . . .-Antigenkomplex sind mit „l“ angegeben und umfassen l, 2 — l, 5 — l, 6 — l, 7 und z_6.

Die H-Antigene der 2. Phase mit dem e, n . . .-Antigenkomplex sind mit „e, n“ bezeichnet und umfassen die Phasen e, n, x — e, n, z_{15} — e, n, z_{15}, z_{16} — e, n, z_{16}, z_{18} usw.

Liste der Salmonella Typen

Salmonella aba = 6, 8 : i : e, n, z_{15}.
Salmonella abaetetuba = 11 : k : 1, 5.
EDWARDS and MCWHORTER: Publ. Health Labor. **10**, 103 (1952).
Salmonella aberdeen = 11 : i : 1, 2.
SMITH: J. of. Hyg. **34**, 351 (1934).
Salmonella abony = 1, 4, 5, 12 : b : e, n, x.
KAUFFMANN: Acta path. scand. (Copenh.) **17**, 1 (1940).
Salmonella abortus bovis = 1, 4, 12, 27 : b : e, n, x.
BERNARD: Z. Hyg. **117**, 352 (1935).
KAUFFMANN: Z. Hyg. **120**, 177 (1937).
Salmonella abortus canis = 4, 5, 12 : b : z_5.
GARD: Z. Hyg. **121**, 139, 432 (1938).
Mit *S. paratyphi B* vereint, da die z_5 Phase eine induzierte Phase ist. KAUFFMANN: Acta path. scand. (Copenh.) **16**, 347 (1939).
Salmonella abortus equi = 4, 12 :—: e, n, x.
Salmonella Subcommittee: J. of Hyg. **34**, 333 (1934).
Salmonella abortus ovis = 4, 12 : c : 1, 6.
KAUFFMANN u. MITSUI: Z. Hyg. **111**, 749 (1930).
LOVELL: J. of Path. **34**, 13 (1931).
Salmonella Subcommittee: J. of Hyg. **34**, 333 (1934).
Salmonella accra = 1, 3, 19 : b : z_6.
Salmonella adabraka = 3, 10 : z_4, z_{23} :—.
Salmonella adelaide = 35 : f, g :—.
ATKINSON: Austral. J. Exper. Biol. a. Med. Sci. **21**, 171 (1943).
Salmonella adeoyo = 16 : g, m :—.
Salmonella aequatoria = 6, 7 : z_4, z_{23} : e, n, z_{15}.
KAUFFMANN and REUL: Acta path. scand. (Copenh.) **26**, 335 (1949).
Salmonella aertrycke = *Salmonella typhi murium.*
Salmonella africana = 4, 12 : r (i) : l, w.
Salmonella agama = 4, 12 : i : 1, 6.
Salmonella agodi = 35 : g, t :—.
Salmonella ahuza = 43 : k : 1, 5.
HIRSCH and SAPIRO-HIRSCH: Acta med. orient. **14**, 298 (1955).
Salmonella ajiobo = 13, 23 : z_4, z_{23} :—.
Salmonella alabama = 9, 12 : c : e, n, z_{15}.
EDWARDS, JOHNSON, MCWHORTER and WEED: Publ. Health Labor. **12**, 15 (1954).

Salmonella alachua = 35 : z_4, z_{23} :—.
LOWERY, SMITH, GALTON and EDWARDS: J. Bacter. **66**, 118 (1953).
Salmonella alagbon = 6, 8 : y : 1, 7.
Salmonella albany = (8), 20 : z_4, z_{24} :—.
WEST and EDWARDS: U. S. Publ. Health Rep. **66**, 1062 (1951).
Salmonella albuquerque = 6, 14, 24 : d : z_6.
EDWARDS and McWHORTER: Publ. Health Labor. **12**, 141 (1954).
Salmonella alexander = 3, 10 : z : 1, 5.
LOWE and BOKKENHEUSER: Im Druck.
Salmonella allandale = 1, 40 : k : 1, 6.
EDWARDS and HERMANN: U. S. Publ. Health Rep. **65**, 547 (1950).
Salmonella altendorf = 4, 12 : c : 1, 7.
HOHN: Zbl. Bakter. I. Orig. **146**, 215 (1940).
Salmonella amager = 3, 10 : y : 1, 2.
KAUFFMANN: Acta path. scand. (Copenh.) **16**, 347 (1939).
Salmonella amersfoort = 6, 7 : d : e, n, x.
HENNING: J. of Hyg. **37**, 561 (1937).
Salmonella amherstiana = (8) : 1, (v) : 1, 6.
EDWARDS and BRUNER: J. of Immun. **44**, 319 (1942).
Salmonella anatum = 3, 10 : e, h : 1, 6.
EDWARDS and RETTGER: J. Bacter. **13**, 73 (1927).
KAUFFMANN u. SILBERSTEIN: Zbl. Bakter. I Orig. **132**, 431 (1934).
Salmonella angoda = 30 : k : e, n, x.
Salmonella aqua = 30 : k : 1, 6.
Salmonella ardwick = 6, 7 : f, g :—.
Salmonella arechavaleta = 4, 5, 12 : a : 1, 7.
HORMAECHE y PELUFFO: Arch. urug. Med. Cir. y Espec. **14**, 217 (1939).
Salmonella atherton = identisch mit *S. waycross*.
ATKINSON, WOODROOFE and MACBETH: Austral. J. Exper. Biol. a. Med. Sci. **28**, 367 (1950).
Salmonella atlanta = 13, 23 : b :—.
SAPHRA and SELIGMANN: J. of Immun. **60**, 557 (1948).
Salmonella austin = 6, 7 : a : 1, 7.
EDWARDS and CHERRY: Publ. Health Labor. **10**, 127 (1952).
Salmonella azteca = 4, 5, 12 : l, v : 1, 5.
OLARTE, McWHORTER, EDWARDS and DE LA TORRE: Im Druck.
Salmonella ball = 1, 4, 12 : y : e, n, x.
ATKINSON, WOODROOFE and MACBETH: Austral J. Exper. Biol. a. Med. Sci. **28**, 377 (1950).
Salmonella bambesa = 9, 12 : a : 1, 5.
LE MINOR: Ann. Inst. Pasteur **91**, 664 (1956).
Biochemisch und serologisch verschieden von *S. miami*.
Salmonella banalia = 6, 8 : b : z_6.
KAUFFMANN, COURTOIS and VAN DEN ABBEELE: Acta path. scand. (Copenh.) **31**, 326 (1952).
Salmonella banana = 4, 5, 12 : m, t :—.
KAUFFMANN, VAN OYE and BALLION: Acta path. scand. (Copenh.) **28**, 43 (1951).

Salmonella bantam = 3, 10 : e, h : l, w.
Identisch mit *Salmonella meleagridis*.
ERBER: Geneesk. Tijdschr. Nederl. Indië **81**, 2123 (1941).

Salmonella baragwanath = 6, 8 : m, t : 1, 5.
BOKKENHEUSER and TOUSSAINT: S. Afric. J. Med. Sci. **20**, 129 (1955).

Salmonella bareilly = 6, 7 : y : 1, 5.
BRIDGES and SCOTT: J. Roy. Army Med. Corps **56**, 241 (1931).

Salmonella batavia = 3, 10 : z_{10} : 1, 5.
Identisch mit *Salmonella lexington*.
ERBER: Meded. Dienst Volksgezdh. Nederl.-Indië **29**, 56 (1940) und Geneesk. Tijdschr. Nederl.-Indië **81**, 2123 (1941).

Salmonella belem = 6, 8 : c : e, n, x.
EDWARDS and FIFE: Publ. Health Labor. **10**, 105 (1952).

Salmonella bergen = 47 : i : e, n, z_{15}.
KAUFFMANN, SAERVOLD, KRISTIANSEN and HENRIKSEN: Acta path. scand. (Copenh.) **37**, 492 (1955).

Salmonella berkeley = 43 : a : 1, 5.
EDWARDS and MCWHORTER: Cornell Veterinarian **43**, 572 (1953).

Salmonella berta = 9, 12 : f, g, t :—.
HORMAECHE, PELUFFO y SALSAMENDI: Arch. urug. Med. Cir. y Espec. **12**, 377 (1938).

Salmonella binza = 3, 15 : y : 1, 5.
KAUFFMANN, VANDEPITTE and VAN GOETHEM: Acta path. scand. (Copenh.) **31**, 431 (1952).

Salmonella birkenhead = 6, 7 : c : 1, 6.
TAYLOR and DOUGLAS: J. Clin. Path. **1**, 237 (1948).

Salmonella bispebjerg = 1, 4, 12 : a : e, n, x.
KAUFFMANN: Z. Hyg. **118**, 540 (1936).

Salmonella bleadon = 17 : (f), g, t :—.
BOYCOTT, TAYLOR and DOUGLAS: J. of Path. **65**, 401 (1953).
DOUGLAS and TAYLOR: Month. Bull. Min. Health, Lond. **13**, 159 (1954).

Salmonella blegdam = 9, 12 : g, m, q :—.
KAUFFMANN: Z. Hyg. **117**, 431 (1935).

Salmonella blockley = 6, 8 : k : 1, 5.
FRIEDMAN, WASSERMANN and SAPHRA: J. Bacter. **70**, 354 (1955).

Salmonella blukwa = 18 : z_4, z_{24} :—.
KAUFFMANN, FAIN and SCHOETTER: Acta path. scand. (Copenh.) **31**, 383 (1952).

Salmonella bodjonegoro = 30 : z_4, z_{24} :—.
ERBER and UTOJO: Hemera Zoa **61**, 171 (1954).

Salmonella boecker = 6, 14 : l, v : 1, 7.
BOECKER and KAUFFMANN: Acta path. scand. **34**, 101 (1954).

Salmonella boksburg = 40 : g, s : e, n, z_{15}.
MORTON, KOORNHOF and BOKKENHEUSER: Im Druck.

Salmonella bolton = 3, 10 : y : e, n, z_{15}.
GREENWOOD, POWIS, DOUGLAS and TAYLOR: Month. Bull. Min. Health, Lond. **12**, 29 (1953).

Salmonella bombay = nicht bestätigt.
White, Bruce: Med. Res. Counc., Spec. Rep. Ser. No 103, 1926. Kauffmann u. Mitsui: Z. Hyg. **111**, 749 (1930).
Salmonella bonariensis = 6, 8 : i : e, n, x.
Monteverde: Rev. Med. y Cienc. afines Buenos Aires **4**, 241 (1942).
Salmonella bonn = 6, 7 : l, v : e, n, x.
Seeliger, v. Vietinghoff-Scheel u. Kall: Z. Hyg. **137**, 192 (1953).
Salmonella borbeck = 13, 22 : l, v : 1, 6.
Hohn: Zbl. Bakter. I Orig. **146**, 215 (1940).
Nicht bestätigt, da die Kultur verlorenging.
Salmonella bovis morbificans = 6, 8 : r : 1, 5.
Basenau: Arch. f. Hyg. **20**, 242 (1894).
Salmonella bracknell = 13, 23, 36 : b : 1, 6.
Salmonella bradford = 4, 12 : r : 1, 5.
Salmonella braenderup = 6, 7 : e, h : e, n, z_{15}.
Kauffmann u. Juel Henningsen: Z. Hyg. **120**, 640 (1937).
Salmonella brancaster = 1, 4, 12 : z_{29} : —.
Macdonald, Sivell, Emms and Douglas: Month. Bull. Min. Health, Lond. **7**, 158 (1948).
Salmonella brandenburg = 4, 12 : l, v : e, n, z_{15}.
Kauffmann u. Mitsui: Z. Hyg. **111**, 740 (1930).
Salmonella brazil = 16 : a : 1, 5.
Edwards and Fife: Publ. Health Labor. **10**, 105 (1952).
Salmonella brazzaville = 6, 7 : b : 1, 2.
Le Minor, Merveille, Bascoulergue et Audebaud: Ann. Inst. Pasteur **87**, 105 (1954).
Salmonella bredeney = 1, 4, 12, 27 : l, v : 1, 7.
Kauffmann: Z. Hyg. **119**, 356 (1937).
Salmonella bristol = 13, 22 : z : 1, 7.
Salmonella bronx = 6, 8 : c : 1, 6.
Saphra, Wassermann and Friedman: J. Bacter. **69**, 477 (1955).
Salmonella broxbourne = identisch mit *Salmonella wien*.
Douglas and Taylor: Month. Bull. Min. Health, Lond. **10**, 198 (1951).
Salmonella budapest = 1, 4, 12 : g, t : —.
Rauss: Z. Immunforsch. **95**, 489 (1939).
Salmonella buenos aires = identisch mit *Salmonella bonariensis*.
Month. Bull. Min. Health, Lond. **13**, 12 (1954).
Salmonella bulawayo = (1), 49 : z : 1, 5.
Kauffmann and Deom: Acta path. scand. (Copenh.). Im Druck.
Salmonella bury = 4, 12, 27 : c : z_6.
Stitt and Walsh: Internat. Bull. Bacter. Nomenclat. a. Taxon. **6**, 51 (1956).
Salmonella butantan = 3, 10 : b : 1, 5.
Peluffo, Bier, Amaral e Biocca: Mem. Inst. Butantan **19**, 211 (1946).
Salmonella buzu = (1), 6, 14, 25 : i : 1, 7.
Salmonella cairina = 3, 10 : z_{35} : z_6.
Kauffmann, Fain and Schoetter: Acta path. scand. (Copenh.) **31**, 383 (1952).

Salmonella cairo = 1, 4, 12, 27 : d : 1, 2.
KAUFFMANN and FLOYD: Acta path. scand. (Copenh.) **32**, 124 (1953).
Salmonella california = 4, 12 : g, m, t : —.
EDWARDS, BRUNER and HINSHAW: J. Inf. Dis. **66**, 127 (1940).
Salmonella cambridge = 3, 15 : e, h : l, w.
WILSON, TAYLOR and ANDERSON: Month. Bull. Min. Health, Lond. **6**, 135 (1947).
Salmonella canastel = 9, 12 : z_{29} : 1, 5.
RANDALL and BRUNER: J. Bacter. **49**, 511 (1945).
Salmonella canoga = (3), (15), 34 : g, s, t : —.
BRUNER and MORAN: J. Bacter. **57**, 135 (1949).
Salmonella caracas = (1), 6, 14, 25 : g, m, s : —.
HORMAECHE y PELUFFO: Noch nicht publiziert.
Salmonella cardiff = 6, 7 : k : 1, 10.
TAYLOR, EDWARD and EDWARDS: Brit. Med. J. **1945 I**, 368.
Vereint mit *Salmonella thompson*, da die 1, 10-Phase eine induzierte Phase ist: Salmonella Subcommittee, Proc. Fourth Internat. Congress Microbiol. 1949, S. 607.
Salmonella carmel = 17 : l, v : e, n, x.
HIRSCH, HIRSCH and SAPIRO-HIRSCH: Acta med. orient. **13**, 42 (1954).
Salmonella carrau = 6, 14, 24 : y : 1, 7.
HORMAECHE, PELUFFO y SALSAMENDI: Arch. urug. Med. Cir. y Espec. **12**, 377 (1938).
HORMAECHE, PELUFFO and PEREYRA: J. Bacter. **47**, 323 (1944).
Salmonella cerro = 18 : z_4, z_{23} : —.
HORMAECHE y PELUFFO: Arch. urug. Med. Cir. y Espec. **19**, 125 (1941).
Salmonella chailey = 6, 8 : z_4, z_{23} : —.
Salmonella champaign = 39 : k : 1, 5.
EDWARDS: Proc. Soc. Exper. Biol. a. Med. **58**, 291 (1945).
Salmonella chandans = 11 : d : e, n, x.
HINSHAW and MCNEIL: J. Bacter. **55**, 870 (1948).
Salmonella charity = 6, 14 : d : e, n, x.
Salmonella chester = 4, 5, 12 : e, h : e, n, x.
KAUFFMANN u. TESDAL: Z. Hyg. **120**, 168 (1937).
Salmonella chiba = gehört nicht zur *Salmonella*-Gruppe.
CARLQUIST: J. Bacter. **54**, 265 (1947).
Salmonella chicago = 28 : r : 1, 5.
SHAUGNESSY, FRIEWER and LESKO: Canad. J. Publ. Health **43**, 300 (1952).
Salmonella chingola = 11 : e, h : 1, 2.
DOUGLAS, TAYLOR and FAIRCHILD: S. Afric. Med. J. **27**, 404 (1953).
Salmonella chinovum = 42 : b : 1, 5.
Salmonella chittagong = (1), 3, 10, (19) : b : z_{35}.
TAYLOR, HAYES, FREEMAN and ANDERSON: J. of Path. **60**, 35 (1948).
Salmonella cholerae suis = 6, 7 : c : 1, 5.
Salmonella Subcommittee: J. of Hyg. **34**, 333 (1934).
Salmonella cholerae suis var. kunzendorf = 6, 7 : — : 1, 5.
Salmonella Subcommittee: J. of Hyg. **34**, 333 (1934).
Proc. Third Internat. Congress Microbiol. 1940, S. 832.

Salmonella christiansborg = 44 : z_4, z_{24} :—.
Salmonella claibornei = 1, 9, 12 : k : 1, 5.
WILCOX and LENNOX: J. of Immun. **49**, 71 (1944).
Salmonella clerkenwell = 3, 10 : z : l, w.
STORY, DOUGLAS and TAYLOR: Month. Bull. Min. Health, Lond. **11**, 22 (1952).
Salmonella clifton = 13, 22 : z_{29} : 1, 5.
DOUGLAS and TAYLOR: Month. Bull. Min. Health, Lond. **13**, 159 (1954).
Salmonella coeln = 4, 5, 12 : y : 1, 2.
SIEVERS: Zbl. Bakt. I Orig. **150**, 52 (1943).
KAUFFMANN: Acta path. scand. Suppl. **54**, 33 (1944).
Salmonella colindale = 6, 7 : r : 1, 7.
PRICE and HOLT: Internat. Bull. Bacter. Nomenclat. a. Taxon. **5**, 1 (1955).
Salmonella colombo = 38 : y : 1, 6.
SCHMID, VELAUDAPILLAI et NILES: Annal. Inst. Pasteur **87**, 106 (1954).
Salmonella colorado = 6, 7 : l, w : 1, 5.
EDWARDS and HERMANN: J. Bacter. **58**, 111 (1949).
Salmonella concord = 6, 7 : l, v : 1, 2.
EDWARDS and HUGHES: J. Bacter. **47**, 574 (1944).
Salmonella coquilhatville = 3, 10 : z_{10} : 1, 7.
KAUFFMANN and LUCASSE: Acta path. scand. (Copenh.) **32**, 335 (1953).
Salmonella corvallis (8), 20 : z_4, z_{23} :—.
EDWARDS and HERMANN: J. Bacter. **58**, 111 (1949).
Salmonella cuba = *Salmonella cubana*.
Month. Bull. Min. Health, Lond. **13**, 12 (1954).
Salmonella cubana = 1, 13, 23 : z_{29} :—.
SELIGMANN, WASSERMANN and SAPHRA: J. Bacter. **51**, 123 (1946).
Salmonella curacao = 6, 8 : a : 1, 6.
EDWARDS and MCWHORTER: Publ. Health Labor. **12**, 17 (1954).
Salmonella dahlem = 48 : k : e, n, z_{15}.
KAUFFMANN, HOFMANN u. PLATZ: Zbl. Bakter. I Orig. **167**, 94 (1956).
Salmonella dakar = 28 : a : 1, 6.
DARRASSE et LE MINOR: Bull. Soc. Path. exot. **48**, 154 (1955).
Salmonella dar-es-salaam = 1, 9, 12 : l, w : e, n.
Salmonella Subcommittee: J. of Hyg. **34**, 333 (1934).
Salmonella daytona = 6, 7 : k : 1, 6.
MORAN and EDWARDS: Proc. Soc. Exper. Biol. a. Med. **62**, 294 (1946).
Salmonella decatur = 6, 7 : c : 1, 5.
KAUFFMANN, EDWARDS and MCWHORTER: Acta path. scand. (Copenh.) **36**, 568 (1955).
Salmonella degania = 40 : z_4, z_{24} :—.
HIRSCH and SAPIRO-HIRSCH: Acta med. orient. **14**, 297 (1955).
„*Salmonella delplata*" = 1, 13, 23 : z_4, z_{23} :—.
Gehört zur *Arizona*-Gruppe.
LEIGUARDA, PESO y KEMPNY: An. Soc. Ci. argent. **148**, 168 (1949).
KAUFFMANN and PETERSEN: Acta path. scand. (Copenh.) **38**, 481 (1956).
Salmonella denver = 6, 7 : a : e, n, z_{15}.
EDWARDS and CHERRY: Publ. Health Labor. **10**, 127 (1952).

Salmonella derby = 1, 4, 12 : f, g :—.
Savage and Bruce White: Med. Res. Counc., Spec. Rep. Ser. No 91, 1925.
Salmonella deversoir = 45 : c : e, n, x.
Watkins, Douglas and Taylor: Internat. Bull. Bacter. Nomenclat. a. Taxon. 5, 5 (1955).
Salmonella djakarta = 31 : z_4, z_{24} :—.
Erber and Kauffmann: Acta path. scand. (Copenh.). Im Druck.
Salmonella djugu = 6, 7 : z_{10} : e, n, x.
Kauffmann, van Oye and Schoetter: Acta path. scand. (Copenh.) **37**, 464 (1955).
Salmonella donna = 30 : l, v : 1, 5.
Watt, de Capito, Hermann and Edwards: U. S. Publ. Health Rep. **65**, 214 (1950).
Salmonella dublin = 1, 9, 12 : g, p :—.
White, Bruce: J. of Hyg. **29**, 443 (1929/30).
Salmonella Subcommittee: J. of Hyg. **34**, 333 (1934) und Proc. Third Internat. Congress Microbiol. 1940, S. 832.
Salmonella duesseldorf = 6, 8 : z_4, z_{24} :—.
Hohn: Zbl. Bakter. I Orig. **146**, 215 (1940).
Kauffmann: Acta path. scand. (Copenh.) **18**, 351 (1941).
Salmonella dugbe = 45 : d : 1, 6.
Salmonella duisburg = 1, 4, 12 : d : e, n, z_{15}.
Korell u. Seeliger: Zbl. Bakter. I Orig. **161**, 421 (1954).
Salmonella durban = 9, 12 : a : e, n, z_{15}.
Henning, Rhodes and Gordon-Johnstone: J. Vet. Sci. a. Animal Industry **16**, 103 (1941).
Salmonella durham = 13, 23 : b : e, n, z_{15}.
Salmonella duval = 1, 40 : b : e, n, z_{15}.
Edwards and West: U. S. Publ. Health Rep. **65**, 839 (1950).
Salmonella eastbourne = 1, 9, 12 : e, h : 1, 5.
Leslie and Shera: J. of Path. **34**, 531 (1931).
Salmonella edinburg = 6, 7 : b : 1, 5.
Watt, de Capito, Edwards and Hermann: U. S. Publ. Health Rep. **65**, 208 (1950).
Salmonella edmonton = 6, 8 : l, v : e, n, z_{15}.
Salmonella elisabethville = 3, 10 : r : 1, 7.
Kauffmann, Delville, Reul and Bouckaert: Acta path. scand. (Copenh.) **27**, 492 (1950).
Salmonella emek = (8), 20 : g, m, s :—.
Hirsch, Henig and Sapiro: J. Bacter. **60**, 213 (1950).
Salmonella emmastad = 38 : r : 1, 6.
Edwards, Rutten and McWhorter: Antonie van Leeuwenhoek **21**, 80 (1955).
Salmonella entebbe = 1, 4, 12, 27 : z : z_6.
Salmonella enteritidis = 1, 9, 12 : g, m :—.
Salmonella Subcommittee: J. of Hyg. **34**, 333 (1934).
Salmonella eschweiler = 6, 7 : z_{10} : 1, 6.
Seeliger u. Pelz: Z. Hyg. **135**, 275 (1952).

Salmonella essen = 4, 12 : g, m : —.
Hohn u. Herrmann: Zbl. Bakter. I Orig. **135**, 505 (1936).
Salmonella ezra = 28 : z : 1, 7.
Hirsch and Sapiro-Hirsch: Acta med. orient. **14**, 298 (1955).
Salmonella falkensee = 3, 10 : i : e, n, z_{15}.
Hofmann u. Pohl: Zbl. Bakter. I Orig. **167**, 413 (1957).
Salmonella fanti = 13, 23 : z_{38} : —.
Salmonella fayed = 6, 8 : l, w : 1, 2.
Anderson, Anderson and Taylor: J. of Path. **59**, 533 (1947).
Salmonella florida = (1), 6, 14, 25 : d : 1, 7.
Cherry, Edwards and Bruner: Proc. Soc. Exper. Biol. a. Med. **52**, 125 (1943).
Salmonella fortune = 4, 12, 27 : z_{10} : z_6.
Salmonella freetown = 38 : y : 1, 5.
Reid, Stitt and Taylor: Internat. Bull. Bacter. Nomenclat. a. Taxon. **6**, 61 (1956).
Salmonella fresno = 9, 12 : z_{38} : —.
Edwards, Browne, McWhorter and Williams: Cornell Veterinarian **44**, 259 (1954).
Salmonella friedenau = 13, 22 : d : 1, 6.
Boecker, Richter u. Winzer: Zbl. Bakter. I Orig. **161**, 45 (1954).
Salmonella frintrop = 1, 9, 12 : b : 1, 5.
Hoffmann, Linzenmeier u. Kall: Zbl. Bakter. I Orig. **165**, 78 (1956).
Salmonella fulica = 4, 5, 12 : a : 1, 5.
Salmonella gallinarum-pullorum = 1, 9, 12 : — : —.
Salmonella Subcommittee: J. of Hyg. **34**, 333 (1934).
Kauffmann: Zbl. Bakter. I Orig. **132**, 337 (1934).
Salmonella gambia = 35 : i : e, n, z_{15}.
Salmonella gaminara = 16 : d : 1, 7.
Hormaeche, Peluffo y Salsamendi: Arch. urug. Med. Cir. y Espec. **12**, 377 (1938).
Salmonella garoli = 6, 7 : i : 1, 6.
le Minor, Devignat et Schoetter: Ann. Inst. Pasteur **85**, 804 (1953).
Salmonella gatuni = 6, 8 : b : e, n, x.
Wilcox and Coates: J. Bacter. **51**, 561 (1946).
Salmonella gelsenkirchen = 6, 7 : l, v : z_6.
Löns u. Seeliger: Im Druck.
Salmonella georgia = 6, 7 : b : e, n, z_{15}.
Seligmann, Saphra and Wassermann: Amer. J. Hyg. **40**, 227 (1944).
Salmonella germiston = 6, 8 : m, t : e, n, x.
Bokkenheuser: S. Afric. J. Med. Sci. **20**, 5 (1955).
Salmonella give = 3, 10 : l, v : 1, 7.
Kauffmann: Z. Hyg. **120**, 177 (1937).
Salmonella glostrup = 6, 8 : z_{10} : e, n, z_{15}.
Kauffmann u. Juel Henningsen: Acta path. scand. (Copenh.) **16**, 99 (1939).
Salmonella gloucester = 1, 4, 12, (27) : i : l, w.
Waite. Taylor and Davey: Month. Bull. Min. Health, Lond. **16**, 14 (1957).

Salmonella godesberg = 30 : g, m : —.
KAUFFMANN, LUND and SEELIGER: Acta path. scand. (Copenh.) **36**, 96 (1955).
Salmonella goerlitz = 3, 15 : e, h : 1, 2.
WILDFÜHR u. HUDEMANN: Z. Hyg. **141**, 129 (1955).
Salmonella goettingen = 9, 12 : l, v : e, n, z_{15}.
HOHN: Zbl. Bakter. I Orig. **146**, 215 (1940).
KAUFFMANN: Acta path. scand. (Copenh.) **18**, 351 (1941).
Salmonella gold coast = 6, 8 : r : l, w.
Salmonella greenside = 50 : z : e, n.
SCHRIRE, KAUFFMANN and EDWARDS: Acta path. scand. (Copenh.). Im Druck.
Salmonella grumpensis = 13, 23 : d : 1, 7.
HORMAECHE, PELUFFO and RICAUD DE PEREYRA: J. Bacter. **47**, 323 (1944).
Salmonella guinea = 44 : z_{10} : —
Salmonella gwaai = 21 : z_4, z_{24} : —.
Salmonella haarlem = (9), 46 : z : e, n, x.
VINK, CLARENBURG, DE NOVY and BEKKER: Antonie van Leeuwenhoek **21**, 367 (1955). KAUFFMANN: Acta path. scand. (Copenh.) **38**, 69 (1956).
Salmonella habana = s. *Salmonella havana*.
Salmonella hadar = 6, 8 : z_{10} : e, n, x.
HIRSCH, GERICHTER, BREGMAN, LUBLING and ALTMAN: Acta med. orient. **13**, 41 (1954).
Salmonella haifa = 1, 4, 5, 12 : z_{10} : 1, 2.
SAPIRO and HIRSCH: J. Bacter. **60**, 101 (1950).
Salmonella hamburg = 1, 9, 12 : g, t : —.
ROHDE u. HOFMANN: Zbl. Bakter. I Orig. Im Druck.
Salmonella hannover = 9, 12 : m, t : —.
Salmonella harrisonburg = (3), (15), 34 : z_{10} : 1, 6.
EDWARDS and MCWHORTER: Cornell Veterinarian **43**, 110 (1953).
Salmonella hartford = 6, 7 : y : e, n, x.
EDWARDS and BRUNER: J. Inf. Dis. **69**, 220 (1941).
Salmonella hato = 4, 5, 12 : g, m, s : —.
EDWARDS, RUTTEN and MCWHORTER: Antonie van Leeuwenhoek **21**, 80 (1955).
Salmonella havana = 1, 13, 23 : f, g : —.
SCHIFF and SAPHRA: J. Inf. Dis. **68**, 125 (1941).
Salmonella heidelberg = 4, 5, 12 : r : 1, 2.
HABS: Zbl. Bakt. I Orig. **130**, 367 (1933).
Salmonella hessarek = 4, 12 : a : 1, 5.
NEEL, LE MINOR et KAWEH: Ann. Inst. Pasteur **85**, 271 (1953).
Salmonella heves = 6, 14, 24 : d : 1, 5.
RAUSS: Z. Immunforsch. **103**, 220 (1943).
Salmonella hidalgo = 6, 8 : r : e, n, z_{15}.
WATT, DE CAPITO, EDWARDS and MORAN: U. S. Publ. Health Rep. **63**, 223 (1948).
Salmonella hillbrow = 17 : b : e, n, z_{15}.
SCHRIRE: S. Afric. J. Med. Sci. **21**, 11 (1956).
Salmonella hindmarsh = (8) : r : 1, 5.
ATKINSON: Noch nicht publiziert.

Salmonella hirschfeldii = *Salmonella paratyphi C*.
Salmonella Subcommittee: J. of Hyg. **34**, 333 (1934).
„*Salmonella holstein*" = gehört nicht zur *Salmonella*-Gruppe.
ROELCKE: Zbl. Bakter. I Orig. **137**, 454 (1936).
KAUFFMANN: Z. Hyg. **119**, 352 (1937).
Salmonella homosassa = (1), 6, 14, 25 : z : 1, 5.
EDWARDS and FIFE: U. S. Publ. Health Rep. **66**, 1060 (1951).
Salmonella horsham = (1), 6, 14, 25 : l, v : e, n, x.
Salmonella hull = 16 : b : 1, 2.
ALEXANDER, DOUGLAS and TAYLOR: Month. Bull. Min. Health, Lond. **13**, 117 (1954).
Salmonella huvudsta = 3, 10 : b : 1, 7.
ALIN and MALMBERG: Acta path. scand. (Copenh.) **39**, 160 (1956).
Salmonella hvittingfoss = 16 : b : e, n, x.
TESDAL: Z. Hyg. **118**, 533 (1936).
KAUFFMANN: Z. Hyg. **118**, 540 (1936).
Salmonella ibadan = 13, 22 : b : 1, 5.
STEVENSON: J. of Path. **66**, 574 (1953).
Salmonella ilala = 28 : k : 1, 5.
MACKEY: East Afr. Med. J. **32**, 1 (1955).
Salmonella illinois = (3), (15), 34 : z_{10} : 1, 5.
EDWARDS and BRUNER: Proc. Soc. Exper. Biol. a. Med. **48**, 240 (1941).
Salmonella indiana = 1, 4, 12 : z : 1, 7.
HAJNA, EDWARDS, MCWHORTER and DAMON: Publ. Health Labor. **13**, 4 (1955).
Salmonella infantis = 6, 7 : r : 1, 5.
WHEELER and BORMAN: J. Bacter. **46**, 481 (1943).
Salmonella inverness = 38 : k : 1, 6.
EDWARDS and HUGHES: Proc. Soc. Exper. Biol. a. Med. **56**, 33 (1944).
Salmonella irumu = 6, 7 : l, v : 1, 5.
KAUFFMANN, COURTOIS and VAN OYE: Acta path. scand. (Copenh.) **24**, 588 (1947).
Salmonella isangi = s. *S. mission var. isangi*.
Salmonella israel = 9, 12 : e, h : e, n, z_{15}.
EDWARDS, ALTMANN and MCWHORTER: Publ. Health Labor. **11**, 141 (1953).
Salmonella italiana = 9, 12 : l, v : 1, 11.
BRUNER and EDWARDS: Proc. Soc. Exper. Biol. a. Med. **58**, 289 (1945).
Mit *S. panama* vereinigt, da die 1,11-Phase eine induzierte Phase ist.
Salmonella Subcommittee: Proc. Fourth Internat. Congress Microbiol. 1949, S. 607.
Salmonella ituri = 1, 4, 12 : z_{10} : 1, 5.
KAUFFMANN and FAIN: Acta path. scand. (Copenh.) **32**, 513 (1953).
Salmonella iwo-jima = identisch mit *Salmonella kentucky*.
CARLQUIST: J. Bacter. **54**, 265 (1947).
Salmonella jacksonville = 16 : z_{29} : —.
GALTON, EDWARDS, FIFE and LEWIS: Public Health Labor. **12**, 147 (1954).
Salmonella jaffna = 1, 9, 12 : d : z_{35}.
SCHMID, EDWARDS, MCWHORTER and VELAUDAPILLAI: J. of Path. **69**, 337 (1955).

Salmonella jangwani = 17 : a : 1, 5.
MACKEY: East Afr. Med. J. **32**, 1 (1955).
Salmonella java = 1, 4, 5, 12 : b : [1, 2].
KAUFFMANN: Acta path. scand. (Copenh.) **33**, 409 (1953).
Salmonella javiana = 1, 9, 12 : l, z_{28} : 1, 5.
EDWARDS and BRUNER: J. of Immun. **44**, 319 (1942).
Von ERBER als „*S. panama var.*" beschrieben: Meded. Dienst Volksgezdh. Nederl.-Indië **29**, 56 (1940).
Salmonella jerusalem = 6, 7 : z_{10} : l, w.
KAUFFMANN and SILBERSTEIN: Acta path. scand. (Copenh.) **27**, 78 (1950).
Salmonella johannesburg = 1, 40 : b : e, n, x.
KAUFFMANN and HENNING: Acta path. scand. (Copenh.) **31**, 586 (1952).
Salmonella kaapstad = 4, 12 : e, h : 1, 7.
HENNING, RHODES and GORDON-JOHNSTONE: J. Vet. Sci. a. Animal Industry **16**, 103 (1941).
Salmonella kaduna = 6, 7 : c : e, n, z_{15}.
Salmonella kalamu = 4, 12 : z_4, z_{24} : —.
VAN OYE, DELCOUR et VAN GOETHEM: Ann. Inst. Pasteur **89**, 587 (1955).
Salmonella kalina = 3, 10 : b : 1, 2.
VAN OYE, GHYSELS et VAN GOETHEM: Ann. Inst. Pasteur: Im Druck.
Salmonella kaltenhausen = 28 : b : z_6.
HOFMANN u. POHL: Zbl. Bakter. I Orig. **162**, 543 (1955).
Salmonella kampala = 1, 42 : c : z_6.
Salmonella kanda = identisch mit *Salmonella meleagridis*.
CARLQUIST: J. Bacter. **54**, 265 (1947).
Salmonella kaneshie = 1, 42 : i : l, w.
Salmonella kaolack = 47 : z : 1, 6.
LE MINOR, DARRASSE et NAZAUD: Ann. Inst. Pasteur **91**, 400 (1956).
Salmonella kapemba = 9, 12 : l, v : 1, 7.
KAUFFMANN, DELVILLE, BOUCKAERT and BALLION: Acta path. scand. (Copenh.) **35**, 307 (1954).
Salmonella kaposvár = 4, 5, 12 : e, (h) : 1, 5.
RAUSS: Zbl. Bakter. I Orig. **147**, 253 (1941).
Mit *Salmonella reading* vereint.
Salmonella karamoja = 40 : z_{41} : 1, 2.
Salmonella kasenyi = 38 : : e, h : 1, 5.
KAUFFMANN and FAIN: Acta path. scand. (Copenh.) **32**, 513 (1953).
Salmonella kentucky = (8), 20 : i : z_6.
EDWARDS: J. of Hyg. **38**, 306 (1938).
Salmonella kenya = 6, 7 : l, z_{13} : e, n, x.
Salmonella kiambu = 4, 12 : z : 1, 5.
Salmonella kibusi = 28 : r : e, n, x.
KAUFFMANN, VAN OYE and VANDEPITTE: Acta path. scand. (Copenh.) **26**, 337 (1949).
Salmonella kikoma = 16 : y : e, n, x.
Salmonella kilwa = 4, 12 : l, w : e, n, x.
MACKEY: East Afr. Med. J. **32**, 1 (1955).

Salmonella kimuenza = 1, 4, 12, 27 : l, v : e, n, x.
KAUFFMANN, VANDEPITTE and VAN GOETHEM: Acta path. scand. (Copenh.) **31**, 431 (1952).

Salmonella kingabwa = 43 : y : 1, 5.
KAUFFMANN and VANDEPITTE: Acta path. scand. (Copenh.) **35**, 71 (1954).

Salmonella kingston = 1, 4, 12, 27 : g, s, t :—.
BRUNER: Cornell Veterinarian **41**, 339 (1951).

Salmonella kinondoni = 17 : a : e, n, x.
MACKEY: East Afr. Med. J. **32**, 1 (1955).

Salmonella kinshasa = 3, 15 : l, z_{13} : 1, 5.
KAUFFMANN and VAN OYE: Acta path. scand. (Copenh.) **27**, 519 (1950).

Salmonella kirkee = 17 : b : 1, 2.
BRIDGES and DUNBAR: J. Roy. Army Med. Corps **67**, 289 (1936).

Salmonella kisangani = 1, 4, 5, 12 : a : 1, 2.
KAUFFMANN and VAN OYE: Acta path. scand. (Copenh.) **24**, 614 (1947).

Salmonella kisarawe = 11 : k : e, n, x.
MACKEY: East Afr. Med. J. **32**, 1 (1955).

Salmonella korovi = 38 : g, m, s :—.
KAUFFMANN and VAN OYE: Acta path. scand. (Copenh.) **36**, 352 (1955).

Salmonella kottbus = 6, 8 : e, h : 1, 5.
KAUFFMANN: Zbl. Bakter. I Orig. **132**, 161 (1934).
Salmonella Subcommittee: Proc. Third Internat. Congress Microbiol. 1940, S. 832.

Salmonella kotte = 6, 7 : b : z_{35}.

„*Salmonella kralendyk*" = 6, 7 : z_4, z_{24} :—.
EDWARDS, RUTTEN and MCWHORTER: Antonie van Leeuwenhoek **21**, 80 (1955).
Zur *Arizona*-Gruppe gehörend: KAUFFMANN and PETERSEN: Acta path. scand. (Copenh.) **38**, 481 (1956).

Salmonella krefeld = 1, 3, 19 : y : l, w.
SEELIGER: Z. Hyg. **136**, 379 (1953).

Salmonella kunduchi = 1, 4, 12, 27 : l, z_{28} : 1, 2.
MACKEY: East Afr. Med. J. **32**, 1 (1955).

Salmonella labadi = 6, 8 : d : z_6.

Salmonella landau = 30 : i : 1, 2.
GÜNTHER u. HANSER: Zbl. Bakter. I Orig. **161**, 363 (1954).

Salmonella langford = 28 : b : e, n, z_{15}.
DOUGLAS and TAYLOR: Month. Bull. Min. Health, Lond. **13**, 159 (1954).

Salmonella lanka = 3, 15 : r : z_6.

Salmonella leoben = 28 : l, v : 1, 5.
ROSCHKA u. ROTTER: Mitt. österr. Sanitätsverw. **57**, 125 (1956).

Salmonella leopoldville = 6, 7 : b : z_6.
KAUFFMANN, VAN OYE and EVENS: Acta path. scand. (Copenh.) **27**, 32 (1950).

Salmonella lexington = 3, 10 : z_{10} : 1, 5.
EDWARDS, BRUNER and RUBIN: Proc. Soc. Exper. Biol. a. Med. **44**, 395 (1940).
Identisch mit *Salmonella batavia*, ERBER: Meded. Dienst Volksgezdh. Nederl.-Indië **29**, 56 (1940) und Geneesk. Tijdschr. Nederl.-Indië **81**, 2123 (1941).

Salmonella lille = 6, 7 : z_{33} : —.
Kauffmann, Buttiaux and Gaumont: Acta path. scand. (Copenh.) **34**, 99 (1954).

Salmonella limete = 1, 4, 12, 27 : b : 1, 5.
Kauffmann and Vandepitte: Acta path. scand. (Copenh.) **34**, 97 (1954).

Salmonella lindenburg = 6, 8 : i : 1, 2.
Lempfrid: Zbl. Bakter. I Orig. **160**, 486 (1954).

Salmonella lindi = 38 : r : 1, 5.
Telling, Taylor and Douglas: Month. Bull. Min. Health, Lond. **10**, 251 (1951).

Salmonella lisboa = 16 : z_{10} : 1, 6.
Figueiredo and Sampaio: J. Soc. Ci. Med. Lisboa **118**, 162 (1954).

Salmonella litchfield = 6, 8 : l, v : 1, 2.
Edwards and Bruner: J. Inf. Dis. **66**, 218 (1940).

Salmonella liverpool = 1, 3, 19 : d : e, n, z_{15}.

Salmonella livingstone = 6, 7 : d : l, w.
Picton, Stirrup, Price and Taylor: J. of Path. **66**, 310 (1953).

Salmonella llandoff = 1, 3, 19 : z_{25} : —.

Salmonella loma-linda = 9, 12 : a : e, n, x.
Edwards: Proc. Soc. Exper. Biol. a. Med. **57**, 104 (1944).

Salmonella lomita = 6, 7 : e, h : 1, 5.
Edwards, Moran, Watt and de Capito: U. S. Publ. Health Rep. **65**, 210 (1950).

Salmonella london = 3, 10 : l, v : 1, 6.
White, Bruce: Med. Res. Counc., Spec. Rep. Ser. No 103, 1926.
Salmonella Subcommittee: J. of Hyg. **34**, 333 (1934).

Salmonella luanshya = 13, 23 : g, s, (t) : —.
Taylor, Edwards, Nicewonger and Nagington.

Salmonella luciana = 11 : a : e, n, z_{15}.
Moran, Edwards and Bruner: Proc. Soc. Exper. Biol. a. Med. **64**, 89 (1947).

Salmonella macallen = 3, 10 : z_{33} : —.
Edwards, Hermann, Watt and de Capito: U. S. Publ. Health Rep. **65**, 212 (1950).

Salmonella madelia = (1), 6, 14, 25 : y : 1, 7.
Cherry, Edwards and Bruner: Proc. Soc. Exper. Biol. a. Med. **52**, 125 (1943).

Salmonella magwa = 21 : d : e, n, x.

Salmonella makiso = 6, 7 : l, z_{28} : z_6.
Kauffmann, van Oye and Vandepitte: Acta path. scand. (Copenh.) **26**, 337 (1949).

Salmonella manchester = 6, 8 : l, v : 1, 7.
Taylor and Douglas: Month. Bull. Min. Health, Lond. **7**, 117 (1948).

Salmonella manhattan = 6, 8 : d : 1, 5.
Edwards and Bruner: Amer. J. Hyg. **34**, 121 (1941).

Salmonella manila = 3, 15 : z_{10} : 1, 5.
Edwards, Kauffmann and McWhorter: Acta path. scand. (Copenh.) **35**, 67 (1954).

Salmonella maracaibo = 11 : l, v : 1, 5.
LE MINOR et FOSSAERT: Ann. Inst. Pasteur **87**, 104 (1954).
Salmonella maritza, Wesselinoff = s. *Salmonella salford var. maritza.*
Salmonella marseille = 11 : a : 1, 5.
MORAN, EDWARDS and BRUNER: Proc. Soc. Exper. Biol. a. Med. **64**, 89 (1947).
Salmonella marylebone = 9, 12 : k : 1, 2.
BOYCOTT, TAYLOR and DOUGLAS: J. of Path. **65**, 401 (1953).
DOUGLAS and TAYLOR: Month. Bull. Min. Health, Lond. **13**, 159 (1954).
Salmonella matadi = 17 : k : e, n, x.
KAUFFMANN and VANDEPITTE: Acta path. scand. (Copenh.) **34**, 97 (1954).
Salmonella matopeni = 30 : y : 1, 2.
TELLING, TAYLOR and DOUGLAS: Month. Bull. Min. Health, Lond. **10**, 251 (1951).
Salmonella mbandaka = 6, 7 : z_{10} : e, n, z_{15}.
KAUFFMANN and REUL: Acta path. scand. (Copenh.) **26**, 335 (1949).
Salmonella meleagridis = 3, 10 : e, h : l, w.
BRUNER and EDWARDS: Amer. J. Hyg. **34**, 82 (1941).
Identisch mit *Salmonella bantam*, ERBER: Geneesk. Tijdschr. Nederl.-Indië **81**, 2123 (1941).
Salmonella memphis = 18 : k : 1, 5.
MORAN, EDWARDS and BRUNER: J. Bacter. **55**, 285 (1948).
Salmonella mendoza = 9, 12 : l, v : 1, 2.
LEIGUARDA, PESO, DE PELAZZOLO and ANSIAUME: U. S. Publ. Health Rep. **66**, 1478 (1951).
Salmonella menston = 6, 7 : g, s, t :—.
COLBECK, DOUGLAS and TAYLOR: J. of Path. **63**, 754 (1951).
Salmonella merseyside = 16 : g, t : 1, 5.
Salmonella mexicana = 6, 8 : d : 1, 2.
VARELA, ZOZAYA y OLARTE: Rev. Inst. salub. y enferm. trop **3**, 209 (1942).
Vereinigt mit *Salmonella muenchen.*
Salmonella mgulani = 38 : i : 1, 2.
TELLING, TAYLOR and DOUGLAS: Month. Bull. Min. Health, Lond. **10**, 251 (1951).
Salmonella miami = 1, 9, 12 : a : 1, 5.
EDWARDS and MORAN: J. Bacter. **50**, 257 (1945).
Salmonella michigan = 17 : l, v : 1, 5.
JUENKER and CALDWELL: Publ. Health Labor. **13**, 66 (1955).
Salmonella mikawasima = 6, 7 : y : e, n, z_{15}.
Isoliert und unvollständig bestimmt von HATTA: Jap. J. Exper. Med. **16**, 201 (1938) *(Salmonella bareilly var. mikawasima).* Bestimmt und benannt von HORMAECHE, PELUFFO and RICAUD DE PEREYRA: J. Bacter. **47**, 323 (1944).
Salmonella milwaukee = 43 : f, g :—.
EDWARDS and FIFE: U. S. Publ. Health Rep. **66**, 1059 (1951).
Salmonella minneapolis = (3), (15), 34 : e, h : 1, 6.
EDWARDS, KAUFFMANN and MCWHORTER: Acta path. scand. (Copenh.) **35**, 67 (1954).

Salmonella minnesota = 21 : b : e, n, x.
Edwards and Bruner: J. of Hyg. **38**, 716 (1938).
Salmonella mishmar-haemek = 1, 13, 23 : d : 1, 5.
Silberstein, Gerichter and Reitler: Acta med. orient. **13**, 40 (1954).
Salmonella mission = 6, 7 : d : 1, 5.
Watt, de Capito, Edwards and Moran: U. S. Publ. Health Rep. **63**, 223 (1948).
Salmonella mission var. isangi = 6, 7 : d : 1, 5.
Kauffmann, Courtois and van Oye: Acta path. scand. (Copenh.) **24**, 588 (1947).
Serologisch und biochemisch verschieden von *Salmonella mission*.
Salmonella mississippi = 1, 13, 23 : b : 1, 5.
Edwards, Cherry and Bruner: Proc. Soc. Exper. Biol. a. Med. **54**, 263 (1943).
Salmonella mjimwema = 1, 9, 12 : b : e, n, x.
Mackey: East Afr. Med. J. **32**, 1 (1955).
Salmonella mobeni = 16 : g, s, (t) :—.
Salmonella monschaui = 35 : m, t :—.
Carlquist and Coates: J. Bacter. **53**, 249 (1947).
Salmonella montevideo = 6, 7 : g, m, s :—.
Hormaeche y Peluffo: Arch. urug. Med. Cir. y Espec. **9**, 673 (1936).
Salmonella montreal = identisch mit *Salmonella wien*.
Laidley, Bailey and Bynoe: Canad. J. Publ. Health **42**, 99 (1951).
Salmonella morehead = 30 : i : 1, 5.
Edwards and McWhorter: Publ. Health Labor. **10**, 103 (1952).
Salmonella morotai = 17 : l, v : 1, 2.
Aktinson, Woodroofe and Macbeth: Austral. J. Exper. Biol. a. Med. Sci. **28**, 377 (1950).
Salmonella moroto = 28 : z_{10} : l, w.
Salmonella moscow = 9, 12 : g, q :—.
Salmonella Subcommittee: J. of Hyg. **34**, 333 (1934) und Proc. Third Internat. Congress Microbiol. 1940, S. 832.
Salmonella muenchen 6, 8 : d : 1, 2.
Mandelbaum: Zbl. Bakter. I Ref. **105**, 377 (1932).
Silberstein: Z. Hyg. **114**, 124 (1932).
Salmonella muenster = 3, 10 : e, h : 1, 5.
Kauffmann u. Silberstein: Zbl. Bakter. I Orig. **132**, 431 (1934). Salmonella Subcommittee: Proc. Third Internat. Congress Microbiol. 1940, S. 832.
Salmonella mundonobo = 28 : d : 1, 7.
Edwards, Rutten and McWhorter: Antonie van Leeuwenhoek **21**, 80 (1955).
Salmonella nachshonim = 1, 13, 23 : z : 1, 5.
Kauffmann, Silberstein and Gerichter: Acta path. scand. (Copenh.) **33**, 79 (1953).
Salmonella nagoya = 6, 8 : b : 1, 5.
Nakajima, Naito, Nakaya and Fukumi: Jap. J. Med. Sci. a. Biol. **6**, 179 (1953).
Salmonella nairobi = 42 : r :—.

Salmonella napoli = 1, 9, 12 : l, z_{13} : e, n, x.
Bruner and Edwards: Proc. Soc. Exper. Biol. a. Med. **58**, 289 (1945).
Salmonella narashino = 6, 8 : a : e, n, x.
Nakaguro u. Yamashita: Literatur unbekannt.
Salmonella nashua = 28 : l, v : e, n, z_{15}.
Salmonella nchanga = 3, 10 : l, v : 1, 2.
Doulas, Taylor and Fairchild: S. Afric. Med. J. **27**, 404 (1953).
Salmonella nchanga var. venusberg = 3, 10 : l, v : 1, 2.
Seeliger: Berl. u. Münch. tierärztl. Wschr. **1955**, 69.
Salmonella ndolo = 9, 12 : d : 1, 5.
Kauffmann, van Oye and Evens: Acta path. scand. (Copenh.) **27**, 32 (1950).
Salmonella neasden = 9, 12 : g, s, t : e, n, x.
Douglas, Taylor and McMath: Month. Bull. Min. Health, Lond. **10**, 250 (1951).
Salmonella ness-ziona = 6, 7 : l, z_{13} : 1, 5.
Kauffmann, Silberstein and Gerichter: Acta path. scand. (Copenh.) **27**, 829 (1950).
Salmonella neukoelln = 6, 7 : l, z_{28} : e, n, z_{15}.
Hofmann u. Henze: Zbl. Bakter. I Orig. **163**, 572 (1955).
Salmonella new brunswick = 3, 15 : l, v : 1, 7.
Edwards: J. of Hyg. **37**, 384 (1937).
Salmonella new-haw = 3, 15 : e, h : 1, 5.
Salmonella newington = 3, 15 : e, h : 1, 6.
Edwards: J. of Hyg. **37**, 384 (1937).
Salmonella Subcommittee: Proc. Third Internat. Congress Microbiol. 1940, S. 832.
Salmonella newington var. tim = 3, 15 : e, h : 1, 6.
Kauffmann: Z. Hyg. **120**, 177 (1937).
Serologisch und biochemisch verschieden von *Salmonella newington*.
Salmonella newlands = 3, 10 : e, h : e, n, x.
Greening, Price and Taylor: Month. Bull. Min. Health, Lond. **12**, 89 (1953).
Salmonella newport = 6, 8 : e, h : 1, 2.
Schütze: Lancet **1920 I**, 93.
Salmonella newport var. puerto rico = 6, 8 : — : 1, 2.
Kauffmann: Zbl. Bakter. I Orig. **132**, 161 (1934).
Salmonella Subcommittee: Proc. Third Internat. Congress Microbiol. 1940, S. 832.
Salmonella new york = identisch mit *Salmonella javiana*.
Kauffmann: Acta path. scand. (Copenh.) Suppl. **54**, 33 (1944).
Salmonella Subcommittee: Proc. Fourth Internat. Congress Microbiol. 1949, S. 608.
Salmonella niarembe = 44 : a : l, w.
Kauffmann and Fain: Acta path. scand. (Copenh.) **32**, 513 (1953).
Salmonella nigeria = 6, 7 : r : 1, 6.
Salmonella niloese = 1, 3, 19 : d : z_6.
Kauffmann: Acta path. scand. (Copenh.) **16**, 347 (1939).
Salmonella nima = 28 : y : 1, 5.

Salmonella nipponbasi = gehört nicht zur *Salmonella*-Gruppe.
Carlquist: J. Bacter. **54**, 265 (1947).
Salmonella norwich = 6, 7 : e, h : 1, 6.
Taylor, Macdonald and Sivell: Month. Bull. Min. Health, Lond. **10**, 76 (1951).
Salmonella nottingham = 16 : d : e, n, z_{15}.
Ludlam, Taylor and Douglas: Month. Bull. Min. Health, Lond. **12**, 29 (1953).
Salmonella nyborg = 3, 10 : e, h : 1, 7.
Kauffmann: Z. Hyg. **120**, 177 (1937).
Salmonella Subcommittee: Proc. Third Internat. Congress Microbiol. 1940, S. 832.
Salmonella oahu = nicht bestätigt.
Carlquist: J. Bacter. **54**, 265 (1947).
Salmonella okatie = 13, 23 : g, s, t : —.
Taylor, Edwards, Nicewonger and Nagington: Im Druck.
Salmonella okefoko = 3, 10 : c : z_6.
Salmonella ona = 28 : g, s, t : —.
Salmonella onarimon = 1, 9, 12 : b : 1, 2.
Kisida: Kitasato Arch. Exper. Med. **17**, 1 (1940).
Salmonella onderstepoort = (1), 6, 14, 25 : e, (h) : 1, 5.
Henning: J. of Hyg. **36**, 525 (1936).
Salmonella oranienburg = 6, 7 : m, t : —.
Kauffmann: Z. Hyg. **111**, 221 (1930).
Salmonella oregon = 6, 8 : d : 1, 2.
Edwards and Bruner: Amer. J. Hyg. **34**, 121 (1941).
Salmonella Subcommittee: Proc. Fourth Internat. Congress Microbiol. 1949, S. 607.
Mit *Salmonella muenchen* vereint.
Salmonella orientalis = 16 : k : e, n, z_{15}.
Carlquist and Conte: Bull. U. S. Army Med. Dept. **1946**, 343.
Salmonella orion = 3, 10 : y : 1, 5.
Barnes, Cherry and Myers: J. Bacter. **50**, 577 (1945).
Salmonella oslo = 6, 7 : a : e, n, x.
Tesdal: Z. Hyg. **119**, 451 (1937).
Salmonella oxford = 3, 10 : a : 1, 7.
Jebb, Douglas and Taylor: Month. Bull. Min. Health, Lond. **10**, 200 (1951).
Salmonella panama = 1, 9, 12 : l, v : 1, 5.
Kauffmann: Zbl. Bakter. I Orig. **132**, 160 (1934).
Salmonella papuana = 6, 7 : r : e, n, z_{15}.
Wilcox, Edwards and Coates: J. Bacter. **49**, 514 (1945).
Salmonella paratyphi = *Salmonella paratyphi A*.
Salmonella paratyphi A = 1, 2, 12 : a : —.
Salmonella Subcommittee: J. of Hyg. **34**, 333 (1934).
Salmonella paratyphi A var. durazzo = 2, 12 : a : —.
Kauffmann u. Silberstein: Zbl. Bakter. I Orig. **132**, 431 (1934).
Salmonella paratyphi B = 1, 4, 5, 12 : b : 1, 2.
Salmonella Subcommittee: J. of Hyg. **34**, 333 (1934).

Salmonella paratyphi B var. java = *Salmonella java.*
KRISTENSEN u. KAUFFMANN: Z. Hyg. **120**, 149 (1937).
KAUFFMANN: Z. Hyg. **141**, 546 (1955).
Salmonella paratyphi B var. odense = 1, 4, 12 : b : 1, 2.
KAUFFMANN: Z. Hyg. **116**, 368 (1934).
Salmonella paratyphi C = 6, 7, Vi : c : 1, 5.
Salmonella Subcommittee: J. of Hyg. **34**, 333 (1934).
Salmonella patience = 28 : d : e, n, z_{15}.
Salmonella penarth = 9, 12 : z_{35} : z_6.
Salmonella pensacola = 9, 12 : g, m, t :—.
MORAN and EDWARDS: Proc. Soc. Exper. Biol. a. Med. **59**, 52 (1945).
Salmonella pharr = 11 : b : e, n, z_{15}.
WATT, DE CAPITO, HERMANN and EDWARDS: U. S. Publ. Health Rep. **65**, 214 (1950).
Salmonella pomona = 28 : y : 1, 7.
EDWARDS: Proc. Soc. Exper. Biol. a. Med. **58**, 291 (1945).
Salmonella poona = 13, 22 : z : 1, 6.
BRIDGES and SCOTT: J. Roy. Army Med. Corps **65**, 221 (1935).
Salmonella portland = 9, 12 : z_{10} : 1, 5.
EDWARDS and MCWHORTER: Internat. Bull. Bacter. Nomenclat. a. Taxon. **5**, 151 (1955).
Salmonella potsdam = 6, 7 : l, v : e, n, z_{15}.
KAUFFMANN u. MITSUI: Z. Hyg. **111**, 740 (1930).
Salmonella praha = 6, 8 : y : e, n, z_{15}.
SEDLÁK and SOLAR: Biol. Listy **30**, No 2 (1949).
Salmonella preston = 1, 4, 12 : z : l, w.
Salmonella pretoria = 11 : k : 1, 2.
HENNING, RHODES and GORDON-JOHNSTONE: J. Vet. Sci. a. Animal Industry **16**, 103 (1941).
Salmonella pueris = 6, 8 : e, h : 1, 2.
WHEELER and BORMAN: J. Bacter. **46**, 481 (1943).
Salmonella Subcommittee: Proc. Fourth Internat. Congress Microbiol. 1949, S. 607.
Mit *Salmonella newport* vereint.
Salmonella pullorum = *Salmonella gallinarum-pullorum.*
Salmonella quiniela = 6, 8 : c : e, n, z_{15}.
STUCKER, GALTON, EDWARDS and FIFE: U. S. Publ. Health Rep. **66**, 1058 (1951).
Salmonella rand = 42 : z : e, n, z_{15}.
BLELOCH and SCHRIRE: Im Druck.
Salmonella reading = 4, 5, 12 : e, h : 1, 5.
SCHÜTZE: Lancet **1920 I**, 93 (s. *Salmonella kaposvár*).
Salmonella redhill = 11 : e, h : l, z_{13}, z_{28}.
Salmonella richmond = 6, 7 : y : 1, 2.
MORAN and EDWARDS: Proc. Soc. Exper. Biol. a. Med. **62**, 294 (1946).
Salmonella ridge = 9, 12 : c : z_6.

Salmonella riogrande = 40 : b : 1, 5.
EDWARDS, MORAN, WATT and DE CAPITO: U. S. Publ. Health Rep. **65**, 210 (1950).
Salmonella roan = 38 : l, v : e, n, x.
Salmonella roodepoort = 13, 22 : z_{10} : 1, 5.
BROMFIELD and BOKKENHEUSER: South Afr. J. Med. Sci. Im Druck.
Salmonella rostock = 1, 9, 12 : g, p, u :—.
KAUFFMANN: Z. Hyg. **111**, 221 (1940).
KAUFFMANN u. MITSUI: Z. Hyg. **111**, 749 (1930).
Salmonella Subcommittee: Proc. Third Internat. Congress Microbiol. 1940, S. 832.
Salmonella rowbarton = 16 : m, t :—.
BOYCOTT, TAYLOR and DOUGLAS: J. of Path. **65**, 401 (1953).
DOUGLAS and TAYLOR: Month. Bull. Min. Health, Lond. **13**, 159 (1954).
Salmonella rubislaw = 11 : r : e, n, x.
SMITH and KAUFFMANN: J. of Hyg. **40**, 122 (1940).
Salmonella ruiru = 21 : y : e, n, x.
Salmonella rutgers = 3, 10 : l, z_{40} : 1, 7.
McWHORTER and EDWARDS: Cornell Veterinarian **46**, 509 (1956).
Salmonella saarbruecken = 1, 9, 12 : a : 1, 7.
GÜNTHER u. GAUL: Zbl. Bakter. I Orig. **161**, 544 (1954).
Salmonella saint paul = 1, 4, 5, 12 : e, h : 1, 2.
EDWARDS and BRUNER: J. Inf. Dis. **66**, 218 (1940).
Salmonella saipan = nicht bestätigt.
CARLQUIST: J. Bacter. **54**, 265 (1947).
Salmonella sakai = identisch mit *Salmonella potsdam*.
Salmonella salford = 16 : l, v : e, n, x.
DOUGLAS, TAYLOR, GREENWOOD and IBBOTSON: Month. Bull. Min. Health, Lond. **10**, 75 (1951).
Salmonella salford var. maritza = biochemisch und serologisch verschieden von *Salmonella salford*.
Salmonella salinatis = 4, 12 : d, e, h : d, e, n, z_{15}.
EDWARDS and BRUNER: J. Bacter. **44**, 289 (1942).
Salmonella sandiego = 4, 5, 12 : e, h : e, n, z_{15}.
KAUFFMANN: Acta path. scand. (Copenh.) **17**, 429 (1940).
Salmonella sanga = (8) : b : 1, 7.
KAUFFMANN, VAN OYE and EVENS: Acta path. scand. (Copenh.) **27**, 262 (1950).
Salmonella san juan = 6, 7 : a : 1, 5.
WATT, DE CAPITO, EDWARDS and HERMANN: U. S. Publ. Health Rep. **65**, 208 (1950).
Salmonella sarajane = 4, 12, 27 : d : e, n, x.
DOUGLAS and TAYLOR: Month. Bull. Min. Health, Lond. **13**, 159 (1954).
Salmonella schleissheim = 4, 12, 27 : b :—.
KAUFFMANN u. TESDAL: Z. Hyg. **120**, 168 (1937).
Salmonella schoeneberg = 1, 3, 19 : z : e, n, z_{15}.
BOECKER u. KAUFFMANN: Zbl. Bakter. I Orig. **158**, 531 (1952).
Salmonella schottmuelleri = *Salmonella paratyphi B*.

Salmonella schwarzengrund = 1, 4, 12, 27 : d : 1, 7.
KAUFFMANN: Acta path. scand. (Copenh.) Suppl. **54**, 33 (1944).
ZIMMERMANN: Zbl. Bakter. I Orig. **153**, 178 (1949).

Salmonella seattle = 28 : a : e, n, x.
EDWARDS and MCWHORTER: Internat. Bull. Bacter. Nomenclat. a. Taxon **5**, 151 (1955).

Salmonella seegefeld = 3, 10 : r (i) : 1, 2.
HOFMANN u. POHL: Zbl. Bakter. I Orig. **167**, 413 (1957).

Salmonella selandia = 3, 15 : e, h : 1, 7.
KAUFFMANN: Z. Hyg. **120**, 177 (1937).

Salmonella sendai = 1, 9, 12 : a : 1, 5.
AOKI u. SAKAI: Zbl. Bakter. I Orig. **95**, 152 (1925).
Salmonella Subcommittee: J. of Hyg. **34**, 333 (1934).

Salmonella senegal = 11 : r : 1, 5.
HINSHAW and MCNEILL: J. Bacter. **52**, 349 (1946).

Salmonella senftenberg = 1, 3, 19 : g, s, t :—.
KAUFFMANN: Z. Hyg. **111**, 221 (1930).

Salmonella senftenberg var. newcastle = biochemische Variante von *Salmonella senftenberg*.
KAUFFMANN u. MITSUI: Z. Hyg. **111**, 749 (1930).
Salmonella Subcommittee: Proc. Third Internat. Congress Microbiol. 1940, S. 832.

Salmonella seremban = 9, 12 : i : 1, 5.
FRISBY and HOLLOS: J. Roy. Army Med. Corps **100**, 55 (1954).

Salmonella shamba = 16 : c : e, n, x.

Salmonella shangani = 3, 10 : d : 1, 5.
KAUFFMANN: Acta path. scand. (Copenh.) **16**, 347 (1939).

Salmonella shanghai = 16 : l, v : 1, 6.
FOURNIER: Rapport sur le fonctionnement technique de l'Institut Pasteur de Changhai **1**, 45 (1948).

Salmonella sheffield = 38 : c : 1, 5.

Salmonella shipley = (8), 20 : b : e, n, z_{15}.
NICEWONGER, TAYLOR and TOMLINSON: Im Druck.

Salmonella shoreditch = 9, 12 : r : e, n, z_{15}.

Salmonella shubra = 4, 5, 12 : z : 1, 2.

Salmonella siegburg = 6, 14, 18 : z_4, z_{23} :—.
KAUFFMANN, EDWARDS, SEELIGER and LUND: Acta path. scand. (Copenh.) **36**, 353 (1955).

Salmonella simi = 3, 10 : r : e, n, z_{15}.
KAUFFMANN and VANDEPITTE: Acta path. scand. (Copenh.) **27**, 181 (1950).

Salmonella simsbury = 1, 3, 19 :—: z_{27} (= 2. Phase von *Salmonella senftenberg*).
BRUNER and EDWARDS: Proc. Soc. Exper. Biol. a. Med. **50**, 174 (1942).

Salmonella singapore = 6, 7 : k : e, n, x.
SEN: J. of Path. **59**, 505 (1947).

Salmonella solna = 28 : a : 1, 5.
ALIN: Acta path. scand. (Copenh.) **38**, 71 (1956).

Salmonella solt = 11 : y : 1, 5.
Rauss: Z. Immunforsch. **103**, 220 (1943).

Salmonella souza = 3, 10 : d : e, n, x.

Salmonella springs = 40 : a : z_{39}.
Bokkenheuser and Edwards: J. of Path. **72**, 687 (1956).

Salmonella stanley = 4, 5, 12 : d : 1, 2.
Schütze: Lancet **1920 I**, 93.
Salmonella Subcommittee: J. of Hyg. **34**, 333 (1934).

Salmonella stanleyville = 4, 5, 12 : z_4, z_{23} : [1, 2].
Kauffmann, Courtois and van Oye: Acta path. scand. (Copenh.) **24**, 588 (1947).
Kauffmann and Defrenne: Acta path. scand. (Copenh.) **29**, 72 (1951).

Salmonella stockholm = 3, 10 : y : z_6.
Lagergren and Oesterling: Acta path. scand. (Copenh.) **29**, 386 (1951).

Salmonella strasbourg = (9), 46 : d : 1, 7.
Kauffmann and Lutz: Acta path. scand. (Copenh.) **36**, 179 (1955).
Kauffmann: Acta path. scand. (Copenh.) **38**, 69 (1956).

Salmonella suipestifer = *Salmonella cholerae suis*.

Salmonella sundsvall = (1), 6, 14, 25 : z : e, n, x.
Olin and Alin: Acta path. scand. (Copenh.) **20**, 607 (1943).

Salmonella szentes = 16 : k : 1, 2.
Rauss: Z. Immunforsch. **103**, 220 (1943).

Salmonella taihoku = identisch mit *Salmonella meleagridis*.
Carlquist: J. Bacter. **54**, 265 (1947).

Salmonella takoradi = 6, 8 : i : 1, 5.
Rewell, Taylor and Douglas: Month. Bull. Min. Health, Lond. **7**, 266 (1948).

Salmonella taksony = 1, 3, 19 : i : z_6.
Rauss: Z. Immunforsch. **103**, 220 (1943).

Salmonella tallahassee = 6, 8 : z_4, z_{32} :—.
Moran and Edwards: Proc. Soc. Exper. Biol. a. Med. **62**, 294 (1946).

Salmonella tananarive = 6, 8 : y : 1, 5.
Le Minor, le Minor et Neel: Ann. Inst. Pasteur **77**, 198 (1949).

Salmonella taunton = 28 : k : e, n, x.
Boycott, Taylor and Douglas: J. of Path. **65**, 401 (1953).
Douglas and Taylor: Month. Bull. Min. Health, Lond. **13**, 159 (1954).

Salmonella teddington = 4, 12, 27 : y : 1, 7.

Salmonella teko = (1), 6, 14, 25 : d : e, n, z_{15}.

Salmonella tel-aviv = 28 : y : e, n, z_{15}.
Kauffmann: Acta path. scand. (Copenh.) **17**, 1 (1940).

Salmonella tel-el-kebir = 13, 23 : d : e, n, z_{15}.

Salmonella tel-hashomer = 11 : z_{10} : e, n, x.
Kauffmann, Silberstein and Gerichter: Acta path. scand. (Copenh.) **27**, 888 (1950).

Salmonella tennessee = 6, 7 : z_{29} :—.
Bruner and Edwards: Proc. Soc. Exper. Biol. a. Med. **50**, 174 (1942).

Salmonella texas = 4, 5, 12 : k : e, n, z_{15}.
WATT, DE CAPITO and MORAN: U. S. Publ. Health Rep. **62**, 808 (1947).
Salmonella thiaroye = 38 : e, h : 1, 2.
LE MINOR, LE MINOR et DARRASSE: Ann. Inst. Pasteur **91**, 935 (1956).
Salmonella thomasville = (3), (15), 34 : y : 1, 5.
EDWARDS, DE CAPITO and FIFE: U. S. Publ. Health Rep. **66**, 1061 (1951).
Salmonella thompson = 6, 7 : k : 1, 5.
SCOTT: J. of Hyg. **25**, 398 (1926).
Salmonella thompson var. berlin = 6, 7 : — : 1, 5.
KAUFFMANN: Z. Hyg. **110**, 537 (1929).
Salmonella Subcommittee: Proc. Third Internat. Congress Microbiol. 1940, S. 832.
Salmonella tim = s. *Salmonella newington var. tim.*
Salmonella tinda = 1, 4, 12, 27 : a : e, n, z_{15}.
KAUFFMANN and JANSSENS: Acta path. scand. (Copenh.) **26**, 719 (1949).
Salmonella tokyo = nicht bestätigt.
WHITE, BRUCE: Med. Res. Counc. Spec. Rep. Ser. No 103, 1926.
KAUFFMANN u. MITSUI: Z. Hyg. **111**, 749 (1930).
Salmonella tosamanga = 6, 7 : z : 1, 5.
MACKEY: East Afr. Med. J. **32**, 1 (1955).
Salmonella tshiongwe = 6, 8 : e, h : e, n, z_{15}.
VAN OYE, DEOM, VERCRUYSSE et FASSEAUX: Ann. Inst. Pasteur. Im Druck.
Salmonella tudu = 4, 12 : z_{10} : 1, 6.
Salmonella tuebingen = 3, 15 : y : 1, 2.
WUNDT u. KLEIN: Z. Hyg. **140**, 481 (1954).
Salmonella tunis = 1, 13, 23 : y : z_6.
LE MINOR, HUET et DRÉAN: Ann. Inst. Pasteur **91**, 936 (1956).
Salmonella typhi = 9, 12, Vi : d :—.
Salmonella Subcommittee: J. of Hyg. **34**, 333 (1934).
Salmonella typhi murium = 1, 4, 5, 12 : i : 1, 2.
Salmonella Subcommittee: J. of Hyg. **34**, 333 (1934).
Salmonella typhi murium var. binns = 1, 4, 5, 12 : — : 1, 2.
SCHÜTZE: Lancet **1920 I**, 93.
Salmonella Subcommittee: Proc. Third Internat. Congress Microbiol. 1940, S. 832.
Salmonella typhi murium var. copenhagen = 1, 4, 12 : i : 1, 2.
KAUFFMANN: Z. Hyg. **116**, 368 (1934).
Salmonella typhi suis = 6, 7 : c : 1, 5.
Salmonella Subcommittee: J. of Hyg. **34**, 333 (1934).
Salmonella typhi suis var. voldagsen = 6, 7 : — : 1, 5.
Salmonella Subcommittee: Proc. Third Internat. Congress Microbiol. 1940, S. 832.
Salmonella uganda = 3, 10 : l, z_{13} : 1, 5.
KAUFFMANN: Acta path. scand. (Copenh.) **17**, 189 (1940).
Salmonella umbilo = 28 : z_{10} : e, n, x.
Salmonella umhlali = 6, 7 : a : 1, 6.
Salmonella umhlatazana = 35 : a : e, n, z_{15}.

Salmonella uphill = 42 : b : e, n, z_{15}.
BOYCOTT, TAYLOR and DOUGLAS: J. of Path. **65**, 401 (1953).
DOUGLAS and TAYLOR: Month. Bull. Min. Health, Lond. **13**, 159 (1954).
Salmonella uppsala = 4, 12, 27 : b : l, 7.
HEILBORN and LAURELL: Acta path. scand. (Copenh.) **40**, 39 (1957).
Salmonella urbana = 30 : b : e, n, x.
EDWARDS and BRUNER: J. Inf. Dis. **69**, 220 (1941).
Salmonella usumbura = 18 : d : l, 7.
KAUFFMANN and FAIN: Acta path. scand. (Copenh.) **33**, 112 (1953).
Salmonella utah = 6, 8 : c : l, 5.
EDWARDS and MCWHORTER: Publ. Health Labor. **12**, 62 (1954).
Salmonella uzaramo = (1), 6, 14, 25 : z_4, z_{24} :—.
TELLING, TAYLOR and DOUGLAS: Month. Bull. Min. Health, Lond. **10**, 251 (1951).
Salmonella vancouver = 16 : c : l, 5.
DOLMAN, RANTA, HUDSON, BYNOE, BAILEY and LAIDLEY: Canad. J. Publ. Health **41**, 23 (1950).
Salmonella vejle = 3, 10 : e, h : l, 2.
HARHOFF: Zbl. Bakter. I Orig. **147**, 194 (1941).
Salmonella veneziana = 11 : i : e, n, x.
BRUNER and JOYCE: J. Bacter. **50**, 371 (1945).
Salmonella venusberg = *S. nchanga var. venusberg.*
Salmonella verity = 17 : e, n, z_{15} : l, 6.
Salmonella victoria = 1, 9, 12 : l, w : l, 5.
Salmonella victoriaborg = 17 : c : l, 6.
Salmonella vinohrady = 28 : m, t :—.
Salmonella virchow = 6, 7 : r : l, 2.
KAUFFMANN: Z. Hyg. **111**, 221 (1930).
Salmonella virginia = (8) : d :—.
SAPHRA and SELIGMANN: Proc. Soc. Exper. Biol. a. Med. **58**, 50 (1945).
Salmonella vom = 4, 12, 27 : l, z_{13}, z_{28} : e, n, z_{15}.
Salmonella wagenia = 1, 4, 12, 27 : b : e, n, z_{15}.
KAUFFMANN, COURTOIS and VAN DEN ABBEELE: Acta path. scand. (Copenh.) **28**, 150 (1951).
Salmonella wandsworth = 39 : b : l, 2.
Salmonella wangata = 9, 12 : z_4, z_{23} :—.
KAUFFMANN and LUCASSE: Acta path. scand. (Copenh.) **32**, 335 (1953).
Salmonella warragul = (1), 6, 14, 25 : g, m :—.
Salmonella waycross = 41 : z_4, z_{23} :—.
SELIGMANN and SAPHRA: J. Bacter. **55**, 561 (1948).
Salmonella welikada = 16 : l, v : l, 7.
Salmonella weltevreden = 3, 10 : r : z_6.
ERBER: Geneesk. Tijdschr. Nederl.-Indië **81**, 2123 (1941).
Salmonella wentworth = 11 : z_{10} : l, 2.
Salmonella weslaco = 42 : z_{36} :—.
EDWARDS, HERMANN, WATT and DE CAPITO: U. S. Publ. Health Rep. **65**, 212 (1950).

Salmonella westhampton = 3, 10 : g, s, t : —.
Edwards and McWhorter: Cornell Veterinarian **43**, 110 (1953).
Salmonella weston = 16 : e, h : z_6.
Boycott, Taylor and Douglas: J. of Path. **65**, 401 (1953).
Douglas and Taylor: Month. Bull. Min. Health, Lond. **13**, 159 (1954).
Salmonella weybridge = 3, 10 : d : z_6.
Salmonella wichita = 1, 13, 23 : d : [z_{37}].
Schiff and Strauss: J. Inf. Dis. **65**, 125 (1939).
Salmonella wien = 4, 12 : b : l, w.
Roschka: Mitt. österr. Sanitätsverw. **52**, 1 (1951).
Salmonella windhoek = 45 : g, t : 1, 5.
Salmonella womba = 4, 12, 27 : c : 1, 7.
Salmonella worcester = 1, 13, 23 : m, t : e, n, x.
Kauffmann and Henning: Acta path. scand. (Copenh.) **31**, 586 (1952).
Salmonella worthington = 1, 13, 23 : z : l, w.
Edwards and Bruner: J. of Hyg. **38**, 716 (1938).
Salmonella wuerzburg = 9, 12 : a : 1, 5.
Le Minor: Ann. Inst. Pasteur **91**, 664 (1956).
Salmonella yodobasi = gehört nicht zur *Salmonella*-Gruppe.
Carlquist: J. Bacter. **54**, 265 (1947).
Salmonella yolo = 35 : c : —.
Van Oye, Kauffmann and Delcour: Acta path. scand. (Copenh.) **37**, 528 (1955).
Salmonella zagreb = 4, 5, 12 : e, h : 1, 2.
Kauffmann: Acta path. scand. (Copenh.) **18**, 351 (1941).
Mit *Salmonella saint-paul* vereint. Salmonella Subcommittee: Proc. Fourth Internat. Congress Microbiol. 1949, S. 607.
Salmonella zanzibar = 3, 10 : k : 1, 5.
Kauffmann: Acta path. scand. (Copenh.) **16**, 347 (1939).
Salmonella zega = 9, 12 : d : z_6.
Fain, Kauffmann and Schoetter: Acta path. scand. (Copenh.) **31**, 325 (1952).

Literatur

Andrewes, F. W.: J. of Path. **25**, 505 (1922); **28**, 345 (1925).
Arkwright, J. A.: J. of Path. **30**, 566 (1927).
Bainbridge, F. A., and R. A. O'Brien: J. of Hyg. **11**, 68 (1911).
Beyer, H. G., and A. L. Reagh: J. Med. Res. **12**, 313 (1904).
Boycott, A. E.: J. of Hyg. **6**, 33 (1906).
Bruner, D. W., and P. R. Edwards: (1) J. Bacter. **42**, 467 (1941).
— — (2) J. Bacter. **53**, 359 (1947).
Castellani, A.: Z. Hyg. **40**, 1 (1902).
Edwards, P. R.: (1) Proc. Soc. Exper. Biol. a. Med. **59**, 49 (1945).
— (2) J. Bacter. **51**, 523 (1946).
—, and D. W. Bruner: (1) J. of Hyg. **38**, 716 (1938).
— — (2) J. of Immun. **44**, 319 (1942).
— — (3) J. Inf. Dis. **69**, 220 (1941).
—, and W. H. Ewing: *Identification of Enterobacteriaceae.* Minneapolis: Burgess Publishing Company 1955.
—, and F. Kauffmann: Amer. J. Clin. Path. **22**, 692 (1952).

Edwards, P. R., F. Kauffmann and E. van Oye: Acta path. scand. (Copenh.) **31**, 5 (1952).
— A. C. McWhorter and M. A. Fife: J. Bacter. **67**, 346 (1954).
—, and A. B. Moran: Proc. Soc. Exper. Biol. a. Med. **61**, 242 (1946).
— — and D. W. Bruner: Proc. Soc. Exper. Biol. a. Med. **66**, 230 (1947).
Hormaeche, E., y C. A. Peluffo: Arch. urug. Med. Cir. y espec. **14**, 217 (1939).
— — y R. Salsamendi: Arch. urug. Med. Cir. y espec. **12**, 377 (1938).
Kauffmann, F.: (1) Z. Hyg. **111**, 221 (1930).
— (2) Zbl. Bakter. I Ref. **96**, 519 (1930).
— (3) Z. Hyg. **111**, 233 (1930).
— (4) Zbl. Hyg. **25**, 273 (1931).
— (5) *Enterobacteriaceae*, 2. Aufl. bei E. Munksgaard. Kopenhagen 1954.
— (6) Acta path. scand. (Copenh.) **39**, 299 (1956).
— (7) J. Bacter. **41**, 127 (1941).
— (8) Acta path. scand. (Copenh.) **18**, 225 (1941).
— (9) Acta path. scand. (Copenh.) Suppl. **54**, 33 (1944).
— (10) Z. Hyg. **117**, 431 (1935).
— (11) Z. Hyg. **119**, 103 (1936).
— (12) Z. Hyg. **120**, 177 (1937).
— (13) Acta path. scand. (Copenh.) **17**, 429 (1940).
— (14) Acta path. scand. (Copenh.) **17**, 1 (1940).
— (15) Z. Hyg. **119**, 356 (1937).
— P. R. Edwards and A. C. McWhorter: Acta path. scand. (Copenh.) **36**, 568 (1955).
—, u. Ch. Mitsui: (1) Z. Hyg. **111**, 749 (1930).
— — (2) Z. Hyg. **111**, 740 (1930).
—, and A. Petersen: Acta path. scand. (Copenh.) **38**, 481 (1956).
—, u. M. Tesdal: Z. Hyg. **120**, 168 (1937).
— u. G. Vahlne: Acta path. scand. (Copenh.) **22**, 119 (1945).
Minor, L. le: (1) Ann. Inst. Pasteur 88, 76 (1955).
— (2) Ann. Inst. Pasteur **91**, **664** (1956).
Rauss, K.: Z. Immunforsch. **95**, 489 (1939).
Salmonella Subcommittee: J. of Hyg. **34**, 333 (1934).
Schütze, H.: (1) J. of Hyg. **20**, 330 (1921).
— (2) Lancet **1920 I**, 93.
Smith, Th., and A. L. Reagh: J. Med. Res. **9**, 270 (1903).
Weil, E., u. A. Felix: Wien. klin. Wschr. **1918**, 637, 986. — Z. Immunforsch. **29**, 24 (1920).
White, P. Bruce: Med. Res. Counc., Spec. Rep. Ser. No 103, 1926.
— A System of Bacteriology, Bd. 4. 1929. London: Majesty's Stationery Office.

V. Activites Bactériolytiques des Microorganismes

Par

MAURICE WELSCH[1]

Table des Matières

Les activités bactériolytiques des microorganismes se manifestent par deux phénomènes, à première vue très différents: l'autolyse et l'hétérolyse.

Aussi longtemps que l'intégrité structurelle du microorganisme est conservée, de multiples barrières intra-cellulaires, anatomiques et physiologiques, assurent une localisation précise des enzymes et en régularisent le jeu harmonieux. Une lésion vient-elle désorganiser cet édifice complexe, voici les enzymes libres d'agir de façon anarchique et de décomposer les divers constituants cellulaires: c'est l'autolyse. Œuvre des enzymes bactériens, elle sera souvent déclenchée, cependant, par un agent extérieur à la cellule, un inducteur d'autolyse, qui la lèse sans toutefois inactiver tous ses systèmes enzymatiques. La dissolution des germes peut souvent, d'autre part, être favorisée par des conditions physico-chimiques déterminées du milieu ambiant: elle est sous la dépendance d'une série d'effecteurs de l'autolyse.

Les phénomènes d'hétérolyse consistent dans la dissolution de certaines bactéries par des agents élaborés par d'autres microorganismes. Ces agents peuvent avoir une action bactériolytique primaire. Mais ils peuvent aussi n'être que des inducteurs d'autolyse. La distinction entre ces deux types d'action est souvent difficile. Toutefois, la capacité de dissoudre des corps microbiens chauffés, non susceptibles de s'autolyser, peut être utilisée comme un des critères permettant parfois de faire cette distinction. L'étude de la bactériolyse des germes chauffés est donc justifiée. Mais on ne peut méconnaître que le chauffage est de nature à modifier l'état physique et chimique, la structure même, des microorganismes étudiés.

[1] Professeur à l'Université de Liège, Directeur des Laboratoires de Microbiologie Générale et Médicale.

Si les manifestations d'hétérolyse sont sans doute fréquemment accompagnées d'une intervention des systèmes autolytiques, inversement, l'autolyse d'une population bactérienne n'est généralement pas un processus strictement endogène. Au contraire, ce dernier, libérant des enzymes lytiques, s'accompagnera d'une dissolution d'origine exogène.

Les phénomènes de bactériolyse ont retenu l'attention des chercheurs pour des raisons très diverses. On attribue à l'autolyse *in vivo* de nombreux germes pathogènes, la libération intra-tissulaire de constituants toxiques, origine de manifestations morbides. Rien d'étonnant donc à ce que l'on ait d'abord étudié l'autolyse dans le but de caractériser et d'isoler les constituants microbiens responsables de l'action pathogène. Corrélativement, on a recherché les propriétés antigéniques des lysats et l'on a envisagé leur utilisation à titre de vaccins.

Phénomènes d'autolyse et d'hétérolyse ont plus tard été étudiés dans le but de mieux connaître la structure bactérienne. Aussi, les voit-on étroitement associés à l'étude de la réaction de Gram et de la structure des enveloppes bactériennes, en particulier de la paroi cellulaire.

La bactériolyse consécutive à l'action de certains antiseptiques, d'antibiotiques, ainsi qu'à l'infection par les virus bactériens, pose d'autres problèmes encore.

Les orientations très diverses qui ont conduit d'innombrables chercheurs à se pencher sur les problèmes de la bactériolyse, ne permettent guère de passer en revue, de façon complète, tous les travaux consacrés à la lyse et aux lysats. Nous limiterons notre exposé à quelques aspects seulement des phénomènes de bactériolyse, en nous excusant d'avoir choisi, de préférence, ceux auxquels notre Laboratoire a pu apporter une contribution.

Nous examinerons tout d'abord le système autolytique du pneumocoque (I) dont l'étude détaillée, poursuivie surtout par l'Ecole d'Avery, a fourni les notions de base relatives au mécanisme de l'autolyse en général.

Nous étudierons ensuite les activités bactériolytiques des staphylocoques (II) qui nous montreront, à l'occasion des phénomènes d'autophagie et d'isophagie, la dualité, endogène et exogène, du phénomène d'autolyse. Cette étude nous conduira ensuite à examiner le mécanisme de l'autolyse en relation avec la réaction de Gram (III).

Nous passerons ensuite en revue successivement les activités bactériolytiques de quelques germes gram-négatifs (IV) et des Bacillacées (V), qui nous amèneront à discuter le rôle des systèmes autolytiques dans la lyse par certains antibiotiques, et le mode d'action des inducteurs d'autolyse à propriétés tensioactives.

Nous terminerons, enfin, en examinant les activités bactériolytiques des actinomycètes (VI). Elles nous révéleront l'existence d'enzymes qui assurent la désintégration bactérienne par un mécanisme semblable à celui du lysozyme, en attaquant, de façon très spécifique, certains constituants de la paroi cellulaire.

I. Le Système Autolytique du Pneumocoque

Le pneumocoque est l'un des microorganismes dont l'activité bactériolytique et les autolysats ont été le plus étudiés. Les premières recherches sur ce sujet, effectuées entre 1910 et 1930, étaient orientées surtout, d'une part, en vue de faciliter l'identification du pneumocoque et sa distinction d'avec les espèces voisines, d'autre part, en vue de découvrir, dans ses lysats, les agents responsables de son activité pathogène. Plus tard, on s'est attaché à élucider le mécanisme même de l'autolyse.

Bürgers (1911) semble avoir été le premier à signaler que les cultures et suspensions de pneumocoques se clarifient très rapidement lorsque se trouvent réunies des conditions physico-chimiques favorables. Toutefois, si les germes ont été chauffés au préalable, ils ne s'autolysent pas.

Selon Mair (1929), un chauffage à 47° C, pendant 60 minutes, réduit déjà considérablement l'intensité et la vitesse de la lyse, tandis qu'un séjour des germes à 56° C, pendant 30 minutes, l'abolit complètement.

La clarification des suspensions pneumococciques se manifeste entre des valeurs de p_H allant depuis 5,0 jusqu'à 8,5, avec un optimum au voisinage de p_H 7,0 (Lord et Nye 1919, Mair 1929). Le phénomène peut être observé entre les températures de 0° et 50° C environ, avec un optimum vers 42° C (Mair 1929). Cet optimum thermique exprime l'état d'équilibre qui s'établit entre les vitesses de lyse et de destruction des agents lytiques, croissant, l'une et l'autre, avec l'élévation de la température.

L'autolyse pneumococcique est entravée en présence de sérum. Elle est, par contre, accélérée par addition d'un autolysat de pneumocoques, lequel, cependant, n'exerce aucune action lytique sur les streptocoques ou les staphylocoques. Ces observations de Lord et Nye (1922a, b, c) soulignent la spécificité du système autolytique du pneumocoque. Elles ont été confirmées et étendues, peu après, par Avery et Cullen (1923). Utilisant des suspensions en solution phosphatée de bactéries préalablement chauffées à 60° C pendant 30 minutes, ou à 120° C pendant 20 minutes, ces auteurs ont établi que les autolysats pneumococciques stériles sont capables de dissoudre les corps de pneumocoques tués, quel qu'en soit le type sérologique. Ils exercent aussi une action lytique évidente, quoique nettement plus faible, sur les corps tués des streptocoques du type *viridans*, apparentés aux pneumocoques, mais aucune sur les corps staphylococciques. L'activité pneumolytique d'un autolysat est proportionnelle à la quantité qu'on en introduit dans la suspension microbienne. Utilisé à faible concentration, insuffisante pour provoquer une dissolution appréciable des corps microbiens, il exerce néanmoins une action manifeste sur la cellule sensible. Sous son action, en effet, celle-ci perd la propriété de conserver le Gram. L'agent responsable de l'activité lytique est thermolabile: un chauffage de 30 minutes à 60° C le détruit complètement et définitivement. Il est, d'autre part, inactivé par oxydation. Mais ce processus est réversible, l'agent lytique pouvant être régénéré par réduction (Neill et Fleming 1927).

Les caractères de l'agent autolytique sont compatibles avec l'idée qu'il serait de nature enzymatique. Avery et Cullen avaient d'ailleurs démontré auparavant l'existence, dans les lysats de pneumocoques, d'enzymes protéolytiques (1920a), lipolytiques (1920b) et saccharolytiques (1920c). Plus tard, Goebel et Avery (1929) ont démontré que l'autolyse pneumococcique s'accompagne d'une protéolyse, mise en évidence par l'augmentation progressive d'azote aminé et d'azote non protéique, et d'une lipolyse dont témoigne la libération d'acides gras éthérosolubles. Des phénomènes identiques sont observés lors de la lyse des pneumocoques chauffés par les autolysats stériles. Ces derniers renferment d'ailleurs des ferments capables d'hydrolyser les protéines et les lipoïdes éthéro-solubles extraits des corps pneumococciques. Meyer, Dubos et Smyth (1936, 1937) y ont ultérieurement décelé une polysaccharidase dégradant les polysaccharides acides respectivement isolés du corps vitré, du cordon ombilical et des streptocoques du groupe A.

L'intervention de ces divers enzymes ne paraît toutefois pas indispensable pour que la bactériolyse se manifeste. En effet, Goebel et Avery (1929) ont montré qu'en présence de désoxycholate sodique, la pneumolyse est accompagnée de lipolyse, mais non de protéolyse. Si, de plus, elle est réalisée à basse température, la lipolyse elle-même est absente. La désintégration des corps microbiens, aboutissant à la clarification des suspensions primitivement troubles, ne semble donc pas exiger une dégradation poussée des constituants chimiques bactériens. Elle peut, au contraire, se limiter à une désorganisation de la structure cellulaire dont les organites constitutifs inframicroscopiques sont alors dispersés dans le milieu, sans subir nécessairement une décomposition chimique profonde. Antérieurement déjà, Jobling et Strouse (1912, 1913) avaient conclu de leurs expériences sur la clarification des suspensions pneumococciques, soit en présence de chloroforme, soit sous l'influence des leucoprotéases, que l'autolyse pouvait se dérouler tout à fait indépendamment de la protéolyse.

Le problème qui se pose dès lors est d'établir en quoi consiste la lésion cellulaire primitive, celle qui provoque la désorganisation structurelle de la bactérie et déclenche ainsi la lyse qu'accompagne, mais seulement si les circonstances le permettent, une dégradation enzymatique plus ou moins profonde des protéines, nucléoprotéines, lipides et polysaccharides microbiens. En particulier, il faudra rechercher la nature et le mode d'action de l'agent responsable de cette lésion primitive.

Ce problème nous amène à considérer, tout d'abord, l'action de diverses substances susceptibles de déclencher et d'accélérer la clarification des suspensions pneumococciques.

Dès 1900, Neufeld avait signalé que la bile provoque une clarification rapide des suspensions de pneumocoques. On sait que cette activité, étroitement spécifique, est utilisée pour identifier les pneumocoques et les distinguer des espèces voisines (Levy 1907, Avery et alii 1917). Après d'innombrables controverses, portant sur l'existence d'une éventuelle corrélation entre la solubilité du pneumocoque dans la bile et sa virulence (Bayer 1907, Mandelbaum 1907, Libman et Rosenthal 1908, Truche et alii, 1911, 1913, Kelly 1920, Malone 1924a, b, Harkins 1925, Falk et Jacobson 1926, Reimann 1927), l'accord semble s'être fait à présent. On admet que tous les pneumocoques sont solubles dans la bile, mais qu'ils présentent, à ce point de vue, des différences quantitatives considérables d'une souche à l'autre, mais sans corrélation étroite avec leur degré de virulence (Neufeld et Schnitzer 1928, Mair 1929).

Neufeld attribuait l'action pneumolytique de la bile aux glycocholate et taurocholate sodiques qu'elle contient. Il l'interprétait comme la conséquence d'une action sur la «membrane» bactérienne. Celle-ci subirait, sous leur influence, un gonflement suivi de précipitation. Diverses substances, les unes présentes dans la bile, les autres pas, possèdent également une activité pneumolytique. C'est ainsi que Nicolle et Adil Bey (1907) ont signalé les effets lytiques des sels d'acides quinolique, cholique, choléique, taurocholique, glycocholique et hypocholique; KOZLOWSKI (1925), celle des acides gras non saturés de la bile; ZIEGLER (1930, 1931, 1932), celle des acides déhydrocholique et désoxydéhydrocholique. LAMAR (1911a, b) a observé l'activité pneumolytique des oléates, linoléates et linolénates alcalins, tandis que DOWNIE, STENT et WHITE (1931) ont étudié celle de la saponine. Selon KLEIN et STONE (1931), cette dernière substance ne serait pneumolytique qu'entre d'étroites limites de concentration et exclusivement en présence de cholestérol ou (KLEIN 1933) d'ergostérol. Plus récemment, POLONOVSKI et COTONI (1947) ont étudié l'action lytique des savons cationiques en général, et SOLOMIDÈS et HIRSCH (1947), celle des savons des acides gras de l'huile de foie de morue. La quinine (ASH et SOLIS-COHEN 1929), le chloroforme (BÜRGERS 1911, STURDZA 1938d), le phosphate tri-sodique (FALK et YANG 1926a, b), l'hydroxyde de sodium 0,01 à 0,2 N (SELLARDS 1918), sont également pneumolytiques.

L'action de tous ces agents, au moins lorsque les conditions expérimentales sont bien choisies, paraît s'exercer spécifiquement sur le pneumocoque, au même titre que celle de la bile. On est donc en droit de se demander quelles relations peuvent exister entre la lyse qu'ils provoquent et l'autolyse spontanée.

NEUFELD et ETINGER-TULCZYNSKA (1930) ont observé que divers agents, capables d'inhiber la dissolution des pneumocoques par la bile ou les sels biliaires, empêchent également leur autolyse. C'est ainsi, par exemple, que les pneumocoques tués par chauffage ne subissent pas davantage la lyse en milieu bilié que l'autolyse spontanée. De même, l'acide phénique à 1%, le formol à 1%, le sulfite ammonique dilué, l'acétone, le chlorure mercurique à 0,05%, empêchent également l'autolyse et la dissolution biliaire des pneumocoques.

D'autre part, ATKIN (1926) a constaté que les colonies secondaires, développées sous forme de papilles au niveau des colonies primaires ayant subi l'autolyse sur gélose au sérum, sont constituées de germes insolubles dans la bile et n'ayant aucune tendance à s'autolyser. Cette résistance n'est toutefois pas une propriété stable, car les subcultures, obtenues à partir de ces germes, fournissent des pneumocoques solubles dans la bile et spontanément autolysables.

WOLLMAN et AVERBUSCH (1932a, b) ont montré que les suspensions de pneumocoques tués, insolubles dans la bile, subissent néanmoins une clarification complète en milieu bilié si l'on y ajoute un petit nombre de pneumocoques vivants. Ce phénomène que les auteurs décrivent sous le nom d'autolyse transmissible, est vraisemblablement identique, dans son mécanisme, à l'isophagie staphylococcique (GRATIA et RHODES 1924, 1925) que nous étudierons plus loin. Il est vraisemblable, en effet, que les pneumocoques vivants, dissous par la bile, libèrent des enzymes autolytiques dont l'action sur les pneumocoques chauffés est bien établie (LORD et NYE 1922a, b, c).

STURDZA (1938a, b, c, d) a souligné les analogies, mais aussi les différences, existant entre l'autolyse pneumococcique, spontanée ou favorisée par congélations et décongélations alternées (AVERY et NEILL, 1924, a), et la dissolution de ces germes sous l'influence des sels biliaires ou du chloroforme. Il montre, par exemple, que les limites de p_H compatibles avec la lyse sont pratiquement les mêmes dans tous les cas. Iode et sérum anti-pneumolysat inhibent aussi bien l'autolyse que la dissolution bactérienne en présence de chloroforme. Par contre, $MgSO_4$, favorable à la lyse biliaire, entrave la lyse chloroformique.

Les observations de Dubos, elles aussi, montrent que la pneumolyse provoquée par les agents les plus divers n'a lieu que si les conditions expérimentales sont telles qu'elles permettent l'action des enzymes autolytiques (Dubos 1937a). Par exemple, des cellules pneumococciques, précipitées par l'acétone ou par l'alcool, à basse température, puis séchées à l'éther, peuvent être remises en suspension dans des milieux dont l'action inhibitrice sur l'autolyse est bien connue: eau bouillante, solution iodée ou formolée. Dans ces conditions, elles conservent leur intégrité morphologique et leur caractère gram-positif. Si, par contre, elles sont remises en suspension dans une solution saline neutre, à la température de 30° à 40° C, elles y subissent une lyse explosive en l'espace de quelques secondes. Acétone ou alcool ont tué les germes. Faute d'apport énergétique, l'architecture cellulaire qui assure la localisation précise de toutes les substances fonctionnelles, ne peut être maintenue. La cellule morte n'est plus guère qu'un minuscule sac dans lequel sont mélangées les substances les plus diverses, et notamment les enzymes. Si ceux-ci n'ont pas été inactivés par le réactif bactéricide, ils entrent en action, pour autant que les conditions physico-chimiques le leur permettent. Dans ce cas, on assiste à la dissolution des germes avec clarification complète de la suspension (Dubos 1937b).

Ces diverses observations amènent à la conclusion que les divers agents soi-disant pneumolytiques ne sont, en réalité, que des inducteurs d'autolyse. Ils exercent sur le germe une action bactéricide plus ou moins étroitement spécifique, en conséquence de laquelle la structure cellulaire intime est bouleversée, ce qui libère toute une série d'enzymes. Ceux-ci s'attaquent alors aux divers constituants cellulaires sensibles à leurs activités respectives. L'inducteur d'autolyse peut, en outre, selon les cas, entraver le jeu de certains enzymes ou favoriser celui de certains autres, soit de façon directe, soit en modifiant l'état physique du substrat. Pauli (1927) a même suggéré que l'autolyse spontanée du pneumocoque serait, en dernière analyse, attribuable à l'auto-production par cet organisme de peroxydes qui joueraient ainsi le rôle d'inducteur naturel de l'autolyse. On comprend, dès lors, que les modifications chimiques associées à l'autolyse, spontanée ou amorcée par des inducteurs divers, puissent être quelque peu différentes selon les cas considérés (Goebel et Avery 1929). On voit aussi qu'une étude de la lyse, amorcée par des inducteurs variés, pourrait peut-être nous révéler l'existence d'une lésion primitive et générale de la cellule en voie d'autolyse.

On sait que les pneumocoques sont des germes gram-positifs et que leur capacité de prendre le Gram persiste après qu'ils ont été traités par des enzymes protéolytiques tels la pepsine ou la trypsine (Wilke, 1929). Nous avons vu, d'autre part, que les pneumocoquesc chauffés, traités par une faible quantité d'autolysat pneumococcique, perdent leur caractère gram-négatif sans modification appréciable de leur intégrité morphologique. De même, des pneumocoques vivants, séparés de leur milieu de culture par centrifugation et repris dans une solution saline dont le p_H est voisin de la neutralité, se transforment progressivement en diplocoques gram-négatifs. Il en est de même encore des pneumocoques traités par de faibles concentrations de bile ou de sels biliaires. Dubos (1938a) a démontré que les pneumocoques tués par le formol, l'iode ou l'acide acétique à p_H 4,5, conservent leur aspect morphologique normal et leur caractère gram-positif aussi longtemps qu'on les maintient en présence de l'agent qui a servi à les tuer. Si, par contre, les germes sont sédimentés et repris en solution saline neutre, ils y perdent rapidement leur capacité de conserver le Gram, en même temps qu'une ribonucléohistone est mise en solution.

On interprète ces observations en supposant que les agents bactéricides utilisés non seulement tuent le pneumocoque, mais inactivent encore la plupart des enzymes qui interviennent dans l'autolyse spontanée. A l'exception, toutefois, de celui qui est responsable de la négativation du Gram. Ce dernier phénomène peut être ainsi considéré comme la première étape de l'autolyse.

L'agent d'origine pneumococcique responsable de la négativation du Gram n'a pas été isolé à l'état de pureté et, par conséquent, si l'on a tout lieu de croire qu'il s'agit d'un enzyme, il est néanmoins impossible d'affirmer avec certitude la nature du substrat auquel il s'attaque. On sait, toutefois, que le pneumocoque peut être converti en diplocoque gram-négatif de taille quelque peu réduite par des enzymes thermostables présents dans divers tissus: leucocytes, pancréas, etc.... (Dubos et McLeod 1938), dont la seule activité connue par ailleurs est la capacité de dépolymériser l'acide ribonucléique (Dubos 1937d, Dubos et Thompson, 1938). On sait aussi que la conversion du caractère gram-positif du pneumocoque peut être obtenue après traitement des germes chauffés par une ribonucléase cristalline (Bartholomew et Umbreit 1944). Tout ceci suggère, par conséquent, mais sans le démontrcr de façon absolue et rigoureuse, que la première lésion cellulaire, initiatrice de l'autolyse du pneumocoque, consiste en la destruction enzymatique d'une structure responsable du caractère gram-positif dans laquelle intervient peut-être un acide ribonucléique. Nous verrons plus loin comment cette hypothèse a été généralisée ultérieurement pour rendre compte des phénomènes de lyse observés avec les germes gram-positifs les plus divers.

Revenons à présent, pour en terminer avec le pneumocoque, aux propriétés des lysats obtenus soit par autolyse naturelle, soit par lyse induite.

En plus des divers ferments hydrolytiques dont l'existence a déjà été signalée, on trouve, dans les lysats pneumococciques, plusieurs autres enzymes et substances d'intérêt biologique.

On y a reconnu, par exemple, un système catalysant la transformation de l'oxyhémoglobine en méthémoglobine (Butterfield et Peabody 1913). Son existence, mise en doute par Cole (1914b), a été confirmée par Morgan et Neill (1924). Cet enzyme est thermolabile et rapidement inactivé par oxydation (Neill et Avery 1924a, b). Avery et Neill ont d'autre part décrit, dans les lysats pneumococciques, un système d'oxydo-réduction dont l'activité se manifeste, selon les conditions expérimentales, par la formation de peroxydes (1924a) ou par la réduction du bleu de méthylène (1924b). Ce système comporte un donneur d'hydrogène non spécifique, thermostable, facilement éliminé par simple lavage des germes, et un enzyme spécifique, thermolabile, inactivé par oxydation (Neill et Avery 1925), plus abondant si l'autolysat provient d'une culture développée en anaérobiose (Neill et Avery 1924c).

La lyse des pneumocoques, spontanée ou artificiellement induite, libère une hémolysine (Cole 1914a, Hewitt et Famulener 1922, Cotoni et Chambrin 1928), thermolabile, antigénique (Neill, Fleming et Gaspari 1927a, b), inactivable par oxydation (Avery et Neill 1924c), mais dont l'activité est restaurée par réduction, celle-ci pouvant être assurée, par exemple, par incubation en présence de bactéries anaérobies (Neill 1926).

La présence de substances toxiques diverses dans les lysats pneumococciques a été signalée par plusieurs auteurs, mais l'accord n'est pas fait, dans tous les cas, sur leur signification. Pour les uns, comme Mair (1928), il s'agit de matériaux toxiques intra-cellulaires, préexistants et libérés par la lyse des germes. Pour d'autres, comme Rosenow (1910, 1912a, b, c) et Cole (1912), il s'agit, au contraire, de substances toxiques formées seulement au cours même de l'autolyse, par action des enzymes sur les constituants normaux du pneumocoque. Le rôle que peuvent jouer ces agents toxiques au cours de l'infection pneumococcique elle-même a été âprement discuté mais reste douteux (Sabin 1931).

Quoi qu'il en soit, rappelons que l'on a mis en évidence, dans les lysats, un principe purpurogène (Julianelle et Reimann 1926, 1927, Pittmann et Falk 1930, Goodner 1933, Goodner et Horsfall 1937), probablement responsable également d'une destruction des plaquettes (Reimann et Julianelle 1926). On y a décrit aussi un agent dermo-nécrosant, très sensible à l'oxydation et à la chaleur (Parker 1928, 1929a, b), considéré comme identique à un principe toxique pour le poumon, produit en plus grande abondance dans les lysats obtenus en anaérobiose (Parker et Pappenheimer 1928, Pittman et Southwick 1930) et ayant qualité d'antigène (Parker et McCoy, 1929). Enfin, on y a décelé encore une toxine déterminant un shock anaphylactoïde (Cole, 1912; Rosenow, 1912a, b, c, Clough 1915, Boehncke et Mouriz-Riesgo 1915), appelée «pneumotoxine» par Weiss (1918), mais qui pourrait n'être rien d'autre que l'histamine (Weiss et Kolmer 1918).

Il est donc bien établi que les autolysats pneumococciques, résultant d'une lyse naturelle ou induite, contiennent diverses substances antigéniques. D'ailleurs, Bigelow (1922) a montré qu'ils pouvaient évoquer une réaction cutanée allergique et Mackenzie et Woo (1925), puis Sharp et Blake (1930), ont établi qu'ils pouvaient agir en qualité d'allergène. Cope et Howell (1931) ont obtenu un phénomène de Shwartzman caractéristique au moyen de lysats biliaires que Reid et Iverson (1941) ont utilisé pour obtenir un sérum anti-pneumococcique.

Avery et Neill (1925) ont toutefois montré que les lysats de pneumocoques encapsulés ne peuvent induire, chez le lapin, la formation d'anticorps type-spécifiques, dirigés contre les polysaccharides capsulaires, et ceci, en dépit du fait que les polysaccharides en question sont présents dans les lysats et parfaitement susceptibles de précipiter *in vitro* en présence de l'anticorps correspondant. Ce fait, reconnu par la plupart des auteurs, n'a cependant pas empêché les essais d'utilisation des lysats à titre de vaccins, tant prophylactiques que thérapeutiques et, semble-t'il, non sans un certain succès (Brotzu 1924, Horder et Ferry 1926, Meyer 1927, Meyer et Sukneff 1928, Barrach 1928, Day 1930, 1934 Ziegler 1933, Blackman 1934), au même titre d'ailleurs que les germes soumis à une digestion par la trypsine (Warden 1912, Hirschfelder 1912) ou à une extraction par acide chlorhydrique (Harley 1934, 1935).

Dubos a montré que l'autolysat pneumococcique stérile, agissant sur des germes encapsulés, tués par chauffage, fait disparaître leur capacité d'induire la production d'anticorps type-spécifiques chez le lapin. Cet effet est observé quel que soit le type auquel appartiennent respectivement le pneumocoque soumis à la lyse et celui qui fournit l'autolysine. Il n'est même pas nécessaire que le germe subisse une désintégration évidente. Il suffit, au contraire, qu'il perde son caractère gram-positif, que ceci soit réalisé par traitement à l'autolysine des germes chauffés, ou par traitement des bactéries vivantes au formol ou à l'acide acétique, suivi de reprise en solution saline neutre. Par contre, après dissolution des cellules chauffées par les enzymes protéolytiques, tels la trypsine ou la chymotrypsine, aboutissant à la formation de débris morphologiquement non reconnaissables, mais gram-positifs, l'antigénicité de type est parfaitement conservée (DUBOS 1937c, 1938b, 1939b).

L'enzyme qui assure la négativation du Gram, considérée comme le premier temps de l'autolyse, entraîne donc, d'autre part, une transformation de l'antigène type-spécifique complet en un simple haptène. Il ne semble altérer en rien le polysaccharide responsable du type et paraît le séparer simplement du complexe mal connu dans lequel il se trouve engagé par des liaisons qui résistent aux protéases, mais non à divers enzymes ayant en commun la capacité de dépolymériser l'acide ribonucléique (DUBOS 1945).

II. Les Activités Bactériolytiques des Staphylocoques

L'autolyse des cultures et suspensions de staphylocoques est certes un phénomène d'observation banale, mais, moins spectaculaire que celle des pneumocoques, elle a été beaucoup moins étudiée.

JAUMAIN, en 1922, a signalé que la clarification des cultures de staphylocoques était singulièrement accélérée lorsqu'on les maintenait en atmosphère confinée, par exemple en tubes scellés. Cette observation, quoique vérifiée à maintes reprises, n'a pas été davantage étudiée pendant longtemps.

Peu après, GRATIA et RHODES (1923) faisaient l'observation suivante. Après addition d'un bactériophage spécifique, les suspensions en bouillon de staphylocoques, tués par chauffage à 60° C, subissent une clarification appréciable après incubation de quelques semaines à la température du laboratoire. Cette clarification est, toutefois, non seulement très lente, mais aussi très irrégulière. Cependant, si l'on ajoute à la suspension de germes chauffés un petit nombre de staphylocoques vivants, sensibles au bactériophage utilisé, on observe une lyse plus rapide et plus constante. Ce phénomène fut tout d'abord considéré comme une attaque des bactéries chauffées par le bactériophage «naissant». Des observations identiques furent faites plus tard, indépendamment, par TWORT (1925) qui les interprétait de la même manière.

Pourtant, quelques mois après leur communication précitée (1923), GRATIA et RHODES (1924) publiaient de nouvelles expériences, démontrant que la clarification des suspensions staphylococciques observée par eux était due, non pas au bactériophage, mais bien à une activité bactériolytique des staphylocoques vivants. Le rôle du virus bactérien n'est que secondaire. Attaquant les staphylocoques vivants, il s'oppose à leur multiplication excessive qui, en son absence, masquerait, par le trouble résultant, la clarification due à la lyse des germes tués. En effet, si le mélange de staphylocoques chauffés et vivants est mis en suspension, non plus en bouillon, mais en soluté physiologique, milieu incapable d'assurer une multiplication de cet organisme, la clarification est évidente, même en l'absence de bactériophage. Il en est de même encore si le mélange de bactéries chauffées et vivantes, bien que fait en bouillon, est placé en tubes scellés. Dans ces conditions, comme l'a montré JAUMAIN (1922), le staphylocoque ne peut se multiplier, il s'autolyse, au contraire, et ses autolysines dissolvent les corps staphylococciques chauffés. Les auteurs ont aussi réalisé des expériences analogues en milieu gélosé. Si quelques staphylocoques vivants sont inoculés à la surface d'une gélose minérale, opacifiée au moyen de corps microbiens staphylococciques tués par chauffage, ils y forment des colonies, et celles-ci s'entourent chacune d'une zone de clarification très nette, explicable par la diffusion d'un agent staphylolytique. GRATIA et RHODES n'ont observé ce phénomène qu'avec des paires de souches convenables, l'une lytique, l'autre sensible. Ils l'ont nommé: isophagie staphylococcique. Leurs observations (GRATIA et RHODES 1926) ont été confirmées par celles de DURAN-REYNALS (1926), de WOLLMAN et DURAN-REYNALS (1926) qui, ainsi que TWORT (GRATIA 1926), se sont ralliés à leur interprétation de la soi-disant action lytique du bactériophage sur les staphylocoques chauffés (WOLLMAN 1926).

GRATIA et RHODES (1924) constatent que les staphylocoques ensemencés en soluté physiologique ne sont pas capables de s'y multiplier et meurent; par contre, ensemencés dans ce même liquide auquel on a ajouté des corps staphylococciques chauffés, ils dissolvent ces derniers et se multiplient jusqu'à un certain point. L'isophagie leur apparaît donc comme une activité nutritive, grâce à laquelle les germes, bien que placés en milieu déficient, peuvent néanmoins survivre aux dépens de corps microbiens morts. L'isophagie, pour autant qu'elle se réalise dans la Nature, serait donc un des facteurs qui assurent la destruction des cadavres bactériens et, ainsi, la remise en circulation des éléments biogènes qui les constituent. Ce serait l'un des multiples chaînons du cycle naturel de la matière organisée (WELSCH 1949a, IMSHENETSKIY et KUZYURINA 1951).

Les activités bactériolytiques des staphylocoques ont été fortuitement redécouvertes puis étudiées dans notre Laboratoire par SALMON (1949a). Celui-ci préparait des géloses enrichies en corps microbiens chauffés pour y cultiver des Myxobactéries. Dans une gélose nutritive, à base de bouillon peptoné, il avait introduit quelques ml d'une épaisse suspension staphylococcique préalablement maintenue au bain d'eau à 65° C pendant 30 minutes. Or, ce chauffage avait été insuffisant pour assurer la mort de tous les germes. Les survivants, se multipliant dans l'épaisseur de la gélose nutritive, y formèrent des colonies. Celles-ci étaient entourées d'une large zone de clarification. Il s'agit, manifestement, d'un phénomène très semblable à celui qu'avaient décrit GRATIA et RHODES, avec la différence, toutefois, que nous sommes ici en présence d'une lyse de staphylocoques chauffés par des individus vivants appartenant non seulement à la même espèce, mais encore à la même souche. Il s'agit donc d'une autophagie plutôt que d'une isophagie.

Après cette observation fortuite, SALMON s'est attaché à l'étude systématique des activités bactériolytiques que le staphylocoque exerce, en milieu gélosé, sur les germes tués par chauffage.

La technique générale utilisée consiste à préparer une suspension bactérienne épaisse, en récoltant, dans 10 ml d'eau distillée, le croît microbien développé en 24 heures à 37° C sur une boîte de Roux contenant 50 ml de bouillon gélosé (peptone: 20 g; extrait de viande: 7,5 g; NaCl: 5 g; agar: 20 g; eau distillée: pour faire 1 litre). Le titre de ces suspensions est de l'ordre de 10^9 germes viables/ml. Elles sont lavées trois fois à l'eau, puis chauffées, généralement à 100° C pendant 3 minutes. On en introduit 1 ml dans 25 ml du milieu gélosé étudié, maintenu liquéfié à 45° C. On ajoute enfin une suspension d'environ 100 staphylocoques vivants et on coule le mélange dans une boîte de Petri. Les zones de lyse peuvent être observées autour des colonies déjà après 24 ou 48 heures d'incubation à 37° C, dans les conditions les plus favorables, mais beaucoup plus tardivement dans d'autres circonstances.

L'examen d'une série de souches diverses, comprenant des staphylocoques blancs et dorés, les uns fraîchement isolés, les autres provenant de collections de culturs, a montré que, contrairement à l'opinion de GRATIA et RHODES (1924), les activités autophagique et isophagique sont des propriétés générales du staphylocoque. En effet, toutes les souches examinées se sont montrées capables de donner lieu à l'autophagie et ne se distinguent entre elles que par de faibles différences quantitatives de cette activité. De même, chacune des souches étudiées provoque la lyse (isophagie) de n'importe quel autre staphylocoque chauffé. Bien plus, l'activité lytique des staphylocoques peut encore s'exercer sur les corps tués de quelques autres espèces microbiennes (hétérophagie): *Sarcina lutea* et *Micrococcus lysodeikticus*, et, à un moindre degré, diverses souches d'*Escherichia coli* et de *Proteus vulgaris*. Dans le cas de ces deux dernières espèces, les zones de lyse observées sont souvent très étendues, mais leur clarification reste incomplète. Diverses corynébactéries, *Streptococcus haemolyticus*, *Pseudomonas aeruginosa* et *Serratia marcescens*, résistent à l'action lytique des staphylocoques.

Les activités bactériolytiques sont inhibées par addition au milieu gélosé de 10% de sérum normal de bœuf ou de 2% de glucose. Elles sont, par contre, tout aussi nettes en gélose maintenue à l'air libre que placée en stricte anaérobiose, dans une atmosphère d'H_2 ou dans l'appareil de FILDES et MCINTOSH (1921). L'introduction dans la gélose d'un réducteur, par exemple 0,1% de thioglycolate de sodium, n'entrave en rien le phénomène. D'ailleurs, les colonies développées dans la profondeur de la gélose s'entourent d'un halo de clarification plus net et plus étendu que celles qui se forment à la surface du milieu. Il semble donc bien qu'une tension d'O_2 quelque peu réduite soit plutôt favorable à l'activité bactériolytique des staphylocoques.

Si l'on recherche l'autophagie staphylococcique dans des géloses où la concentration en bouillon nutritif est progressivement diminuée, on constate que les phénomènes cessent de se manifester bien avant que la possibilité d'un développement bactérien ait disparu. Ainsi, SALMON (1949b) constate que, dans les géloses staphylococciques préparées au moyen de bouillon nutritif dilué au $^1/_{10}$, les colonies s'entourent encore d'une zone de clarification évidente, bien que de dimensions réduites. Par contre, si le bouillon de base est dilué au $^1/_{100}$, les colonies de staphylocoques qui se forment encore n'exercent plus aucune action lytique visible, même au cours d'une incubation de 15 jours. On est donc amené à conclure que la dilution du bouillon nutritif élimine d'abord un élément nécessaire aux activités lytiques du staphylocoque, mais non indispensable à sa multiplication.

D'autre part, si des suspensions de staphylocoques chauffés, faites en eau distillée ou en solution minérale (KNO_3: 2 g; K_2HPO_4: 1 g; KCl: 0,5 g; $MgSO_4$: 0,5 g; $FeSO_4$: 0,01 g; eau distillée: pour faire 1 litre), sont gélosées et ensemencées avec une centaine de staphylocoques vivants, on pourra observer, mais irrégulièrement, la formation de colonies, dont la taille est d'ailleurs réduite. Mais ce développement, en réalité, est attribuable aux minimes quantités de substances nutritives qui sont enlevées à la gélose nutritive lors de la récolte des germes sensibles et fortuitement introduites avec eux dans le milieu (SALMON 1949b). En effet, si les staphylocoques récoltés sont soigneusement lavés à l'eau, à trois reprises, avant d'être chauffés puis introduits dans l'eau gélosée ou la gélose minérale, les germes vivants, inoculés dans ce milieu, ne sont plus capables d'y former des colonies visibles. Inversement, l'addition d'un ml d'une eau distillée ayant servi à laver une gélose nutritive vierge, à raison de 10 ml pour 50 ml en boîte de Roux, peut suffire à rendre propice au développement du staphylocoque 25 ml d'eau gélosée ou de gélose minérale.

Dans l'eau distillée gélosée, opacifiée au moyen de corps staphylococciques lavés et chauffés, contenant en outre 1% de bouillon nutritif, les staphylocoques vivants forment des colonies, mais celles-ci ne s'entourent pas d'une zone clarification, même après addition d'autolysine (WELSCH 1949c) staphylococcique qui y dissout cependant les corps bactériens. Par contre, lors d'une expérience identique, mais réalisée cette fois dans la solution minérale gélosée, on voit que les colonies y exercent leur activité bactériolytique. De toute évidence, en plus des éléments nutritifs apportés par la petite quantité de bouillon, et nécessaires pour assurer la croissance du staphylocoque, il faut un élément additionnel, apporté par la solution minérale, indispensable pour que s'exerce l'activité lytique. En éliminant à tour de rôle les sels constitutifs de la solution minérale utilisée, SALMON a montré que cet élément est K_2HPO_4, encore efficace à la concentration de 0,01 à 0,02%. Son rôle ne s'explique pas essentiellement par son action sur le p_H. En effet, les tampons de phosphates primaire et secondaire assurant des valeurs de p_H comprises entre 6,0 et 7,6 sont parfaitement convenables pour permettre l'activité lytique, alors que l'eau gélosée staphylococcique, dans laquelle la lyse ne se manifeste pas, possède un p_H de 7,3. D'ailleurs, les géloses microbiennes, dépourvues de phosphates, mais ajustées à des valeurs de p_H comprises entre 6,0 et 7,6, ne permettent pas l'activité lytique.

WELSCH, SALMON et HEUSGHEM (1949) ont confirmé l'importance du phosphate pour le phénomène d'autophagie en préparant des géloses à partir de bouillon dont ce sel avait été éliminé par précipitation au moyen d'un excès de $BaCl_2$. Ces géloses permettent la formation de colonies de staphylocoques, mais, lorsqu'elles sont opacifiées au moyen de corps staphylococciques chauffés, elles ne sont pas propices à la manifestation des activités bactériolytiques. L'addition de 0,1% K_2HPO_4 à ces milieux suffit cependant à permettre l'autophagie, et ce, même lorsque le phosphate ainsi ajouté est en quantité insuffisante pour neutraliser complètement l'excès de $BaCl_2$. Il ne semble pourtant pas que la bactériolyse soit alors possible à la faveur du $BaHPO_4$ précité, car l'addition directe de ce sel à une gélose staphylococcique contenant 1% de bouillon, ne permet pas aux colonies de s'entourer d'une zone de lyse. On arrive ainsi à penser que des composés phosphorés organiques, présents dans le bouillon, pourraient éventuellement jouer, comme l'orthophosphate, le rôle d'effecteur de l'autophagie.

En bon accord avec cette interprétation est le fait que les dilutions limites respectives auxquelles trois peptones différentes permettent encore l'activité bactériolytique sont, au moins grossièrement, proportionnelles à leur teneur en phosphore. On ne peut d'ailleurs s'attendre à une proportionnalité rigoureuse, car le degré et la vitesse d'utilisation de composés distincts peuvent naturellement être quelque peu différents. Cinq cents ml d'un bouillon dix fois plus concentré que normalement, ont été séparés, par dialyse à travers cellophane et contre 5 litres d'eau distillée, en deux fractions: un dialysat et un indialysé, amenés ensuite, l'un et l'autre, au volume de 5 litres par addition d'eau distillée. La teneur en P total et la dilution limite permettant encore l'autophagie ont été déterminées pour le bouillon normal et chacune des deux fractions, dialysable et indialysable. On constate que cette dernière est un effecteur moins bon que les deux autres, ce qui semble indiquer que les composés phosphorés organiques à grosses molécules sont plus difficilement utilisés par le staphylocoque.

SALMON (1950c) a ensuite montré, de façon directe, que des composés organiques définis, tels le glycérophosphate sodique, l'acide ribonucléique, la lécithine animale, pouvaient agir comme effecteurs de l'autophagie.

WELSCH et SALMON (1949b, 1950a) ont montré, d'autre part, que l'orthophosphate peut être remplacé par certains sels minéraux dérivés du phosphore. En eau gélosée, contenant 1% de bouillon ainsi que des corps staphylococciques lavés et chauffés, les staphylocoques vivants se développent et exercent leur activité lytique pour autant que l'on ait introduit dans le milieu 0,05 à 0,2% de phosphate dipotassique, 0,05 à 0,4% de métaphosphate sodique, ou 0,0125 à 0,025% de pyrophosphate sodique. La clarification, visible après 24 ou 48 heures dans les cas de K_2HPO_4, n'apparaît toutefois qu'après 4 à 7 jours dans le cas des autres sels et reste moins étendue. Avec les phosphite et hypophosphite, aucune lyse n'a été observée, même après 15 jours ou plus d'incubation, les concentrations variant de 0,004 à 1%. Quant au pyrophosphate il exerce une action inhibitrice nette sur la croissance du staphylocoque, même à des dilutions où son effet sur le p_H ne peut être mis en cause.

On constate par ailleurs que si ces divers succédanés, minéraux ou organiques, sont mis en incubation avec des staphylocoques, il y a libération d'orthophosphate (SALMON 1950c). Le rôle effecteur de ces substances s'exerce donc probablement, dans tous les cas, par l'intermédiaire de l'orthophosphate dont l'activité paraît être ainsi étroitement spécifique.

On doit évidemment se demander si l'orthophosphate favorise la production des autolysines par le staphylocoque, s'il assure leur diffusion à partir des germes, ou s'il est nécessaire pour permettre leur action sur les corps bactériens chauffés. Il n'est pas possible de répondre de façon certaine à ces questions. Toutefois, les expériences suivantes sont peu compatibles avec l'idée que le rôle principal du phosphate porterait sur l'élaboration des lysines. En effet, les staphylocoques, inoculés en eau gélosée, contenant 1% de bouillon et des corps staphylococciques, lavés et chauffés, forment des colonies, mais ne provoquent pas de bactériolyse. Après 5 jours d'incubation, ces colonies ont atteint, depuis longtemps, leur développement maximal. Il paraît logique d'admettre que les germes qui les constituent sont alors en phase de repos et ne métabolisent plus de façon active. Or, il suffit, à ce moment, d'étaler sur la gélose renfermant ces colonies, une mince couche d'une solution gélosée de phosphate (ortho, méta ou pyro) pour qu'après quelques heures, chaque colonie soit entourée d'un halo de clarification très net (Welsch et Salmon 1949b). Etant donné que la présence du phosphate n'est pas indispensable à l'autolyse du staphylocoque (Welsch 1949b), ces observations nous font croire que, très vraisemblablement, le rôle effecteur de ce sel dans les phénomènes d'autophagie, isophagie et hétérophagie, est surtout d'assurer la libération des agents lytiques.

Une expérience d'exécution très analogue suggère, d'autre part, que l'agent bactériolytique n'est pas élaboré, comme un enzyme adaptatif, par suite d'une induction attribuable aux corps staphylococciques chauffés, substrat sensible à son action. En effet, si on laisse se développer des colonies de staphylocoques dans une mince couche de bouillon gélosé, et, qu'après 5 ou 6 jours d'incubation, moment où leurs germes constitutifs sont très vraisemblablement en phase de repos, on les recouvre avec une eau gélosée opacifiée par des corps staphylococciques lavés et chauffés, on observe, quelques heures plus tard, une clarification évidente au niveau de chaque colonie (Salmon 1949a).

La démonstration que les staphylocoques ne peuvent se développer en gélose minérale opacifiée au moyen de corps microbiens chauffés, en tout cas lorsque ceux-ci ont été bien lavés, n'est évidemment pas compatible avec l'interprétation primitive de Gratia et Rhodes (1924), selon laquelle l'isophagie serait une fonction permettant aux germes de se multiplier, en milieu nutritivement déficient, aux dépens des cadavres de leurs congénères. Les activités bactériolytiques exercées en milieu gélosé par les staphylocoques nous apparaissent, au contraire, comme un phénomène contingent, accompagnant éventuellement la multiplication bactérienne, mais qui n'en est nullement la source. L'autophagie n'est observée, en effet, que dans les milieux où la croissance des germes est possible, même en l'absence d'une addition de staphylocoques lavés et chauffés.

Salmon (1950a) a tenté de démontrer de façon formelle que les corps staphylococciques chauffés ne peuvent apporter aux staphylocoques vivants tout ce qui est nécessaire pour assurer leur développement. Dans ce but, il les a incorporés non plus à des géloses, mais à des gels de silice, c'est-à-dire à des milieux dépourvus de toute autre source de matière organique. Or, les staphylocoques vivants, inoculés dans ces gels, en l'absence comme en présence d'orthophosphate, n'y forment pas de colonies. Il en est de même encore si le gel de silice est enrichi par addition d'un hydrolysat de caséine, privé de vitamines par adsorptions répétées sur charbon activé (Norit A). Des expériences de contrôle montrent, bien entendu, qu'en gel de silice contenant du bouillon, les staphylocoques forment des colonies aussi volumineuses qu'en gélose nutritive et à la même vitesse.

Dans ces conditions expérimentales, les corps staphylococciques lavés et tués, ne peuvent donc être utilisés ni comme source exclusive de matière et d'énergie, ni comme simple source des quelques facteurs de croissance requis pour le développement du staphylocoque.

Malheureusement, ces observations ne permettent pas d'affirmer qu'il en est toujours ainsi. En effet, dans un gel de silice contenant assez de bouillon pour que les staphylocoques y forment des colonies, renfermant en outre du phosphate et des corps bactériens lavés et chauffés, il n'y a, néanmoins, aucune trace de bactériolyse, contrairement à ce qui se passe, dans les mêmes conditions, en milieu gélosé. L'absence d'utilisation des corps staphylococciques comme source nutritive est donc peut-être tout simplement attribuable au fait que les germes vivants, introduits dans le milieu silicé, sont incapables de les digérer. Cette absence de bactériolyse pourrait résulter du manque d'un facteur favorisant hypothétique normalement apporté par l'agar. L'expérience montre qu'il n'en est cependant rien: l'addition d'agar aux gels de silice ne permet pas à l'autophagie de se manifester. L'absence de bactériolyse pourrait, d'autre part, résulter d'une action inhibitrice de la silice. Tel est bien le cas, en effet. L'addition de silice colloïdale à des géloses permettant l'autophagie, entrave partiellement (0,02%) ou totalement (0,12%) l'activité lytique selon sa concentration. Cet effet inhibiteur ne porte pas sur l'élaboration des lysines, mais bien sur leur activité. Il est la conséquence de leur adsorption sur la silice colloïdale (SALMON 1950b) dont la capacité de fixer d'autres agents bactériolytiques avait été signalée déjà par CUMMINS et WEATHERALL (1931).

Pour défendre l'interprétation de GRATIA et RHODES, on pourrait avancer que, même en milieu gélosé, où l'action des lysines est possible, une cellule staphylococcique isolée serait néanmoins incapable de se développer aux dépens des seuls corps staphylococciques, lavés et chauffés, parce-qu'elle ne dispose pas, en quantité suffisante, des enzymes nécessaires pour les digérer. La présence d'une minime quantité de bouillon nutritif lui permettrait toutefois de commencer à se multiplier, assurerait ainsi l'élaboration des lysines, grâce auxquelles les corps microbiens seraient ensuite désintégrés et utilisés comme matériaux nutritifs. Cette explication ne peut cependant être retenue, croyons-nous. En effet, l'addition de corps staphylococciques, lavés et chauffés, à un milieu déterminé, propice à l'autophagie, n'améliore cependant en rien la croissance des staphylocoques vivants qu'on y introduit. D'autre part, ces derniers, ensemencés dans un autolysat staphylococcique (WELSCH 1949d), stérilisé et gélosé, enrichi ou non en phosphate, n'y forment pas de colonies (SALMON 1949b).

WOLLMAN et DURAN-REYNALS (1926), DURAN-REYNALS (1926), WOLLMAN (1926) ont interprété l'isophagie staphylococcique comme une «autolyse transmissible». WOLLMAN et AVERBUSCH (1932a, b) ont avancé cette même interprétation plus tard, pour expliquer un phénomène analogue observé avec des pneumocoques. Selon ces auteurs, le processus d'autolyse se «transmettrait», de façon à vrai dire assez mystérieuse, des germes vivants aux germes morts auxquels ils sont mélangés. Il convient cependant de signaler que les colonies staphylococciques responsables de la clarification des corps chauffés qui les entourent ne manifestent, à aucun moment, de signes quelconques d'autolyse.

Pour nous, les phénomènes d'autophagie, isophagie et hétérophagie, sont la conséquence d'une sécrétion ou excrétion d'un agent lytique diffusible par des cellules parfaitement saines, cette libération exigeant la présence d'une concentration minimale d'un orthophosphate ou d'un succédané, ainsi qu'un p_H compris entre 6,0 et 7,6. Le phénomène de dissolution des corps bactériens chauffés par ces lysines est purement fortuit et ne joue aucun rôle biologique particulier pour le staphylocoque actif lui-même. Ces agents lytiques peuvent toutefois être libérés aussi lors de l'autolyse des staphylocoques et sont vraisemblablement responsables, au moins en partie, de la clarification des suspensions de ces germes en milieu liquide.

Ces considérations nous amènent à examiner à présent le phénomène de l'autolyse staphylococcique.

Etendant les observations de JAUMAIN (1922), WELSCH (1949b) a montré que la mise des staphylocoques en anaérobiose stricte ou en atmosphère confinée en tubes scellés, n'est pas indispensable pour qu'on obtienne leur autolyse. Des cultures en bouillon, développées avec aération forcée sur un agitateur approprié, subissent, en effet, une clarification très appréciable si on les transvase simplement ensuite dans des tubes, de préférence étroits et hauts. L'autolyse staphylococcique est donc favorisée par une simple réduction de l'oxygénation, une anaérobiose relative. On constate d'ailleurs que les cultures agitées, placées ensuite en atmosphère d'hydrogène ou dans l'appareil de FILDES et MCINTOSH, ne s'y clarifient ni plus vite, ni davantage que par simple transfert en tubes. Mais leur clarification reste insignifiante si on les laisse immobiles, dans un large flacon d'ERLENMEYER, condition qui assure de meilleurs échanges gazeux avec l'air ambiant.

WELSCH n'a pas observé d'effet favorable du chloroforme sur l'autolyse des cultures staphylococciques, et SALMON (1950b) a montré qu'elle n'était pas entravée par addition de 0,008 à 0,5% de silice colloïdale. La lyse est également observée avec des cultures faites en bouillon contenant 10% de sérum normal de bœuf, 2% de glucose ou 0,5% de silice, agents qui, par ailleurs, entravent l'autophagie (SALMON 1949a). Par contre, le formol et l'acide thioglycolique inhibent l'autolyse, cet effet étant, au moins partiellement, neutralisé par addition de sérum normal.

WELSCH, SALMON et HEUSGHEM (1949) ont obtenu l'autolyse de cultures staphylococciques faites en bouillon dont les phosphates minéraux avaient été préalablement précipités par un excès de $BaCl_2$. Il semble donc que la présence de ce sel ne soit pas indispensable à l'autolyse ou qu'il puisse être remplacé par des composés phosphorés organiques du bouillon.

L'autolyse ne se limite pas aux cultures de staphylocoques en bouillon, mais elle peut être obtenue également au moyen de germes lavés, mis en suspension dans des milieux définis divers. Les bactéries provenant de cultures jeunes, en phase de croissance exponentielle, sont plus favorables.

Prélevés d'une culture sur gélose nutritive et remis en suspension dans de l'eau distillée, les staphylocoques y subissent une lyse, généralement peu importante, et d'ailleurs irrégulière. Si l'on a soin de laver les germes avant de les abandonner à 37° dans l'eau, on constate qu'ils ne s'autolysent pas. Mais ces germes lavés se dissolvent si on les remet en suspension dans une solution de K_2HPO_4 à 0,1% ou dans du bouillon dilué au $^1/_{10}$. Dans ce dernier cas, on peut d'ailleurs observer que la suspension passe alternativement par des phases de clarification et d'opacification, indice d'une succession de lyses et de croissances post-lytiques (WELSCH 1949d).

Au même titre que l'orthophosphate, les métaphosphate, pyrophosphate et glycérophosphate sodiques favorisent, à concentration convenable, l'autolyse des suspensions staphylococciques lavées. Les phosphite et hypophosphite sodiques sont indifférents ou entravent l'autolyse en eau distillée, selon leur concentration. Par contre, ils inhibent toujours nettement l'autolyse en solution phosphatée. Contrairement aux observations faites pour l'autophagie, certains sels, non dérivés du phosphore, peuvent jouer le rôle d'effecteurs de l'autolyse (WELSCH 1950a, WELSCH et SALMON 1950a): NaCl, KCl, $NaNO_3$ et $MgSO_4$. Contrairement à celui de $MgSO_4$, l'effet favorable de KCl et $NaNO_3$ sur l'autolyse ne se manifeste pas en milieu phosphaté, où il fait parfois place à une action inhibitrice. Inversement, la présence de silice colloïdale n'empêche pas l'autolyse des staphylocoques en solution phosphatée, alors qu'elle entrave leur autophagie (SALMON 1950b).

Dans certaines conditions, on peut observer une clarification des suspensions de staphylocoques tués par chauffage (WELSCH, 1949e). Les suspensions épaisses de ce germe, soumises à une température de 56° C pendant 3 heures, ce qui ne provoque pas d'altérations morphologiquement reconnaissables, de même que les suspensions autoclavées à 120° C pendant 30 minutes, ce qui ne fait pas perdre aux microbes leur forme générale, mais diminue considérablement leur colorabilité et leur capacité de conserver le Gram, sont, les unes et les autres, très stables, aussi bien en eau distillée qu'en solution phosphatée ou en bouillon au $^1/_{10}$. Par contre, les staphylocoques ayant subi une «flash-sterilization» (DUBOS et MCLEOD 1938) à 80° C, réalisée en introduisant la suspension bactérienne épaisse dans 50 fois son volume d'eau portée à 80° C et maintenue ensuite à cette température pendant 30 minutes, subissent une dissolution lente en solution phosphatée, plus rapide en bouillon au 1/10, bien que leur morphologie et leur colorabilité n'aient pas été sensiblement modifiées par le traitement thermique. Les suspensions staphylococciques portées à 100° C pendant 3 à 5 minutes sont moins stables encore: leur dissolution lentement progressive est apparente non seulement en solution phosphatée et en bouillon au 1/10, mais aussi en eau distillée. Dans ce cas, les altérations morphologiques sont comparables à celles des bactéries autoclavées.

Utilisant les suspensions staphylococciques les plus stables, c'est-à-dire tuées par chauffage à 56° C prolongé pendant 3 heures, WELSCH (1950a) a constaté qu'elles ne subissaient pas de lyse appréciable en solution de métaphosphate, pyrophosphate, glycérophosphate, phosphite ou hypophosphite alcalins, aux concentrations comprises entre 0,1 et 1%.

Lorsque des cultures de staphylocoques, faites en bouillon, sur agitateur, sont transférées en tubes où elles se clarifient, l'autolysat, débarrassé des corps microbiens et des débris par une centrifugation énergique, puis stérilisé par addition de quelques gouttes de chloroforme et séjour à basse température pendant 2 ou 3 jours, se montre capable d'accélérer l'autolyse des staphylocoques vivants et de dissoudre les suspensions aqueuses de staphylocoques chauffés et de quelques autres germes: *Sarcina lutea* et *Micrococcus lysodeikticus*, mais non de certains autres: streptocoques, corynebactéries, bâtonnets gram-négatifs (WELSCH, 1949c). L'activité d'une telle préparation, dénommée autolysine staphylococcique, se manifeste encore, selon les cas, après dilution entre 1/40 et 1/320. L'allure cinétique de la lyse qu'elle provoque dépend de sa concentration et de la richesse en germes de la suspension sensible utilisée.

L'autolysine dissout mal les staphylocoques tués au formol et repris dans l'eau distillée. Elle lyse, comme les staphylocoques chauffés, ces germes traités à l'acide thioglycolique et devenus ainsi gram-négatifs (WELSCH 1949b).

L'autolysine peut être obtenue à partir de cultures faites en bouillon dont les phosphates minéraux ont été éliminés par précipitation au moyen d'un excès de $BaCl_2$. De même, une autolysine, obtenue à partir de cultures en bouillon normal, conserve son entière activité si l'on en précipite les phosphates (WELSCH, SALMON et HEUSGHEM 1949).

L'activité de l'autolysine se manifeste le mieux sur les suspensions bactériennes en eau distillée. L'addition d'orthophosphate ou d'autres sels, dérivés ou non du phosphore, inhibe légèrement son activité (WELSCH 1950a); l'addition de bouillon (WELSCH 1949a) ou de silice (SALMON 1949b), l'entrave très nettement ou, à concentration suffisante, l'abolit totalement. Les autolysats, obtenus à partir de cultures en bouillon contenant de la silice, ont une activité bactériolytique réduite ou nulle, selon la concentration (SALMON 1950b).

L'autolysine staphylococcique n'hydrolyse pas les préparations d'acide ribonucléique de levure. Elle ne permet pas davantage de mettre en évidence les éléments nucléaires d'*Escherichia coli* fixée au liquide de Chabaud ainsi que le fait la ribonucléase pancréatique.

Il est évident que les conditions expérimentales n'influencent pas de la même façon les diverses activités bactériolytiques du staphylocoque (WELSCH et SALMON 1950b, c).

L'action de l'autolysine est entravée par la silice, le bouillon, le sérum normal; elle n'exige point la présence d'orthophosphate ou de l'un de ses succédanés; elle se manifeste en aérobiose comme en anaérobiose.

Les activités autophagique, isophagique, hétérophagique, exercées sur des germes chauffés par les colonies staphylococciques développées en gélose, sont également inhibées par la silice et le sérum normal, mais non par le bouillon; elles exigent la présence d'orthophosphate ou de l'un de ses succédanés; elles sont favorisées en milieu modérément réducteur.

Or, ces conditions sont celles qui permettent la libération de l'autolysine. Aussi, ces activités peuvent-elles être attribuées à une production d'autolysine par les colonies croissant dans la gélose. L'absence d'inhibition par le bouillon pourrait s'expliquer par le fait que l'autolysine, en milieu gélosé, reste concentrée au voisinage des colonies productrices, alors qu'elle est fortement diluée lorsqu'elle est libérée en milieu liquide. Il importe, toutefois, de souligner que si ces activités s'exercent sur les germes que peut dissoudre l'autolysine, elles se manifestent aussi, dans une certaine mesure, à l'égard de quelques souches de bâtonnets gram-négatifs insensibles à l'autolysine.

L'autolyse des cultures et suspensions de staphylocoques, par contre, n'est entravée ni par le bouillon, ni par le sérum, ni par la silice; elle n'exige pas la présence d'orthophosphate ou de ses succédanés, mais peut être observée en solution de sels minéraux non dérivés du phosphore. Est-ce à dire que ce phénomène est indépendant de toute intervention de l'autolysine? Nous pensons plutôt qu'un même système autolytique intervient dans toutes ces activités bactériolytiques, mais n'agit pas, dans tous les cas, au même endroit. La dissolution des suspensions bactériennes chauffées par l'autolysine, de même que les phénomènes d'autophagie, isophagie et hétérophagie, représentent une destruction des cellules sensibles par un agent qui leur est extérieur. L'autolyse d'une suspension bactérienne, au contraire, est un processus d'origine mixte, à la fois exogène et endogène. Cette dernière, relativement indépendante des conditions qui prévalent dans le milieu, est favorisée par une anaérobiose partielle: les suspensions staphylococciques ne s'autolysent rapidement que si elles passent d'un état de bonne aération à des conditions de moins bonne oxygénation. La dissolution cellulaire d'origine endogène libère les lysines qui, de l'extérieur, agissent alors sur d'autres bactéries. Les inhibiteurs de l'autolysine, silice et sérum, empèchent la dissolution d'origine exogène, mais n'ont aucun effet sur la dissolution d'origine endogène.

La dualité du processus intime de l'autolyse est souligné encore par le fait que la clarification est plus poussée lorsque les conditions permettent une multiplication microbienne limitée. Elle est, par exemple, meilleure en bouillon au 1/10 qu'en eau distillée ou en solution phosphatée. Dans ces conditions, en effet, l'autophagie exogène vient s'ajouter à l'autolyse strictement endogène. A l'appui de cette interprétation, on peut rappeler que divers antibiotiques, pénicilline, actinomycine, streptomycine, n'ayant aucune action sur l'activité de l'autolysine vis-à-vis des germes chauffés, réduisent nettement, à dose bactériostatique, l'intensité de l'autolyse des suspensions vivantes, en bouillon au 1/10 ou en eau distillée contenant 10% d'autolysine. Dans le premier cas, l'inhibition peut même être complète (WELSCH 1949d).

III. Mecanisme de l'Autolyse des Bactéries Gram-positives

En passant en revue les études relatives à l'autolyse du pneumocoque, nous avons vu que la perte du caractère gram-positif de cet organisme avait été souvent observée, tant au cours de son autolyse spontanée que de son autolyse induite par divers agents. Ce phénomène a été généralement considéré comme le résultat d'une activité enzymatique représentant la première étape de l'autolyse. Dans le cas de l'isophagie des staphylocoques, GRATIA et RHODES (1924) ont également signalé que la clarification reste toujours incomplète. «Les staphylocoques tués et dissous», écrivent-ils, «sont réduits à l'état de petits granules qui, chose curieuse, ne prennent plus le Gram; il ne reste qu'un squelette Gram-négatif que les microbes vivants ne peuvent dissoudre.»

Les recherches effectuées pour élucider le mécanisme de la coloration de Gram (BARTHOLOMEW et MITTWER 1952, WELSCH 1948a) ont clairement mis en évidence la participation, dans cette réaction, d'un complexe périphérique dans lequel l'acide ribonucléique, sous forme de sel magnésien, joue un rôle important. On sait que les cellules de diverses bactéries gram-positives, traitées par la bile ou le cholate sodique à 60° C, sont transformées en cytosquelettes gram-négatifs, en même temps que des hydrates de carbone, une faible quantité de protéine, et du ribonucléate de magnésium sont mis en liberté. Ce dernier constituant peut être «replaqué» sur les cytosquelettes, homologues ou hétérologues, à condition qu'ils aient été conservés en présence de formol. Il leur rend alors leur caractère gram-positif (HENRY et STACEY 1946). Des observations identiques ont été faites avec des germes tués par chauffage puis traités par la ribonucléase cristallisée (BARTHOLOMEW et UMBREIT 1944).

La négativation du Gram a été observée aussi lors de la lyse bactérienne par le lysozyme et après action de cet enzyme sur des germes chauffés (MEYER 1946, CUTINELLI et LA MANNA 1947, TETI et FIORE 1949, CALIFANO 1949, PETRONELLI 1950). WEBB (1948), utilisant une préparation de lysozyme purifié, en tout cas dépourvue de toute activité nucléasique, a montré que tous les germes gram-positifs étudiés (les levures exceptées), tués par chauffage puis soumis à son action, deviennent gram-négatifs en libérant des substances réductrices en faible quantité et de l'acide ribonucléique. Celui-ci peut être «replaqué» sur des cytosquelettes obtenus par action de la bile, mais non sur ceux que fournit le lysozyme. Il est donc certain que la libération d'acide nucléique, avec perte du caractère gram-positif, peut être réalisée par des enzymes très différents (TETI 1952, TETI et ALBANO 1950).

D'autre part, si ces observations montrent, sans discussion possible, que la région superficielle des germes gram-positifs est un substrat de la réaction de Gram, elles n'excluent nullement que d'autres portions de la bactérie, l'ensemble de son cytoplasme par exemple, ne puissent l'être aussi (LAMANNA et MALETTE 1953).

Lors de la conversion des pneumocoques en cytosquelettes gram-négatifs, THOMPSON et DUBOS (1938) ont mis en évidence la libération de ribonucléoprotéine et d'acide ribonucléique. HENRY, STACEY et TEECE (1945) ont de même constaté que *Clostridium welchii*, soumis à une autolyse de durée relativement courte, perd son caractère gram-positif avec solubilisation de ribonucléoprotéine (JONES, MUGGLETON et STACEY 1950).

Sur la base de ces diverses observations, STACEY et WEBB (1948), JONES, STACEY et WEBB (1949), ont développé une théorie générale, destinée à rendre compte du mécanisme de l'autolyse des germes gram-positifs.

Selon ces auteurs, l'autolyse comporterait deux stades successifs. Le premier consiste en une perte du caractère gram-positif. Il est attribuable à des enzymes ayant en commun la capacité d'hydrolyser l'acide ribonucléique, mais qui possèdent, d'autre part, une étroite spécificité, en vertu de laquelle ils ne peuvent convertir en squelettes gram-négatifs que les représentants de l'espèce bactérienne dont ils proviennent («spécificité d'espèce»). Ces enzymes agissent au p_H optimal de 8,0, sont irréversiblement détruits à p_H 4,0 ou à la température de 80° C, et réversiblement inactivés par oxydation. Le second stade de l'autolyse consiste dans la désintégration des cytosquelettes gram-négatifs. Elle est attribuée à un ou plusieurs enzymes protéolytiques, ne manifestant pas de spécificité étroite, agissant au p_H optimal de 8,0 à 8,5, mais inactifs en milieu réducteur.

La présence d'enzymes possédant les propriétés précitées a été démontrée chez diverses bactéries gram-positives. On sait, par ailleurs, que si les germes gram-positifs intacts résistent à l'action des protéases, telle la trypsine, leurs cytosquelettes gram-négatifs, au contraire, sont rapidement dissous par de tels ferments (WEBB 1948).

Dans un cas particulier, la succession des étapes de l'autolyse spontanée a été suivie par emploi de plusieurs méthodes de coloration. Il s'agit de l'autolyse de *Bacillus subtilis* en eau distillée, étudiée par BARTHOLOMEW et MITTWER (1951). Ces auteurs ont constaté que les bacilles gram-positifs abandonnent d'abord du ribonucléate de magnésium et se transforment en germes gram-négatifs tout en conservant leur taille normale. A ce stade, la paroi cellulaire et l'ectoplasme restent présents et le traitement des cytosquelettes par le ribonucléate magnésien peut leur rendre le caractère gram-positif. Ensuite, la cellule présente une diminution de taille, associée à la disparition de la membrane cellulaire et de l'ectoplasme, ainsi qu'à l'impossibilité de lui conférer à nouveau son caractère gram-positif par « replaquage » de ribonucléate magnésien. Enfin, la cellule, complètement désintégrée, n'est plus reconnaissable.

On peut toutefois se demander s'il est justifié de considérer la lésion primitive, caractéristique de l'autolyse débutante, comme une perte *spécifique* de la capacité de conserver le Gram (Welsch et Salmon 1950c).

En effet, si l'on examine des staphylocoques en voie d'autolyse, non seulement après les avoir colorés au Gram, mais aussi avec des colorants basiques usuels, le bleu de toluidine par exemple, on constate que la perte du caractère gram-positif est toujours associée à une diminution considérable de la basophilie en général (Welsch 1949b, e). Or, si la propriété d'être gram-positif exige, sans aucun doute, l'existence d'une affinité pour les colorants basiques, elle est, d'autre part, sous la dépendance d'autres facteurs plus étroitement spécifiques. Par conséquent, seule l'incapacité de prendre le Gram, tout en conservant la basophilie, serait l'indice certain d'une lésion portant spécifiquement sur le substrat de la réaction de Gram.

Il convient de remarquer, à ce propos, que divers agents dont la capacité de négativer le Gram a été démontrée, sont, par ailleurs, utilisés aussi pour mettre en évidence les noyaux bactériens, et ce, aussi bien chez les germes gram-négatifs que gram-positifs. Or, il est généralement admis à présent que la mise en évidence de ces noyaux est le résultat d'un démasquage, réalisé par l'élimination de l'acide ribonucléique abondant du cytoplasme. Vendrely et Lipardy (1947) ont, en effet, démontré, par des analyses chimiques directes, que la méthode d'hydrolyse chlorhydrique, originellement utilisée par Piekarsky (1937), puis développée par Robinow (1942, 1944), consiste effectivement en une extraction sélective des acides ribonucléiques. Il y a longtemps déjà, Deussen (1920) avait observé que la capacité de retenir le Gram peut être perdue après traitement des germes par des acides ou des alcalis. Ces derniers d'ailleurs, à concentration convenable, peuvent également réaliser l'extraction sélective des acides ribonucléiques (Sulkin 1951) et démasquer les noyaux bactériens (Welsch et Nihoul 1948).

La ribonucléase cristallisée transforme les germes gram-positifs en cytosquelettes gram-négatifs, mais elle enlève aussi l'acide ribonucléique cytoplasmique des bactéries, gram-positives ou gram-négatives, laissant des noyaux, basophiles grâce à leur acide désoxyribonucléique, dans un cytoplasme qui ne retient plus guère les colorants basiques (Boivin, Vendrely et Tulasne 1947, Tulasne et Vendrely 1947a, b).

Le traitement au lysozyme qui peut négativer le Gram, est également susceptible, seul ou associé à une nucléase (Petronelli, 1951), de mettre en évidence les noyaux de *Micrococcus lysodeikticus* (Caselli 1947, Caselli et Carta 1951) ou du pneumocoque (Cutinelli et la Manna 1950). En associant au lysozyme la streptomycine qui se fixe aux acides nucléiques et empêche ainsi leur désintégration, on peut conserver intacts, dans les lysats de *Micrococcus lysodeikticus*,

Sarcina lutea et *Bacillus megaterium*, des formations morphologiquement identiques aux noyaux, donnant une réaction de Feulgen positive (WELSCH 1952a).

Le sulfate de lauryl, inducteur de la lyse pneumococcique (BAYLISS 1937), est, d'autre part, utilisé aussi pour démasquer les noyaux bactériens (DELMOTTE 1953a, b), ainsi d'ailleurs que la bile (CUTINELLI 1949).

On sait encore que les organismes normalement gram-positifs peuvent ne pas présenter ce caractère lorsqu'ils ont été cultivés en milieu déficient. Ici encore, l'absence du caractère gram-positif est liée à une diminution générale d'affinité pour les colorants basiques, grâce à laquelle les noyaux peuvent souvent être mis directement en évidence, et qui est associée à une réduction de la teneur en acide ribonucléique (DE LEY 1951).

Au cours de l'autolyse des staphylocoques, l'inaptitude à conserver le Gram se manifeste à la périphérie de la cellule, mais il persiste une région centrale gram-positive plus ou moins étendue (WELSCH 1949b, e). C'est uniquement avec les staphylocoques d'abord maintenus en milieu acide, puis abandonnés à l'autolyse en solution neutre ou légèrement alcaline, que l'on observe, après ce stade, une seconde étape où la cellule apparaît toute entière gram-négative. Dans tous les autres cas, la désintégration bactérienne complète intervient avant que ce stade de cytosquelette complètement gram-négatif puisse être observé (WELSCH et DELCAMBE, 1951). En conformité avec les observations de QUERSIN (1950), nous interprétons les granules gram-positifs, démasqués par l'autolyse, comme des noyaux bactériens. Cet auteur a montré, en effet, qu'après hydrolyse chlorhydrique ou traitement à la ribonucléase, les noyaux, colorables au Giemsa, ont le caractère gram-positif.

Dans le cas de l'autolyse des staphylocoques en tout cas, probablement même des bactéries gram-positives en général, il nous paraît donc plus exact de dire que la première étape de l'autolyse consiste en une perte d'acide ribonucléique et de ribonucléoprotéine, se traduisant par une diminution générale de l'affinité pour les colorants basiques et, par conséquent, de la capacité de prendre le Gram.

Cette perte d'acides nucléiques se traduit d'ailleurs aussi par le comportement des germes traités par l'auramine et observés en lumière ultraviolette. Les staphylocoques en voie d'autolyse, examinés de la sorte, ne montrent qu'une faible fluorescence verdâtre, au lieu de la fluorescence jaune vif caractérisant les cellules intactes (WELSCH 1950b). Or, selon OSTER (1951), la fluorescence normale de l'auramine est verdâtre, mais devient jaune vif en présence d'acide nucléique.

La libération d'acide nucléique et de ribonucléoprotéine au début de l'autolyse staphylococcique est-elle exclusivement le résultat d'un phénomène enzymatique ? Lorsque des suspensions staphylococciques denses sont chauffées, soit à 60° C ou 80° C pendant 60 minutes, soit à 100° C pendant 10 minutes, diluées ensuite en eau distillée ou en solution phosphatée, abandonnées enfin, pendant 2 ou 3 heures, à 20° C, on constate qu'elles perdent ainsi 5 à 10% de leur azote et 10 à 15% de leur phosphore (WELSCH et DELCAMBE 1951). Bien que cette perte soit pratiquement la même avec les trois modes de chauffage utilisés, dans les deux premiers cas, il n'y a pas de diminution appréciable de la basophilie et du caractère gram-positif, alors que ces propriétés sont nettement modifiées dans le troisième cas. On sait, cependant, que le chauffage des bactéries à des températures même modérées, en extrait des acides nucléiques (CALIFANO 1950, CANTELMO 1950).

Il est donc probable que les liaisons rompues à 100° C d'une part, à 60° C ou 80° C d'autre part, ne sont pas identiques, puisque, pour une libération quantitativement égale de P et N, on observe néanmoins des degrés de basophilie très différents.

D'autre part, les suspensions de bactéries, chauffées à 80° C pendant une heure ou à 100° C pendant 10 minutes, puis diluées en eau distillée ou en solution de phosphate à 37° C, subissent une clarification progressive modérée, de l'ordre de 30 à 40%. Au cours de celle-ci, la solubilisation du P et de l'N se déroule comme dans le cas de l'autolyse des germes non chauffés, où la clarification peut atteindre jusqu'à 90% et plus. Mais elle reste, bien entendu, quantitativement beaucoup plus faible. Dans ces diverses conditions, le phosphore est libéré surtout sous forme organique, mais non précipitable par l'acide trichloracétique. L'azote, au contraire, est d'abord libéré sous forme de substances protéiques dont l'hydrolyse a lieu ensuite. La libération du P est relativement plus rapide que celle de l'N ce qui se traduit par une élévation progressive de la valeur du rapport N_b/P_b, exprimant le quotient du taux d'azote lié au résidu bactérien centrifugeable à celui du P correspondant.

Si la solubilisation des constituants staphylococciques, et en particulier ceux renfermant du phosphore, est un phénomène exclusivement enzymatique, les ferments en cause ne possèdent certes pas la thermolabilité que STACEY et WEBB (1948) attribuent à la nucléinase responsable de la négativation du Gram. Toutefois, la libération de P et d'N ayant encore lieu dans les suspensions maintenues à 0°, 60°, 80° et 100° C, nous croyons plutôt que ces phénomènes ne sont que partiellement d'origine enzymatique.

Des staphylocoques non chauffés, ayant préalablement séjourné à 0° C pendant 40 heures, soit en eau distillée, soit dans des solutions tampons de p_H 4,0, 5,0 ou 7,5, sont repris, d'une part, en eau distillée, d'autre part, en solution de K_2HPO_4 à 0,25%, et mis en incubation à 37° C. Les suspensions en eau distillée des germes provenant de l'eau ou du tampon à p_H 7,5 se clarifient normalement, en libérant progressivement P et N selon la cinétique habituelle. Les suspensions en eau distillée provenant des tampons de p_H 4,0 et 5,0, loin de se clarifier, s'opacifient. Mais, tandis que dans la première, dont le p_H est de 5,0 environ, il n'y a aucune solubilisation de P et d'N, celle-ci est, par contre, nette dans la seconde dont le p_H est voisin de 6,0. Les suspensions correspondantes en solution phosphatée se comportent de façon différente. Celles qui proviennent de l'eau ou du tampon à p_H 7,5 s'autolysent dans les délais habituels, mais la libération du P est cette fois relativement plus lente que celle de l'N. Celles qui proviennent des tampons acides subissent une autolyse à tous points de vue normale. Le séjour prolongé à p_H 4,0 ou 5,0 n'a donc nullement détruit les enzymes responsables de l'autolyse et, en particulier, n'a pas irréversiblement inactivé le processus responsable de la libération des acides nucléiques. Celui-ci ne peut donc pas être attribué à un enzyme identique à la nucléase de STACEY et WEBB qui est définitivement détruite à p_H 4,0. Le processus responsable de la libération des composés phosphorés paraît avoir été altéré, au contraire, par séjour prolongé en milieu neutre ou légèrement alcalin, et, d'ailleurs, sans que cela n'entrave l'autolyse.

L'intervention d'un enzyme agissant au niveau des ribonucléoprotéines superficielles est-elle donc bien une étape préalable *nécessaire* pour que la désintégration cellulaire puisse avoir lieu ensuite?

Remarquons tout d'abord que si cette suite d'évènements peut être valablement postulée pour rendre compte du processus de dissolution exogène, elle n'est certainement pas nécessaire, *a priori*, lorsqu'on envisage l'autolyse stricte, d'origine endogène.

Or, la dissolution par l'autolysine des staphylocoques rendus gram-négatifs après traitement à l'acide thioglycolique, n'est nullement plus rapide que celle des germes chauffés restés gram-positifs (WELSCH 1949b). Bien plus, compte tenu de leur clarification spontanée, les suspensions de staphylocoques chauffés à 56° C ou soumis à une «flash-sterilization» à 80° C, restés gram-positifs, sont plus sensibles à l'autolysine que les suspensions de germes chauffés à 100° ou à 120° C, rendus ainsi largement gram-négatifs (WELSCH 1949e). L'autolysine staphylococcique n'agit d'ailleurs pas sur l'acide ribonucléique extrait de la levure, ni sur l'acide ribonucléique d'*Escherichia coli* fixée (WELSCH 1949c). Ajoutée à une suspension de germes chauffés à 100° C, dont elle favorise la dissolution, elle augmente nettement la libération de l'N, mais non celle du P (WELSCH et DELCAMBE 1951).

A la suite de ces diverses observations, nous pensons que la nucléinase étroitement spécifique, thermolabile à 80° C et irréversiblement détruite à p_H 4,0 ne peut être considérée comme un élément constitutif indispensable du système autolytique des bactéries gram-positives.

Les observations de nombreux auteurs s'accordent, d'autre part, pour reconnaître que l'autolyse du staphylocoque est favorisée par des conditions réductrices appropriées. WELSCH et DELCAMBE (1951) ont constaté que l'autolyse d'un staphylocoque s'effectuait de façon tout à fait identique, tant au point de vue de la clarification que de la solubilisation du P et de l'N, d'une part, en solution tamponnée de p_H 7,5, d'autre part, dans ce même milieu, additionné d'H_2S, ou placé dans un récipient où une pression positive de ce gaz était maintenue de façon constante. La lyse observée dans ces conditions aboutit à une désintégration complète des cellules, et non à une simple transformation des germes en cytosquelettes gram-négatifs. On ne peut donc admettre comme un fait de portée générale que la désintégration des cellules en cours d'autolyse soit attribuable à l'action de protéinases inactives en milieu réducteur.

IV. Activités Autophagique, Isophagique et Autolytique des Germes Gram-négatifs

L'importance de l'autolyse dans la pathogénie des infections par germes gram-négatifs est implicitement admise de façon générale. Pourtant, ce phénomène n'a guère été étudié de façon aussi approfondie que pour certain germes gram-positifs.

On sait que les coques gram-négatifs, gonocoques et méningocoques, s'autolysent aisément. On sait également que des bâtonnets gram-négatifs libèrent certains de leurs constituants lorsqu'on les abandonne à la macération en présence de chloroforme (LINZ 1948). La production, par certaines souches d'*Escherichia coli*, de substances spécifiquement lytiques pour des bactéries voisines a été signalée, par exemple, par GUÉLIN (1943) et par ZAMENHOF (1944). BERNARD, GUILLERM et GALLUT (1937) ont étudié la digestion des vibrions cholériques chauffés à 56° ou 100° C par les filtrats de culture homologues. PETROVANU

(1925) a montré que l'eau oxygénée, à concentration convenable, induit l'autolyse du vibrion cholérique et de divers autres germes, surtout gram-négatifs. Les autolysats ainsi obtenus dissolvent les suspensions homologues correspondantes, à la condition qu'elles soient préparées au moyen de jeunes cultures.

Plusieurs chercheurs italiens ont signalé que certains acides aminés, particulièrement la nor-valine, la nor-leucine et la glycine, induisent des anomalies morphologiques chez des bactéries gram-négatives. Les colonies microbiennes, sous l'influence de ces acides aminés, et tout particulièrement lorsqu'ils sont associés à des extraits bactériens, notamment de *Bacillus subtilis*, subissent une lyse plus ou moins prononcée (AMATI 1947, COLOMBO 1947, LORIZIO et RISMONDO 1947, MASERA 1947, RISMONDO 1946, 1947, STRANO 1950, VACIRCA 1946, 1947a, b, c, 1948).

La lyse des suspensions bactériennes sous l'influence de la glycine a été étudiée plus particulièrement, d'abord par MACULLA et COWLES (1948), puis par GORDON et ses collaborateurs (1949—1954).

Dans notre laboratoire, SALMON (1951a, b) a recherché les activités bactériolytiques éventuelles des Entérobactéries en utilisant les techniques antérieurement choisies pour l'étude correspondante faite chez le staphylocoque.

En gélose contenant des corps microbiens homologues chauffés, les colonies d'*Escherichia coli* φ s'entourent d'une zone de clarification qui reste partielle, mais très étendue (autophagie). Ces colonies dissolvent aussi les corps tués d'autres souches d'*E. coli* (isophagie), mais non ceux des autres espèces microbiennes étudiées (hétérophagie): *Micrococcus pyogenes*, *Sarcina lutea*. Ces phénomènes d'autophagie et d'isophagie n'ont pas été observés avec les autres souches d'*E. coli* examinées. Ils semblent donc être particuliers à ce microorganisme qui servit de souche sensible à GRATIA (1932), lors de sa découverte des principes responsables de l'iso-antagonisme entre bacilles coli, les colicines (GRATIA et FREDERICQ 1946).

L'autophagie d'*E. coli* φ est partiellement inhibée par un excès de bouillon. Mais, à condition que la concentration en orthophosphate soit suffisante (0,025%), elle peut être obtenue dans tous les milieux étudiés qui permettent la formation de colonies. Toutefois, contrairement à ce qui se passe pour le staphylocoque, l'orthophosphate ne peut être remplacé ici par d'autres sels. Le pyrophosphate sodique permet une certaine autophagie, mais elle reste peu étendue et n'apparaît que tardivement.

Les procédés qui sont efficaces pour obtenir l'autolyse des staphylocoques, mise des cultures, après développement en aérobiose, soit en tubes scellés, soit en atmosphère d'hydrogène, soit dans l'appareil de FILDES et MCINTOSH, ne provoquent pas d'autolyse appréciable d'*Escherichia coli* ou d'autres bactéries voisines.

Les diverses Entérobacteriacées examinées, récoltées à partir de gélose nutritive et lavées, subissent toutefois une lyse modérée en solution d'orthophosphate potassique à 0,2%, et une dissolution beaucoup plus prononcée en pyrophosphate sodique de même concentration ou en solution M de glycine. L'action du pyrophosphate ne peut être essentiellement attribuée à son alcalinité. En effet, certaines souches seulement se dissolvent dans des solutions de bicarbonate de p_H comparable, et, même dans ces cas, la clarification ne débute alors que très tardivement. Les Entérobactériacées préalablement tuées par chauffage ne subissent pas de lyse, ni en orthophosphate, ni en pyrophosphate, ni en glycine. Il paraît donc logique de considérer ces agents comme des inducteurs d'autolyse.

Le système lytique d'*E. coli* est plus sensible à la chaleur que celui du staphylocoque. Il est partiellement inactivé par le chloroforme, mais se retrouve intact dans les bactéries tuées par le formol.

Les lysats d'Entérobactériacées, débarrassés des germes résiduels par centrifugation, ne possèdent pas d'activité lytique pour les suspensions homologues vivantes ou chauffées. On ne peut donc obtenir, à partir de ces germes, l'équivalent de l'autolysine staphylococcique.

La comparaison des activités lytiques du pyrophosphate et de la glycine (Salmon 1952) montre l'existence de différences de cinétique importantes. De plus, tandis que la glycine assure la dissolution des bactéries provenant de cultures d'âges très différents, le pyrophosphate, au contraire, agit nettement mieux lorsqu'on utilise des germes prélevés dans de jeunes cultures. La solubilisation des protéines microbiennes par la glycine est plus élevée, pour un même degré de clarification, que par le pyrophosphate. L'activité du pyrophosphate, mais non celle de la glycine, diminue lorsqu'on utilise des suspensions bactériennes plus denses. Le résidu de la lyse en glycine se dissout en pyrophosphate, mais l'inverse n'a pas lieu. Enfin, l'association de la glycine et du pyrophosphate exerce une synergie lytique évidente.

Glycine et pyrophosphate, à concentration convenable, inhibent le développement d'*E. coli*. Dans l'un comme dans l'autre cas, par cultures répétées en présence de concentrations croissantes de ces réactifs, on peut obtenir des variants résistants. Ceux-ci se montrent alors spécifiquement moins sensibles à l'action lytique de l'inducteur correspondant, mais ils ne présentent aucune résistance croisée.

Salmon (1957) a également procédé à une analyse des constituants antigéniques présents dans les lysats de *Salmonella schottmuelleri* respectivement obtenus en glycine, en pyrophosphate, en solution mixte de glycine et pyrophosphate, en orthophosphate.

Les trois premiers lysats sont nettement plus riches en protéines que le quatrième. Par électrophorèse en gélose, suivie de coloration à l'amidoschwarz, une seule fraction protéique a été décelée dans le lysat en pyrophosphate. Cette même fraction peut être reconnue aussi dans les lysats en glycine et en glycine-pyrophosphate. Elle y fournit également des réactions positives avec le noir Soudan et avec le réactif de Schiff. Le lysat en glycine montre en outre trois autres fractions protéiques et le lysat en glycine-pyrophosphate, deux.

Tous ces lysats, injectés au lapin, induisent la formation d'anticorps qui agglutinent *S. schottmuelleri* et précipitent certains de ses lysats. Par la méthode de précipitation sérologique en gélose, Salmon a décelé quatre fractions antigéniques distinctes pour l'ensemble des quatre lysats étudiés. L'une, b, est commune aux quatre préparations; une autre, a, se retrouve dans les lysats en glycine et en glycine-pyrophosphate; les deux dernières, c et d, dans les lysats en pyrophosphate et en glycine-pyrophosphate. La technique de l'immunoélectrophorèse confirme l'essentiel de ces observations.

V. Activités Bactériolytiques des Bacillacées

Les phénomènes d'autolyse accompagnant la sporulation sont fréquents chez les germes des genres *Bacillus* et *Clostridium.* Le système autolytique de *Clostridium welchii*, particulièrement étudié, a fourni des arguments à la théorie générale de l'autolyse des germes gram-positifs que nous avons déjà discutée.

MONOD (1942) a signalé la brusque autolyse de *Bacillus subtilis* consécutive à l'inanition carbonée en milieu de culture contenant, au départ, certains sucres bien déterminés.

Tout récemment, GREENBERG et HALVORSON (1955) ont étudié la production, par *Bacillus terminalis*, d'une autolysine thermostable, non dialysable, active entre les valeurs de p_H 5,0 et 5,5. Cette substance lyse également les cellules de *Bacillus cereus* qui, d'ailleurs, ne représente probablement qu'une variété de *B. terminalis.* Elle est inactive sur les autres espèces du genre *Bacillus.* Des autolysines différentes, possédant notamment une autre spécificité d'action, seraient produites par *Bacillus polymyxa.*

LWOFF, SIMINOVITCH et KJELGAARD (1950) ont signalé que la souche lysogène, *Bacillus megaterium* 899, subit une autolyse rapide et intense lorsqu'elle est placée en anaérobiose au cours de sa phase de croissance exponentielle. Ce phénomène a été confirmé par WELSHIMER (1951) qui a souligné l'existence d'une corrélation entre le pouvoir lysogène de *B. megaterium* et son aptitude à s'autolyser en anaérobiose. Dans notre Laboratoire, SMOLIAR (1951) a étudié les activités bactériolytiques de trois souches de *B. megaterium*: la souche lysogène 899 et deux souches non lysogènes, 377 et 36, en utilisant les méthodes antérieurement mises au point pour une étude similaire chez le staphylocoque et les Entérobactériacées (SALMON, 1949a, 1951a). Il a constaté que l'aptitude à l'autolyse en eau distillée, en solution de phosphate ou en bouillon nutritif, est plus marquée pour la souche lysogène 899 que pour la non lysogène 377. Elle est pratiquement nulle pour la souche 36. L'autolyse n'est favorisée ni par l'orthophosphate, ni par le pyrophosphate, mais bien par $CaCl_2$. Ce sel permet même une lyse modérée de la souche 36. L'autolysat de *B. megaterium* 899 accélère la lyse des suspensions homologues, mais n'a que peu d'effet sur les germes chauffés. Il est pratiquement inactif sur les suspensions, vivantes ou chauffées, des souches 377 et 36. Il est à remarquer que des mutants streptomycino-résistants de *B. megaterium* 36 se sont montrés extrêmement sensibles à l'autolyse, contrairement à la souche mère. Les souches 899 et 377, mais non 36, donnent lieu au phénomène d'autophagie en gélose. Chacune des trois souches manifeste une activité isophagique à l'égard des deux autres. Toutes trois, enfin, exercent une action hétérophagique sur les corps chauffés de *Micrococcus lysodeikticus*, de *Sarcina lutea* et, dans une certaine mesure, d'*Escherichia coli.*

L'activité bactériolytique de plusieurs espèces du genre *Bacillus* est connue depuis longtemps déjà. Dès 1907, en effet, NCOLLE signalait l'action dissolvante qu'exercent les filtrats de culture de *B. subtilis*, de *B. mesentericus* et de divers *Tyrothrix* isolés par DUCLAUX, sur des bactéries gram-positives (pneumocoques) ou gram-négatives *(Escherichia coli, Salmonella typhosa, Vibrio comma, Pfeifferella mallei).* Il attribuait la bactériolyse à une action enzymatique et constatait que les lysats de pneumocoques, mais non ceux du bacille de la morve, vaccinaient le lapin.

Des observations analogues ont été faites à de multiples reprises par la suite. PRINGSHEIM (1920) a rapporté la dissolution de *Corynebacterium diphtheriae* par un *Bacillus mesentericus vulgatus* dont l'activité bactériolytique, non attribuable à des protéases, se manifeste aussi vis-à-vis de quelques autres germes. Ces constatations ont été confirmées plus tard par WEILAND (1936). D'après cet auteur, le principe lytique en cause, étroitement spécifique pour le bacille de la diphtérie, ne serait que partiellement détruit après séjour pendant une heure à 100° C.

KIMMELSTIEL (1923) a observé la dissolution des cultures en bouillon de divers germes: staphylocoque, bacille du côlon, bacille de Flexner, etc...., par un *Bacillus mycoïdes* dont les filtrats ne sont toutefois pas bactériolytiques. MÜCH et SARTORIUS (1924), puis MÜCH (1925), ont souligné que la propriété bactériolytique n'appartenait qu'à un petit nombre de souches de *B. mycoïdes* dont ils font une variété particulière: *B. cytolyticus.* Ils constatent que des souches distinctes de *B. cytolyticus* manifestent des spectres d'activité lytique différents, observation qui fut confirmée ultérieurement par FRANKE et ISMET (1926). A la suite des constatations de KIMMELSTIEL (1924) et de SARTORIUS (1924), selon lesquelles les lysats obtenus à l'intervention de *B. cytolyticus* sont de bons antigènes, des préparations de ce genre ont été utilisées pour l'usage thérapeutique, notamment sous le nom de «Sentocym».

D'après KIMMELSTIEL (1924), VAN CANNEYT (1926) et MAGARAO *et alii* (1944), *B. mycoïdes* et *B. subtilis* seraient capables de lyser le bacille de Koch, et les filtrats de culture de ces germes auraient une action favorable sur l'évolution de la tuberculose expérimentale.

ROSENTHAL (1925a, b, d), puis ROSENTHAL et ILITCH (1926), ont étudié l'action bactériolytique de *B. mesentericus*, et surtout d'une série de *Tyrothrix* thermophiles, sur une série d'espèces microbiennes. Ayant observé que ces organismes sont capables de détruire de nombreux germes, tant chauffés que vivants, ROSENTHAL et DURAN-REYNALS (1926a, b) ont tenté de les acclimater au milieu intestinal, avec l'espoir de les y voir exercer une action antagoniste bienfaisante sur la flore digestive normale.

L'activité lytique de *B. megaterium*, *B. subtilis* et *B. mycoïdes* a été plus récemment signalée à nouveau par BITTER (1941), ARK et HUNT (1941), ANZAI *et alii* (1948).

RAMON, RICHOU et RAMON (1945a, b), DELAUNAY 1948, ont étudié l'activité lytique des filtrats de culture de *B. subtilis* comparativement avec ses propriétés gélatinolytiques, antibiotiques et antidotiques.

OLIVIER et ses collaborateurs (1945, 1946) ont décrit, sous le nom d'endosubtilysine, une préparation extraite de *B. subtilis* (OLIVIER 1946) qui possède des propriétés lytique, bactériostatique et bactéricide pour divers germes, parmi lesquels *Mycobacterium tuberculosis* (OLIVIER et DRUTEL, 1947).

VALLÉE (1945a), enfin, a étudié une préparation, la subtilysine, provenant également de cultures de *B. subtilis*, qui lyse lentement les cultures vivantes de divers germes et résiste au formol (1945b). Avec BOREL et SAUDINOS (1946, 1947), il a comparativement étudié les activités bactériolytique, gélatinolytique et amylolytique de cette préparation.

Ces observations établissent que, sans aucun doute, diverses souches appartenant au genre *Bacillus* possèdent la capacité de provoquer la lyse de bactéries très diverses, tantôt vivantes, tantôt tuées par chauffage. Elles ne nous apprennent toutefois que peu de chose au sujet de la nature des agents lytiques et de leur mode d'action.

Schiller (1914) avait observé qu'un *Lactobacillus acidophilus* était capable de rendre gram-négatifs, puis de dissoudre, en 18 heures à 37° C, les streptocoques ajoutés à ses cultures. Les filtrats de culture de ce lactobacille ne sont nullement bactériolytiques, mais les filtrats des cultures mixtes, où le streptocoque lui est associé, sont spécifiquement streptolytiques. Généralisant ces observations, Schiller a introduit le concept de l'antagonisme induit ou provoqué. Celui-ci consiste en une sorte d'adaptation enzymatique, en vertu de laquelle certains germes, en culture mixte, élaboreraient des bactériolysines spécifiquement actives contre le microbe associé (Schiller et Adamovskaja 1931). Ainsi, en l'absence d'aliment, mais en présence de bactéries diverses, y compris le bacille de la tuberculose (1925b, 1930, 1933), les levures (Schiller 1924a) et les bactéries protéolytiques élaboreraient des bactériolysines spécifiques (Schiller, 1924b, c 1925a).

Les observations de Schiller n'ont jamais été confirmées. Il est évidemment regrettable que cet auteur ait choisi, pour étudier l'antagonisme soi-disant induit, des espèces microbiennes capables de manifester des activités bactériolytiques même après culture pure. Welsch (1941) n'a pu constater que les activités bactériolytiques de plusieurs *Bacillus* spp. fussent exaltées ou spécifiquement orientées après leur culture en présence de bactéries diverses.

Après que la production d'antibiotiques par les cultures de diverses Bacillacées eût été démontrée, Rosenthal (1941) a réexaminé ses souches de *Tyrothrix* dont plusieurs sont, en fait, identifiables à l'espèce *B. subtilis*. Il a constaté que, sur 31 de ces microorganismes, 6 possédaient des propriétés antagonistes pour les germes gram-positifs seulement, et 18 pour les germes gram-positifs et gram-négatifs. En ce qui concerne l'activité bactériolytique, toutes les souches examinées étaient capables de lyser les bactéries gram-négatives chauffées, mais aucune n'avait une action appréciable sur les germes gram-positifs chauffés.

Plusieurs observations anciennes, notamment celles de Kimmelstiel (1924) et de Callerio et Dolcini (1932), montrent que les activités lytiques des Bacillales sont parfois attribuables à un agent filtrable et thermostable à 100° C. Ceci permet de se demander, par conséquent, s'il s'agit bien là d'une activité de nature enzymatique ou si, au contraire, les phénomènes de lyse ne seraient pas, directement ou indirectement, attribuables à des substances bactéricides qui induiraient une autolyse ou déclencheraient une désintégration d'origine physico-chimique de la cellule sensible. Cette dissolution pourrait être éventuellement complétée, d'ailleurs, soit par les enzymes des filtrats bruts, soit par le système autolytique bactérien lui-même.

Une intervention de substances lipoïdiques a été souvent soupçonnée (Zukerman et Minkewitsch 1925, Dresel 1926, Gundel 1930, Gundel et Wagner 1930, Besta et Kuhn 1934, Auerswald 1938), puis, Hettche et Weber (1939) ont montré que les activités antibiotiques de certains *Bacillus* spp. sont probablement attribuables à la production d'acides gras, tels les acides iso-valérique et oléique. Des substances de ce genre pourraient sans doute se fixer à la paroi cellulaire et y provoquer une désorganisation initiatrice de lyse. Tel est, en effet, le mode d'action que l'on attribue généralement aujourd'hui aux agents tensioactifs ou détergents (HOTCHKISS 1946), conformément aux suggestions originales de BAKER, HARRISON et MILLER (1941).

Les observations au microscope électronique de bactéries traitées par le bromure de cétyltriméthylammonium (SALTON, HORNE et COSSLETT 1951, DAWSON, LOMINSKI et STERN 1953) sont en bon accord avec cette interprétation. On sait aussi que cette substance tensio-active provoque une solubilisation rapide de divers constituants bactériens (SALTON et ALEXANDER 1950, SALTON 1951).

Certains des antibiotiques élaborés par des *Bacillus* manifestent une certaine activité bactériolytique et se comportent comme des détergents. C'est le cas, par exemple, pour la tyrothricine de *B. brevis* (DUBOS 1939a) dont l'activité à cet égard est attribuable à l'un de ses constituants, la tyrocidine, l'autre composant, la gramicidine, n'ayant aucune action lytique (DUBOS et HOTCHKISS 1941).

Les bactéries, mises en présence de tyrocidine, libèrent certains de leurs constituants intra-cellulaires (HOTCHKISS 1944, FONG et KRUEGER 1950, GALE et TAYLOR 1951, GROS et MACHEBOEUF 1948) même en l'absence de bactériolyse au sens habituel du mot. Leur examen au microscope électronique révèle des lésions de la paroi cellulaire (JOHNSON 1944, MITCHELL et CROWE 1948).

La polymyxine de *B. polymyxa* possède également une activité bactériolytique (WHITE, ALVERSON, BAKER et JACKSON 1949). Cet antibiotique est un agent tensio-actif puissant et des constituants cellulaires bactériens sont rapidement solubilisés sous son action (NEWTON 1953). La localisation de cette substance au niveau d'une membrane cytoplasmique, sous jacente à la paroi cellulaire, a été élégamment démontrée par NEWTON (1955, 1856), grâce à l'emploi d'un dérivé fluorescent de l'antibiotique. FEW (1954) a étudié, par microscopie électronique, les lésions qui résultent de son action.

Bien que tyrocidine et polymyxine puissent solubiliser des constituants bactériens chez de multiples espèces sensibles à leur action antibiotique ,elles ne provoquent la lyse complète ou importante que de certaines d'entre elles seulement, celles qui, précisément, sont capables de subir une autolyse rapide dans des conditions adéquates. On les considère donc, assez généralement (HOTCHKISS 1944), comme des inducteurs de l'autolyse.

Ces considérations nous amènent à envisager, pour terminer cette section, le mécanisme de l'action bactériolytique d'un antibiotique d'origine toute différente, la pénicilline. Ici encore, la lyse n'est observée qu'avec certaines des espèces sensibles et TODD (1945) considère qu'il s'agit d'une bactériolyse secondaire, induite par l'antibiotique certes, mais dont le système autolytique bactérien est le principal responsable. A l'appui de son opinion, il a montré que l'autolysine pneumococcique favorise la dissolution des pneumocoques en milieu pénicilliné. Son observation n'a toutefois pu être confirmée par LEVADITI et HENRY (1946). D'autre part, ni TODD, ni WELSCH, n'ont pu accélérer ou intensifier la lyse pénicillinique des staphylocoques par addition d'autolysine. Enfin, selon LEVADITI et VAISMAN (1945), les altérations morphologiques de la lyse pénicillinique diffèrent de celles de l'autolyse.

La lyse pénicillinique pose d'ailleurs un autre problème. C'est l'observation d'une dissolution de vieilles colonies staphylococciques par un *Penicillium notatum* qui est à l'origine de la découverte de FLEMING (1929). Ce dernier a répété volontairement cette expérience dans la suite (1946). Mais les tentatives effectuées par d'autres (JENNINGS 1944, FISHER 1946) pour mettre en évidence une action lytique du *Penicillium*, de son filtrat de culture, ou de la pénicilline purifiée, sur des cultures arrivées à la phase stationnaire de leur cycle de croissance, n'ont guère été couronnées de succès. WELSCH (1947c) a cependant observé une lyse, mais très lente et très incomplète, des suspensions de staphylocoques lavés en présence de 0,04 à 40 u/ml. Il rapporte, d'autre part (1947d), que, lors d'expériences effectuées en 1937, il n'avait pu obtenir la lyse de suspensions bactériennes chauffées par les filtrats de culture du *Penicillium* de FLEMING et que cet organisme ne pouvait s'y développer comme le font, cependant, de nombreux streptomycètes.

L'absence d'activité lytique appréciable de la pénicilline purifiée sur les cultures ayant atteint leur complet développement suggérait que la pénicilline brute originale, simple filtrat de culture, contenait peut-être, en plus de l'antibiotique, un agent directement lytique. HUTTER crut avoir découvert cet agent auquel elle donna le nom de notalysine (1945, 1946). Elle constatait, en effet, qu'une souche de *Penicillium*, cultivée dans des conditions où elle ne produisait, croyait-on, pas de pénicilline, exerçait néanmoins une action lytique sur les staphylocoques en multiplication active. Cette action lytique, d'autre part, n'était pas détruite après traitement à la pénicillinase. Plus tard, cependant, dans notre Laboratoire, HUTTER (1948, 1950) a constaté que sa souche, même dans les conditions de culture particulières choisies, produisait une petite quantité de pénicilline. Celle-ci résistait à la pénicillinase par suite de la présence, dans les cultures, d'un inhibiteur de cet enzyme. En fait, les activités lytiques observées étaient donc attribuables à la pénicilline elle-même.

VI. Activités Bactériolytiques des Actinomycètes

Dans ce groupe d'organismes, les phénomènes d'autolyse sont fréquents (DMITRIEFF et SOUTEEFF 1936, KRASSILNIKOV 1938, KRASSILNIKOV et KORENIAKO 1938, KATZNELSON 1940), mais c'est l'activité hétérolytique surtout qui a fait l'objet d'études approfondies.

Après la découverte fortuite de l'isophagie staphylococcique (GRATIA et RHODES, 1924), GRATIA, interprétant ce phénomène comme un mode de nutrition bactérienne, s'est demandé s'il n'existerait point d'autres activités «bactériophagiques» analogues qui, s'ajoutant à l'action prédatrice bien connue des Protozoaires et des Myxomycètes, contribueraient à maintenir l'équilibre de la flore microbienne dans la Nature.

Pour aborder l'étude de cette question, il utilisa la gélose microbienne: eau ou solution saline gélosée dans laquelle sont incorporés des corps bactériens, tantôt chauffés, tantôt vivants, que l'on expose ensuite simplement à la contamination par les poussières de l'air, ou que l'on inocule avec des dilutions convenables de matériaux naturels variés: eau de rivière ou d'étang, terre, macération de produits végétaux, etc. . . . Dans ces conditions, les corps bactériens constituant l'unique source appréciable de matière organique, seuls se développent, théoriquement, les germes capables de les dissoudre et de les utiliser à des fins nutritives. De plus, la gélose microbienne, milieu sélectif (BOUNAMEAUX 1953), révèle également, de façon directe, l'activité bactériolytique: les colonies actives s'entourent, en effet, d'une zone de clarification évidente.

A la suite d'une recherche systématique, GRATIA et DATH (1924) ont reconnu que les germes bactériolytiques les plus fréquemment isolés par cette méthode appartiennent au genre *Streptothrix*, plus généralement connu à présent sous le nom de *Streptomyces*.

Certes, avant eux, GASPERINI (1890) avait vu que le *Streptothrix* se développe souvent à la surface des bactéries et des hyphomycètes, et il croyait que ses filaments peuvent en digérer la paroi. Mais son observation est restée isolée et n'a donné lieu à aucune expérience. De même, LIESKE (1921), au cours de ses études systématiques sur les «champignons rayonnés», avait reconnu que certains streptomycètes peuvent former, sur gélose contenant des bactéries, chauffées à 60° ou 100° C, ou tuées par le chloroforme, des colonies qui clarifient le milieu autour d'elles. LIESKE avait même suggéré que les agents responsables de cette bactériolyse pourraient avoir des applications en thérapeutique et que le phénomène pourrait jouer un rôle biologique important dans la Nature. Toutefois, ses observations sont restées perdues dans son importante monographie qui attira surtout l'attention des morphologistes et des taxonomistes (GRATIA et DATH 1925c).

GRATIA et DATH (1924) ont montré que les streptomycètes actifs, isolés sur la gélose bactérienne, peuvent être inoculés dans des suspensions bactériennes très diverses (GRATIA et ALEXANDER 1931), tuées par chauffage et diluées en eau distillée: ils s'y développent et en provoquent la dissolution. Ces lysats, stérilisés par filtration, possèdent une activité bactériolytique, à la vérité très faible et plus ou moins spécifique pour les germes homologues chauffés. L'activité lytique existe aussi, nettement plus élevée d'ailleurs, dans les filtrats de culture en bouillon des streptomycètes (GRATIA et DATH 1925a). Son intensité varie avec l'âge de la culture et dépend de l'état de sporulation des germes (GRATIA et DATH 1925b). L'activité lytique peut se manifester à l'égard de certaines espèces non préalablement tuées par chauffage (GRATIA et DATH 1924). Elle est inhibée par une augmentation de la concentration saline (DATH 1927).

GRATIA et DATH (1925b, 1926) ont signalé que les lysats bactériens, obtenus à l'intervention des streptomycètes ou de leurs filtrats, bien que pauvres en substances protéiques, conservent néanmoins un pouvoir antigénique important. De plus, ils présentent une toxicité moindre que les suspensions bactériennes intactes. Les auteurs n'ont toutefois publié que peu de détails sur ce sujet; ils concernent l'activité antigénique d'un «mycolysat» de vibrions cholériques. Néanmoins, à la suite de cette observation, GRATIA (1930b, 1931, 1934) préconisa l'emploi des «mycolysats» pour la vaccinothérapie de certaines infections, en particulier des staphylococcies. Il a rapporté des résultats thérapeutiques non dépourvus d'intérêt, mais ses tentatives pour mettre en évidence une élaboration d'anticorps restèrent négatives (GRATIA 1930a).

La démonstration que certains antigènes bactériens sont conservés, après lyse par les streptomycètes ou leurs agents actifs, ne vint que beaucoup plus tard seulement. La libération du polysaccharide C, antigène caractéristique des groupes de LANCEFIELD (1933), lors de la dissolution des streptocoques par les bactériolysines d'actinomycètes, a été mise en évidence par MAXTED (1948). Le procédé de cet auteur a été utilisé ensuite par plusieurs chercheurs, pour préparer l'antigène de groupe préalablement à l'identification sérologique des streptocoques (LONGFELLOW *et alii* 1949, WANNAMAKER *et alii* 1950, KLINGE 1955).

Dans notre Laboratoire, Castermans a d'abord tenté, mais sans succès, de mettre en évidence la production d'opsonines spécifiques chez les animaux vaccinés par un mycolysat de staphylocoques. Il a montré ensuite (1950) que la dissolution des streptocoques libère un matériel pleinement antigénique, capable d'induire, chez le lapin, la formation d'anticorps précipitants spécifiques. Il a constaté aussi que l'antigène streptococcique de groupe, extrait selon la technique de Wadsworth (1939), reste inaltéré après incubation prolongée en présence d'un *Streptomyces* activement bactériolytique.

Castermans s'est ensuite attaché à l'étude du sort des antigènes somatiques de *Salmonella typhimurium* au cours de la dissolution de cet organisme par les streptomycètes ou leurs filtrats actifs. L'antigène somatique, extrait selon la méthode décrite par Boivin et ses collaborateurs (1933), conserve sa capacité de précipiter en présence de l'anticorps spécifique après avoir été mis en incubation prolongée avec un streptomycète bactériolytique. La dissolution de la salmonelle libère un matériel qui précipite sous l'influence du sérum anti-0 correspondant et qui peut être extrait par la méthode à l'acide trichloracétique de Boivin et collaborateurs (Castermans 1950b). Ce matériel est bien l'antigène 0 complet, pleinement antigénique, car, injecté au lapin, il peut fonctioner comme agglutinogène, précipitogène et anaphylactogène (Castermans 1951). De plus, il confère à la souris une protection efficace contre l'infection expérimentale homologue subséquente (Castermans, 1954). Des observations analogues ont été faites pour *Salmonella typhosa* (Castermans 1956). Elles permettent de conclure que la dissolution, par les streptomycètes ou leurs agents actifs, d'Entérobactériacées préalablement chauffées, constitue un procédé efficace et commode pour l'isolement des antigènes 0. D'autre part, Salmon (1957) a montré, à la suite d'une étude comparative par électrophorèse, séro-précipitation en milieu gélosé et immuno-électrophorèse, que plusieurs antigènes distincts sont présents dans les mycolysats et les autolysats de *Salmonella schottmuelleri*, mais ne sont pas identiques dans chacun d'eux.

L'activité bactériolytique des actinomycètes a été occasionnellement décrite par quelques autres chercheurs. Rosenthal (1925c) a isolé un *Streptothrix*, contaminant accidentel d'une culture de *Corynebacterium diphtheriae*, qui dissout les suspensions de ce germe en eau gélosée et inhibe sa croissance en milieu nutritif. Borodulina (1935), puis Krassilnikov et Koreniako (1939), ont étudié les modalités de dissolution de diverses espèces bactériennes sous l'influence d'actinomycètes. Ils suggèrent que le principe lytique en cause pourrait être apparenté au lysozyme (Fleming 1922). Kriss (1940) a isolé un agent bactériolytique des cultures d'un *Streptomyces violaceus* et le considère comme identique ou très voisin du lysozyme.

L'étude du mécanisme de la bactériolyse effectuée par les actinomycètes a été entreprise par Welsch, à partir de 1936, et poursuivie par lui et ses collaborateurs jusqu'à présent (Welsch et Salmon 1949a; Ghuysen et Welsch 1952, 1953, Welsch et Ghuysen 1957). Les principaux résultats des recherches effectuées jusqu'ici sont réunis dans deux mémoires (Welsch 1947e, Ghuysen, 1957).

Utilisant quelques souches de streptomycètes, et, particulèrement, un organisme dénommé *Streptomyces albus* souche G, Welsch (1936, 1937b, 1939a) a montré qu'au moment de la sporulation, il apparaît, dans les cultures stationnaires, un agent bactériolytique capable de dissoudre les suspensions de bactéries préalablement stérilisées par chauffage ou par voie chimique, et même, pour quelques espèces, sans stérilisation préalable. L'addition d'un peu de liquide possédant cette activité, que ce soit le filtrat d'une culture en bouillon ou l'extrait aqueux d'une culture en milieu gélosé, favorise la multiplication du streptomycète dans les suspensions bactériennes en eau distillée ou en solution saline. Elle permet aussi la dissolution par le streptomycète des suspensions de nombreuses espèces gram-positives, non préalablement stérilisées et insensibles à l'action lytique du filtrat de culture stérile (WELSCH 1937a, e, 1938a, 1939e).

Les propriétés physico-chimiques de l'agent responsable de ces divers effets ont fait admettre sa nature protéique: taille voisine de celle de l'ovalbumine d'après les données de la filtration fractionnée (WELSCH et ELFORD 1937), caractères de stabilité, solubilité, précipitabilité et adsorbabilité (WELSCH 1937d, 1939c), inactivation par la chaleur avec incrément thermique critique de valeur élevée (WELSCH 1938e), inactivation par irradiation ultra-violette (WELSCH 1939d). L'étude cinétique de la bactériolyse qu'il provoque, étudiée par la méthode néphélométrique (WELSCH 1938b, 1939b), a fait admettre, d'autre part, sa nature enzymatique (WELSCH 1938c, d). Pour des raisons de commodité, les filtrats de culture de streptomycètes possédant ces propriétés ont été nommés «actinomycétine» (WELSCH 1937c), par analogie avec le terme pénicilline que FLEMING (1929) avait utilisé pour désigner un filtrat de culture brut de *Penicillium notatum* possédant un ensemble d'activités anti-microbiennes caractéristique.

Un streptomycète possède donc une activité bactériolytique du «type actinomycétine» lorsqu'il se comporte de la façon suivante.

1. Inoculé dans les suspensions, en eau ou en solution saline appropriée, des bactéries les plus diverses, préalablement inactivées par chauffage ou moyens chimiques convenables, il s'y développe et en provoque la dissolution rapide.

2. Il se comporte de la même manière lorsqu'il est inoculé dans les suspensions de bactéries gram-positives diverses, même non tuées au préalable. Cette activité est très nettement favorisée par addition à l'inoculat d'une petite quantité de l'actinomycétine qu'il produit.

3. Son filtrat de culture stérile, ou actinomycétine, dissout les suspensions bactériennes chauffées, et, plus irrégulièrement, les suspensions de bactéries gram-positives vivantes.

L'activité bactériolytique du type actinomycétine s'exerçant non seulement sur des corps bactériens tués, mais aussi sur des germes vivants, elle peut, à juste titre, être considérée comme une activité antibiotique (WELSCH, 1947b). Elle est très répandue chez les actinomycètes (WELSCH 1942a, b, 1947e).

L'activité lytique d'une actinomycétine obtenue après culture d'un streptomycète dans une suspension de bactéries en milieu non nutritif reste toujours très faible. D'autre part, l'addition de corps microbiens aux milieux de culture n'augmente pas l'activité de l'actinomycétine produite et ne favorise pas sélectivement son action lytique pour l'espèce utilisée. Les ferments bactériolytiques des actinomycètes n'ont donc pas le caractère des enzymes adaptatifs. Ils ne peuvent davantage être considérés comme des substances dont l'élaboration serait la conséquence d'un antagonisme induit ou provoqué, au sens de SCHILLER (1914, 1924—1933).

L'activité des actinomycétines sur les germes chauffés, d'une part, leur spectre d'action très étendu, d'autre part, ne permettent guère de les rapprocher du lysozyme. En particulier, l'activité des actinomycétines n'est pas plus marquée pour *Micrococcus lysodeikticus*, substrat type du lysozyme, que pour le staphylocoque, espèce insensible à cet enzyme (WELSCH 1942b).

GRATIA et DATH (1924) avaient constaté l'activité protéolytique des filtrats de cultures d'actinomycètes. Toutefois, l'action bactériolytique des actinomycétines ne paraît pas devoir être attribuée à des protéases banales. Les méthodes de purification par précipitation, classiquement utilisées en chimie des protéines, ont permis d'obtenir des actinomycétines concentrées et partiellement purifiées (WELSCH 1941). Leur activité bactériolytique spécifique, par mg de matière sèche, était bien plus élevée que celle de préparations cristallisées de trypsine ou de chymotrypsine, alors qu'inversément, leur activité protéolytique, sur la gélatine ou la caséine, était beaucoup plus faible. Pour insister sur la spécificité de l'enzyme responsable de la lyse des germes chauffés, WELSCH (1947a) lui a donné le nom d'actinozyme. La spécificité de l'agent bactériolytique des actinomycètes qui dissout les bactéries chauffées a d'ailleurs été soulignée ultérieurement aussi par TAI et VAN HEYNINGEN (1951).

La corrélation habituellement observée entre l'activité lytique d'une actinomycétine sur les germes chauffés et son aptitude à lyser les germes gram-positifs vivants, ou, tout au moins, à favoriser leur dissolution par le streptomycète en voie de développement, faisait penser que l'actinozyme intervenait aussi dans la désintégration des bactéries non chauffées. D'ailleurs, les modifications des milieux de culture qui aboutissaient à une amélioration de l'activité lytique des actinomycétines pour les germes vivants, entraînaient toujours, simultanément, une augmentation du rendement en actinozyme (WELSCH 1947a, e).

L'inverse n'était cependant pas nécessairement vrai (WELSCH 1946). De plus, des préparations d'actinomycétine concentrées par précipitation au sulfate ammonique, bien que pouvant être 100 fois plus actives sur les germes gram-négatifs chauffés, restaient néanmoins sans action appréciable sur les staphylocoques vivants (WELSCH, 1941). Enfin, la cinétique de la lyse des germes gram-négatifs chauffés est très différente de celle des germes gram-positifs chauffés ou vivants. Dans le premier cas, la clarification se déroule selon une cinétique d'ordre un. Dans le second, la réaction procède de façon analogue, mais seulement après une phase de latence plus ou moins prolongée, la dissolution des germes non chauffés se déroulant, mais à vitesse réduite, de la même façon que celle des germes correspondants chauffés (WELSCH 1947e).

L'ensemble de ces faits conduisait à attribuer la lyse des germes gram-positifs, chauffés ou non, à l'actinozyme, ferment qui, à lui seul, peut lyser les bactéries gram-négatives chauffées, mais à postuler, en outre, l'intervention indispensable d'un ou plusieurs autres agents additionnels. WELSCH (1947e) s'est alors demandé si la phase de latence précédant la lyse des germes gram-positifs, morts ou vifs, ne pourrait être, au moins en partie, l'expression d'une réaction préparatoire, au terme de laquelle les bactéries sensibles seraient transformées en cytosquelettes gram-négatifs. Il n'a cependant pu mettre en évidence un stade gram-négatif au cours de la lyse des germes gram-positifs. De plus, il a constaté que les staphylocoques, rendus gram-négatifs par traitement au moyen de bile à 60° C, sont dissous par l'actinomycétine, exactement de la même manière que les germes témoins restés gram-positifs. Enfin, il n'a pu déceler de ribonucléase dans ses actinomycétines (WELSCH 1948b).

Ces résultats négatifs n'ont toutefois qu'une importance secondaire. En effet, les actinomycétines dissolvant les germes gram-négatifs chauffés sont également actives sur les germes gram-positifs chauffés. Leurs effets respectifs sur ces deux types de substrats ne différant que du point de vue de la cinétique, il ne serait sans doute pas impossible de les expliquer par l'intervention d'un seul et même enzyme.

Par contre, des actinomycétines actives sur germes chauffés peuvent être absolument dépourvues de toute action lytique sur les bactéries gram-positives vivantes. La dissolution de ces dernières ne peut donc être l'œuvre du seul actinozyme, à moins d'admettre que les germes gram-positifs vivants, placés dans les milieux favorables aux manifestations de la bactériolyse, doivent d'abord y mourir progressivement et ne sont dissous qu'après leur mort.

SWERTZ (1949), étudiant, dans notre Laboratoire, la formation du halo de clarification partielle (WELSCH 1947e) qui entoure parfois la zone de bactériolyse complète provoquée par les actinomycètes, a pu montrer, par numération directe des germes survivants, respectivement dans les régions totalement clarifiées, partiellement clarifiées et restées troubles, que la dissolution des staphylocoques se fait à un moment où, en l'absence de produits diffusant à partir du streptomycète, ils seraient encore parfaitement viables. Notons en passant que l'étude du phénomène de halo (SWERTZ 1949, SMOLIAR 1950) révèle une participation probable du système autolytique du staphylocoque lui-même.

La production éventuelle d'un agent bactéricide par les streptomycètes bactériolytiques a donc été recherchée. L'examen d'un grand nombre de souches a tout d'abord montré l'existence d'une corrélation entre l'activité lytique et la capacité d'inhiber, faiblement d'ailleurs, le développement des germes gram-positifs, en particulier des Bacillacées. Il y a, par contre, corrélation négative entre la possession d'une activité lytique et la capacité de produire des substances antibiotiques hautement actives (WELSCH 1942b, WAKSMAN, HORNING, WELSCH et WOODRUFF, 1942). D'autre part, les solvants des lipoïdes ont effectivement permis d'extraire, à partir des précipités ammoniques d'actinomycétine, un matériel acidique manifestant, à forte dilution, une action bactéricide élective sur les espèces gram-positives (WELSCH 1941). Toutefois, des substances apparemment identiques peuvent être extraites, de la même manière, à partir des milieux de culture vierges, bien qu'en plus faible quantité cependant. Mais, de toute façon, leur intervention dans la dissolution des germes vivants ne peut être considérée comme nécessaire. En effet, les actinomycétines délipoïdées conservent inchangée leur activité à l'égard des staphylocoques non chauffés (WELSCH 1947e).

Etendant à la lyse par les actinomycètes leur théorie de l'autolyse des germes gram-positifs, JONES, SWALLOW et WEBB (1948) ont expliqué celle-ci en supposant que les germes sensibles sont d'abord tués par un agent bactéricide, perdent ensuite leur capacité de retenir le Gram sous l'influence d'une nucléase spécifique, sont enfin dissous à l'intervention de plusieurs protéases qui furent étudiées par MUGGLETON et WEBB (1952a, b). Applicable peut-être au *Streptomyces* de ces auteurs, cette explication n'est certainement pas valable pour les actinomycètes étudiés par WELSCH.

Pourtant, la possibilité d'intervention d'un agent bactéricide, au moins dans certaines circonstances, reste ouverte, notamment à la suite des observations faites dans notre Laboratoire par SMOLIAR (1949). Celui-ci a constaté que la lyse des staphylocoques autour des colonies du streptomycète est plus étendue en solution phosphatée gélosée qu'en eau distillée gélosée. La numération directe des germes viables, montre que la survie est plus longue en milieu phosphaté qu'en eau distillée, mais que, dans les deux cas, il existe, autour de chaque colonie d'actinomycète, une zone où la mortalité est très importante et précoce. Cette zone de léthalité déborde largement les limites de la zone de clarification. Il semble donc que, dans ces conditions expérimentales en tout cas, l'actinomycète élabore un agent bactéricide qui diffuse plus vite et plus loin que le principe bactériolytique. Il n'est toutefois pas démontré que son action soit indispensable, ni même simplement favorable à la lyse.

Les investigations que nous venons de passer en revue montrent clairement que l'actinozyme ne peut, isolément, lyser les bactéries gram-positives vivantes. Elles n'ont cependant pu mettre en évidence un agent additionnel qui, concuremment avec cet enzyme, rendrait compte de la bactériolyse des germes vivants. Il convenait donc de se demander si les corrélations observées entre activité lytique sur germes chauffés et vivants ne sont pas purement accidentelles et si la lyse des germes vivants n'est pas totalement indépendante de l'actinozyme.

S'attachant à cette question de l'unicité ou de la multiplicité des agents lytiques de l'actinomycétine, GHUYSEN (1952a), dans notre Laboratoire, a tout d'abord étudié, de façon comparative les activités de l'actinomycétine sur des suspensions d'*Escherichia coli* tuée par chauffage à 100° C pendant 3 minutes, d'une part, de *Micrococcus pyogenes aureus* non chauffé, d'autre part. Ces deux types d'activité seront respectivement désignés sous les noms d'activité colilytique et staphylolytique. L'une et l'autre apparaissent régulièrement dans les cultures de *Streptomyces albus* G, en milieu contenant les sels de la solution de Czapek et une peptone convenable, lorsque l'incubation est faite à la température optimale de 28° C, en flacons de Roux de 1 litre de capacité, contenant 250 ml de milieu, maintenus sur un agitateur adéquat.

L'évolution des deux types d'activité est toutefois bien différente. L'agent colilytique est produit lorsque la concentration en peptone atteint 0,1%. Sa production augmente progressivement avec une élévation du taux de peptone jusqu'à 1%. Si ce dernier est encore augmenté, le rendement en principe colilytique ne diminue pas, mais sa production est plus lente et le titre maximum est atteint plus tardivement. L'agent colilytique est stable, son titre ne diminuant que lentement et légèrement au cours d'une incubation prolongée.

L'activité staphylolytique, par contre, n'apparaît dans les cultures que si le taux de peptone atteint 0,5%. Sa production est optimale pour une concentration en peptone de 1% et diminue pour des concentrations plus élevées. Quelle que soit cette dernière, l'activité maximale est obtenue après une incubation de

96 heures environ. L'agent staphylolytique, instable, disparaît assez rapidement au cours d'une incubation prolongée.

A concentration convenable (GHUYSEN 1953a), tous les cations divalents étudiés peuvent favoriser la production du principe colilytique, au moins dans des conditions expérimentales déterminées. Les résultats les meilleurs et les plus constants sont obtenus avec le Co^{++} à la concentration 10^{-3} M. L'influence particulière de cet ion est d'ailleurs bien mise en évidence par les faits suivants. Les actinomycétines obtenues à température anormalement élevée (36° C), ou en présence soit d'1% d'agar-agar, soit de divers cations, tels Cu^{++}, Ni^{++}, Fe^{++}, produisent, après avoir été dialysées, une bactériolyse d'*E. coli* chauffée qui n'obéit pas à la cinétique habituelle d'ordre un. Or, les actinomycétines préparées dans les mêmes conditions, mais, avec addition de Co^{++} au milieu de culture, ont un comportement tout à fait normal (GHUYSEN, 1957).

A concentration convenable, tous les cations divalents examinés, mais à l'exception de Ca^{++} cependant, peuvent également favoriser, dans des conditions expérimentales déterminées, la production du principe staphylolytique. C'est le Mg^{++}, à la concentration $2{,}5 \times 10^{-4}$ M, qui, cette fois, fournit les résultats les meilleurs et les plus constants.

En milieu contenant du Co et du Mg, les deux types d'activité sont obtenus en quantité appréciable, aussi bien en flacons de Roux agités qu'en cuves de grande capacité (100 ou 500 litres), pourvues d'un dispositif d'agitation mécanique et d'aération forcée (WELSCH et DELCAMBE 1954). Dans ces conditions, l'augmentation du taux d'aération favorise la production de l'activité colilytique, mais non celle de l'activité staphylolytique. Celle-ci est maximale, en effet, pour une aération de 0,2 litre d'air par litre de culture par minute, la pression étant maintenue à 0,5 kg/cm^2.

Les agents colilytique et staphylolytique se distinguent non seulement par les conditions requises pour assurer leur production, mais également par les conditions qui permettent leur activité optimale (GHUYSEN 1952a, 1953a).

La lyse par l'actinomycétine d'*E. coli* chauffée est, dans de larges limites (0,015 à 0,08 ou davantage), indépendante de la force ionique. Elle est, d'autre part, inhibée par l'ion Mn^{++} (BORN 1952). Sauf exceptions que nous avons mentionnées plus haut, elle se déroule, en solution M/30 de phosphates à p_H 8,0, selon une cinétique d'ordre un.

La lyse du staphylocoque vivant présente, au contraire, un optimum très net lorsque la force ionique est voisine de 0,02. En plus de cet effet général, il existe d'ailleurs une action spécifique, liée à la nature même des ions choisis. L'association d'orthophosphate et de sulfate ammonique, à p_H 7,0, fournit les meilleurs résultats. On peut y ajouter du Mn^{++}, pour déprimer l'activité colilytique, sans que cet ion altère l'activité staphylolytique. Dans ces conditions, la dissolution du staphylocoque obéit généralement, elle aussi, mais après une phase de latence, à une cinétique d'ordre un.

Ces observations expérimentales suggèrent l'existence d'une dualité des principes respectivement actifs sur *E. coli* chauffée et *M. pyogenes* vivant. GHUYSEN (1952b) tenta de séparer ces agents l'un de l'autre par adsorption spécifique. Divers adsorbants usuels, charbon activé, alumine, amidon, fixent énergiquement les agents bactériolytiques de l'actinomycétine, mais leur élution ultérieure n'a pu être obtenue. Le traitement d'une actinomycétine par la silice colloïdale, dans des conditions physico-chimiques bien déterminées, a toutefois fourni des préparations dans lesquelles les proportions des deux types d'activité étaient profondément modifiées. En utilisant les corps microbiens eux-mêmes comme matériel d'adsorption, on constate que les activités colilytique et staphylolytique sont, l'une et l'autre, fixées par les corps, chauffés ou non, d'*E. coli* et de *M. pyogenes*. Les conditions d'adsorption et d'élution sont cependant bien différentes pour chacun des deux types d'activité. Optimale dans tous les cas à p_H voisin de 9,2, l'adsorption est relativement peu influencée par la valeur de la force ionique pour l'agent colilytique, mais elle n'est appréciable qu'à très faible force ionique pour l'agent staphylolytique. Chaque substrat bactérien manifeste, d'autre part, une capacité d'adsorption plus élevée pour le principe lytique homologue qu'il fixe de façon préférentielle. L'élution qui, dans tous les cas, est obtenue à force ionique élevée et p_H soit très alcalin (10,6), soit très acide (3,0), fournit des liquides dans lesquels les proportions des deux types d'activité sont fortement modifiées. Mais leur séparation complète n'a pu être obtenue et le rendement de l'opération est faible.

En utilisant comme adsorbant des colonnes de sable sur lequel les bactéries avaient été préalablement fixées, GHUYSEN (1954b) obtint des actinomycétines concentrées et partiellement purifiées. Au cours de ses observations, il a constaté que certains échantillons de sable, non artificiellement recouverts de bactéries, adsorbaient eux-mêmes non seulement les agents lytiques, mais encore divers enzymes protéolytiques de l'actinomycétine, digérant la caséine, la kératine, l'épidermine (GHUYSEN 1954c, GHUYSEN et LÉGER, 1954), le fibrinogène et la fibrine (GHUYSEN 1954d), la mucine (BERGAMINI 1954), etc. . . . En choisissant des conditions physico-chimiques appropriées pour l'adsorption et l'élution, il a obtenu des préparations dans lesquelles les proportions de ces diverses activités étaient profondément modifiées. Ces constatations sont en bon accord avec l'idée, autérieurement avancée, que les propriétés bactériolytiques sont indépendantes des activités protéolytiques banales.

Ni la précipitation fractionnée, ni l'adsorption n'ayant permis de séparer les agents lytiques de l'actinomycétine, GHUYSEN eut recours à la méthode d'extraction fractionnée (GHUYSEN 1957). L'actinomycétine est fixée sur des particules de verre pyrex qui sont ensuite disposées en une colonne. On fait successivement passer à travers celle-ci des solutions de sulfate ammonique de p_H déterminé et de concentration progressivement décroissante. Les fractions successives sont séparément recueillies et analysées. Dans ces conditions, l'activité colilytique est retrouvée dans les fractions à concentration saline élevée, en même temps d'ailleurs que les enzymes protéolytiques. Toutefois, le rendement respectif de ces divers agents varie de façon différente selon que le p_H choisi est 9,5 ou 5,5. Ceci confirme donc leur individualité respective. L'activité staphylolytique, par contre, est retrouvée dans les fractions moins riches en sel, dont certaines sont totalement dépourvues d'activité colilytique. La séparation des deux types d'activité bactériolytique, réalisée par ce procédé, démontre donc, de façon directe et définitive, que la dissolution des staphylocoques vivants est possible sans intervention du principe colilytique ou actinozyme.

Des études portant sur la cinétique de la bactériolyse (GHUYSEN 1953b, 1957) montrent, d'autre part, que la dissolution des *E. coli* chauffées, ainsi que celle des staphylocoques chauffés et vieillis pendant quelques semaines à basse température, est attribuable à l'actinozyme. Les suspensions de staphylocoques non chauffés sont dissoutes par un système staphylolytique totalement distinct de l'actinozyme. Enfin, les suspensions de staphylocoques, chauffés et immédiatement utilisés, sont normalement lysées par l'actinozyme, mais, en présence de Mn^{++}, inhibiteur spécifique de cet agent lytique, elles peuvent être clarifiées par le système staphylolytique.

Le rendement en agent staphylolytique, lors de l'extraction fractionnée, n'atteint toutefois que 90%, après extraction à p_H 9,5, et seulement 45% à p_H 5,5. Dans les deux cas, cependant, on retrouve 100% de l'activité staphylolytique si, lors du dosage de chaque fraction, on lui ajoute une petite quantité de la fraction extraite par la solution ammonique la plus concentrée. Cette observation indique donc que l'agent responsable de la staphylolyse est probablement complexe, qu'il comprend, à côté d'un agent proprement lytique pour le staphylocoque, un activateur dont le mode d'action reste à préciser.

La méthode utilisée pour démontrer la dualité des principes lytiques n'est malheureusement pas applicable à leur préparation. Elle est, en effet, très longue et elle aboutit à une dilution considérable des substances actives.

Utilisant alors un milieu de culture où la peptone était remplacée par un hydrolysat de caséine et de kératine, plus favorable à la production de l'agent staphylolytique, GHUYSEN (1957) a purifié et isolé le système staphylolytique par chromatographie sur échangeur d'ions. En employant l'amberlite IRC—50, convenablement traitée au préalable par NaOH, il a pu retenir isolément sur la colonne la seule activité staphylolytique qui fut ensuite récupérée par élution et désignée sous le nom de fraction 1. Le rendement de l'opération est apparemment assez pauvre, mais l'activité complète est restaurée par addition d'une petite quantité du matériel non adsorbé initialement (fraction 2), ce qui confirme la complexité du système staphylolytique. L'agent staphylolytique de la fraction 1, ou F 1, a été hautement purifié par chromatographie sur amberlite XE—64. L'agent staphylolytique présent dans la fraction 2, ou F 2, est, par contre, beaucoup plus difficile à isoler, ses propriétés physico-chimiques étant proches de celles de l'actinozyme et des protéases auxquels il est associé. GHUYSEN l'a néanmoins obtenu, en petite quantité, dans un état de pureté satisfaisante, dépourvu, en tout cas, des agents colilytique et protéolytiques.

Les deux agents, F 1 et F 2, constituant le système staphylolytique, sont thermolabiles et non dialysables. Le premier dissout les suspensions de staphylocoques non chauffés, le second n'a qu'une activité très faible. Leur association, cependant, exerce une action bactériolytique hautement synergique sur le staphylocoque dont la dissolution se déroule alors selon une cinétique d'ordre un. Ces deux agents ne sont d'ailleurs pas spécifiquement staphylolytiques. L'un et l'autre peuvent assurer une lyse plus ou moins rapide et étendue, selon les cas, de diverses autres bactéries gram-positives: *Micrococcus lysodeikticus, Sarcina lutea, Sporosarcina ureae, Clostridium welchii*, divers *Bacillus* spp. Toujours, cependant, leur association fournit une lyse plus rapide et plus régulière.

Salton et Ghuysen (1957) ont montré que ces deux agents dissolvent les suspensions de parois cellulaires, isolées et purifiées (Salton et Horne, 1951b; Salton 1952a, 1953a) provenant des bactéries sensibles. Ici encore, selon l'espèce considérée, F 1 est tantôt plus actif que F 2 (staphylocoque), tantôt moins *(S. lutea)*. Mais, dans tous les cas, F 1 et F 2 agissant ensemble, sont considérablement plus efficaces. La dissolution des parois cellulaires se fait de façon tout à fait parallèle à celle des corps microbiens correspondants. Elle est notamment influencée de la même manière par les variations de la force ionique. La bactériolyse peut donc, à bon droit, être attribuée à l'attaque spécifique d'un constituant de la paroi cellulaire des germes sensibles.

L'analyse chimique des produits libérés lors de la digestion des parois, tant par F 1 que par F 2 ou par les deux agents réunis, montre qu'il n'apparaît ni sucres, ni sucres aminés, mais uniquement, selon les cas, un ou deux acides aminés libres (glycine et alanine) et des peptides de petite taille, non identifiés.

L'activité staphylolytique de l'actinomycétine est donc attribuable à l'action synergique de deux peptidases, agissant d'ailleurs également sur d'autres bactéries gram-positives. Elles seront désignées sous les noms d'actinolysopeptidases 1 et 2. La bactériolyse de ces germes par l'actinomycétine ressemble donc très fort à celle que provoque le lysozyme. Cet enzyme, en effet, agit également sur la paroi cellulaire (Salton 1952b, 1953, Tomcsik et Guez-Holzer, 1952, Weibull 1953, Welshimer 1953, Grula et Hartsell 1954a, b), mais le substrat de son action est très différent: c'est un mucopolysaccharide (Meyer, Palmer, Thompson et Khorazo 1936, Epstein et Chain 1940).

L'actinozyme et les actinolysopeptidases constituant le système staphylolytique de l'actinomycétine ne rendent toutefois pas compte de toutes ses activités bactériolytiques. En effet, étudiant la dissolution de streptocoques par les actinomycètes, selon la méthode de Maxted (1948), McCarty (1952a, b) a séparé des filtrats de culture de son streptomycète un enzyme qui agit sur la paroi cellulaire (Salton 1952c, Salton 1953a) de ces germes et qui est considéré comme une polysaccharidase. Cet enzyme diffère du lysozyme car il est inactif sur les bactéries sensibles à ce dernier. Les observations de McCarty ont incité Welsch et Ghuysen (1953) à examiner l'activité streptolytique de l'actinomycétine de *S. albus* G. L'étude cinétique de la streptolyse, dans diverses conditions expérimentales, montre que l'activité streptolytique de l'actinomycétine n'est exclusivement attribuable ni à l'actinozyme, ni au complexe staphylolytique. Lors de l'extraction de l'actinomycétine au sulfate ammonique, l'agent streptolytique est extrait en même temps que l'agent staphylolytique. Le rendement de l'opération est extrêmement pauvre et l'activite originale ne peut être restaurée par addition de fractions extraites à salinité plus élevée. L'agent staphylolytique F 1 purifié n'exerce qu'une action lytique très modérée sur le streptocoque non chauffé. La fraction F 2 est totalement inactive et n'a, dans ce cas, aucun effet potentiateur sur F 1. La recherche des activités staphylolytique et streptolytique chez une série de streptomycètes montre d'ailleurs qu'il n'y a pas de corrélation étroite entre ces deux propriétés (Welsch 1954). Il existe donc, dans l'actinomycétine de *Streptomyces albus* G, un agent spécifiquement streptolytique dont l'étude est à faire et qui, d'ailleurs, est peut-être identique à la polysaccharidase de McCarty.

GHUYSEN (1954a) a recherché ensuite l'activité de l'actinomycétine sur le pneumocoque. Une actinomycétine ayant traversé une colonne de sable approprié, et ainsi débarrassée de toute activité protéolytique, colilytique, staphylolytique et streptolytique, provoque néanmoins la dissolution des suspensions de pneumocoques, chauffés ou vivants. L'analyse par extraction fractionnée montre l'existence de deux agents distincts, l'un lysant les pneumocoques chauffés ou vivants, l'autre n'agissant que sur les pneumocoques non chauffés. La recherche des activités bactériolytiques d'une série de streptomycètes sur plusieurs espèces microbiennes montre d'ailleurs l'absence de corrélation étroite entre les activités staphylolytique, streptolytique, pneumolytique, caséinolytique, kératinolytique et mucasique (BERGAMINI, GHUYSEN et WELSCH 1954). Il existe donc, dans l'actinomycétine, au moins deux agents spécifiquement pneumolytiques dont l'étude est à faire.

La multiplicité des agents lytiques de l'actinomycétine est donc à présent parfaitement établie (WELSCH, GHUYSEN et CASTERMANS 1955). Ce complexe bactériolytique comporte: l'actinozyme, responsable de la dissolution des germes gram-négatifs tués et habituellement aussi des germes gram-positifs tués; le système dit staphylolytique, formé de deux actinolysopeptidases, conjointement responsables de la dissolution de divers germes gram-positifs vivants; un agent streptolytique et deux agents pneumolytiques.

Il est remarquable qu'aucune activité n'ait été décelée à l'égard des germes gram-négatifs vivants. Les nombreux streptomycètes examinés par WELSCH (1942a, b, 1947e) se sont tous montrés incapables de provoquer la dissolution des suspensions non chauffées d'*E. coli*. D'autre part, SALTON (1955) a isolé des actinomycètes bactériolytiques au moyen d'agar opacifié par des parois cellulaires de *Streptococcus faecalis*, mais non d'*Escherichia coli*. Aucun des organismes ainsi isolés n'attaquaient les parois de germes gram-négatifs.

Etant donné que les agents bactériolytiques dont le mode d'action a pu être élucidé semblent bien exercer leur effet initial sur la paroi cellulaire, on est en droit de se demander si la résistance à la lyse des germes gram-négatifs n'est pas liée à la nature de leur membrane. Les analyses de HOLDSWORTH (1952), de SALTON (1952a, b, c, 1953a), de CUMMINS et HARRIS (1956) ont montré qu'il existe, en effet, des différences de composition très nettes entre les parois de ces deux types d'organismes. L'hydrolyse des parois des bactéries gram-positives ne fournit qu'un nombre limité d'acides aminés, à l'exclusion des acides cycliques et à groupe thiol. Par contre, l'hydrolyse des parois de bactéries gram-négatives fournit l'ensemble des acides aminés habituels aux protéines naturelles; elles sont en outre riches en lipoïdes. Notons que, selon MITCHELL et MOYLE (1951), les parois du staphylocoque auraient une composition assez particulière: elles seraient constituées d'un complexe glycérophosphoprotéique.

On connaît d'ailleurs plusieurs cas où une modification convenable de la paroi cellulaire rend une bactérie sensible à l'action lytique d'une substance à laquelle elle était préalablement résistante. L'altération des parois cellulaires de germes gram-négatifs provoquée par le chauffage est mise en évidence au moyen du microscope électronique (SALTON et HORNE 1951a). On lui attribue la sensibilité à la trypsine des germes chauffés, contrastant avec leur résistance à l'état frais (SALTON 1953b). De même, un chauffage modéré ou un traitement par l'acétone peuvent rendre sensibles au lysozyme des bactéries normalement résistantes à son action (Peterson et Hartsell 1955, Warren, Gray et Bartell 1955). Le lysozyme lui-même, agissant sur des bactéries dont il ne provoque pas la lyse, peut les rendre sensibles à la trypsine (Webb 1948). Des mécanismes de ce genre rendront peut-être compte de deux observations faites dans notre laboratoire. Smoliar (1952) a constaté que l'actinomycétine agit en synergie avec le pyrophosphate et la glycine pour assurer la dissolution de suspensions non chauffées d'*Escherichia coli*. Castermans (1955) a constaté, d'autre part, que les suspensions en solution phosphatée de *Salmonella typhosa* et de *Salmonella typhimurium*, non préalablement chauffées, subissent une clarification rapide, sous l'influence de l'actinomycétine, lorsque l'incubation est faite à 60 ou 65° C, le phénomène se déroulant fréquemment en un temps trop court pour assurer la stérilisation de suspensions témoins maintenues à cette température. Nous ignorons si l'agent lytique intervenant dans ces deux circonstances est l'actinozyme ou s'il s'agit d'un autre principe encore.

Conclusions

Beaucoup de bactéries, sinon toutes, possèdent un système autolytique dont l'activité varie toutefois considérablement d'une espèce à l'autre. Il n'est pas étonnant que l'on ait davantage étudié l'autolyse des germes faciles à cultiver et dont les cultures âgées ou les suspensions lavées se clarifient de la façon la plus manifeste. Ce sont, avant tout, le pneumocoque, le staphylocoque, divers représentants des genres *Bacillus* et *Clostridium*. Mis à part les gonocoque et méningocoque, dont la culture malaisée fait un matériel expérimental plus difficile à manier, les germes gram-négatifs usuels ne présentent pas d'autolyse très spectaculaire. L'existence d'un système autolytique est toutefois mise en évidence chez eux par des inducteurs d'autolyse, tels le pyrophosphate, la glycine, et, selon Wollman (1932a), divers antiseptiques dilués.

On a décelé, dans les autolysats bactériens les plus divers, des antigènes et des substances toxiques. La supériorité des lysats-vaccins sur les suspensions bactériennes intactes, affirmée par certains, reste toutefois à démontrer. Il convient de souligner que des lysats d'un même germe, obtenus à l'intervention d'inducteurs d'autolyse différents, peuvent renfermer des antigènes distincts. L'origine des toxines présentes dans les lysats est variable. Tantôt elles représentent un matériel endocellulaire préexistant que libère simplement l'autolyse, tantôt un matériel formé *ex novo* au cours de la lyse, par dégradation enzymatique des constituants cellulaires. Leur rôle éventuel dans la pathogénie des infections, consécutive à la bactériolyse *in vivo*, est généralement admis dans certains cas (endotoxine des Salmonelles), mais nié dans d'autres (pneumocoque).

L'autolyse, comme les procédés de destruction mécanique des germes, est un moyen parfois utilisé avec succès pour libérer et étudier certains systèmes enzymatiques endocellulaires. En particulier, les suspensions bactériennes lysées ont parfois la capacité d'activer l'autolyse des germes homologues vivants et de les solubiliser après qu'ils ont été tués par chauffages. Considérées sous cet aspect, les suspensions lysées sont souvent désignées sous le nom d'autolysine.

Les conditions qui favorisent l'autolyse, l'activité des autolysines, les phénomènes d'autophagie, isophagie et hétérophagie, ont été étudiées chez le pneumocoque, le staphylocoque, *Bacillus megaterium* et la souche φ d'*Escherichia coli*. Elles sont quelque peu différentes pour chacune des espèces considérées. Dans le cas du staphylocoque, où toutes ces activités ont pu être étudiées, on peut conclure que l'autophagie, l'isophagie et l'hétérophagie, sont attribuables à la diffusion, à partir des colonies microbiennes, des enzymes autolytiques. Cette libération exige la présence d'un orthophosphate ou d'un succédané qui, vraisemblablement, n'est d'ailleurs actif qu'après avoir été converti lui-même en orthophosphate.

L'intervention du système autolytique des bactéries lors de la dissolution que provoque leur infection par un bactériophage est probable, mais n'a cependant guère été étudiée (WEINBERG et AZNARD 1922, WOLLMAN et WOLLMAN 1932 HUPPERT et PANIJEL 1954, BROWN 1956). Certes, les lésions nucléaires reconnues lors de la lyse batériophagique diffèrent de celles que l'on voit lors de la simple autolyse (BEUMER et QUERSIN 1947, QUERSIN 1948). Mais il est bien certain que, dans le premier cas, l'infection virale qui précède puis accompagne la désintégration cellulaire est de nature à provoquer des altérations cytologiques spécifiques additionnelles.

Des phénomènes tels que la lyse bactériophagique par entrainement (WAHL et JOSSE-GOICHOT, 1949), la lyse en glacière des staphylocoques lysogènes irradiés (CAVALLO et CANTELMO 1951, WELSCH 1952b, WELSCH, CAVALLO et CANTELMO, 1953), la lyse collatérale (WELSCH 1953a, b), ne sont guère explicables que par l'intervention d'agents autolytiques libérés par la lyse bactériophagique d'une fraction de la population.

Il semble, d'autre part, que la bactérie infectée de phage, le complexe, puisse élaborer des autolysines que ne synthétise pas la bactérie vierge d'infection virale. On connait, en effet, la grande sensibilité des complexes à l'action lytique d'agents tels le cyanure de sodium (DOERMANN, 1952) ou le chloroforme (SÉCHAUD et KELLENBERGER 1956) qui en permettent la lyse prématurée. On sait aussi que, dans certains lysats bactériophagiques, on peut mettre en évidence des lysines spécifiques, d'abord considérées comme des enzymes élaborés par le virus lui-même (SERTIC 1929, SERTIC et BULGAKOV 1937). Leur biosynthèse résulte toutefois, plus vraisemblablement, d'un phénomène d'induction de la bactérie par le virus qui l'infecte (RALSTON *et alii*, 1955). On sait que ces lysines, ou prolysines (PANIJEL et HUPPERT, 1954, 1956), des lysats bactériophagiques, responsables des halos qui entourent parfois les zones de bactériophagie, peuvent solubiliser des germes tués. Les unes s'attaquent spécifiquement à des constituants bactériens superficiels (GRATIA 1937, HUMPHRIES 1948), de nature probablement polysaccharidique, les autres, par contre, à des constituants protéiques, identiques aux récepteurs de phage (ANDERSON 1945, QUERSIN et DIRKX 1951).

Le mécanisme de l'autolyse reste encore mal connu. Il est certain que ce phénomène, tel qu'il se déroule spontanément, met en jeu un nombre considérable d'enzymes très divers. Mais on sait aussi que, par un choix approprié des conditions expérimentales, la clarification d'une suspension bactérienne peut être obtenue sans que protéines, lipides et polysaccharides subissent une dégradation poussée.

Des observations portant sur plusieurs espèces bactériennes ont montré que la première étape de l'autolyse, chez les germes gram-positifs, se traduit par l'incapacité de conserver le Gram.

On en a conclu que l'autolyse était nécessairement déclenchée par un enzyme très spécifique, une nucléase, agissant sur le complexe gram-positif, protéine-polysaccharide-ribonucléate de Mg, dont beaucoup admettent qu'il est localisé au niveau de la paroi cellulaire. Il convient toutefois de remarquer que si le caractère gram-positif de la paroi cellulaire semble bien démontré, il n'est nullement établi que la réaction de Gram soit localisée exclusivement à cette structure. D'autre part, le rôle prépondérant accordé au ribonucléate de Mg dans la réaction de Gram positive n'est pas unanimement admis. Selon MITCHELL et MOYLE (1951, 1954), par exemple, la présence d'un excès de polyglycérophosphates chez les espèces gram-positives aurait une signification plus grande. En tout cas, la négativation du Gram que l'on arrive à mettre en évidence comme premier temps de l'autolyse, ne nous semble pas être nécessairement une lésion spécifique du substrat gram-positif. En effet, elle s'accompagne toujours d'une diminution considérable de la basophilie générale du cytoplasme, alors que, au moins dans certains cas, les noyaux, restés basophiles et gram-positifs, deviennent bien apparents. La manifestation initiale de l'autolyse nous parait être une solubilisation d'acides ribonucléiques, qui parfois, d'ailleurs, n'est que partiellement d'origine enzymatique. La série d'enzymes dont STACEY et ses collaborateurs postulent l'intervention, dans un ordre bien déterminé, pour expliquer les stades successifs de l'autolyse, n'ont certainement pas la signification générale que l'on veut leur attribuer. En particulier, l'autolyse du staphylocoque peut être réalisée dans des conditions incompatibles avec les propriétés attribuées à ces enzymes.

La capacité de clarifier les suspensions d'une autre espèce, l'hétérolyse, a été rencontrée chez plusieurs bactéries, en particulier de nombreuses Bacillacées et Streptomycétacées.

Dans certains cas, cette activité hétérolytique est sans doute attribuable à la production d'un agent antibiotique jouant le rôle d'inducteur d'autolyse. Lipoides variés, tyrocidine, polymyxine, provoqueraient ainsi la mort des bactéries sensibles à leur action respective, et induiraient la bactériolyse de celles d'entre elles qui possèdent un système autolytique puissant. Douées de propriétés tensioactives, ces substances se fixeraient soit à la paroi cellulaire, soit à la membrane cytoplasmique, et y provoqueraient une désorganisation, aboutissant à la désintégration complète de la cellule bactérienne, éventuellement suivie de dégradation enzymatique des constituants microbiens.

Dans d'autres cas, l'activité hétérolytique est attribuable à des enzymes qui digèrent un constituant bactérien déterminé. Il est certain que de nombreuses bactéries, en particulier des *Streptomyces*, élaborent des enzymes, distincts des protéases banales, par exemple l'actinozyme, qui dissolvent spécifiquement les corps microbiens tués par chauffage ou par des agents chimiques. D'autre part, on connait deux agents bactériolytiques, produits par des *Streptomyces*, qui provoquent la dissolution des bactéries sensibles en détruisant un constituant de leur paroi cellulaire. L'un est l'enzyme streptolytique décrit par McCARTY. C'est une polysaccharidase. L'autre est le système dit staphylolytique de l'actinomycétine. GHUYSEN a montré qu'il comprenait deux facteurs agissant en synergie. Avec SALTON, il a montré que chacun de ces facteurs est une peptidase capable de solubiliser la paroi cellulaire des germes sensibles (actinolysopeptidases).

Etant donné leur mode d'action, ces agents doivent être rapprochés du lysozyme, que produisent d'ailleurs certains microorganismes. Il convient également de les rapprocher des enzymes qui détruisent un constituant bactérien superficiel, sans toutefois provoquer une bactériolyse. Ainsi, agissent, notamment, les enzymes d'origine bactérienne qui digèrent spécifiquement les polysaccharides capsulaires type-spécifiques des pneumocoques (AVERY et DUBOS 1930, DUBOS et AVERY 1931, DUBOS 1932, 1935, DUBOS et BAUER 1935, SICKLES et SHAW, 1933, 1934a, b, 1935; SHAW, 1937). Ceux-ci ont été utilisés, avec le succès que l'on sait, pour démontrer le rôle important de la capsule dans la virulence de ce germe (AVERY et DUBOS 1931, GOODNER, DUBOS et AVERY, 1932, GOODNER et DUBOS 1932, FRANCIS, TERRELL, DUBOS et AVERY 1934). Ainsi agissent également les protéases qui digèrent la protéine M des streptocoques (LANCEFILD 1943), l'hyaluronidase qui détruit la capsule de ces mêmes organismes (MCLEAN 1942, MAXTED 1952), et plusieurs autres enzymes microbiens capables de digérer les capsules d'organismes divers (MORGAN et THAYSEN, 1933, MEYER et MORGAN 1935): streptocoques (HIRST 1941), *Klebsiella pneumoniae* (THOENNISSEN 1921), *Bacillus subtilis* (GREEN et STAHMANN 1952, KREAM et alii 1954, THORNE et alii 1954), *Bacillus anthracis* (NORDBERG et THORSELL 1955).

Résumé

M. WELSCH. Activités bactériolytiques des Microorganismes.

On passe en revue les travaux consacrés: (1) à l'autolyse du pneumocoque; (2) aux activités bactériolytiques des staphylocoques (autophagie, isophagie, hétérophagie, autolyse); (3) au mécanisme de l'autolyse des germes gram-positifs; (4) aux activités autophagiques, isophagiques et autolytiques des bâtonnets gram-négatifs; (5) aux activités bactériolytiques des Bacillacées, et (6) des Actinomycètes.

Le mécanisme de l'autolyse et les enzymes qui en sont responsables sont encore très mal connus. En particulier, les enzymes dont l'intervention a été postulée par STACEY et ses collaborateurs pour rendre compte de l'autolyse des bactéries gram-positives, n'ont pas la signification générale que l'on a voulu leur donner. La disparition du caractère gram-positif, souvent observée au début de l'autolyse, parait être la conséquence d'une perte générale d'acides ribonucléiques, plutôt qu'une atteinte spécifique d'un substrat gram-positif localisé. L'autolyse d'une suspension bactérienne est un phénomène complexe, à la fois endogène (autolyse vraie) et exogène (autophagie).

L'activité hétérolytique des microorganismes est souvent attribuable à la production d'un agent antibiotique jouant le rôle d'inducteur d'autolyse. Dans d'autres cas, elle peut être attribuée à des enzymes directement lytiques. Parmi ceux-ci, les seuls actuellement bien individualisés sont: (1) le lysozyme, produit par divers microorganismes, mais surtout étudié sous la forme présente dans le blanc d'œuf; (2) l'enzyme streptolytique, isolé par MCCARTY des cultures d'un *Streptomyces* de MAXTED; (3) les deux actinolysopeptidases isolées par GHUYSEN de l'actinomycétine de WELSCH. Tous ces enzymes doivent leur action bactériolytique à leur capacité de digérer un constituant de la paroi cellulaire bactérienne, de nature polysaccharidique dans le cas des deux premiers, protidique dans le cas des deux derniers.

Références

AMATI, A.: Grandezza della carica batterica, peptoni e decorso della lisi da amino-acidi e da estratto di *Subtilis*. Boll. Soc. ital. Biol. sper. **23**, 210 (1947).

ANDERSON, T. F.: On a bacteriolytic substance associated with a purified bacterial virus. J. Cellul. a. Comp. Physiol. **25**, 1 (1945).

ANZAI, H., D. NAGAKI, K. DATE, N. TSUKAKOSHI, T. MATORI and M. NAKAMURA: On the antagonistic action and the enzymatic action of spore-forming bacilli. Kitasato Arch. Exper. Med. **21**, 20 (1948).

ARK, P. A., and M. L. HUNT: Saprophytes antagonistic to phytopathogenic and other microorganisms. Science (Lancaster, Pa.) **93**, 354 (1941).

ASH, R., and M. SOLIS-COHEN: Influence of quinin on proliferation and autolysis of pneumococci. J. Inf. Dis. **45**, 449 (1929).

ATKIN, E. E.: The rationale of bile solubility of pneumococcus. Brit. J. Exper. Path. **7**, 167 (1926).

AUERSWALD, H.: Welche Mikroorganismen wirken auf Diphtherie- und Pseudodiphtheriebazillen antagonistisch Zbl. Bakter. I Orig. **142**, 32 (1938).

AVERY, O. T., H. T. CHICKERING, R. COLE and A. R. DOCHEZ: Acute lobar pneumonia. Prevention and serum treatment. Monograph No 7 of the Rockefeller Institute for Medical Research, New York, 1917.

—, and G. E. CULLEN: Studies on the enzymes of pneumococcus. I. Proteolytic enzymes. J. of Exper. Med. **32**, 547 (1920a).

— — Studies on the enzymes of pneumococcus. II. Lipolytic enzymes: esterase. J. of Exper. Med. **32**, 571 (1920b).

— — Studies on the enzymes of pneumococcus. III. Carbohydrate-splitting enzymes: invertase, amylase, inulase. J. of Exper. Med. **32**, 583 (1920c).

— — Studies on the enzymes of pneumococcus. IV. Bacteriolytic enzyme. J. of Exper. Med. **38**, 199 (1923).

—, and R. J. DUBOS: The specific action of a bacterial enzyme on pneumococci of type III. Science (Lancaster, Pa.) **72**, 151 (1930).

— — The protective action of a specific enzyme against type III pneumococcus infection in mice. J. of Exper. Med. **54**, 73 (1931).

—, and J. M. NEILL: Studies on oxidation and reduction by pneumococcus. II. The production of peroxide by sterile extracts of pneumococcus. J. of Exper. Med. **39**, 357 (1924a).

— — Studies on oxidation and reduction by pneumococcus. III. Reduction of methylene blue by sterile extracts of pneumococcus. J. of Exper. Med. **39**, 543 (1924b).

— — Studies on oxidation and reduction by pneumococcus. IV. Oxidation of hemotoxin in sterile extracts of pneumococcus. J. of Exper. Med. **39**, 745 (1924c).

— — The antigenic properties of solutions of pneumococcus. J. of Exper. Med. **42**, 355 (1925).

BAKER, Z., R. W. HARRISON and B. F. MILLER: Inhibition of action of detergents by phospholipids. J. of Exper. Med. **74**, 621 (1941).

BARACH, A. L.: Factors involved in the production of immunity with pneumococcus vaccine. I. Active and passive immunity during the first seven days after the injection of antigen. J. of Exper. Med. **48**, 83, 567 (1928).

BARTHOLOMEW, J. W., and T. MITTWER: Cell structure in relation to the Gram reaction as shown during lysis of *Bacillus subtilis*. J. Gen. Microbiol. **5**, 39 (1951).

— — The Gram stain. Bacter. Rev. **16**, 1 (1952).

—, and W. W. UMBREIT: Ribonucleic acid and the Gram stain. J. Bacter. **48**, 567 (1944).

BAYER, G.: Untersuchungen über die Gallenhämolyse. I. Über die Hemmungswirkung normaler Sera. Biochem. Z. **5**, 368 (1907).

BAYLISS, M.: Visible action of sodium lauryl sulphate on microorganisms. J. Labor. a. Clin. Med. **22**, 701 (1937).

BERGAMINI, L.: Recherches sur les mucases des *Streptomyces*. Rev. belge Path. Méd. expér. **23**, 370 (1954).

— J. M. GHUYSEN et M. WELSCH: L'activité antibiotique du type «actinomycétine» chez les *Streptomyces*. C. r. Soc. Biol. Paris **148**, 733 (1954).

BERNARD, N. P., J. GUILLERM et J. GALLUT: Action d'une protéidase du vibrion cholérique sur ce même vibrion. C. r. Soc. Biol. Paris **126**, 478 (1937).

Besta, B., u. H. Kuhn: Untersuchungen über Antagonismus zwischen Diphtheriebazillen und anderen Bakterien. Z. Hyg. **116**, 520 (1934).

Beumer, J., et L. Quersin: Etude microscopique de la lyse bactériophagique par coloration des noyaux bactériens. C. r. Soc. Biol. Paris **141**, 1280 (1947).

Bigelow, G. H.: Intracutaneous reactions in lobar pneumonia. Arch. Int. Med. **29**, 221 (1922).

Bitter, R. C.: Bacterial antibiosis. J. Colorado-Wyoming Acad. Sci. **3**, 16 (1941).

Blackman, S. S. Jr.: Pneumococcal lipoïd nephrosis and the relation between nephrosis and nephritis. II. Experimental studies. Bull. Johns Hopkins Hosp. **55**, 85 (1934).

Boehncke, K. E., u. J. Mouriz-Riesgo: Zum Mechanismus der Pneumokokkenimmunität. Z. Hyg. **79**, 355 (1915).

Boivin, A., L. Mesrobeanu et I. Mesrobeanu: Extraction d'un complexe toxique et antigénique à partir du bacille d'Aertrycke. C. r. Soc. Biol. Paris **114**, 307 (1933).

— R. Vendrely et R. Tulasne: Le rôle des acides nucléiques dans la constitution et dans la vie de la cellule bactérienne, et plus généralement de touts les cellules vivantes. 6e Congr. Int. Cytol. expérim., Stockholm, p. 208, 1947.

Born, G. V. R.: The extracellular bacteriolytic enzymes of a species of *Streptomyces*. J. Gen. Microbiol. **6**, 344 (1952).

Borodulina, J. S.: Interrelations of soil *Actinomyces* and *B. mycoïdes*. Microbiologia URSS. **4**, 561 (1935).

Bounameaux, Y.: Influence de la nature du milieu de culture sur l'isolement de germes staphylolytiques et staphylostatiques à partir du sol. C. r. Soc. Biol. Paris **147**, 716 (1953).

Brotzu, G.: Über die Herstellung einer polyvalenten Pneumokokkenvaccine. (Experimentelle Untersuchungen.) Z. Hyg. **103**, 139 (1924).

Brown, A.: A study of lysis in bacteriophage-infected *Escherichia coli*. J. Bacter. **71**, 482 (1956).

Bürgers, T. J.: Über Auflösungserscheinungen an Bakterien. Zbl. Bakter. Beil. zu Abt. I, Ref. **50**, 125 (1911).

Butterfield, E. E., and F. W. Peabody: The action of pneumococcus on blood. J. of Exper. Med. **17**, 587 (1913).

Califano, L.: Action of lysozyme on the liberation of nucleic acid of bacteria. Boll. Ist. sieroter. milan. **28**, 366 (1949).

— Azione del calore sulla liberazione di acido nucleinico dei bacteri. Atti Accad. naz. Lincei **9**, 142 (1950).

Callerio, C., e C. Dolcini: Sull'azione litica di alcuni germi del gruppo del' *B. mycoïdes* su altre specie batteriche. Boll. Soc. med.-chir. Pavia **46**, 849 (1932).

Canneyt, J. van: Action du *Bacillus subtilis* et de ses sécrétions sur le bacille de la tuberculose. C. r. Soc. Biol. Paris **95**, 878 (1926).

Cantelmo, P.: Ricerche sulla liberazione da calore di acido nucleinico dei germi, in rapporto alla eta delle culture. Atti Accad. naz. Lincei **9**, 104 (1950).

Caselli, P.: Il comportamento del nucleo batterico nella lisi da lisozima. Boll. Soc. ital. Biol. sper. **23**, 1176 (1947).

—, e R. Carta: Indagini quantitative sulla perdita di materiali citoplasmatici da parte della cellula batterica durante la lisi da lisozima. Riv. Biol. **43**, 547 (1951).

Castermans, A.: Antigène streptococcique et bactériolyse par les actinomycètes. C. r. Soc. Biol. Paris **144**, 428 (1950a).

— Antigène 0 de *Salmonella typhimurium* et bactériolyse par les actinomycètes. C. r. Soc. Biol. Paris **144**, 1250 (1950b).

— Antigène 0 complet de *Salmonella typhimurium* et bactériolyse par les actinomycètes. C. r. Soc. Biol. Paris **145**, 156 (1951).

— Vaccination de la souris contre *Salmonella typhimurium* au moyen de suspensions bactériennes lysées par l'actinomycétine. C. r. Soc. Biol. Paris **148**, 176 (1954).

— Influence du phosphate et de la température sur la lyse des salmonelles par l'actinomycétine. C. r. Soc. Biol. Paris **149**, 204 (1955).

— Activités antigéniques des salmonelles lysées par les actinomycètes ou l'actinomycétine. Rev, belge Path. Méd. expér. **25**, 5 (1956).

Cavallo. G., et P. Cantelmo: Induction par les rayons ultra-violets de la lyse bactériophagique des staphylocoques lysogènes. C. r. Soc. Biol. Paris **145**, 1419 (1951).

CLOUGH, P. W.: Some observations on hypersensitiveness to pneumococcus protein, with special reference to its relation to immunity. Bull. Johns Hopkins Hosp. **26**, 37 (1915).
COLE, R. I.: Toxic substances produced by pneumococcus. J. of Exper. Med. **16**, 644 (1912).
— Pneumococcus hemotoxin. J. of Exper. Med. **20**, 346 (1914a).
— The production of methemoglobin by pneumococci. J. of Exper. Med. **20**, 363 (1914b).
COLOMBO, C.: Temperatura di incubazione e decorso della lisi della colonie batteriche da amino-acidi et da estratti batterici. Boll. Soc. ital. Biol. sper. **23**, 212 (1947).
COPE, E. J., and K. M. HOWELL: Local skin reactivity to "filtrates" of pneumococcus. J. Inf. Dis. **48**, 570 (1931).
COTONI, L., et N. CHAMBRIN: Recherches sur l'hémolysine pneumococcique. Ann. Inst. Pasteur **42**, 1536 (1928).
CUMMINS, C. S., and H. HARRIS: The chemical composition of the cell-wall in some gram-positive bacteria and its possible value as a taxonomic character. J. Gen. Microbiol. **14**, 583 (1956).
CUMMINS, S. L., and C. WEATHERALL: The retardation of lytic processes by colloïdal silica solutions. Brit. J. Exper. Path. **12**, 239 (1931).
CUTINELLI, C.: Lytic action of the bile on bacterial nuclei. Riv. Ist. sieroter. ital. **24**, 157 (1949).
—, e N. LA MANNA: Modificazioni morfologiche et chimiche dello pneumococco per azione del lisozima. Boll. Soc. ital. Biol. sper. **23**, 1157 (1947).
— — Comportamento del nucleo dello pneumococco nella lisi da lisozima. Riv. Ist. sieroter. ital. **25**, 1 (1950).
DATH, S.: Sur l'action bactériolytique du *Streptothrix*. C. r. Soc. Biol. Paris **97**, 1194 (1927).
DAWSON, I. M., I. LOMINSKI and H. STERN: An electron-microscope study of the action of cetyltrimethylammonium bromide on *Staphylococcus aureus*. J. of Path. **66**, 513 (1953).
DAY, H. B.: The preparation of antigenic specific substance from streptococci and pneumococci (Type I). Brit. J. Exper. Path. **11**, 164 (1930).
— The preparation of pneumococcal species antigen. J. of Path. **38**, 171 (1934).
DELAUNAY, A.: Action bactéricide et bactériolytique *in vitro* des filtrats de culture du *B. subtilis* sur la bactéridie charbonneuse. Etude macroscopique et microscopique. Revue d'Immunol. **12**, 222 (1948).
DELMOTTE, N.: Action du lauryl-sulfate et du lauryl-sulfonate de soude sur l'aspect microscopique du colibacille. C. r. Soc. Biol. Paris **147**, 173 (1953a).
— Mise en évidence des noyaux bactériens par traitement des frottis au lauryl-sulfate ou lauryl-sulfonate de soude. Antonie van Leeuwenhoek **19**, 117 (1953b).
DEUSSEN, E.: Die Gramsche Bakterienfärbung ihr Wesen und ihre Bedeutung. Biochem. Z. **103**, 123 (1920).
DMITRIEFF, S., et G. SOUTEEFF: Sur les phénomènes de dissociation et de lyse observés dans les cultures de l'*Actinomyces bovis* Bostroem. Essais d'application des filtrats de cultures lysées au traitement de l'actinomycose. Ann. Inst. Pasteur **56**, 470 (1936).
DOERMANN, A. H.: The intracellular growth of bacteriophage. I. Liberation of intracellular bacteriophage T 4 by premature lysis with another phage or with cyanide. J. Gen. Physiol. **35**, 645 (1952).
DOWNIE, A. W., L. STENT and S. M. WHITE: The bile solubility of pneumococcus, with special reference to the chemical structure of various bile salts. Brit. J. Exper. Path. **12**, 1 (1931).
DRESEL, E. G.: Bakteriolyse durch Fettsäuren und deren Abkömmlinge. Zbl. Bakter. I Orig. **97**, 178 (1926).
DUBOS, R. J.: Factors affecting the yield of specific enzyme in cultures of the bacillus decomposing the capsular polysaccharide of type III pneumococcus. J. of Exper. Med. **55**, 377 (1932).
— Studies on the mechanism of production of a specific bacterial enzyme which decomposes the capsular polysaccharide of type III pneumococcus. J. of Exper. Med. **62**, 259 (1935).
— The autolytic system of pneumococci. J. of Exper. Med. **65**, 873 (1937a).
— Mechanism of the lysis of pneumococci by freezing, thawing, bile and other agents. J. of Exper. Med. **66**, 101 (1937b).
— The effect of the bacteriolytic enzyme of pneumococcus upon the antigenicity of encapsulated pneumococci. J. of Exper. Med. **66**, 113 (1937c).

Dubos, R. J.: The decomposition of yeast nucleic acid by a heat-resistant enzyme. Science (Lancaster, Pa.) **85**, 549 (1937d.)
— The effect of formaldehyde on pneumococci. J. of Exper. Med. **67**, 389 (1938a).
— Immunization of experimental animals with a soluble antigen extracted from pneumococci. J. of Exper. Med. **67**, 799 (1938b).
— Studies on a bactericidal agent extracted from a soil bacillus. I. Preparation of the agent, its activity *in vitro*. J. of Exper. Med. **70**, 1 (1939a).
— Enzymatic analysis of the antigenic structure of pneumococci. Erg. Enzymforsch. **135**, (1939b).
— The bacterial cell. Cambridge, Massachusetts: Harvard University Press 1945.
—, and O. T. Avery: Decomposition of the capsular polysaccharide of pneumococcus type III by a bacterial enzyme. J. of Exper. Med. **54**, 51 (1931).
—, and J. H. Bauer: The use of graded collodion membranes for the concentration of a bacterial enzyme capable of decomposing the capsular polysaccharide of type III pneumococcus. J. of Exper. Med. **62**, 271 (1935).
—, and R. D. Hotchkiss: The production of bactericidal substances by aerobic sporulating bacilli. J. of Exper. Med. **73**, 629 (1941).
—, and C. M. McLeod: The effect of a tissue enzyme upon pneumococci. J. of Exper. Med. **67**, 791 (1938).
—, and R. H. S. Thompson: The decomposition of yeast nucleic acid by a heat-resistant enzyme. J. of Biol. Chem. **124**, 501 (1938).
Duran-Reynals, F.: Bactériophage et microbes tués. C. r. Soc. Biol. Paris **94**, 242 (1926).
Epstein, L. A., and E. Chain: Some observations on the preparation and properties of the substrate of lysozyme. Brit. J. Exper. Path. **21**, 339 (1940).
Falk, I. S., and M. A. Jacobson: Studies on respiratory diseases. XXVIII. The electrophoretic potentials, the dissolution and the serum agglutination of pneumococci in the presence of sodium oleate. J. Inf. Dis. **38**, 188 (1926).
—, et S. Y. Yang: Studies on respiratory diseases. XXV. The influence of certain electrolytes and non electrolytes on the bile solubility of pneumococci. J. Inf. Dis. **38**, 1 (1926a).
— — Studies on respiratory diseases. XXVI. The lysis of pneumococci by sodium oleate. J. Inf. Dis. **38**, 8 (1926b).
Few, A. V.: Electron microscopy of disrupted bacteria treated with polymyxin E. J. Gen. Microbiol. **10**, 304 (1954).
Fildes, P., and J. McIntosh: An improved form of McIntosh and Fildes anaerobic jar. Brit. J. Exper. Path. **2**, 153 (1921).
Fisher, A. M.: A study of the mechanism of action of penicillin as shown by its effect on bacterial morphology. J. Bacter. **52**, 539 (1946).
Fleming, A.: On a remarkable bacteriolytic element found in tissues and secretions. Proc. Roy. Soc. Lond. Ser. B **93**, 306 (1922).
— On the antibacterial action of cultures of a *Penicillium*, with special reference to their use in the isolation of *B. influenzae*. Brit. J. Exper. Path. **10**, 226 (1929).
— Penicillin. Its practical application. London: Butterworth & Co. 1946.
Fong, J., and A. P. Krueger: The lytic action of tyrothricin and its derivatives on *Staphylococcus aureus*. J. Gen. Physiol. **33**, 311 (1950).
Francis jr., T., E. E. Terrell, R. J. Dubos and O. T. Avery: Experimental Type III pneumococcus pneumonia in monkeys. II. Treatment with an enzyme which decomposes the specific capsular polysaccharide of pneumococcus Type III. J. of Exper. Med. **59**, 641 (1954).
Franke, H., u. A. Ismet: Über Cytolyse. Zbl. Bakter. I Orig. **99**, 570 (1926).
Gale, E. F., and E. S. Taylor: The assimilation of amino-acids by bacteria. II. The action of tyrocidin and some detergent substances in releasing amino-acids from the internal environment of *Streptococcus faecalis*. J. Gen. Microbiol. **1**, 76 (1947).
Gasperini, G.: Recherches morphologiques et biologiques sur un microorganisme de l'atmosphère, *Streptothrix foersteri*. Cohn. Ann. Microgr. **10**, 449 (1890).
Ghuysen, J. M.: L'activité staphylolytique de l'actinomycétine. C. r. Soc. Biol. Paris **146**, 1268 (1952a).
— Analyse et purification de l'actinomycétine par adsorption. C. r. Soc. Biol. Paris **146**, 1812 (1952b).

GHUYSEN. J. M.: Influence des ions bivalents et de la température sur la production et les activités des principes bactériolytiques de l'actinomycétine. C. r. Soc. Biol. Paris **147**, 1502 (1953a).
— Dualité des principes bactériolytiques de l'actinomycétine. Arch. internat. Physiol. et Biochim. 259 (1953b).
— Activité pneumolytique de l'actinomycétine. C. r. Soc. Biol. Paris **148**, 729 (1954a).
— Concentration et purification des principes bactériolytiques de l'actinomycétine par chromatographie sur corps microbiens. C. r. Soc. Biol. Paris **148**, 1283 (1954b).
— Epilage enzymatique par des filtrats de culture de *Streptomyces* spp. 27e Congr. int. Chim. ind., Bruxelles, vol. 3, p. 760, 1954c.
— Action de l'actinomycétine sur le fibrinogène et la fibrine. C. r. Soc. Biol. Paris **148**, 1694 (1954d).
— Activités bactériolytiques de l'actinomycétine de *Streptomyces albus* G. Arch. internat. Physiol. et Biochim. **65**, 173 (1957).
—, et G. LÉGER: Activités protéolytiques et épilantes de l'actinomycétine. C. r. Soc. Boil. Paris **148**, 1691 (1954).
—, and M. WELSCH: On the staphylolytic action of actinomycetin. J. Gen. Microbiol. **7**, iv (1952).
— — Les activités bactériolytiques de l'actinomycétine. 6e Congr. int. Microbiol., Rome, Atti, vol. 1, p. 599, 1953.
GOEBEL, W. F., and O. T. AVERY: A study of pneumococcus autolysis. J. of exper. Med. **49**, 267 (1929).
GOODNER, K.: The effect of pneumococcus autolysates upon pneumococcus dermal infection in the rabbit. J. of Exper. Med. **58**, 153 (1933).
—, and R. J. DUBOS: Studies on the quantitative action of a specific enzyme in Type III pneumococcus infection in rabbits. J. of Exper. Med. **56**, 521 (1932).
— — and O. T. AVERY: The action of a specific enzyme upon the dermal infection of rabbit with Type III pneumococcus. J. of Exper. Med. **55**, 393 (1932).
—, and F. L. HORSFALL jr.: The purpuric reaction produced in animals by derivatives of the pneumococcus. Proc. Soc. Exper. Biol. a. Med. **37**, 178 (1937).
GORDON, J., and R. A. HALL: The sensitivity of glycine-resistant organisms to the bactericidal action of serum. J. of Path. **62**, 274 (1950).
— — et L. H. STICKLAND: A comparison of the degree of lysis by glycine of normal and glycine-resistant organisms. J. of Path. **61**, 581 (1949).
— — — Observations on resistance of *Bacterium coli* to glycine. J. of Path. **63**, 285 (1951).
— — — The inhibition by various agents of the lysis of *Bacterium coli* by glycine. J. of Hyg. **51**, 215 (1953).
— — — The lysis of *Bacterium coli* by amino-acids. J. of Hyg. **52**, 510 (1954).
GRATIA, A.: L'action lytique des staphylocoques vivants sur les staphylocoques tués. Au sujet d'une note récente de WOLLMAN et DURAN-REYNALS. C. r. Soc. Biol. Paris **95**, 734 (1926).
— Sur l'immunité antistaphylococcique par les mycolysats. C. r. Soc. Biol. Paris **104, 1058** (1930a).
— Le traitement des infections à staphylocoques par les bactériophages et les mycolysats staphylococciques. Bull. Mém. Soc. nat. Chir. Paris **56, 344** (1930b).
— Mise au point de quelques notions de bactériophagie. Brux. méd. **11**, 697 (1931).
— Antagonisme microbien et bactériophagie. Ann. Inst. Pasteur **48**, 413 (1932).
— La dissolution des bactéries et ses applications pratiques. Bull. Acad. roy. Méd. Belg. **14**, 285 (1934).
— Le phénomène du halo et la synergie des bactériophages. C. r. Soc. Biol. Paris **126**, 418 (1937).
—, et J. ALEXANDER: Sur la mycolyse par le *Streptothrix* de divers microbes pathogènes. C. r. Soc. Biol. Paris **106**, 1288 (1931).
—, et S. DATH: Propriétés bactériolytiques de certaines moisissures. C. r. Soc. Biol. Paris **91**, 1442 (1924).
— — Moisissures et microbes bactériophages. C. r. Soc. Biol. Paris **92**, 461 (1925a).
— — De l'action bactériolytique des *Streptothrix*. C. r. Soc. Biol. Paris **92**, 1125 (1925b).
— — A propos de l'action bactériolytique du *Streptothrix*. C. r. Soc. Biol. Paris **93**, 451 (1925c).

Gratia, A., et S. Dath: Propriétés bactériolytiques des *Streptothrix*. C. r. Soc. Biol. Paris **94**, 1267 (1926).
—, et P. Fredericq: Diversité des souches antibiotiques de *B. coli* et étendue variable de leur champ d'action. C. r. Soc. Biol. Paris **140**, 1032 (1946).
—, et B. Rhodes: Action du principe lytique sur les émulsions de staphylocoques vivants et de staphylocoques tués. C. r. Soc. Biol. Paris **89**, 1171 (1923).
— — De l'action lytique des staphylocoques vivants sur les staphylocoques tués. C. r. Soc. Biol. Paris **91**, 640 (1924).
— — The Twort-d'Herelle phenomenon. Lancet **1926**, 204.
Green, M., and M. A. Stahmann: Synthesis and enzymatic hydrolysis of glutamic acid polypeptides. J. of Biol. Chem. **197**, 771 (1952).
Greenberg, R. A., and H. O. Halvorson: Studies on an autolytic substance produced by an aerobic spore-forming bacterium. J. Bacter. **69**, 45 (1955).
Gros, F., et M. Macheboeuf: Recherches chimiques sur les antibiotiques: Tyrothricine. C. r. Soc. Biol. Paris **142**, 738 (1948).
Grula, E. A., and S. E. Hartsell: Lysozyme and morphological alterations induced in *Micrococcus lysodeikticus*. J. Bacter. **68**, 171 (1954a).
— — Lysozyme action and its relation to the Nakamura effect. J. Bacter. **68**, 302 (1954b).
Guelin, A.: Lyse bactérienne provoquée par une souche de *B. coli*. Ann. Inst. Pasteur **69**, 382 (1943).
Gundel, M.: Über Bakterienlipoïde. Zbl. Bakter. I Orig. **116**, 81 (1930).
—, u. W. Wagner: Weitere Studien über Bakterienlipoïde. Z. Immunforsch. **69**, 63 (1930).
Harkins, M. J.: Studies on the pneumococcolytic activity of the bile from experimental bacterial cholecystitis of rabbits. J. Labor. a. Clin. Med. **10**, 716 (1925).
Harley, D.: The immunizing action of extracts of pneumococci (Types 1 and 2) in mice and rabbits. Brit. J. Exper. Path. **15**, 161 (1934).
— Species immunity to pneumococcus. Brit. J. Exper. Path. **16**, 14 (1935).
Henry, H., et M. Stacey: Histochemistry of the gram staining reaction for microorganisms. Nature (Lond.) **151**, 671 (1943).
— — and E. G. Teece: Nature of the Gram-positive complex in microorganisms. Nature (Lond.) **156**, 720 (1945).
Hettche, H. O., u. B. Weber: Die Ursache der bakteriziden Wirkung von Mesentericus-filtraten. Arch. f. Hyg. **123**, 69 (1939).
Hewitt, J. A. W., et L. W. Famulener: The hemolytic properties of the pneumococcus. Proc. Soc. Exper. Biol. a. Med. **19**, 368 (1922).
Hirschfelder, J. O.: The production of active and passive immunity to the pneumococcus with a soluble vaccine. A preliminary report. J. Amer. Med. Assoc. **59**, 1373 (1912).
Hirst, G. K.: The effect of a polysaccharide-splitting enzyme on streptococcal infection. J. of Exper. Med. **73**, 493 (1941).
Holdsworth, E. S.: The nature of the cell-wall of *Corynebacterium diphtheriae*. Biochim. et Biophysica Acta 8, 110 (1952).
Horder, T., and N. S. Ferry: A search for an ideal antigen for therapeutic immunization. Brit. Med. J. **1926**, 177.
Hotchkiss, R. D.: Gramicidin, tyrocidin and tyrothricin. Adv. Enzymol. **4**, 153 (1944).
— The nature of the bactericidal action of surface active agents. Ann. New York Acad. Sci. **46**, 479 (1946).
Humphries, J. C.: Enzymic activity of bacteriophage culture lysates. I. A capsule lysin active against *Klebsiella pneumoniae* type A. J. Bacter. **56**, 683 (1948).
Huppert, J., et J. Panijel: Influence des conditions de préparation de la suspension bactérienne sur la sensibilité de *E. coli* Fb tué à l'action d'une substance lytique non reproductible en série (lysine). C. r. Acad. Sci. Paris **238**, 1163 (1954).
Hutter, S.: Les substances antibiotiques du *Penicillium notatum* Westling. Contribution à l'étude de la notatine et mise en évidence d'un principe bactériolytique nouveau: la notalysine. Schweiz. med. Wschr. **1945**, 411.
— Nouvelles recherches sur la notalysine, principe distinct de la pénicilline. Schweiz. med. Wschr. **1946**, 1138.
— Rôle d'une antipénicillinase dans les phénomènes attribués à la notalysine. C. r. Soc. Biol. Paris **142**, 560 (1948).

HUTTER, S.: Un inhibiteur de la pénicillinase dans les filtrats de *Penicillia* producteurs de pénicilline. Biochim. et Biophysica Acta **4**, 572 (1950).

IMSHENETSKIY, A. A., i L. A. KUZYURINA: Bacteriotrophic microorganisms. The evolution of cannibalism and parasitism. Microbiologia URSS. **20**, 3 (1951).

JAUMAIN, D.: Autolyse microbienne en tubes scellés. C. r. Soc. Biol. Paris **87**, 790 (1922).

JENNINGS, M. A.: Inédit (1944). Cité d'après FLOREY, H. W., E. CHAIN, N. G. G. HEATLEY, M. A. JENNINGS, A. G. SANDERS, E. P. ABRAHAM et M. E. FLOREY, Antibiotics, vol. 2, p. 1152. Oxford: University Press 1949.

JOBLING, J. W., and S. STROUSE: Studies on ferment action. Immunization with proteolytic clivage products of pneumococci. J. of Exper. Med. **16**, 860 (1912).

— — Studies on fermentation. IX. A note on the relation between lysis and proteolysis of pneumococci. J. of Exper. Med. **18**, 597 (1913).

JOHNSON, F. H.: Observations on the electron microscopy of *B. cereus* and tyrothricin action. J. Bacter. **47**, 551 (1944).

JONES, A. S., P. W. MUGGLETON and M. STACEY: The Gram-complex in *Clostridium welchii*. Nature (Lond.) **166**, 650 (1950).

— A. J. STACEY and M. WEBB: Studies on the autolytic system of gram-positive micro-organisms. I. The lytic system of staphylococci. Biochim. et Biophysica Acta **3**, 383 (1949).

— A. J. SWALLOW and M. WEBB: The exocellular bacteriolytic system of soil actinomycetes. 1. The nature and properties of the lytic system. Biochim. et Biophysica Acta **2**, 167 (1948).

JULIANELLE, L. A., and H. A. REIMANN: The production of purpura by derivatives of pneumococcus. I. General considerations of the reaction. J. of Exper. Med. **43**, 87 (1926).

— — The production of purpura by derivatives of pneumococcus. III. Further studies on the nature of the purpura-producing principle. J. of Exper. Med. **45**, 609 (1927).

KATZNELSON, H.: Autolysis of a thermophilic *Actinomyces*. Soil Sci. **49**, 83 (1940).

KELLY, F. B.: On the solution of pneumococci by bile. Amer. J. Publ. Health **10**, 708 (1920).

KIMMELSTIEL, P.: Über eine biologische Eigenschaft eines Wurzbazillus. Zbl. Bakter. I Orig. **89**, 113 (1923).

— Weitere Versuche über die bakteriolytischen Fähigkeiten des *B. mycoïdes*. Med. Klin. **1924**, 419.

KLEIN, S. J.: Studies on the solubility of pneumococcus in saponine. II. Sensitization by ergosterol. J. Bacter. **26**, 215 (1933).

—, et F. M. STONE: The lysis of pneumococcus by saponine. J. Bacter. **22**, 387 (1931).

KLINGE, K.: Die serologische Gruppenbestimmung der Streptokokken durch fermentative Antigenbereitung. Z. Hyg. **14**, 402 (1956).

KOZLOWSKI, A.: Comparative studies of the action on the pneumococcus of bile acids and unsaturated fatty acids found in bile in the form of soaps. J. of Exper. Med. **42**, 453 (1925).

KRASSILNIKOV, N. A.: The phenomenon of autolysis in actinomycetales. I. Cultural and morphological picture of autolysis. Microbiologia URSS. **7**, 708 (1938).

—, i A. I. KORENIAKO: The phenomenon of autolysis in actinomycetales. II. Influence of environmental conditions upon autolysis of actinomycetes and proactinomycetes. Microbiologia URSS. **7**, 829 (1938).

— — The bactericidal substance of the actinomycetes. Microbiologia URSS. 8, 673 (1939).

KREAM, J., B. A. BOREK, C. J. DI GRADO and M. BOVARNICK: Enzymatic hydrolysis of glutamyl polypeptide and its derivatives. Arch. of Biochim. a. Biophysics **53**, 333 (1954).

KRISS, A. E.: On the lysozyme of *Actinomycetes*. Microbiologia URSS. **9**, 32 (1940).

LAMANNA, C., and F. MALLETTE: Basic bacteriology, p. 122—135. Baltimore: Williams & Wilkins Co. 1953.

LAMAR, R. V.: Chemo-immunological studies on localized infections. I. Action on the pneumococcus and its experimental infections of combined oleate and antipneumococcus serum. J. of Exper. Med. **13**, 1 (1911a).

— Chemo-immunological studies on localized infections. III. Some further observations upon the action of certain soaps on the pneumococcus and its experimental infections. J. of Exper. Med. **14**, 256 (1911b).

LANCEFIELD, R. C.: A serological differentiation of human and other groups of hemolytic streptococci. J. of Exper. Med. **57**, 571 (1933).
— Studies on the antigenic composition of group A hemolytic streptococci. I. Effects of proteolytic enzymes on streptococcal cells. J. of Exper. Med. **78**, 465 (1943).
LEVADITI, C., et J. HENRY: Pénicilline et autolysine. C. r. Soc. Biol. Paris **140**, 734 (1946).
—, et A. VAISMAN: Mécanisme de la lyse pénicillique. Bull. Acad. Méd. Paris **129**, 564 (1945).
LEVY, R.: Differentialdiagnostische Studien über Pneumokokken und Streptokokken. Virchows Arch. **187**, 327 (1907).
LEY, J. DE: Cytochemical observations on nucleic acids and nucleoproteins in nitrogen deficient microorganisms. Antonie van Leeuwenhoek **17**, 15 (1951).
LIBMAN, E., and J. ROSENTHAL: The action of bile on the pneumococcus, streptococcus and *Streptococcus mucosus*. Proc. New York Path. Soc. 8, 40 (1908).
LIESKE, R.: Morphologie und Biologie der Strahlenpilze. Leipzig: Gebr. Boerntrager 1921.
LINZ, R.: Lyse chloroformique et antigène somatique. Ann. Inst. Pasteur **75**, 544 (1948).
LONGFELLOW, D., S. G. CARY and L. R. KUHN: Use of Maxted's streptomyces filtrate for rapidly extracting group carbohydrates from hemolytic streptococci. Amer. J. Clin. Path. **19**, 880 (1949).
LORD, F. T., and R. N. NYE: The relation of the pneumococcus to hydrogen-ion concentration, acid death-point and dissolution of the organism. J. of Exper. Med. **30**, 389 (1919).
— — Studies on the pneumococcus. II. Dissolution of pneumococci at varying hydrogen-ion concentrations. Effect of temperature, previous killing of the organisms, and fresh human serum on the phenomenon. Behavior of other organisms. J. of Exper. Med. **35**, 689 (1922a).
— — Studies on the pneumococcus. III. Dissolution of pneumococci in pneumonic cellular material at varying hydrogen-ion concentrations. Resistance of certain other organisms to dissolution. J. of Exper. Med. **35**, 699 (1922b).
— — Studies on the pneumococcus. IV. Effect of bile at varying hydrogen-ion concentrations on dissolution of pneumococci. J. of Exper. Med. **35**, 703 (1922c).
LORIZIO, V., e R. RISMONDO: Alterazioni morfologiche e colturali del *B. pneumoniae* di Friedländer da amino-acidi e da estratti di *subtilis*. Giorn. Batter **37**, 53 (1947).
LWOFF, A., L. SIMINOVITCH et N. KJELDGAARD: Induction de la production de bactériophage chez une bactérie lysogène. Ann. Inst. Pasteur **79**, 815 (1950).
MACKENZIE, G. M., and S. T. WOO: The production and significance of cutaneous allergy to pneumococcus protein. J. of Exper. Med. **41**, 65 (1925).
MACULLA, E. S., and P. B. COWLES: The use of glycine in the disruption of bacterial cells. Science (Lancaster, Pa.) **107**, 376 (1948).
MCCARTY, M.: The lysis of group A hemolytic streptococci by extracellular enzymes of *Streptomyces albus*. I. Production and fractionation of the lytic enzymes. J. of Exper. Med. **96**, 555 (1952a).
— The lysis of group A hemolytic streptococci by extracellular enzymes of *Streptomyces albus*. II. Nature of the cellular substrate attacked by the lytic enzymes. J. of Exper. Med. **96**, 569 (1952b).
MCLEAN, D.: The in vivo decapsulation of streptococci by hyaluronidase. J. of Path. **54**, 284 (1942).
MAGARAO, M. F., A. ARRIAGADA et S. THALES: Lytic action of *Bacillus subtilis* on *Mycobacterium tuberculosis*. Rev. Braz. Med. **1944**, 556.
MAIR, W.: The purpura-producing substance in pneumococcus and the heritable susceptibility of mice. J. of Path. **31**, 215 (1928).
— The Pneumococcus, in: A system of bacteriology in relation to medicine. Med. Res. Council, H. M. S. Office **2**, 164 (1929).
MALONE, R. H.: An examination of the factors concerned in Neufeld's solubility test. Indian. J. Med. Res. **11**, 867 (1924a).
— Degree of bile solubility as a means of differentiating pathogenic from non-pathogenic capsulated diplococci. Indian. J. Med. Res. **11**, 877 (1924b).
MANDELBAUM, M.: Über die Wirkung von taurocholsaurem Natrium und tierischer Galle auf den Pneumokokkus, *Streptococcus mucosus* und auf die andern Streptokokken. Münch. med. Wschr. **54**, 1431 (1907).

MASERA, A.: Potenziale ossido-reduttivo del terreno e decorso della lisi delle colonie batteriche da amino-acidi e da estratti batterici. Boll. Soc. ital. Biol. sper. **23**, 209 (1947).

MAXTED, W. R.: Preparation of streptococcal extracts for Lancefield grouping. Lancet **1948**, 255.

— Enhancement of streptococcal bacteriophage lysin by hyaluronidase. Nature (Lond.) **170**, 1020 (1952).

MEYER, H.: Über aktive Immunisierung von Mäusen durch in Natrium taurocholicum gelöste Pneumokokken. Z. Hyg. **107**, 416 (1927).

—, u. W. W. SUKNEFF: Über die immunisierende Wirkung von avirulenten Pneumokokken und von Kulturfiltraten und Auflösungen virulenter Pneumokokken. Z. Hyg. **109**, 78 (1928).

MEYER, K., R. J. DUBOS et E. M. SMYTH: Action of the lytic principle of pneumococcus on certain tissue polysaccharides. Proc. Soc. Exper. Biol. a. Med. **34**, 816 (1936).

— — — The hydrolysis of the polysaccharide acids of vitreous humor, of umbilical cord, and of streptococcus by the autolytic enzyme of pneumococcus. J. of Biol. Chem. **118**, 71 (1937).

—, and W. T. J. MORGAN: The relationship between the heterophile hapten and the specific polysaccharide hapten of the smooth form of bact. Shigae. Brit. J. Exper. Path. **16**, 476 (1935).

—, T. W. PALMER, R. THOMPSON and D. KHORAZO: On the mechanism of lysozyme action. J. of Biol. Chem. **113**, 479 (1936).

MITCHELL, P. D., and G. R. CROWE: A note on electron micrographs of normal and tyrocidin-lysed streptococci. J. Gen. Microbiol. **1**, 85 (1948).

—, and J. MOYLE: The glycerophospho-protein complex envelope of *Micrococcus pyogenes*. J. Gen. Microbiol. **5**, 981 (1951).

— — The Gram-reaction and cell composition: nucleic acids and other phosphate fractions. J. Gen. Microbiol. **10**, 533 (1954).

MONOD, J.: Sur un phénomène de lyse lié à l'inanition carbonée. Ann. Inst. Pasteur **68**, 444 (1942).

MORGAN, H. J., and J. M. NEILL: Methemoglobin formation by sterile culture filtrates of pneumococcus. J. of Exper. Med. **40**, 269 (1924).

MORGAN, W. T. J., and A. C. THAYSEN: Decomposition of specific bacterial polysaccharides by a species of Myxobacterium. Nature (Lond.) **132**, 604 (1933).

MÜCH, H.: Über den *B. cytolyticus*. Münch. med. Wschr. **1925**, 374.

—, u. P. SARTORIUS: Über die neuartige Lysine des *Mycoïdes* „Müch". Med. Klin. **1924**, 345.

MUGGLETON, P. W., and M. WEBB: The exocellular bacteriolytic system of soil actinomyces. II. A further investigation of the lytic system. Biochim. et Biophysica Acta **8**, 431 (1952a).

— — The exocellular bacteriolytic system of soil actinomyces. III. The separation and characterization of the proteolytic enzymes. Biochim. et Biophysica Acta **8**, 526 (1952b).

NEILL, J. M.: Studies on oxidation and reduction of immunological substances. I. Pneumococcus hemotoxin. J. of Exper. Med. **44**, 199 (1926).

—, and O. T. AVERY: Studies on oxidation and reduction by pneumococcus. V. The destruction of oxyhemoglobin by sterile extracts of pneumococcus. J. of Exper. Med. **39**, 757 (1924a).

— — Studies on oxidation and reduction by pneumococcus. VI. The oxidation of enzymes in sterile extracts of pneumococcus. J. of Exper. Med. **40**, 405 (1924b).

— — Studies on oxidation and reduction by pneumococcus. VII. Enzyme activity of sterile filtrates of aerobic and anaerobic cultures of pneumococcus. J. of Exper. Med. **40**, 423 (1924c).

— — Studies on oxidation and reduction by pneumococcus. VIII. Nature of the oxidation-reduction systems in sterile pneumococcus extracts. J. of Exper. Med. **41**, 285 (1925).

—, and W. L. FLEMING: Studies on the oxidation and reduction of immunological substances. VI. The "reactivation" of the bacteriolytic activity of oxidized pneumococcus extracts. J. of Exper. Med. **46**, 263 (1927).

— — and E. L. GASPARI: Studies on the oxidation and reduction of immunological substances. VIII. The antigenic properties of hemolytically active and hemolytically inactive modifications of pneumococcus hemotoxin. J. of Exper. Med. **46**, 735 (1927a).

NEILL. J. M., W. L. FLEMING and E. L. GASPARI: Studies on the oxidation and reduction of immunological substances. X. Immunological distinctions between the hemotoxin and the "protein fraction" of the pneumococcus cell. J. of Exper. Med. **46**, 777 (1927b).

NEUFELD, F.: Über eine specifische bakteriolytische Wirkung der Galle. Z. Hyg. **34**, 454 (1900).

—, u. R. ETINGER-TULCZYNSKA: Untersuchungen zur Gallenlösung der Pneumokokken. Arch. f. Hyg. **103**, 107 (1930).

—, u. R. SCHNITZER: Pneumokokken, in W. KOLLE, R. KRAUSS u. P. UHLENHUTH, Handbuch der pathogenen Mikroorganismen, 3e ed., Vol. 4, p. 913. Jena: Gustav Fischer; Berlin u. Wien: Urban & Schwarzenberg 1928.

NEWTON, B. A.: The release of soluble constituents from washed cells of *Pseudomonas aeruginosa* by the action of polymyxin. J. Gen. Microbiol. **9**, 54 (1953).

— A fluorescent derivative of polymyxin: its preparation and use in studying the site of action of the antibiotic. J. Gen. Microbiol. **12**, 226 (1955).

— The properties and mode of action of the polymyxins. Bacter. Rev. **20**, 14 (1956).

NICOLLE, M.: Action du *Bacillus subtilis* sur diverses bactéries. Ann. Inst. Pasteur **21**, 613 (1907).

—, et ADIL BEY: Action de la bile sur le pneumocoque et diverses autres bactéries. Ann. Inst. Pasteur **21**, 20 (1907).

NORDBERG, B. K., and W. THORSELL: The effect of certain enzyme systems on the capsule of *Bacillus anthracis*. J. Bacter. **69**, 367 (1955).

OLIVIER, H. R.: Biological properties of *Bacillus subtilis*. Nature (Lond.) **157**, 238 (1946).

—, et P. DRUTEL: Etude comparée de l'action exercée sur le bacille de Koch homogène par la pénicilline, la streptomycine et l'endosubtilysine. C. r. Soc. Biol. Paris **141**, 1002, 1040 (1947).

—, and L. DE SAINT RAT: Action du bacille subtil et de l'endosubtilysine sur le bacille de Koch (souche ARLOING et COURMONT) et dans la tuberculose expérimentale. Rev. Tbc. **10**, 50 (1946).

— — P. BONÈT-MAURY et E. BLANCHON: Etude de l'action biologique de *Bacillus subtilis*. Bull. Acad. Méd. Paris **129**, 358 (1945).

— — — — Pouvoir lytique du *Bacillus subtilis* sur le colibacille, les bacilles typhique et paratyphiques. C. r. Soc. Biol. Paris **139**, 464 (1945).

OSTER, G.: Fluorescence de l'auramine 0 en présence d'acide nucléique. C. r. Acad. Sci. Paris **232**, 1708 (1951).

PANIJEL, J., et J. HUPPERT: Spécificité bactérienne et mode d'action de la substance lytique non reproductible en série (lysine) accompagnant la reproduction d'un bactériophage. C. r. Acad. Sci. Paris **238**, 1452 (1954).

— — Recherches sur les prolysines. La prolysine du phage Fcz. Ann. Inst. Pasteur **90**, 619 (1956).

PARKER, J. T.: The production of skin necrosis by certain autolysates of pneumococcus (Types I and II). J. of Exper. Med. **47**, 531 (1928).

— Differentiation between some toxic substances in anaerobically produced autolysates of pneumococci (Types I and II). J. of Exper. Med. **49**, 695 (1929a).

— Experimental pneumonia in guinea-pigs. II. Effect of anti-autolysates sera on pneumococcus pneumonia in guinea-pigs. J. of Exper. Med. **50**, 161 (1929b).

—, and M. VAN S. MCCOY: The production and titration of potent horse anti-pneumotoxin. J. of Exper. Med. **50**, 103 (1929).

—, and A. M. PAPPENHEIMER: Experimental pneumonia in guinea-pigs. I. The effect of certain toxic autolysates of pneumococci. J. of Exper. Med. **48**, 695 (1928).

PAULI, P.: Sulla importanza dell'acqua ossigenata nel metabolismo del pneumococco. Boll. Ist. sierot. milan **6**, 225 (1927).

PETERSON, R. G., and S. E. HARTSELL: The lysozyme spectrum of the gram-negative bacteria. J. Inf. Dis. **96**, 75 (1955).

PETRONELLI, A.: Modificazione della colorabilita indotte del lisozima su alcune specie microbie. Boll. Ist. sieroter. milan. **29**, 252 (1950).

— Una semplice tecnica di evidenziazione dei nuclei nei microbi. Boll. Ist. sieroter. milan. **30**, 499 (1951).

PETROVANU, G.: L'action de l'eau oxygénée sur quelques groupes de microbes. Catalases microbiennes. Phénomènes d'autolyse. C. r. Soc. Biol. Paris **92**, 459 (1925).

PIEKARSKI, G.: Cytologische Untersuchungen an Paratyphus- und Coli-Bakterien. Arch. Mikrobiol. **8**, 428 (1937).

PITTMAN, M., and I. S. FALK: Studies on respiratory diseases. XXXIV. Some relations between extracts, filtrates and virulence of pneumococci. J. Bacter. **19**, 327 (1930).

—, and M. A. SOUTHWICK: Studies on respiratory diseases. XXXV. The pathology of pneumococcus infection in mice. J. Bacter. **19**, 363 (1930).

POLONOVSKI, J., et L. COTONI: Lyse du pneumocoque par les savons cationiques. Ann. Inst. Pasteur **73**, 1155 (1947).

PRINGSHEIM, E. G.: Über die gegenseitige Schädigung und Förderung von Bakterien. Zbl. Bakter. II **51**, 72 (1920).

QUERSIN, L.: Images nucléaires au cours de divers phénomènes de lyse microbienne. Ann. Inst. Pasteur **75**, 522 (1948).

— Images nucléaires et coloration de Gram. Ann. Inst. Pasteur **79**, 767 (1950).

—, et J. DIRKX: Etude du facteur lytique d'Anderson. Ann. Inst. Pasteur **80**, 378 (1951).

RALSTON, D. J., B. S. BAER, M. LIEBERMAN and A. P. KRUEGER: Virolysin: a virus-induced lysin from staphylococcal phage lysates. Proc. Soc. Exper. Biol. a. Med. **89**, 502 (1955).

RAMON, G., R. RICHOU et P. RAMON: Sur les propriétés diastasiques, bactériostatiques, bactéricides et bactériolytiques des filtrats de culture de *Bacillus subtilis*. C. r. Acad. Sci. Paris **220**, 543 (1945).

— — — De l'antagonisme microbien en général et en particulier des propriétés bactériostatiques, bactéricides et bactériolytiques des filtrats de culture du bacille subtilis. Bull. Acad. Méd. Paris **129**, 277 (1945).

REID, R. D., et H. A. IVERSON: The use of sodium desoxycholate-lysed pneumococci as antigen for the production of anti-pneumococcal serum. J. Bacter. **41**, 69 (1941).

REIMANN, H. A.: Studies concerning the relationship between pneumococci and streptococci. J. of Exper. Med. **45**, 1 (1927).

—, and L. A. JULIANELLE: The production of purpura by derivatives of pneumococcus. II. The effect of pneumococcus extract on the blood platelets and corpuscles. J. of Exper. Med. **43**, 97 (1926).

RISMONDO, R.: Sul fenomeno di lisi batterica da amino-acidi e da estratti batterici. Boll. Soc. ital. Biol. sper. **22**, 1272 (1946).

— L'azione litica degli amino-acidi e dell estratto di *B. subtilis* sul *Bacterium pneumoniae* di Friedlander. Giorn. Batter. **36**, 105, (1947).

ROBINOW, C. F.: A study of the nuclear apparatus of bacteria. Proc. Roy. Soc. Lond., Ser. B **130**, 299 (1942).

— Cytological observations on *Bact. coli*, *Proteus vulgaris* and various aerobic spore-forming bacteria, with special reference to the nuclear structures. J. of Hyg. **43**, 413 (1944).

ROSENOW, E. C.: The autolysis of pneumococci and the effect ot the injection of autolyzed pneumococci. J. Amer. Med. Assoc. **54**, 1943 (1910).

— Further studies of the toxic substances obtainable from pneumococci. J. Inf. Dis. **11**, 94 (1912a).

— On the nature of the toxic substance from pneumococci. J. Inf. Dis. **11**, 235 (1912b).

— On the toxicity of broth, of pneumococcus broth culture filtrates, and on the nature of the proteolytic enzyme obtainable from pneumococci. J. Inf. Dis. **11**, 286 (1912c).

— Partially autolyzed pneumococci in the treatment of lobar pneumonia. Results in two hundred cases. J. Amer. Med. Assoc. **70**, 759 (1918).

ROSENTHAL, L.: Microbes bactériolytiques (lysobactéries). C. r. Soc. Biol. Paris **92**, 78 (1925a).

— Mécanisme de l'action des lysobactéries. C. r. Soc. Biol. Paris **92**, 472 (1925b).

— La lyse des bacilles diphtériques effectuée par un *Streptothrix*. C. r. Soc. Biol. Paris **93**, 77 (1925c).

— Sur des lysobactéries thermophiles. C. r. Soc. Biol. Paris **93**, 1569 (1925d).

— Action of sporogenic aerobic bacilli on living and dead bacteria. Proc. Soc. Exper. Biol. a. Med. **46**, 448 (1941).

—, et F. DURAN-REYNALS: *Tyrothrix scaber* et flore intestinale. C. r. Soc. Biol. Paris **94**, 309 (1926a).

Rosenthal, L., et F. Duran-Reynals: Le sort du *Tyrothrix scaber* dans l'organisme après son introduction parentérale. C. r. Soc. Biol. Paris **94**, 1059 (1926b).

—, et Z. Ilitch: Sur le pouvoir cytolytique des filtrats de *Tyrothrix scaber*. C. r. Soc. Biol. Paris **95**, 10 (1926).

Sabin, A. B.: Role of certain anaerobic toxins in pneumococcus infection. Proc. Soc. Exper. Biol. a. Med. **28**, 457 (1931).

Salmon, J.: Activité bactériolytique directe des staphylocoques. C. r. Soc. Biol. Paris **143**, 9 (1949a).

— Mécanisme de l'action bactériolytique directe des staphylocoques. C. r. Soc. Biol. Paris **143**, 993 (1949b).

— Activité bactériolytique de *Staphylococcus aureus*. Sa signification. Son inhibition par la silice. C. r. Soc. Biol. Paris **144**, 715 (1950a).

— Action inhibitrice de la silice sur l'autolysine de *Staphylococcus aureus*. C. r. Soc. Biol. Paris **144**, 978 (1950b).

— Transformation microbiologique en orthophosphate de substances qui peuvent remplacer ce sel pour l'autophagie staphylococcique. C. r. Soc. Biol. Paris **144**, 1289 (1950c).

— Autophagie et autolyse chez *Escherichia coli*. C. r. Soc. Biol. Paris **145**, 776 (1951a).

— Autolyse des entérobactériacées. C. r. Soc. Biol. Paris **145**, 1413 (1951b).

— Comparaison des activités bactériolytiques du pyrophosphate et de la glycine. C. r. Soc. Biol. Paris **146**, 792 (1952).

— Constitution antigénique des lysats de *Salmonella schottmuelleri*. Rev. belge Path. Méd. exper. **1957**.

Salton, M. R. J.: The adsorption of cetyltrimethylammonium bromide by bacteria, its action in releasing cellular constituents and its bactericidal effects. J. Gen. Microbiol. **5**, 391 (1951).

— Studies on the bacterial cell-wall. III. Preliminary investigation of the chemical constitution of the cell-wall of *Streptococcus faecalis*. Biochim. et Biophysica Acta **8**, 510 (1952a).

— Cell-wall of *Micrococcus lysodeikticus* as the substrate of lysozyme. Nature (Lond.) **170**, 746 (1952b).

— The nature of the cell-walls of some Gram-positive and Gram-negative bacteria. Biochim. et Biophysica Acta **9**, 334 (1952c).

— Studies on the bacterial cell-wall. IV. The composition of the cell-walls of some Gram-positive and Gram-negative bacteria. Biochim. et Biophysica Acta **10**, 512 (1953a).

— Cell structure and the enzymic lysis of bacteria. J. Gen. Microbiol. **9**, 512 (1953b).

— Isolation of *Streptomyces* spp. capable of decomposing preparations of cell-walls from various microorganisms and a comparison of their lytic activities with those of certain actinomycetes and myxobacteria. J. Gen. Microbiol. **12**, 25 (1955).

— Cellular structure and enzymatic bacteriolysis. Conf. et Rapports, 3e Congr. Int. Biochim., Bruxelles. Imprimerie Vaillant-Carmanne S. A., Liège, p. 404, 1956.

—, and A. E. Alexander: The release of cellular constituents from *Staphylococcus aureus* treated with cetyltrimethylammonium bromide. J. Gen. Microbiol. **4**, ii (1950).

—, and J. M. Ghuysen: Action de l'actinomycétine sur les parois cellulaires bactériennes. Biochim. et Biophysica Acta **24**, 160 (1957).

—, and R. W. Horne: Studies on the bacterial cell-wall. I. Electron microscopical observations on heated bacteria. Biochim. et Biophysica Acta **7**, 19 (1951a).

— — Studies of the bacterial cell-wall. II. Methods of preparation and some properties of cell-walls. Biochim. et Biophysica Acta **7**, 177 (1951b).

— — et V. E. Cosslett: Electron microscopy of bacteria treated with cetyltrimethylammonium bromide. J. Gen. Microbiol. **5**, 405 (1951).

Sartorius, F.: Neuartige Lysine bei *Mykoïdes*-Bakterien. Zbl. Bakter. I Orig. **93**, 162 (1924).

Schiller, I.: Sur les produits des microbes en association. Zbl. Bakter. I Orig. **73**, 123 (1914).

— Sur l'antagonisme provoqué. C. r. Soc. Biol. Paris **91**, 152 (1924a).

— Über erzwungene Antagonisten. Zbl. Bakter. I Orig. **91**, 68 (1924b); **92**, 124 (1924c); **94**, 64 (1925a); **96**, 54 (1925b); **103**, 304 (1927); **129**, 139 (1933).

— Sur l'antagonisme provoqué. Bactéries et bacille de Koch. C. r. Soc. Biol. Paris **105**, 423 (1930).

SCHILLER, I., et F. ADAMOVSKAJA: Sur l'antagonisme provoqué. Procédé pour obtenir de grandes quantités de bactériolysines anti-diphtériques et anti-streptococciques. C. r. Soc. Biol. Paris **107**, 480 (1931).

SÉCHAUD, J., et E. KELLENBERGER: Lyse précoce, provoquée par le chloroforme chez les bactéries infectées par du bactériophage. Ann. Inst. Pasteur **9**, 102 (1956).

SELLARDS, A. W.: Selective action of dilute sodium hydroxide on certain races of the pneumococcus. J. Amer. Med. Assoc. **71**, 1301 (1918).

SERTIC, V.: Untersuchungen über einen Lysinzonen bildenden Bakteriophagen. I. Der Aufbau der Bakteriophagenkolonien. Zbl. Bakter. **110**, 125 (1929).

—, et N. BULGAKOV: Lysines de bactériophages présentant différentes thermorésistances. C. r. Soc. Biol. Paris **108**, 948 (1931).

SHARP, W. B., and F. G. BLAKE: Host factors in the pathogenesis of pneumococcus pneumonia. J. of Exper. Med. **52**, 501 (1930).

SHAW. M.: Decomposition of pneumococcus carbohydrate by the combined activity of strains of two bacterial species. J. Bacter. **33**, 644 (1937).

SICKLES, G. M., and M. SHAW: Microorganisms which decompose the specific carbohydrate of pneumococcus Types II and III. J. Inf. Dis. **53**, 38 (1933).

— — Action of microorganisms from soil on type-specific and non type-specific pneumococcus Type I carbohydrates. Proc. Soc. Exper. Biol. a. Med. **31**, 443 (1934a).

— — A systematic study of microorganisms which decompose the specific carbohydrates of the pneumococcus. J. Bacter. **28**, 415 (1934b).

— — A microorganism which decomposes the specific carbohydrate of pneumococcus type VIII. Proc. Soc. Exper. Biol. a. Med. **32**, 857 (1935).

SMOLIAR, V.: Nouveaux aspects des activités bactériolytiques et bactéricides des actinomycètes. C. r. Soc. Biol. Paris **143**, 1147 (1949).

— Influence des sels minéraux sur la formation du halo au cours de la bactériolyse par les actinomycètes. C. r. Soc. Biol. Paris **144**, 1256 (1950).

— Activités bactériolytiques de *Bacillus megaterium*. C. r. Soc. Biol. Paris **145**, 1416 (1951).

— Synergie bactériolytique de l'actinomycétine et de la glycine. C. r. Soc. Biol. Paris **146**, 1620 (1952).

SOLOMIDES, J., et A. HIRSCH: Sur la lyse du pneumocoque par les savons et les acides gras de l'huile de foie de morue. C. r. Soc. Biol. Paris **141**, 328 (1947).

STACEY, M., and M. WEBB: Some components of the lytic system of Gram-positive microorganisms. Nature (Lond.) **126**, 11 (1948).

STRANO, A.: Azione batteriolitica degli amino-acidi. Patol. sper. ital. **38**, 273 (1950).

STURDZA, S. A.: Sur le mécanisme de la lyse du pneumocoque par les sels biliaires. Les propriétés antigéniques de l'autolysine pneumococcique. C. r. Soc. Biol. Paris **129**, 410 (1938a).

— Sur le mécanisme de la lyse du pneumocoque par les sels biliaire. La sensibilité de l'autolysine des pneumocoques aux rayons ultra-violets. C. r. Soc. Biol. Paris **129**, 412 (1938b).

— Sur le mécanisme de la lyse des pneumocoques par congélation. C. r. Soc. Biol. Paris **129**, 1227 (1938c).

— Sur la lyse du pneumocoque par le chloroforme. Arch. roum. Path. exper. **11**, 171 (1938d).

SULKIN, N. M.: Histochemical localization of ribonucleoproteins by alkaline hydrolysis. Proc. Soc. Exper. Biol. a. Med. **78**, 32 (1951).

SWERTZ, L.: Etude de l'activité bactériolytique des actinomycètes et particulèrement des phénomènes de halo. Rev. belge Path. Méd. expér. **19**, 214 (1949).

TAI, T. Y., and W. E. van HEYNINGEN: Bacteriolysis by a species of streptomyces. J. Gen. Microbiol. **5**, 110 (1951).

TETI, M.: Isolamento di una ribonucleoproteina nel mezzo liquido della bacteriolisi lisozimica. Riv. Ist. sieroter. ital. **27**, 279 (1952).

—, e V. ALBANO: Modificazioni indotte dal lisozime sulla *S. typhi*. Boll. Ist. sieroter. milan. **29**, 436 (1950).

—, e M. FIORE: Morphological alterations caused by lysozyme on organisms of different sensitivity. Boll. Ist. sieroter. milan. **28**, 160 (1949).

THOMPSON, R. H. S., and R. J. DUBOS: The isolation of nucleic acid and nucleoprotein fractions from pneumococci. J. of Biol. Chem. **125**, 65 (1938).

Thorne, G. B., C. G. Gomez, H. E. Noyes and R. D. Housewright: Production of glutamyl polypeptide by *Bacillus subtilis*. J. Bacter. **68**, 307 (1954).
Todd, E. W.: Function of autolytic enzymes in bacteriolysis by penicillin. Lancet **1945**, 172.
Toenissen, E.: Untersuchungen über die Kapsel (Gummihülle) der pathogenen Bakterien. II. Die chemische Beschaffenheit der Kapsel und ihr dadurch bedingtes Verhalten gegenüber der Fixierung und Färbung. Zbl. Bakter. I Orig. **85**, 225 (1921).
Tomcsik, J., u. S. Guex-Holzer: Änderung der Struktur der Bakterienzelle im Verlauf der Lysozym-Einwirkung. Schweiz. Z. allg. Path. u. Bakter. **15**, 518 (1952).
Truche, C., L. Cotoni et A. Raphael: Etudes sur le pneumocoque. VIII. Action de la bile sur les pneumocoques humains et animaux. Ann. Inst. Pasteur **27**, 886 (1913).
—, A. Cramer et L. Cotoni: Etudes sur le pneumocoque. I. Virulence du pneumocoque humain pour la souris. Ann. Inst. Pasteur **25**, 480 (1911).
Tulasne, R., and R. Vendrely: Demonstration of bacterial nuclei with ribonuclease. Nature (Lond.) **160**, 225 (1947 a).
— — Mise en évidence des noyaux bactériens par la ribonucléase. C. r. Soc. Biol. Paris **141**, 674 (1947).
Twort, F. W.: The transmissible bacterial lysis and its action on dead bacteria. Lancet **1925**, 642.
Vacirca, F.: Un nuovo metodo di lisi batterica da amino-acidi e da estratti batterici. Boll. Soc. ital. Biol. sper. **22**, 1271 (1946).
— Alterazioni morfologiche del corpo batterico da amino-acidi. Med. Ital. **27**, 209 (1947 a).
— Lisi delle colonie batteriche da amino-acidi e da estratti batterici. Boll. Soc. ital. Biol. sper. **23**, 208 (1947 b).
— L'azione degli essudati sul processo di lisi batterica da amino-acidi e da estratti di subtilis. Giorn. Batter. **36**, 297 (1947 c).
— Sul mecanismo del fenomeno di lisi batterica da amino-acidi e da estratti di subtilis. Riv. ital. Ig. **8**, 45 (1948).
Vallée, M.: Etude du pouvoir bactériolytique du *Bacillus subtilis*. C. r. Soc. Biol. Paris **139**, 148 (1945 a).
— Etude du pouvoir bactériolytique du *Bacillus subtilis*. Lysine et analysine. C. r. Soc. Biol. Paris **139**, 648 (1945 b).
—, A. J. Borrel et S. Saudinos: Titrage de la subtilysine. Relation entre son pouvoir bactériolytique et son pouvoir gélatinolytique. C. r. Soc. Biol. Paris **141**, 350 (1947).
—, et S. Saudinos: Titrage de la subtilysine. Relation entre son pouvoir bactériolytique et son pouvoir amylolytique. C. r. Soc. Biol. Paris **140**, 850 (1946).
Vendrely, R., et J. Lipardy: Acides nucléiques et noyaux bactériens. C. r. Acad. Sci. Paris **223**, 342 (1946).
Wadsworth, A. B.: Standard methods, 2nd ed. Baltimore: Williams & Wilkins Co. 1939.
Wahl, R., et Josse-Goichot: La lyse bactériophagique par entrainement. Ann. Inst. Pasteur **76**, 43 (1949).
Waksman, S. A., E. S. Horning, M. Welsch and H. B. Woodruff: Distribution of antagonistic actinomycetes in nature. Soil Sci. **54**, 281 (1942).
Wanamaker, L. W., F. W. Denny, C. H. Rammelkamp jr. and W. R. Brinks: Use of Maxted's method for group classification of hemolytic streptococci. Proc. Soc. Exper. Biol. a. Med. **73**, 467 (1950).
Warden, C. C.: Immunity in pneumococcus infections. J. Amer. Med. Assoc. **59**, 2085 (1912).
Warren, G. H., J. Gray and P. Bartell: The lysis of *Pseudomonas aeruginosa* by lysozyme. J. Bacter. **70**, 614 (1955).
Webb, M.: The action of lysozyme on heat-killed Gram-positive microorganisms. J. Gen. Microbiol. **2**, 260 (1948).
Weibull, C.: The isolation of protoplasts from *Bacillus megaterium* by controlled treatment with lysozyme. J. Bacter. **66**, 688 (1953).
Weiland, P.: Bakterizide Wirkung von Mesentericusfiltraten auf Diphtheriebazillen. Zbl. Bakter. I Orig. **136**, 451 (1936).
Weinberg, M., et P. Aznard: Autobactériolysines et phénomène de d'Hérelle. C. r. Soc. Biol. Paris **86**, 833 (1922).
Weiss, C.: Studies in pneumonia. IX. The properties of pneumotoxin and its probable function in the pathology of lobar pneumonia. J. Med. Res. **39**, 103 (1918).

WEISS, C., and J. A. KOLMER: Studies in pneumonia. VIII. A skin reaction to pneumotoxin. J. of Immun. **3**, 395 (1918).

WELSCH, M.: Propriétés bactériolytiques du *Streptothrix* et sporulation. C. r. Soc. Biol. Paris **123**, 1013 (1936).

— La dissolution des germes vivants par les *Streptothrix*. C. r. Soc. Biol. Paris **124**, 573 (1937a).

— Endo- et exo-bactériolysines des actinomyces (Streptothrix). C. r. Soc. Biol. Paris **124**, 1240 (1937b).

— Influence de la nature du milieu de culture sur la production de lysines par les actinomyces. C. r. Soc. Biol. Paris **126**, 244 (1937c).

— De quelques propriétés du principe bactériolytique des actinomyces. C. r. Soc. Biol. Paris **126**, 247 (1937d).

— Influence des sels minéraux sur la mycolyse. C. r. Soc. Biol. Paris **126**, 1254 (1937e).

— Le mécanisme de la mycolyse des germes vivants. C. r. Soc. Biol. Paris **127**, 347 (1938a).

— Mise au point d'une technique néphélométrique pour l'étude de la mycolyse. C. r. Soc. Biol. Paris **128**, 795 (1938b).

— La cinétique de la mycolyse. C. r. Soc. Biol. Paris **128**, 799 (1938c).

— Dosage néphélométrique du principe bactériolytique des actinomyces. C. r. Soc. Biol. Paris **128**, 1172 (1938d).

— Inactivation par la chaleur du principe bactériolytique des actinomyces. C. r. Soc. Biol. Paris **128**, 1175 (1938e).

— Développement des actinomyces et sécrétion de bactériolysines. C. r. Soc. Biol. Paris **130**, 104 (1939a).

— Mise au point d'une technique néphélométrique pour l'étude de la mycolyse. Influence de la réaction du milieu sur le trouble des suspensions microbiennes. C. r. Soc. Biol. Paris **130**, 797 (1939b).

— Influence de la réaction du milieu sur la production, l'activité et la stabilité des bactériolysines des actinomyces. C. r. Soc. Biol. Paris **130**, 800 (1939c).

— De l'inactivation du principe bactériolytique des actinomyces par les rayons ultra-violets. C. r. Soc. Biol. Paris **131**, 1296 (1939d).

— Bacteriolytic properties of actinomyces. Third int. Congr. Microbiol., New York, Rep. Proc. p. 260, 1939e.

— Bactericidal substances from sterile culture-media and bacterial cultures. J. Bacter. **42**, 801 (1941).

— Incidence of bacteriolytic properties among actinomycetes. J. Bacter. **43**, 10 (1942a).

— Bacteriostatic and bacteriolytic properties of actinomycetes. J. Bacter. **44**, 571 (1942b).

— Production d'actinomycétine et d'actinomycine dans les cultures submergées de divers actinomycètes. Schweiz. Z. allg. Path. u. Bakter. **9**, 379 (1946).

— Activité bactériolytique de certaines actinomycétines sur *Staphylococcus aureus* et *Bacillus megatherium* vivants. Bull. Soc. Chim. biol. Paris **29**, 362 (1947a).

— Actinomycetin. J. Bacter. **53**, 101 (1947b).

— A propos de l'action bactériolytique de la pénicilline. C. r. Soc. Biol. Paris **141**, 436 (1947c).

— La pénicilline, agent chimiothérapique né et développé dans le laboratoire de microbiologie. Scalpel **100**, 939 (1947d).

— Phénomènes d'antibiose chez les actinomycètes. Rev. belge Path. Méd. expér. **18**, Suppl. 2 (1947e).

— La réaction de Gram, son mécanisme, sa signification. Rev. méd. Liège **3**, 257 (1948a).

— De la nature du système bactériolytique des actinomycètes. C. r. Soc. Biol. Paris **142**, 1589 (1948b).

— Les antibiotiques. Armes thérapeutiques nouvelles fournies par les microorganismes. Rev. méd. Liège **4**, 301 (1949a).

— A propos de l'autolyse du staphylocoque. C. r. Soc. Biol. Paris **143**, 719 (1949b).

— Production et activité bactériolytique de l'autolysine staphylococcique. C. r. Soc. Biol. Paris **143**, 721 (1949c).

— Nouvelles recherches sur l'autolyse du staphylocoque. C. r. Soc. Biol. Paris **143**, 997 (1949d).

Welsch, M.: Dissolution spontanée ou par l'autolysine staphylococcique des suspensions de *Staphylococcus aureus* chauffé. C. r. Soc. Biol. Paris **143**, 999 (1949e).

— Sels dérivés du phosphore et autolyse de *Staphylococcus aureus*. C. r. Soc. Biol. Paris **144**, 1253 (1950a).

— Etude de l'autolyse staphylococcique par la microscopie de fluorescence. C. r. Soc. Biol. Paris **144**, 1429 (1950b).

— Streptomycine et lysozyme. Antonie van Leeuwenhoek **18**, 23 (1952a).

— Lyse à basse température de staphylocoques lysogènes irradiés. C. r. Soc. Biol. Paris **146**, 1397 (1952b).

— Lyse collatérale des staphylocoques normaux en présence de staphylocoques lysogènes induits par irradiation ultraviolette. C. r. Soc. Biol. Paris **147**, 537 (1953a).

— Nouvelles recherches sur la lyse collatérale des staphylocoques. C. r. Soc. Biol. Paris **147**, 1486 (1953b).

— Activités staphylolytique et streptolytique des filtrats de culture de *Streptomyces* spp. C. r. Soc. Biol. Paris **148**, 604 (1954).

—, G. Cavallo et P. Cantelmo: Induction lysogénique chez *Micrococcus pyogenes aureus*. Rev. belge Path. Méd. expér. **22**, 241 (1953).

—, et L. Delcambe: Recherches sur la libération de phosphore et d'azote au cours de l'autolyse staphylococcique. Bull. Soc. Chim. biol. Paris **33**, 488 (1951).

— — Installation-pilote pour recherches microbiologiques. 27e Congr. int. Chim. ind., Bruxelles, vol. 3, p. 803, 1954.

— et W. J. Elford: Détermination de la taille des bactériolysines d'actinomyces par ultrafiltration. C. r. Soc. Biol. Paris **125**, 1053 (1957).

—, et J. M. Ghuysen: Activité streptolytique de l'actinomycétine. C. r. Soc. Biol. Paris **147**, 1659 (1953).

— — The antibiotics of actinomycetin. Giorn. Microbiol. **1957**.

— — et A. Castermans: The multiplicity of bacteriolytic agents in actinomycetin. Conf. et Rapports, 3e Congr. int. Biochim., Bruxelles, Imprimerie H. Vaillant-Carmanne, Liège p. 413, 1955.

—, et E. Nihoul: A propos de la mise en évidence du noyau bactérien. C. r. Soc. Biol. Paris **143**, 1449 (1948).

—, and J. Salmon: Bacteriolytic properties of actinomycetes and staphylococci. J. Gen. Microbiol. **3**, xxvii (1949a).

— — Sels dérivés du phosphore et activité bactériolytique du staphylocoque. C. r. Soc. Biol. Paris **143**, 1399 (1949b).

— — Action comparée de divers sels sur l'autophagie et l'autolyse de *Staphylococcus aureus*. C. r. Soc. Biol. Paris **144**, 1292 (1950a).

— — Effecteurs de l'autophagie et de l'autolyse staphylococciques. Ve Congr. Int. Microbiol. Rio de Janeiro, Rés. Comm., p. 36, 1950b.

— — Quelques aspects de la staphylolyse. Ann. Inst. Pasteur **79**, 802 (1950c).

— — et C. Heusghem: Importance des phosphates pour les activités bactériolytiques du staphylocoque. C. r. Soc. Biol. Paris **143**, 1152 (1949).

Welshimer, H. J.: Observations on phage output and on anaerobic lysis of lysogenic strains of *Bacillus megaterium*. J. Bacter. **61**, 153 (1951).

— The action of lysozyme on the cell wall and capsule of *Bacillus megaterium*. J. Bacter. **66**, 112 (1953).

White, H. J., C. M. Alverson, M. W. Baker and E. R. Jackson: Comparative biological studies of polymyxin and aerosporin. Ann. New York Acad. Sci. **51**, 879 (1949).

Wilke, W.: Versuche über Gram-Festigkeit von Pneumokokken und Diphtheriebazillen. Zbl. Bakter. I Orig. **113**, 262 (1929).

Wollman, E.: Action lytique des staphylocoques vivants sur les staphylocoques tués (A propos de la note de A. Gratia). C. r. Soc. Biol. Paris **95**, 679 (1926).

— Autolyse bactérienne et antiseptique. C. r. Soc. Biol. Paris **111**, 578 (1932a).

— Recherches sur le phénomène de Twort-d'Hérelle (bactériophagie). Ann. Inst. Pasteur **49**, 41 (1932b).

—, et M. Averbusch: Transmission de la lyse biliaire des pneumocoques à des formes de ces germes insolubles dans la bile. C. r. Soc. Biol. Paris **110**, 623 (1932).

WOLLMAN, E., et M. AVERBUSCH: Lyse biliaire des pneumocoques. Transmission à des souches normalement non solubles. Application au diagnostic différentiel du pneumocoque. Bull. Acad. Méd.Paris **107**, 126 (1932b).

—, et F. DURAN-REYNALS: Bactériophage et autolyse. C. r. Soc. Biol. Paris **94**, 1330 (1926).

—, et MME. E. WOLLMAN: Lyse secondaire non bactériophagique du staphylocoque. C. r. Soc. Biol. Paris **110**, 636 (1932).

ZAMENHOF, S.: A specific lytic substance in *Escherichia coli*. J. Bacter. **49**, 413 (1944).

ZIEGLER, E. E.: The specific effect of bile salts on pneumococci and pneumococcus pneumonia. Arch. Int. Med. **4**, 644 (1930).

— Sodium dehydrocholate. Its specific effect on pneumococci. J. Labor. a. Clin. Med. **16**, 868 (1931).

— The effect on pneumococci of sodium dehydrocholate, a bile salt derivative. J. Labor. a. Clin. Med. **17**, 317 (1932).

— Some properties of pneumocholin, a biochemical antigen. J. Labor. a. Clin. Med. **18**, 695 (1933).

ZUKERMAN, I., u. I. MINKEWITSCH: Zur Frage des bakteriellen Antagonismus. Cité d'après Zbl. Bakter. I Ref. **80**, 483 (1925).

VI. Die Bakteriencytologie im Vergleich zum Feinbau höher differenzierter Zellen

Von

Helmut Ruska [1]

Stoffwechsel, Bewegung und Genetik der Bakterien entsprechen grundsätzlich gleichen Lebenserscheinungen bei morphologisch höher differenzierten Zellen des Pflanzen- und Tierreichs. Es besteht kein Zweifel, daß diese Funktionen an identische oder chemisch ähnliche Substanzgruppen gebunden sind. Umstritten dagegen ist das Verhältnis des Feinbaus von Bakterienorganellen zum Feinbau von Mitochondrien, Kern und Centrosomen höherer Zellen. Geht man von der Voraussetzung aus, daß die rezenten Bakterien noch die Organisation phylogenetisch ältester Zellen besitzen, so taucht damit das Problem auf, ob auch Zellorganellen eine phylogenetische Entwicklung durchgemacht haben, oder ob Strukturen, die in der höheren Zelle zum gesetzmäßigen Bestand gehören, in Bakterien bereits in vergleichbarer Differenzierung der Struktur präformiert sind. Eine ähnliche Frage wurde für die Ontogenese bereits im 18. Jahrhundert diskutiert, nämlich Präformation des fertigen Organismus in den Keimzellen oder Epigenese.

Wie später noch genauer zu zeigen sein wird, postulieren einige Autoren Homologien zwischen Volutinkörnchen oder Enzymanhäufungen an Zellpolen der Bakterien und Mitochondrien höherer Zellen, außerdem zwischen den Kernäquivalenten der Bakterien und den bei ihrer Teilung auftretenden Strukturen mit dem Kern, den Chromosomen und Centriolen der klassischen Cytologie. Andere dagegen halten nur einfachere Organisationsformen mit weniger komplizierten Verhaltensweisen für nachgewiesen. Aus den einfachen Formen in Spaltpflanzen müßten sich dann die differenzierten Strukturen in Flagellaten und ihren Abkömmlingen im submikroskopischen ebenso wie im mikroskopischen Bereich erst während der Phylogenese entwickelt haben. Mit der alten Diskussion um Präformation und Epigenese in der Ontogenese hat die gegenwärtige Frage nach der Phylogenese innerer Zellstrukturen auch das Problem der Raumbeanspruchung gemeinsam. Wir wissen, daß den höheren Zellen äquivalente Funktionen in der Bakterienzelle untergebracht sind. Wieweit haben aber den Zellkernen und Chromosomen, Centriolen und Kernspindeln, Mitochondrien, cytoplasmatischen Membransystemen und deren feineren Teilstrukturen homologe Strukturen noch Platz in einem Bacterium, das oft selbst nur die Größe von kleinen Mitochondrien oder von Centrosomen besitzt? Und schließlich, wie verhalten sich diese Strukturen, wenn beim Formwandel der Bakterien Kleinformen auftreten, in denen die Raumfrage noch kritischer wird?

[1] Division of Laboratories and Research, New York State Department of Health.

Die gegenwärtig vorliegenden Untersuchungen sind von einer Lösung dieser Fragen noch weit entfernt. Teils weil das Material zu einem vollständigen Vergleich bei weitem nicht ausreicht, teils weil erst sehr wenig davon mit den besten derzeit verfügbaren Methoden untersucht ist. Nur das elektronenmikroskopische Bild des Dünnschnitts von fixierten Zellen oder besser noch Bilder von Serienschnitten geben über die innere Struktur und über die Beziehung dieser Strukturen zueinander genügend Auskunft. Die dazu notwendigen Methoden sind erst vor kurzem so weit verbessert worden, daß reproduzierbare Ergebnisse eine genauere Deutung erlauben (Maaløe und Birch-Andersen; Glauert, Rogers und Glauert; Moore und Grimley).

Überblickt man die Literatur der letzten Jahre, so zeigt sich kein Mangel an zusammenfassenden Arbeiten, die analytisch-chemische, enzymchemische, immunologische und genetische Daten in ihren Beziehungen zur Mikrostruktur der Bakterienzelle darstellen. Hingewiesen sei auf die Aufsätze zur Bakteriencytologie von Marshak, zur Histochemie von Vendrely, zum chemischen Aufbau der Zellwand von Cummins, auf die von Spooner und Stocker herausgegebene, ausgezeichnete Diskussion der Bakterien-Anatomie auf dem 6. Symposium der Society for General Microbiology und aus dem deutschen Schrifttum vor allem auf die Darstellung der Bakterienzelle von Winkler. Ebenso fehlt es nicht an zusammenfassenden Werken über die elektronenmikroskopisch sichtbaren Strukturen höher organisierter Zellen. Wichtig sind die Erkenntnisse des Symposiums über Feinstrukturen der Zelle (Leiden 1954), der von Porter herausgegebene Bericht der Konferenz über Gewebefeinstrukturen (N. Y. Harriman 1956) und die Darstellung der Ultrastruktur der Zelle von Sjöstrand.

Es wäre verfrüht, diese hervorragenden Informationsquellen mit einzelnen Ergebnissen der letzten Monate zu ergänzen. Dagegen dürfte eine Darstellung der derzeitigen Grenzen berechtigter Homologisierungen von Bakterienstrukturen mit Strukturen höher entwickelter Zellen gegensätzliche Auffassungen verdeutlichen und künftige Fragestellungen schärfer formulieren lassen. Den Arbeiten von Bringmann, Mudd und DeLamater bleibt das Verdienst, die Suche nach einem Verständnis der inneren Organisation der Bakterienzelle aktiviert zu haben. Bringmann betonte dabei stets den Gedanken der Entwicklung von Innenstrukturen der niederen Formen zu Kern und Mitochondrien der höheren Formen, während Mudd und DeLamater eine weitgehendere Präformation der Innenstrukturen nachzuweisen suchten. Die Arbeiten begannen in einer Zeit, in der lokalisierte färberische Eigenschaften und histochemische Reaktionen elektronenmikroskopisch mit dem Aussehen der intakten oder ebenfalls gefärbten Bakterienzelle verglichen werden konnten, die ausschlaggebenden Schnittverfahren jedoch noch nicht oder ungenügend entwickelt waren. Ohne diese blieb es bei der Erkennung von Reaktionsorten und Orten der Ablagerung von Reaktionsprodukten. Die gezogenen Folgerungen sind daher zum Teil nicht mehr haltbar oder umstritten.

Metachromatische Körnchen und Mitochondrien

Kurze Zeit wurden Volutinkörnchen der Bakterien (Größe 0,02—0,6 μ) und Mitochondrien höherer Zellen (0,5 bis mehrere μ) als gleichwertige Organellen betrachtet. Im Elektronenmikroskop fallen bei der Betrachtung von Totalpräparaten beide durch scharfe Begrenzung und große Dichte auf, aber erst ähnliche färberische Eigenschaften ließen an eine Homologie denken. Im Schnitt zeigen die Mitochondrien eine charakteristische Struktur, und zwar eine äußere Doppelmembran, deren innere Schicht innere Doppelmembranen bildet. Diese teilen jedes Mitochondrion in kommunizierende Binnenräume auf. Den Volutinkörnchen dagegen fehlt eine vergleichbare Struktur. Sie tritt auch, wie die Schnittbilder von CHAPMANN und HILLIER, von MAALØE, BIRCH-ANDERSEN und SJÖSTRAND, von BRADFIELD und von KIMURA, ITO und OZAKI zeigen, an keiner anderen Stelle des Bakterienkörpers auf. Außerdem sind nach BRADFIELD die mit den Funktionen der Mitochondrien höherer Zellen (Krebscyclus, oxydative Phosphorylierung) verwandten Funktionen in Bakterien an eine Zellfraktion gebunden, die aus 10—20 mμ großen Partikeln besteht. Teilchen dieser Größe können nichts mehr mit der Struktur von Mitochondrien gemeinsam haben, deren Doppelmembranen bereits eine Dicke von 10—20 mμ erreichen. Spekulationen über wesentlich dünnere Membranen erübrigen sich aus Gründen der Raumbeanspruchung molekularer Lipoproteidschichten. Bezeichnungen wie Mitochondrien (MUDD), Mitochondrienäquivalente (BRINGMANN) und Chondrioide (KELLENBERGER) haben somit in der Bakteriologie ihre Berechtigung verloren, gleichgültig ob darunter Volutinkörnchen oder polar gelegene Verdichtungen des Protoplasmas verstanden wurden. Nach BISSET hat MUDD in einer persönlichen Mitteilung die Auffassung von einer wahren Homologie der metachromatischen Körnchen mit den Mitochondrien aufgegeben und spricht nur noch von einer Analogie der Färbungsreaktionen.

Kernäquivalente und Kerne

In der Frage nach dem Kern gehen die Forderungen bezüglich einer Homologie der Ausstattung von Bakterienzelle und höherer Zellen mit Recht weiter als in der Mitochondrienfrage. Hier handelt es sich nicht nur um die Frage, ob enzymatische Systeme durch die Anordnung in einer spezifischen Struktur zusätzliche physikalisch-chemische Wirkungsmöglichkeiten gewonnen haben oder ob diese noch fehlen. Vielmehr muß nach den Erfahrungen der Genetik gefordert werden, daß auch in der Bakterienzelle Erbfaktoren geordnet in die Tochterzelle übertragen werden, und dies erscheint bislang kaum anders denkbar als durch eine lineare, in der Längsrichtung verdopplungsfähige und teilbare Struktur. Über diese Forderung hinausgehend verficht DELAMATER die Auffassung , daß das Kernäquivalent der Bakterien und sein Verhalten bei der Teilung dem Kern höherer Zellen homolog sei. Chromosomen sollen durch die Phasen der Mitose gehen, sich in 2 Chromatiden trennen und durch das Eingreifen der von Centriolen ausgehenden Kernspindel auf die Tochterzellen übertragen werden. Im Gegensatz zu höheren Zellen (abgesehen von einigen Flagellaten) soll die Kernmembran während der Teilung persistieren. Demgegenüber betonen Autoren mit Erfahrungen an elektronenmikroskopisch untersuchten Dünnschnitten, daß eine Kernmembran in den Bakterien fehlt und weder verschiedene Phasen der Chromosomen noch Kernspindel und Centriolen beobachtet worden sind. Die Bezirke der Kernäquivalente bzw. Nucleoide von PIEKARSKI oder der Chromatinkörper von ROBINOW erscheinen vielmehr in Schnittbildern als unscharf begrenzte Areale geringer Substanzdichte. Einsenkungen des intakten Bakterienkörpers, die sich beim Trocknen der Zelle an den Stellen der histochemisch nachgewiesenen

Kernäquivalente bilden, ließen schon vorher auf ihren hohen Wassergehalt schließen. Innerhalb der Kernareale sind in Schnitten, die nach dem gegenwärtigen Urteil als optimal erhalten angesehen werden müssen, dichtere Stränge sichtbar. MAALØE und BIRCH-ANDERSEN wiesen einen tubulären Aufbau der Stränge nach und interpretierten die Wände als schraubig aufgerollte Fäden der Desoxynucleoproteide. Damit wären diese Fäden mit den sehr viel größeren Chromosomen- bzw. Chromatidenspiralen in ihrem maximal kontrahierten Zustand vergleichbar. Streckungen der Wandspirale, wie sie in der Prophase bei den Chromosomen der Kerne höherer Zellen auftreten, sind bis jetzt nicht beobachtet.

Zu den Fragen der Raumbeanspruchung geben MAALØE und BIRCH-ANDERSEN an, daß die gesamte Menge der Desoxyribonucleinsäure ($1{,}13 \cdot 10^{-14}$ g je Zelle nach LARK und MAALØE) einen Faden der etwa 1000fachen Länge der Bakterienzelle ausmachen würde. Die Fadendicke beträgt unter Einschluß von Proteinen 25—50 Å. Dies entspricht nach MAALØE und BIRCH-ANDERSEN der Wanddicke der in elektronenmikroskopischen Schnittbildern festgestellten tubulären Struktur im Bereich der Kernäquivalente. Da der Durchmesser der Röhren rund 500 Å beträgt, kommen etwa 1600 Å auf eine Schraubenwindung und unter günstigen Annahmen auf 1 μ Bakterienlänge noch mehr als 10 μ Spirale, in der die 1000fache Länge des DNS-Fadens unterzubringen wäre. Serienschnitte müßten dies noch belegen. Für eine Kernmembran wäre zwischen dem wenig dichten Areal der Kernäquivalente und dem dichten Cytoplasma noch genügend Raum vorhanden, doch ist sie, wie erwähnt, nicht mit Sicherheit nachgewiesen. Bei einer Gleichheit in den strukturellen Beziehungen zwischen Kern und Cytoplasma müßte außerdem in der umstrittenen Interphase des Bakterienkerns eine Doppelmembran erwartet werden, deren äußere cytoplasmatische Schicht die bei höheren Zellen aufgefundenen Beziehungen zum endoplasmatischen Reticulum hat (WATSON, RUSKA, NOVIKOFF u. a.). Immerhin ist es bemerkenswert, daß das Kernmaterial in der Bakterienzelle in einer verhältnismäßig flüssigen Phase des Cytoplasmas zu liegen scheint, der zum Teil eine ähnliche funktionelle Bedeutung zukommen könnte wie der perinucleären Zysterne höherer Zellen.

Wie Verdopplung und Teilung des Kernmaterials und sein Übergang in die elektronenmikroskopisch sichergestellten Kleinformen (RAETTIG) und Gonidien (BISSET) vor sich gehen und wie es sich in „large bodies" verhält, ist noch nicht geklärt[1]. Diese Fragen werden schwieriger Untersuchungen unter Einschluß elektronenmikroskopischer Aufnahmen von Serienschnitten bedürfen. Die Befunde und Interpretationen von MAALØE und BIRCH-ANDERSEN haben die Struktur des bakteriellen Kernäquivalents in nähere Beziehungen zur Struktur von Chromosomen gebracht, als es je zuvor geschah. Die Streitigkeiten ROBINOWs, BISSETs u. a. einerseits mit DELAMATER und seinem Arbeitskreis andererseits betreffen primär nur die Frage der direkten Teilung und der mitotischen Teilung des Kernmaterials. Hinzu kommt, daß DELAMATER die Mitwirkung von Centriolen und Kernspindel bei der Mitose in Bakterien forderte, deren Anwesenheit nicht unbedingte Voraussetzung für das Zustandekommen einer Mitose zu sein scheint (HARTMANN). Den Chromosomen selbst werden schon Bewegungskräfte zugeschrieben.

---

[1] Nachtrag bei der Korrektur: Siehe Schnittbilder von „large bodies" bei PONTIERI.

Wimpern, Geißeln und Bakteriengeißeln

Das Auftreten von Cilien und Flagellen wie auch der Kernspindeln in höheren Zellen steht mit dem Verhalten des Centrosoms in engem Zusammenhang. Der Teilung des Centrosoms zur Lieferung der Centriolen für die Spindel kann eine weitere Teilung eines Centriols zur Lieferung des Basalkorns bzw. Blepharoplasten folgen. Das zusätzliche Zentrum bildet dann die Geißel. Elektronenmikroskopische Untersuchungen haben diese Befunde in überraschender Weise erweitert. Cilien und Flagellen von pflanzlichen und tierischen Zellen bestehen aus 9 Paaren peripherer und einem Paar zentraler Fibrillen. Die Centriolen der Kernspindel wiederum lassen einen Aufbau aus 9 tubulären Gebilden erkennen, die ihrerseits zu einem kurzen Rohr angeordnet sind (Bernhard und de Harven, de Harven persönliche Mitteilung und Demonstration weiteren Materials; Grassé, Carasso und Favard). Das Centrosom und seine Abkömmlinge, die verschiedenartige motorische Funktionen erfüllen, besitzen also über das gesamte Pflanzen- und Tierreich 9 Primärstrukturen in ringförmiger oder tubulärer Anordnung.

Die Bakteriengeißel läßt einen Aufbau aus 9 Untereinheiten vermissen. Sie erscheint elektronenmikroskopisch in der Regel homogen. Nur selten sind Formen beobachtet worden, in denen die einzelne Geißel aus 2 oder 3 umeinandergewundenen Fäden aufgebaut ist (Starr und Williams; Labaw und Mosley). Verschiedenartige Geißeln können am selben Organismus auftreten (Houwink und van Iterson). Basalkörnchen sind zwar an vielen Bakterien nachgewiesen, doch ist noch nicht gesichert ob sie allen Bakteriengeißeln zukommen (Stocker). Ihre Struktur ist wenig geklärt, vor allem wegen des Fehlens von Schnittbildern. Es wäre überraschend, an den Basalkörnchen 9 Untereinheiten aufzufinden, da die Raumfrage, zumal bei vielgeißeligen Zellen, sehr kritisch wird. Ein solcher Befund würde neue Argumente in die Diskussion um die Centrosomen der Bakterienzelle bringen. Insgesamt fehlt bisher der Nachweis einer Homologie der Bakteriengeißeln mit Geißeln und Cilien höherer Zellen. Es fehlt der Nachweis einer Abstammung der bakteriellen Basalkörnchen von Centrosomen, und es fehlt der Nachweis von 9 Teilstrukturen in Basalkorn und Bakteriengeißel. Nach Bisset entstehen Basalkörnchen und Geißeln in den Tochterzellen der polar begeißelten Bakterien de novo, doch ist auch diese Frage nicht völlig entschieden (Stocker). Die Zusammenfassung aller geißeltragenden Organismen (Pflanzen, Tiere, Pilze und Bakterien) und ihre Gegenüberstellung zu der einzig nicht begeißelten Gruppe der Blaualgen (Bisset) erscheint verfrüht, solange über die Homologie aller Flagellen und die damit zusammenhängenden Fragen von Basalkörnchen und Centrosomen keine Klarheit besteht.

Plasmamembranen

Experimente über Plasmolyse bei Bakterien ließen schon vor langer Zeit auf eine Plasmamembran schließen, die unter der den Bakterien eigentümlichen, stabileren Zellwand liegt. Sie ist elektronenmikroskopisch an intakten Bakterien gesehen worden, erscheint aber deutlicher in guten Dünnschnitten. Den elegantesten Nachweis für das Vorhandensein einer Plasmamembran, zusätzlich zur Zellwand, brachten Versuche, die zum Ziele hatten, die Zellwand mit Lysozym aufzulösen, ohne die Zelle zu vernichten. Der frei werdende, nur von der Plasmamembran bedeckte Protoplast kugelt sich ab, hat die Fähigkeit, eine neue Zellwand zu bilden, verloren, behält aber einen nahezu normalen Stoffwechsel, wächst, teilt sich und bildet unter geeigneten Voraussetzungen Sporen sowie Phagen (WEIBULL, McQUILLEN). Die Geißeln bleiben mit dem isolierten Protoplast verbunden, doch sind noch keine Bedingungen gefunden worden, die auch ihre Beweglichkeit erhalten hätten. Nach Zerfall der Protoplasten können Plasmamembranen isoliert werden, die noch einen erheblichen Anteil von Enzymen enthalten. BRADFIELD führt dies auf anhaftende Granula zurück, und WEIBULL macht darauf aufmerksam, daß die Fähigkeit der bakteriellen Plasmamembranen, Proteine zu synthetisieren, an die gleichen Eigenschaften der Mikrosomenfraktion höherer Zellen erinnert. Diese Fraktion höherer Zellen enthält außer Plasmamembranen auch Bruchstücke der intraplasmatischen Membranen (endoplasmatisches Reticulum), die sich nach PALADE morphogenetisch von der äußeren Plasmamembran ableiten lassen. Das endoplasmatische Reticulum fehlt den Bakterien, scheint aber bei Blaualgen (BRADFIELD) und höheren Algen (HODGE, McLEAN und MERZER) bereits aufzutreten. Form und mechanische Festigkeit erhält die Bakterienzelle im wesentlichen durch die Zellwand, die osmotischen Eigenschaften werden vorwiegend auf die Plasmamembran bezogen (MITCHELL und MOYLE). Damit sind an die Plasmamembranen der Bakterien sowie der höheren Zellen selektive Permeabilität, d. h. Austauschkontrolle zwischen Zelle und Umgebung sowie ähnliche enzymatische Reaktionen gebunden.

Präformation oder Entwicklung

Naturgemäß tauchte das Problem der Entwicklung zunächst bei der Betrachtung individueller Formen auf, wo es jedem aufmerksamen Beobachter ständig entgegentritt. Der Entwicklungsgedanke in der Stammesgeschichte ist viel jüngeren Datums, aber seit DARWIN und seinen Vorgängern in der Biologie nicht mehr zu entbehren. Zur Frage nach der Entwicklung der Organismen kam die Frage nach dem ersten Auftreten von Organen und Organsystemen und der Differenzierung von Geweben und Zellen. Der Streit um Präformation und Epigenese in den mikroskopischen Bereichen konnte nur ausbrechen und mit so ungewöhnlicher Schärfe geführt werden, weil der Entwicklungsgedanke, der sich sonst auf allen biologischen Gebieten durchsetzte, bei der Betrachtung von Organellen kleinster Dimensionen nicht genügend berücksichtigt wurde. Man darf also der mit morphologischen Belegen zu treffenden Entscheidung des Streites um die phylogenetische Präformation des Zellkerns, der Centrosomen und ihrer Abkömmlinge mit einigem Recht vorausgreifen, in der Annahme, daß diese Strukturen nicht dem allgemeinen Gesetz der Entwicklung entzogen sind. Ein Hinweis auf die Virusforschung genügt, um daran zu erinnern, daß nicht alles genetische Geschehen nach den aus mikroskopischen Beobachtungen an höheren Zellen erschlossenen Regeln verläuft. Gemeinsam bleibt allen genetischen Vorgängen die identische Reproduktion von Nucleoproteiden und die Eigenschaft der polymeren Nucleinsäuren, spezifische Proteine zu bilden, die ihrerseits wieder in die biologischen Vorgänge eingreifen. Mit der phylogenetischen Entwicklung geht außerdem insgesamt eine Zunahme des genetischen Materials einher. Die Übertragung identischer Eigenschaften in geordneter Form stellt man sich durch

Verdopplung linearer Anordnungen von eigenschaftsbestimmenden Faktoren vor. Doch hat F. O. SCHMITT vor kurzem die Hypothese ausgesprochen, daß variable Ordnungstendenzen, wie sie organischen Makromolekülen innewohnen, zu einer Ordnung von genetischem Material führen könnten, ohne daß sich dieses als Ganzes verdoppelt und daß die eigenschaftsbestimmenden Faktoren nicht streng lokalisiert, sondern durch die Gesamtanordnung der Teilbausteine gegeben sein könnten. Dann wäre das Wesen der Entwicklung außer in einer Hinzufügung neuartiger Teilbausteine des genetischen Materials in ständigen Änderungen der Gesamtanordnung, in der Massenzunahme des geordneten Materials und im Anwachsen der geordneten Dimensionen zu sehen, wie es in der Tat Mikroskop und Elektronenmikroskop in fortschreitendem Maße enthüllen.

Herrn Dr. R. CAESAR danke ich für kritische Durchsicht der Arbeit.

Literatur

BERNHARD, W., et É. DE HARVEN: Sur la présence dans certaines cellules de mammifères d'un organite de nature probablement centriolaire. C. r. Acad. Sci. Paris **242**, 288—290 (1956).

— — Étude au microscope électronique de l'ultra structure du centriole chez les vertébrés. Z. Zellforsch. **45**, 378—398 (1956).

BISSET, K. A.: Cellular organization in bacteria. In: Bacterial Anatomy, p. 1—18. Cambridge: University Press 1956.

BRADFIELD, J. R. G.: Organization of bacterial cytoplasm. In: Bacterial Anatomy, p. 296 bis 317. Cambridge: University Press 1956.

BRINGMANN, G.: Die Organisation der Kernäquivalente der Spaltpflanzen unter Berücksichtigung elektronenmikroskopischer Befunde. Zbl. Bakter. II, **107**, 40—70 (1952).

CHAPMAN, G. B., and J. HILLIER: Electron microscopy of ultra-thin sections of bacteria. J. Bacter. **66**, 362—373 (1953).

CUMMINS, C. S.: The chemical composition of the bacterial cell wall. Internat. Rev. of Cytol. **5**, 25—50 (1956).

DELAMATER, E. D.: Bacterial chromosomes and their mechanism of division. In: Bacterial Anatomy, p. 215—260. Cambridge: University Press 1956.

GLAUERT, A. M., G. E. ROGERS and R. H. GLAUERT: A new embedding medium for electron microscopy. Nature (Lond.) **178**, 803 (1956).

GRASSÉ, P.-P., N. CARASSO et P. FAVARD: Les ultrastructures cellulaires au cours de la spermiogenèse de l'escargot (Helix pomatia L.). Ann. Sci. nat. zool., II. sér. **18**, 339—380 (1956).

HARTMANN, M.: Allgemeine Biologie, 4. Aufl., 940 S. Stuttgart: Gustav Fischer 1953.

HODGE, A. J., J. D. MCLEAN and F. V. MERZER: A possible mechanism for the morphogenesis of lamellar systems in plant cells. J. Biophys. Biochem. Cytol. **2**, 597—608 (1956).

HOUWINK, A. L., and W. VAN ITERSON: Electron microscopical observations on bacterial cytology. II. A study on flagellation. Biochim. et Biophysica Acta **5**, 10—44 (1950).

KELLENBERGER, E.: Les formes caractéristiques des nucléoides de *E. coli*. Symp. Bact. Cytol. VI. Internat. Congr. Rom, 1953, p. 45—66.

KIMURA, R., Y. ITO and Y. OZAKI: Cytological studies of thiamine decomposing bacteria with the application of ultramicrotomy and electron microscopy. Proc. Jap. Acad. **32**, 63—66 (1956).

LABAW, L. W., and V. M. MOSLEY: Periodic structure in the flagella of *Brucella bronchiseptica*. Biochim. et Biophysica Acta **17**, 322—324 (1955).

LARK, K. G., and O. MAALØE: Nucleic acid synthesis and the division cycle of *Salmonella typhimurium*. Biochim, et Biophysica Acta **21**, 448—458 (1956).

MAALØE, O., and A. BIRCH-ANDERSEN: On the organization of the „nuclear material" in *Salmonella thyphimurium*. In: Bacterial Anatomy, p. 261—278. Cambridge: University Press 1956.

MAALØE, O., and A. BIRCH-ANDERSEN and F. S. SJÖSTRAND: Electron micrographs of sections of *E. coli* cells infected with the bacteriophage T 4. Biochim. et Biophysica Acta **15**, 12—19 (1954).

MARSHAK, A.: Bacterial cytology. Internat. Rev. of Cytol. **4**, 103—114 (1955).

MCQUILLEN, K.: Capabilities of bacterial protoplasts. In: Bacterial Anatomy, p. 127—149. Cambridge: University Press 1956.

MITCHELL, P., and J. MOYLE: Osmotic function and structure in bacteria. In: Bacterial Anatomy, p. 150—180. Cambridge: University Press 1956.

MOORE, D. H., and PH. GRIMLEY: J. Biophys. Biochem. Cytol. **3** (1957) (im Druck).

MUDD, S.: Cytology of bacteria I. Annual Rev. Microbiol. **8**, 1—22 (1954).

NOVIKOFF, A. B.: Electron microscopy: Cytology of cell fractions. Science (Lancaster, Pa.) **124**, 969—972 (1956).

PALADE, G. E.: The endoplasmic reticulum. J. Biophys. a. Biochem. Cytol. **2** (Suppl.), **85—98** (1956).

PIEKARSKI, G.: Die Zellkernäquivalente der Bakterien. Erg. Hyg. **26**, 333—364 (1949).

PONTIERI, G.: Über die sog. „large bodies" von *Escherichia coli*. Zbl. Bakter. I Orig. **165**, 524—530 (1956).

— Osservazioni sul nucleo bacterico in sezioni ultrasottili di E. coli. Giorn. di Microbiologica **1**, 367—374 (1956).

PORTER, K. R.: Proceedings of a conference on tissue fine structure. J. Biophys. a. Biochem. Cytol. **2** (Suppl.), **454** (1956).

RAETTIG, H.: Elektronenmikroskopische Darstellung von Bakterienkleinformen der Enterobacteriaceen. Z. wiss. Mikrosk. **62**, 92—105 (1954).

ROBINOW, C. F.: The chromatin bodies of bacteria. In: Bacterial Anatomy, p. 181—214. Cambridge: University Press 1956.

RUSKA, H.: Die gegenwärtige Entwicklung der morphologischen Grundlagen in Zytologie und Zytopathologie. Zbl. Bakter. I Orig. **166**, 546—553 (1956).

SCHMITT, F. O.: Chromosomes genes and macromolecular systems. Nature (Lond.) **177**, 503—505 (1956).

SJÖSTRAND, F. S.: The ultrastructure of cells as revealed by the electron microscope. Internat. Rev. of Cytol. **5**, 455—533 (1956).

SPOONER, E. T. C., and B. A. D. STOCKER: Bacterial Anatomy, p. 360. Cambridge: University Press 1956.

STARR, M. P., and R. C. WILLIAMS: Helical fine structure of flagella of a motile diphtheroid. J. Bacter. **63**, 701—706 (1952).

STOCKER, B. A. D.: Bacterial flagella: morphology, constitution and inheritance. In: Bacterial Anatomy, p. 19—40. Cambridge: University Press 1956.

Symposium on fine structure of cells, held at the 8th congress of cell biology, Leiden 1954, p. 321. New York: Interscience Publishers, Inc.; Groningen: The Netherlands, P. Noordhoff LTD — Publishers 1955.

VENDRELY, R.: Histochemistry of bacteria. Internat. Rev. of Cytol. **4**, 115—142 (1955).

WATSON, M. L.: The nuclear envelope. Its structure and relation to cytoplasmic membranes. J. Biophys. Biochem. Cytol. **1**, 257—270 (1955).

WEIBULL, C.: Bacterial protoplasts; their formation and characteristics. In: Bacterial Anatomy, p. 111—126. Cambridge: University Press 1956.

WINKLER, A.: Die Bakterienzelle, 124 S. Stuttgart: Gustav Fischer 1956.

VII. Die serologische Diagnostik der Viruserkrankungen des Menschen[1,2]

Von

WALTER HENNESSEN

Mit 2 Abbildungen

Inhalt

A. Allgemeine Grundlagen

Die serologische Virusdiagnostik, wie sie heute bei einer Reihe von Krankheiten üblich ist, wurde durch eine Kette von Erkenntnissen ermöglicht, in denen sich die Besonderheiten der Viruserkrankungen wie die der Virusarten manifestieren.

---

[1] Aus dem Institut für Hygiene und Mikrobiologie der Medizinischen Akademie Düsseldorf (Direktor: Prof. Dr. med. W. KIKUTH).

[2] Herrn Prof. Dr. W. KIKUTH zum 60. Geburtstag am 21. Dezember 1956.

Darüber hinaus kann die SVD (serologische Virusdiagnostik) wohl als typisches Ergebnis einer Synthese zwischen klinischer Empirie und naturwissenschaftlichem Denken gelten, wie eine solche in der medizinischen Mikrobiologie häufig zu finden ist.

Erkenntnisse, die dazu verhalfen, daß es heute für viele Viruserkrankungen eine leistungsfähige SVD gibt, wurden aus fast allen Gebieten der Medizin zusammengetragen; es seien in diesem Zusammenhang nur die Beiträge von Innerer Medizin, Kinderheilkunde, Pathologie, Epidemiologie, Chemie, Physik usw. aufgezählt, um herauszustellen, daß der Fortschritt, welcher schließlich im serologischen Laboratorium erzielt wurde, in Wahrheit ein Erfolg der gesamten Medizin und nicht derjenige einer ihrer Disziplinen ist. Es wird sich im Laufe dieser Ausführungen zeigen lassen, daß tatsächlich kein Fortschritt erfolgen konnte, wenn nur eine Sparte gesondert aufstrebte, ohne daß der notwendige Unterbau in gemeinsamer medizinischer Zusammenarbeit erstellt worden wäre. Als Beispiel sei hier auf die jüngsten Ergebnisse der Virologie verwiesen, die in der Isolierung zahlreicher neuer Virusarten bestehen, für welche in der Humanmedizin kein Krankheitssubstrat bekannt ist, so daß diese Viren „isoliert", ja verwaist dastehen und konsequenterweise auch als Waisen- (Orphan-) Viren bezeichnet worden sind.

Wenn bei der Erörterung der Voraussetzungen einer SVD Selbstverständlichkeiten, ja Binsenwahrheiten erwähnt werden wie die Tatsache, daß nur solche Erkrankungen für eine serologische Diagnose geeignet sind, deren Ätiologie klar ist, so geschieht dies aus der Absicht, vom Grundsätzlichen her die Entwicklung der heutigen Situation auf diesem Gebiet zu verstehen.

Die Vielzahl von Voraussetzungen für eine SVD läßt sich vielleicht am ehesten an den Grundzügen der serologischen Diagnose von Infektionskrankheiten im allgemeinen aufzeigen, um dann für die Viruserkrankungen im besonderen dargelegt zu werden.

Grundzüge der Serodiagnose von Infektionskrankheiten

Die serologische Diagnose von Infektionskrankheiten beruht auf dem Phänomen, daß ein infizierendes Agens den Organismus zur Bildung von Abwehrstoffen, Antikörpern, veranlaßt, daß also der Ansteckungsstoff als Antigen wirksam wird. Nachdem sich herausgestellt hatte, daß dies grundsätzlich der Fall ist (EHRLICH), lag es nahe, den Nachweis dieser spezifischen Antikörper auch diagnostisch auszuwerten, indem man aus bekannten infektiösen Agentien Antigene herstellte und diese in vitro mit dem Serum von Patienten reagieren ließ. Die Art der Reaktion ist hier im Grunde belanglos, obwohl sich bei den Viruskrankheiten zeigen lassen wird, daß ihr in diesem Spezialfall eine entscheidende Bedeutung zukommen kann. Den Versuchen einer serologischen Diagnose lag nun keineswegs nur der Gedanke zugrunde, durch eine relativ einfache in vitro-Reaktion die oft mühselige Suche nach dem Krankheitserreger aus Patientenmaterial zu umgehen, entscheidend war vielmehr die Erkenntnis, daß eine Reihe von Krankheitserregern ubiquitär vorkommt, daß demnach der Nachweis dieser Erreger grundsätzlich nicht schon die ätiologische Klärung eines Krankheitszustandes bedeuten kann, sofern nicht der Nachweis des Erregers aus dem Blut gelingt. Seit es sich gezeigt hatte, daß symptomenfreie Keimträger existieren, mußte der Nachweis auch pathogener Keime aus dem Exkret oder Sekret eines Patienten prinzipiell an Beweiskraft verlieren [HENOCH (*54*)]. Als die Untersuchungen von NICOLLE (*85*) zum Begriff der „inapparenten Infektion" führten, mußte schließlich RIVERS (*89*) die Definition der Infektion neu fassen als Kontakt des Erregers mit dem Wirt unter spezifischer AK-Bildung, wobei der Erregernachweis wie auch die Auslösung eines bestimmten Krankheitsbildes nicht mehr als Beweis für eine stattgehabte Infektion postuliert wurden [vgl. auch GÄDECKE (*38*)]. In der Bildung von AK ist demnach das untrügliche Kriterium dafür zu sehen, daß sich der Wirtsorganismus tatsächlich mit dem Erreger auseinandergesetzt hat. Daraus läßt sich folgern, daß bei Infektionskrankheiten, deren Erreger bekannt sind, der AK-Nachweis, wie er bei der serologischen Diagnostik erfolgt, grundsätzlich eine Infektion mit größerer Sicherheit anzuzeigen vermag als der Nachweis der Erreger selbst.

Aus dem Gesagten geht hervor, daß die Erkenntnisse, welche zur Neufassung des Infektionsbegriffes führten, notwendigerweise den Begriff der serologischen Diagnostik schärfer festlegen. Diese kann nur in dem Sinne erfolgen, daß der Zweck jeder Sero-Diagnostik im Nachweis der Infektion gesehen werden muß und nicht etwa im Nachweis der Erkrankung. Es muß als Überforderung des serologischen Labors betrachtet werden, wenn allein auf Grund des Ausfalls eines Reagensglasversuches die Krankheit eines Menschen diagnostiziert werden soll. Daß hier auf diese selbstverständliche Beschränkung der Sero-Diagnostik verwiesen wird, geschieht unter dem Eindruck der sich häufenden Berichte über Doppelinfektionen [vgl. auch HENGEL (*46*)], die zu besonderer Vorsicht bei der Bewertung von serologischen Reaktionsausfällen mahnen.

Es ist nicht ohne Reiz, der Verwirklichung der Erkenntnis, daß nur dann eine Infektion vorliegt, wenn es zu einer Wirt—Erreger—Reaktion gekommen ist, in der Praxis der Diagnostik nachzugehen. Dabei zeigt sich, daß dem genannten Prinzip bei keiner der klassischen, bakteriellen Infektionskrankheiten entsprochen werden kann, soweit es sich um die Frühdiagnose handelt. Der Grund hierfür liegt sowohl in der Pathogenese wie auch im rein Technischen. Der Nachweis der Erreger im Blute des Patienten macht bei vielen bakteriellen Infektionen keine nennenswerten, technischen Schwierigkeiten, so daß kein Bedarf für Erleichterung der Diagnostik durch serologische Methoden besteht. Darüber hinaus bringt es die Pathogenese bakterieller Erkrankungen mit sich, daß die Bildung von AK oft erst relativ später erfolgt. Diese Verhältnisse sollen bei der Besprechung der Besonderheiten von Virusinfektionen weiter unten dargelegt werden. Für die Laboratoriumsdiagnose ergibt sich daraus ein Vorgehen, wie es sich z. B. am Typhus abdominalis demonstrieren läßt. Die frühzeitige Laboratoriumsdiagnose beruht hier auf dem Erregernachweis im Blute des Patienten; erst im fortgeschrittenen Stadium gewinnt dann die serologische Erfassung der Infektion (Gruber-Widal-Reaktion) an Bedeutung. Ähnliche Verhältnisse liegen bei Lues-Infektionen vor. Die Frühdiagnose wird hier an Hand des Spirochätennachweises im Primäraffekt gestellt. Erst Wochen danach wird der Fall serologisch mit der Wa.R. faßbar, wobei hier außer acht gelassen werden darf, daß es sich bei der klassischen Diagnostik um unspezifische Antigene handelt.

Gleichartige Beispiele ließen sich noch zahlreich anführen, aber schon aus den genannten scheint genügend klar hervorzugehen, daß eine wesentliche Voraussetzung für die erfolgreiche Frühdiagnose aus dem Patientenserum darin besteht, daß die in Frage stehende Infektion nicht nur grundsätzlich zur AK-Bildung führt, sondern daß diese Reaktion auch frühzeitig im Krankheitsverlauf einsetzt. Wo dies nicht der Fall ist, beschränkt sich der Wert jeder Serodiagnose auf die Klärung epidemiologischer Fragestellungen, wenn nicht der spezifische Ablauf der Erkrankung, wie z. B. bei der Lues, bei langen Ruhepausen ihre Erfassung nur noch serologisch ermöglicht, so daß hier die Serodiagnostik klinisch unentbehrlich wurde.

In diesem Zusammenhange gewinnt eine weitere Eigenschaft der gebildeten AK an Bedeutung. So selbstverständlich es ist, daß nur der bald nach der Infektion auftretende AK eine serologische Frühdiagnose erlaubt, so notwendig erscheint aber auch, daß diese AK nicht zu lange nach der Infektion nachweisbar bleiben. Als Ideal eines diagnostisch verwertbaren AK wäre demnach ein solcher zu bezeichnen, dessen Persistenz mit dem Infektionsablauf synchron verläuft. Ein derartiger AK hat sich bisher nicht auffinden lassen, vielmehr ist in allen Fällen ein längeres Verweilen der AK als der auslösenden Erreger im Wirtsorganismus zu beobachten. Für die Diagnose einer Infektion, welche nicht nur die Art der Ansteckung, sondern tunlichst auch den Zeitpunkt derselben andeuten sollte, ist es daher unzweckmäßig, den Nachweis derjenigen AK zu wählen, die als Indicatoren einer Immunität angesehen werden müssen. Dies spielt vor allem bei denjenigen Krankheiten eine Rolle, nach welchen eine lang dauernde Immunität zu erwarten ist. Hier liegt einer der Gründe für die beschränkte Brauchbarkeit beispielsweise der Gruber-Widal-Reaktion bei der Diagnose des Typhus abdominalis, denn die AK, welche dabei erfaßt werden, stehen in engem Zusammenhang mit der Typhus-Immunität, so daß ein positiver Ausfall nicht nur durch eine akute Infektion, sondern auch durch eine lange zurückliegende Auseinandersetzung des Organismus mit den Erregern hervorgerufen werden kann. Ein Erreger-Wirt-Kontakt kann schließlich auch durch eine aktive Immunisierung ausgelöst worden sein, so daß sich der Wert der Gruber-Widal-Reaktion für die Diagnostik verschlechtert, wenn es sich um Typhus-schutzgeimpfte Personen handelt.

Auch für die Bedeutung der Persistenz von AK für eine serologische Diagnose kann das Beispiel der Typhusinfektion herangezogen werden, da sich hier einige Unterschiede im Verhalten der einzelnen AK nachweisen lassen. Diese Unterschiede der Typhus-AK werden bei der Serodiagnose insofern berücksichtigt, als der Nachweis von Typhus 0-AK und der von Ty.-Vi-AK einen besseren Hinweis auf eine Infektion darstellt als etwa derjenige von Ty.-H-AK. Wie bei den meisten bakteriellen Erkrankungen liegt aber auch beim Typhus in der Serodiagnostik ein erheblicher Unsicherheitsfaktor, der ihren Wert, welcher schon durch das späte Auftreten der AK vermindert wird, weitgehend einschränkt [Schmidt (*104*)].

Die angeführten Beispiele führen zu der Folgerung, daß für eine brauchbare, d. h. frühe und zuverlässige Serodiagnostik der Nachweis von solchen AK zweckmäßig ist, welche frühzeitig auftreten, um möglichst bald nach Abklingen der Infektion aus dem Blute zu verschwinden, die also als *flüchtige* Antikörper bezeichnet werden können.

Bei Infektionen, in welchen keine flüchtigen AK bekannt sind, muß selbstverständlich auf persistierende zurückgegriffen werden. Es soll aber betont werden, daß damit die frühzeitige Serodiagnose grundsätzlich unmöglich wird, weil in diesen Fällen nur die Differenz des AK-Spiegels in zwei zeitlich auseinander liegenden Serumproben für die akute Infektion beweisend ist. Als Beispiel sei hier der Hirst-Test bei Influenza und die KBR bei der Ornithose angeführt, die beide persistierende AK nachweisen und deshalb für die Frühdiagnose nur beschränkt brauchbar sind.

Mit der Gegenüberstellung der Bedeutung von flüchtigen und von persistierenden AK für die Serodiagnostik kann sich die Erörterung von den serodiagnostischen Grundzügen den

Besonderheiten der Viruskrankheiten

in serologischer Beziehung zuwenden. Dabei sei hervorgehoben, daß schon der soeben gestreifte Fragenkomplex der Persistenz von AK erst bei den Virusinfektionen systematisch angegangen worden ist, so daß die Kenntnisse auf diesem Gebiete weitgehend von der Virologie vermittelt wurden. Dieser Umstand erklärt sich aus der Tatsache, daß bisher eigentlich nur die Viruskrankheiten die Möglichkeit boten, der unterschiedlichen Persistenz der verschiedenen AK nachzugehen (s. weiter unten).

Diejenigen Eigentümlichkeiten der Virusinfektionen, welche sie von den bakteriellen Infektionen unterscheiden, sind die gleichen, welche auch ihre serologischen Besonderheiten hervorrufen. Dabei steht die Natur ihrer Erreger im Vordergrund, da der Erreger einer Infektionskrankheit als Antigen die serologisch entscheidende Rolle spielt.

So wenig exakte Daten über die chemische Struktur der Virusarten vorliegen, so sehr ist doch ihre ausgezeichnete antigene Wirksamkeit bekannt. Die Spezifität von Virusantigenen ist so groß, daß es z. B. gelingt, mit dem rohen Preßsaft virusinfizierter Pflanzen im Tierkörper AK hervorzurufen, die den serologischen Nachweis zum empfindlichsten Hilfsmittel beim Studium von pflanzenpathogenen Viren macht [Pollard (*87*)]. Dies ist um so erstaunlicher, als keine Anzeichen dafür bekannt sind, daß sich diese Pflanzenviren etwa im Tier vermehren würden; die antigene Wirkung kann hier nur von der injizierten Menge Virus ausgehen. Die ausgezeichnete antikörperbildende Potenz ließ sich für alle bekannten Virusarten nachweisen [Luria (*74*)], so daß die Grundforderung für eine serologische Diagnose erfüllt ist.

Es bleibt demnach zu untersuchen, wie sich bei Virusinfektionen der zeitliche Ablauf der AK-Bildung zur Entwicklung der Infektion verhält, welche Eigenschaften die AK haben, und welche Eigentümlichkeiten der Virusarten als Antigene von serodiagnostischer Bedeutung sind.

Über den Verlauf einer Virusinfektion ist sehr wenig bekannt, da es der virologischen Technik erst seit kurzer Zeit möglich ist, in das Geschehen vom Zeitpunkt des ersten Zusammentreffens Virus — Organismus bis zum Ausbruch der manifesten Erkrankung, also in die Inkubationszeit, hineinzuleuchten [Burnet (*23*)]. Es erscheint aber berechtigt, die Beobachtungen Fenners (*34*) bei einer Virusinfektion der Maus als typisch für den Ablauf von Viruskrankheiten anzusehen, um so mehr als sie mit Einzelbefunden bei Virusinfektionen des Menschen, besonders aber mit dem von BODIAN (*16*) weitgehend geklärten Infektionsverlauf bei der Poliomyelitis übereinstimmen. Danach kommt es bei Virusinfektionen in der Regel schon im Inkubationsstadium zu einer primären Virämie, die im Falle der Mäuseektromelie und bei der Poliomyelitis 3 Tage post infect. und rund 5 Tage vor Auftreten der klinischen Manifestation nachgewiesen werden konnte. Ohne auf den Ort der AK-Entstehung oder auf denjenigen der Vermehrung des Virus in der Inkubationszeit eingehen zu wollen, läßt sich nach dieser frühen Virämie in jedem Falle auch eine frühzeitige AK-Bildung erwarten.

Es kann daher wohl mit Recht angenommen werden, daß durch die primäre Virämie im Inkubationsstadium noch vor der Manifestation der betreffenden Viruserkrankung die AK-Bildung in Gang gesetzt wird. Abbildung 1 versucht,

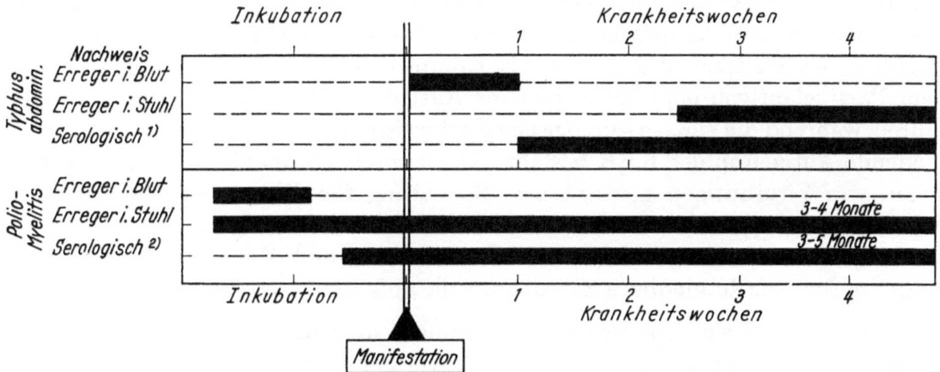

Abb. 1. Schema des zeitlichen Ablaufs von Laboratoriumsbefunden bei bakterieller und bei Virus-Infektion.

diese Verhältnisse für die Poliomyelitis und die S. typhi Infektion klarzustellen, indem der zeitliche Ablauf der Infektionen schematisch verglichen wird, soweit dieser im Laboratorium faßbar ist. Selbstverständlich läßt sich nicht jede Angabe dieses Schemas verallgemeinern, man geht aber wohl in der Annahme nicht fehl, daß in diesem unterschiedlichen Verhalten bei beiden Krankheiten auch Gründe für die verschiedenartige Anwendung diagnostischer Methoden zu suchen sind. Ausnahmen bei der Vielzahl bakteriologischer und Viruserkrankungen verbieten eine kritiklose Übernahme für jede Krankheit ganz von selbst.

Wenn eine Virusinfektion schon früh zu einer AK-Bildung führt, dann werden in diesem Zusammenhang einige Ausführungen über das Virusantigen notwendig. Die ausgezeichneten Antigen-Qualitäten von Virussuspensionen wurden schon erwähnt, sie lassen jedoch einen unmittelbaren Schluß auf die antigenen Eigenschaften bei einer Infektion nicht zu, da man sich eine Infektion so vorzustellen hat, daß kleine Virusdosen die Ansteckung auslösen, daß aber die Infektion erst „angeht" und unterhalten wird, wenn sich die anfänglich kleine Virusmenge in den empfänglichen Zellen des Wirtsorganismus vermehrt. Die serologische Wirkung einer Infektion muß dementsprechend von den antigenen Eigenschaften des Virus im Stadium der Vermehrung abhängen.

Die Vermehrung der infektiösen Viruseinheiten ist auch heute noch weitgehend unklar. Untersuchungen der jüngsten Zeit haben aber doch einige Fakten erbracht, die von serologischer Bedeutung sind. So zeigten die Experimente von BURNET (*23*), SCHÄFER (*101*), (*102*) und anderen Autoren, daß bei der Virusvermehrung neben den eigentlichen Partikeln auch noch kleinere Einheiten auftreten, die anscheinend mit dem „Soluble antigen“ [HOYLE und FAIRBROTHER (*60*)] identisch sind. Es handelt sich dabei um kleinste corpusculäre Elemente, die oft nur $^1/_{10}$ oder noch weniger der Größe des eigentlichen infektiösen Virusteilchens ausmachen — aus diesem Grunde wurden sie anfänglich als „gelöst“ angesehen —, die selbst weder infektiös noch vermehrungsfähig, als Antigene aber hochwirksam sind. Ihre Rolle bei der Virusvermehrung soll hier nicht erörtert werden. Der Klarheit halber sei lediglich erwähnt, daß SCHÄFER (*101*), (*102*) zeigen konnte, daß Viruspartikel nach Ätherbehandlung große Massen der kleinen Elemente freisetzen, wobei die Viruspartikel eine Art von Auflösung erfahren. Auf Grund ihrer Fähigkeit, in Gegenwart von Antikörpern Komplement zu binden, hat es sich eingebürgert, statt des ursprünglichen, „löslichen Antigens“ nun vom komplementbindenden Antigen zu sprechen. Selbstverständlich binden auch die infektiösen Partikel, das Partikelantigen, mit AK Komplement. Es lassen sich jedoch mit dem Partikelantigen eine Reihe anderer Antigen-Antikörper-Reaktionen durchführen, während die einzige serologische Nachweismöglichkeit für das komplementbindende Antigen in der KBR besteht. Die gesonderte Erwähnung dieser beiden Arten von Virusantigen gewinnt ihre Bedeutung für die serologische Diagnose durch weitere Eigenschaften der komplementbindenden Antigene. Es sei vorausgeschickt, daß sie damit diejenigen Forderungen weitgehend erfüllen, welche eingangs für jede brauchbare Serodiagnostik aufgestellt wurden.

Es ließ sich zeigen, daß die komplementbindenden Antigene der verschiedenen Virusarten schon früh im Laufe der Virusvermehrung auftreten, für Influenza konnte HOYLE sogar ihr Auftreten noch vor der Produktion von Viruspartikeln feststellen. Obschon experimentelle Untersuchungen darüber nicht vorliegen, da technische Schwierigkeiten hier eine Grenze setzen, darf wohl angenommen werden, daß auch die komplementbindenden Antigene das Stadium der primären Virämie der Viruspartikel mitmachen, ja daß sie vielleicht darüber hinaus noch im Blute kreisen. Sie rufen daher ebenfalls schon frühzeitig eine AK-Bildung hervor.

Diese AK stehen in keinem Zusammenhang mit der Immunität, so wie auch die sie verursachenden komplementbindenden Antigene selbst nicht infektiös sind. Die komplementbindenden AK verschwinden mehr oder weniger kurze Zeit nach der Infektion aus dem Blute, ohne daß dafür ein Grund angegeben werden könnte, wenn man von hier wohl sicher nicht zweckmäßigen, teleologischen Gesichtspunkten absieht. Die komplementbindenden AK sind also flüchtig und eignen sich deshalb für die Serodiagnose in besonderem Maße.

Als weitere Besonderheit zeigte sich die unterschiedliche Spezifität von Partikel- und komplementbindendem Antigen und entsprechend diejenige der zugehörigen AK. Während die Viruspartikel bei vielen Virusarten stammspezifisch sind, d. h. die Infektion mit einem Stamm schützt nicht oder nur unvollständig gegen eine solche mit einem anderen Stamm gleichen Typs, liegt bei den komplementbindenden Antigenen eine ausgesprochene Typenspezifität vor, die verschiedentlich Anzeichen einer Artspezifität aufweisen [HENNESSEN (*48*)]. Für die Serodiagnostik bedeutet das eine außerordentliche Erleichterung insofern, als damit die oft verwirrende Vielzahl von Virus-Erregerstämmen gleichen Typs durch ein all diesen Stämmen gemeinsames Antigen erfaßt werden kann, wie dies z. B. bei der Influenza der Fall ist.

In Tabelle 1 sind diejenigen Virusarten angeführt, bei welchen bisher der Nachweis von Partikel- und komplementbindendem Antigen als getrennte Substanzen gelungen ist. Obwohl hier noch die eine oder andere seltenere Virusart mit aufzuführen wäre, zeigt die Tabelle doch mit bemerkenswerter Deutlichkeit daß die Klärung dieser für eine serologische Virusdiagnostik entscheidend wichtige Frage bisher längst nicht bei allen Virusarten, erst recht nicht bei all denen mit klinischer Bedeutung, erfolgt ist. Im einzelnen soll darauf bei der speziellen Betrachtung der verschiedenen Viruskrankheiten eingegangen werden. Es muß aber schon jetzt betont werden, daß die theoretischen Erkenntnisse die serologische Virusdiagnostik auf einen sicheren Boden zu stellen vermögen, daß aber praktische Hindernisse auf diesem Gebiete einem befriedigenden Arbeiten noch allzu häufig entgegenstehen.

Tabelle 1. *Struktur einiger Virusantigene*

Virusart	Infektiöse Partikel nachgewiesen	Gesondertes komplementbindendes Antigen nachgewiesen
Influenza	+	+
Mumps	+	+
Psittakose-Ornithose	+	+
Lymphogranuloma inguin.	+	(+)
Trachom	+	(+)
Einschlußblennorrhoe	+	(+)
Katzenkratzkrankheit	+	(+)
Adenoviren (APC)	+	+
Epidemische Keratoconjunctivitis	(+)	?
Variola	+	+
Vacinia	+	+
Varicellen	(+)	?
Herpes zoster	(+)	?
Herpes simplex	+	?
Maul- und Klauenseuche	+	+
Poliomyelitis	+	+
Lymphocytäre Choriomeningitis	+	+
Coxsackie-Viren	+	?
Gelbfieber	+	+
Virusencephalitiden	+	?
Denguevirus	+	?
Pappatacifieber	+	?
Tollwut	+	?
Schnupfen	0	0
Masern	0	0
Röteln	0	0
Primäre atypische Pneumonie	0	0
Hepatitis	0	0

Hindernisse für die serologische Virusdiagnose

Als wichtigster Hemmschuh einer virologischen Serodiagnostik wurden die mangelnden Kenntnisse einer Reihe von Virus-Antigenen schon erwähnt. Es ist dadurch schwierig, wenn nicht unmöglich, brauchbare Antigene herzustellen, da diese die Voraussetzung jeder serologischen Diagnostik sind.

Darüber hinaus aber verdient eine Reihe von Erschwerungen Erwähnung, die mit dem eingangs skizzierten Erfordernis zusammenhängen, nach dem an einem Fortschreiten der Medizin eben alle Disziplinen beteiligt sein müssen. Dies ist selbst bei denjenigen Krankheiten nicht in dem notwendigen Maße der Fall, welche seit langem als Virus-Infektionen erkannt sind. So ist z. B. die Vorstellung von der Entität der Krankheitsbilder weitgehend ungenügend. Es ist nicht in ausreichendem Maße bekannt, daß sich alle Virusinfektionen der gemäßigten Klimazone insofern mit dem Anblick von Eisbergen vergleichen lassen, als bei ihnen der weitaus größte Teil der Infektionen symptomlos verläuft, ein ebenfalls erheblicher Teil derart symptomarm bleibt, daß klinisch keinerlei Abgrenzung erfolgen kann, und daß nur ein verschwindend geringer Prozentsatz schließlich diejenige Manifestation zeigt, welche zum Begriff der einzelnen Erkrankung geführt hat. In diesem „Eisberg-artigen" Verhalten der Virusinfektionen, das auf die meist sehr hohe Kontagiosität der Virusarten zurückgeführt werden muß, liegt nun zwar kein grundsätzliches Hindernis für eine Serodiagnose, jedoch bleibt für die Praxis einer solchen die Gefahr, daß bei einem notwendigerweise häufigen Nachweis von Virusinfektionen im Einzelfall ein Zweifel an der Zuverlässigkeit auftaucht. Daraus ergibt sich erneut die Notwendigkeit darauf hinzuweisen, daß im serologischen Laboratorium niemals eine Krankheitsdiagnose, sondern jeweils nur die Infektions-Diagnose zu stellen ist. Als Beispiel seien in diesem Zusammenhang Untersuchungen von FOX (*35*) erwähnt, der wie andere Autoren erneut nachweisen konnte, daß bei der Poliomyelitis auf einen manifesten Fall fast 1000 symptomlose Infektionen zu erwarten sind.

Als weitere Schwierigkeit der SVD kommt der Booster-Effekt und mit ihm die heterotype Reaktion in Betracht. Wie andere Antigene [SCHMIDT (*104*)], so rufen auch die Virusantigene bei Reinfektion ein schnelles Wiederauftreten von AK hervor, die nach der 1. Infektion verschwunden oder abgesunken waren. Dadurch kann es einerseits zu einer Vortäuschung einer Infektion mit einem bestimmten Virus kommen, welche in Wahrheit durch ein anderes, wenn auch verwandtes Virus verursacht wurde; zum anderen aber ist in der Virusserologie durchaus die Möglichkeit einer „anamnestischen Reaktion" im Sinne BIELINGS (*13*) zu erwarten, wobei ein unspezifischer, parenteraler Reiz zum Wiederauftreten von früher einmal gebildeten AK führen kann. Erschwerungen der SVD, die, über diese allgemeinen Erscheinungen hinausgehend, auftreten können, sollen im speziellen Teil erwähnt werden.

Methodik der SVD

Da sich weiter oben zeigen ließ, daß der Nachweis von flüchtigen, früh auftretenden AK für jede Serodiagnose erwünscht ist, daß weiter die komplementbindenden Virusantigene zur Bildung von AK mit diesen Eigenschaften führen, welche darüber hinaus noch den Vorteil haben, typenspezifisch mit einem Antigen den Nachweis vieler Stämme zu erlauben, läßt sich folgern, daß im Nachweis der komplementbindenden AK mit der KBR die Methode der Wahl für die SVD gegeben ist.

Im Laufe der Zeit wurde beinahe für jede Virusart eine besondere Technik angegeben, ein Umstand, der sich aus der Verschiedenartigkeit der Antigene erklärt. Zur Veranschaulichung sei hier die Abb. 2 angeführt. Diese graphische Darstellung zeigt Komplement-Bindungs-Reaktionen mit einer Anzahl von Antigenen und den zugehörigen Immunseren. Dabei zeigt sich auch im graphischen Bilde, daß die erreichbaren oder tatsächlich erreichten Maximaltiter der AK bei den verschiedenen Krankheiten schwanken, wobei die Extreme der Abb. 2, Ornithose und Poliomyelitis, nicht einmal als Endwerte angesehen werden können [LIPPELT und BRAND (*73*), HENNESSEN (*49*)].

Allen Methoden der Virus-KBR ist jedoch gemeinsam, daß als Antigen eine Virussuspension benutzt wird, die meist aus Gründen der Sicherheit gegen Labor-infektionen inaktiviert wird. Die Virussuspensionen stellen entweder Extrakte aus virusbefallenen Organen dar, oder sie werden aus virusinfizierten Bruteiern gewonnen, oder es handelt sich um Virus aus Gewebekultur. Die Reinigung oder Anreicherung der Antigene erfolgt bei Organantigen durch fraktionierte Aufschließung, durch die der Organanteil im Endprodukt reduziert wird. Ei-Antigene erfordern meist sehr viel weniger Aufbereitung, während Gewebekultur-Antigene entweder fast ohne Vorbereitung oder durch die Ultrazentrifuge konzentriert benutzt werden. In der Praxis erfolgt nur selten eine Trennung in Partikel und komplementbindendes Antigen, so daß die Beurteilung und Bewertung von Virus-KBR einige Erfahrung verlangt. Als Beispiel sei auf die Influenza verwiesen. Hier kann bei Benutzung der üblichen Mischantigene aus Allantoisflüssigkeit von Bruteiern eine positive Reaktion sowohl durch das Partikel-Antigen als auch durch das komplementbindende Antigen hervorgerufen sein, denn selbstverständlich wird bei der Bindung der neutralisierenden AK an die Viruspartikel Komplement verbraucht. Da neutralisierende AK von einer lange zurückliegenden Infektion stammen können, läßt sich nicht unbedingt auf den Zeitpunkt der zu beurteilenden Ansteckung schließen. Bei einiger Erfahrung wird aber die Höhe des AK-Spiegels darüber Auskunft geben, wie die Bewertung zu erfolgen hat.

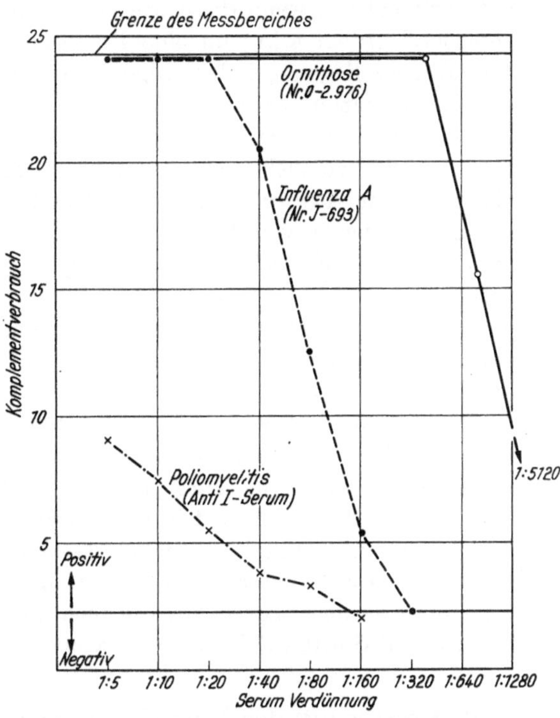

Abb. 2. Quantitative Komplementbindungsreaktionen mit verschiedenen Virusantigenen und den zugehörigen Antiseren.

Ein weiterer Zug, der allen Komplement-Bindungs-Reaktionen mit Virusantigen gemeinsam ist, liegt in der Dauer der Bindungszeit. Diese soll in jedem Fall länger als 30 min betragen und erfolgt am zweckmäßigsten über Nacht bei + 4° C [Kolmer (*67*)]. Es hat sich durchweg eingebürgert, Virus-KBR als Mikro-Reaktion unter Verwendung kleinster Mengen durchzuführen. Der Grund hierfür liegt vor allem in der stets sehr kostspieligen und zeitraubenden Antigenherstellung, die Einsparungen an Material wünschenswert macht. Wie bei jeder serologischen Reaktion ist auch bei Komplementbindungen mit dem Vorkommen von Hemmstoffen (Inhibitoren) in den zu prüfenden Seren zu rechnen. Diese Inhibitoren können den negativen Ausfall der Reaktionen trotz sicherer Infektion herbeiführen, so daß ein gewisser Prozentsatz der Fälle serologisch refraktär bleibt [Schmidt et al. (*105*)]. Die Anzahl der so nicht erfaßbaren Seren ist aber gering [Melnick (*78*)]. Interessanterweise kann ein Serum, das in der Komplementbindung mit dem einen Virusantigen refraktär ist, sehr wohl mit einem anderen Virusantigen reagieren, wie dies bei Doppelinfektionen nachgewiesen werden konnte [HENNESSEN (*51*)].

Neben der KBR finden weitere Reaktionen in der SVD Verwendung. Der Natur der Viren entsprechend, verbieten sich eine Reihe der in der übrigen Serologie möglichen Verfahren. So werden Präcipitationen und Agglutinationen— von Ausnahmen mehr theoretischer Natur abgesehen — mit Viruslösungen dadurch unmöglich, daß man mit Virusaufschwemmungen praktisch nie die Mindestkonzentration reaktionsfähiger Teilchen erreicht, welche für diese Reaktionen gefordert werden müssen. Der Grund für die notwendigen, hohen Konzentrationen liegt in der Tatsache, daß bei diesen Reaktionen die Masse der Partikel entscheidend ist; wenn diese wie bei Viren sehr klein ist, muß ihre Anzahl entsprechend größer sein, um eine Reaktion hervorzurufen [LURIA (*74*)]. Andere Verhältnisse liegen bei denjenigen Virusarten vor, welche die Eigenschaft zur Hämagglutination besitzen. Als Prototyp der Hämagglutinations-Hemmungs-Reaktion ist der Hirst-Test anzuführen [HIRST (*56*)]. Bei dieser Art der Untersuchung wird versucht, die hämagglutinierende Fähigkeit der Viruspartikel, die als spezifische Adsorption zwischen Erythrocyten und Virusteilchen aufzufassen ist, durch Zusatz von Patienten-Serum zu hemmen. Damit ist der H.A.H.T. praktisch eine Methode zur Erfassung von virusneutralisierenden AK. So unentbehrlich er für die Untersuchung der entsprechenden Virusarten ist, so wenig hat er sich für die Diagnose von Virusinfektionen bewährt, da eben ganz allgemein der Nachweis neutralisierender AK für die Diagnose einer Infektion unzweckmäßig ist.

Mit diesen allgemeinen Bemerkungen kann das Gebiet der Methodik verlassen werden. Es wurde bewußt davon abgesehen, die unspezifischen Nachweisverfahren zu erörtern. Ihre mangelnde Spezifität, ihre widersprechende Beurteilung wie die Unklarheit ihrer Zusammenhänge zur Virusinfektion verhindern ihre Zuordnung zum eigentlichen Gebiet der serologischen Virus-Diagnostik. Darüber hinaus darf hier und später auf die Abhandlung der Methodik im einzelnen verzichtet werden. Es sei für die technischen Einzelheiten auf die Bücher von LEPINE und SOHIER (*71*) und von KLÖNE (*65*) verwiesen.

B. Erörterung des Vorgehens bei den einzelnen Virusinfektionen

Bei der Besprechung der Möglichkeiten einer SVD für die verschiedenen Viruserkrankungen taucht eine Schwierigkeit auf, die in ihrer Banalität symptomatisch für unsere Kenntnisse in der Virologie ist. Es zeigt sich, daß für die Einordnung, ja für die Reihenfolge, in welcher die Infektionen abgehandelt werden, ein verbindliches, ordnendes Prinzip nicht existiert. Wie schon erwähnt, macht die Vielzahl von Krankheitsbildern, welche von der gleichen Virusart hervorgerufen werden können, den Krankheitszustand als Bezugssystem für die Virusarten unbrauchbar. Aus dieser Ursache war es bisher unmöglich, dem Versuch einer Systematik der Virusarten das Krankheitsbild zugrunde zu legen [RUSKA (*93*), MINER (*82*)]. Da eine medizinische Einordnung ausgeschlossen ist, resultiert der merkwürdige Umstand, daß die einschlägigen Werke der Weltliteratur sämtlich nach Gesichtspunkten angeordnet sind, die weniger durch die Materie, vielmehr aber durch das jeweilige Interessengebiet der Verfasser bedingt sind [GILDE-

MEISTER (*39*), DOERR-HALLAUER (*30*), v. ROOYEN-RHODES (*91*), RIVERS (*90*), GRUMBACH-KIKUTH (*43*)].

Für die vorliegende Erörterung scheint ein Kompromiß zweckmäßig, das in der Übernahme der Gruppeneinteilung der *Com. on Nomenclature* besteht, wobei im Anschluß an diese Gruppen auch klinisch verwandte Virusarten besprochen werden. Der Anspruch auf logische Gliederung mußte daher zugunsten der größeren Übersichtlichkeit für den medizinischen Betrachter aufgegeben werden.

Influenzagruppe

Die Influenza-Viren können wohl mit Recht als Virusart gelten, deren Erforschung am weitesten vorgetrieben werden konnte. Der Umstand, daß die Züchtung des Influenza-Virus im Brutei keine Schwierigkeiten macht, erlaubt die Gewinnung von qualitativ wie quantitativ befriedigenden Antigenen. Für die Sero-Diagnose hat sich besonders die KBR als nützlich erwiesen, so daß die Hämagglutinations-Hemmung-Reaktion (Hirst-Test) für die Diagnose mehr zurücktritt. Bei der Influenza-KBR ist es von besonderem Vorteil, daß die verwendeten Antigene typenspezifisch reagieren, andernfalls wäre bei diesem anscheinend besonders „mutationsfreudigen" Virus eine Serodiagnose a priori unmöglich. Darüber hinaus ist es bei der hohen Durchseuchung der Bevölkerung von entscheidender Bedeutung, daß die komplementbindenden AK schon wenige Monate post infectionem zur Norm absinken [BURNET (*23*)].

Für die Methodik der Influenza-KBR wurden von zahlreichen Autoren Modifikationen angegeben, nachdem HOYLE und FAIRBROTHER (*60*) die Reaktion beschrieben hatten. Die Auswahl der jeweiligen Technik ist für das Ergebnis selbst unerheblich, sie kann daher dem einzelnen Untersucher überlassen bleiben, sofern die allgemeinen Erfordernisse von Virus-KBRen gewahrt bleiben. Es sei hier lediglich auf das von FULTON und DUMBELL (*37*) angegebene Verfahren hingewiesen, da sich dieses durch maximale Exaktheit auszeichnet, wenn auch die diagnostische Brauchbarkeit in der Praxis durch eine gewisse Umständlichkeit beeinträchtigt wird. Diese Methode wird daher in der ganzen Virologie mit Recht besonders häufig angewandt, wenn es um die Klärung von Reaktionsmöglichkeiten mit neuaufgefundenen Virusantigenen geht. Das Prinzip der Reaktion beruht auf der gleichlaufenden Titrierung von fallenden Serumverdünnungen gegen fallende Komplementdosen bei konstantem Antigen.

Trotz der günstigen Voraussetzungen für eine Serodiagnose der Influenza sind die diagnostischen Ergebnisse oft unbefriedigend. Dies ist in epidemiefreien Intervallen mehr der Fall als zur Zeit einer Epidemie. Bei endemischem Auftreten kommt die hohe Durchseuchung der Bevölkerung nicht nur mit Influenza-Viren, sondern auch mit anderen Virusarten ähnlicher Pathogenität zur Auswirkung. Während bei wirklichen Influenza-Epidemien oder Pandemien ein beträchtlicher Teil der Bevölkerung erkrankt, so daß die anderen Virusinfektionen der Atemwege relativ zurückgehen, ist dies bei endemischem Auftreten nicht der Fall. Es kommen dann vielmehr Patienten zur Untersuchung, die sich klinisch kaum unterscheiden, von denen aber nur ein Teil mit Influenza-Virus infiziert ist. Der Reaktionsausfall verliert dabei scheinbar an Spezifität, während in Wirklichkeit dann nur serodiagnostisch zu klären ist, welche Infektion tatsächlich vorliegt. So fanden z. B. BERGE et al. (*10*) unter klinisch gleichartigen Fällen von Erkrankungen des Respirationstraktes nur 10% mit Influenza, während sich für den Rest andere Ursachen nachweisen ließen.

Bei der Influenzadiagnostik liegt also einer jener Fälle vor, der trotz großer virologischer und serologischer Kenntnisse infolge mangelhafter Abgrenzbarkeit des Krankheitsbildes eine praktische Anwendung der Serodiagnostik behindert. In größeren Untersuchungsreihen nimmt die Influenza daher auch nicht ganz den Platz ein, der auf Grund des häufigen Vorkommens zu erwarten wäre [TERZIN et al. (*114*), HENNESSEN (*51*).]

Es soll hier nur kurz auf die Hämagglutinations-Hemmungs-Reaktion (Hirst-Test) eingegangen werden. Das Verfahren wurde von HIRST (*56*) angegeben und von SALK (*99*) weiter ausgebaut. Es beruht auf der Eigenschaft von Influenzaviruspartikeln, bestimmte Erythrocyten zu agglutinieren. Die Agglutination der Blutkörperchen von Mensch, Meerschweinchen und Hühnern wird durch eine Adsorption der Virusteilchen an die Erythrocyten-Membran verursacht, wobei es unter anderem durch das sog. „Brückenphänomen" (ein Partikel an 2 Erythrocyten) zur Zusammenballung der Blutkörperchen kommt. Diese Fähigkeit der Virusteilchen wird durch spezifische Antiseren gehemmt, so daß sich mit konstanten Virussuspensionen die Höhe der im Serum vorhandenen AK feststellen läßt. Die Natur dieser Reaktion als Neutralisation des Viruseffektes schränkt die praktische Verwendung zur Diagnose ein. Wie sich schon weiter oben zeigen ließ, ist der Nachweis von neutralisierenden AK im Serum wegen des lang anhaltenden Vorkommens und wegen ihres Zusammenhanges mit der Immunität unzweckmäßig. Zum Studium der Eigenschaften von Virusstämmen hat sich diese Reaktion aber außerordentlich bewährt, so daß die großen Kenntnisse auf dem Gebiet der Influenza ohne den Hirst-Test undenkbar wären. Mittels der Hämagglutinations-Hemmungsreaktion ließ sich z. B. die epidemiologisch bedeutsame Feststellung treffen, daß Großepidemien anscheinend deshalb von Influenza-A-Virusstämmen hervorgerufen werden, weil diese eine größere Variationsbreite haben, während die konstanteren Eigenschaften der Influenza-B-Stämme zu ihrem mehr endemischen Auftreten führten [BRANS und MULDER (*18*), BOZZO (*17*), HENNESSEN (*47*)].

Zusammenfassend läßt sich feststellen, daß die Serodiagnose der Influenza gut ausgebaut ist. Sie beruht vornehmlich auf der KBR, gegenüber welcher die Hämagglutinations-Hemmungs-Reaktion serodiagnostisch zurücktritt. Ihre Bedeutung für Epidemien ist klar, in epidemiefreien Intervallen vermag sie leicht denjenigen zu verwirren, der sich nicht Rechenschaft gibt über den Formenreichtum, welchen das Influenza-Virus, aber auch andere Virusarten den zugehörigen Krankheitsbildern zu geben vermag.

Mumps

Die Verwandtschaft zwischen dem Mumpsvirus und dem der Influenza führte zu ihrer Einordnung in die gleiche Virusgruppe (Myxovirus), so daß ihre Besprechung an dieser Stelle erfolgen soll. Ihrer Verwandtschaft entsprechend ist auch die Serodiagnose bei Mumps derjenigen bei Influenza sehr ähnlich. Die diagnostische Situation bei Mumps ist sogar insofern noch günstiger als bei der

Grippe, als die Virusart Mumps anscheinend keine verschiedenen Serotypen aufweist [MARIS et al. (*76*)]. Die Reaktionen können daher mit nur einem Antigen durchgeführt werden. Unter diesen für eine Serodiagnose förderlichen Umständen ist es schwer verständlich, daß von ihr so relativ wenig Gebrauch gemacht wird. Als Erklärung dafür kann lediglich angeführt werden, daß die Kenntnis der Krankheitsbilder, welche das Mumpsvirus hervorrufen kann, nicht alle jene Zustände umfaßt, mit denen tatsächlich gerechnet werden muß. Bei näherer Beschäftigung mit dieser Frage konnte erst kürzlich wieder bestätigt werden (HENNESSEN), was ENDERS (*31*) betont, daß nämlich die Mumps „sine parotite" sehr viel häufiger ist, als allgemein angenommen wird. Von den Organmanifestationen, die ohne Schleimdrüsenerkrankungen einhergehen können, seien hier nur aufgeführt: Zentralnervensystem-Befall, Entzündungen des Genitalapparates (männlich und weiblich) und fast aller inneren Organe einschließlich des Myokards u. a. m.

Von Bedeutung scheinen die Untersuchungen von HENLE et al. (*53*), in denen festgestellt werden konnte, daß bei Mumps die komplementbindenden AK spätestens nach einem Jahr, meist aber schon nach 3—6 Monaten unter die Grenze der Nachweisbarkeit sinken, so daß ein positiver Befund mit entsprechender Titerhöhe als sicherer Indicator einer kürzlich stattgehabten Infektion angesehen werden darf.

Wie bei der Influenza hat sich auch für Mumps die KBR als diagnostische Methode bewährt. Hier wie dort wird das Antigen aus Bruteiern gewonnen. Einzelheiten der Antigen-Bereitung finden sich unter anderem bei MÜLLER und BRAND (*84*). Auch hier ist ein früher Anstieg der komplementbindenden Antikörper bei Infektionen zu erwarten [HENLE et al. (*52*)]. Demgegenüber tritt der Hämagglutinations-Hemmungs-Test [LEVENS und ENDERS (*72*)] an diagnostischer Bedeutung zurück, so daß auch in dieser Hinsicht große Ähnlichkeit zwischen Mumps- und Influenzavirus besteht. Auch die Fähigkeit zur Hämagglutination bezieht sich bei beiden Viren auf Erythrocyten fast der gleichen Species [BEVERIDGE und LIND (*12*)].

Psittakose — Lymphogranuloma ing.-Gruppe

Obwohl nach Ansicht einer Reihe von Autoren die Erreger der hier zu besprechenden Krankheiten nicht unter die eigentlichen Viruskrankheiten zu rechnen sind [ANDREWES (*1*)], müssen sie trotzdem vom Standpunkt der klinischen Virologie an dieser Stelle ihre Berücksichtigung finden. Ihrem von anderen Virusarten abweichenden Verhalten — Empfindlichkeit gegen Chemotherapeutica und Antibiotica, eigene Stoffwechselvorgänge, Größe, Vermehrungscyclus — steht ein serologischer Unterschied zur Seite, der sich serodiagnostisch erschwerend auswirkt. Während bei den sog. echten Viren die komplementbindenden Antikörper flüchtig und damit diagnostisch geeignet auftreten, ist dies bei den großen Virusarten nicht der Fall. Vielmehr ist bei ihnen eine Persistenz der komplementbindenden Antikörper zu beobachten, die sich über Jahre erstreckt. Aus diesem Grunde ist für die Diagnose unbedingt eine 2. Serumprobe zu untersuchen, da nur daraus — und selbst dann nur schwer — auf die Akuität des Krankheitsprozesses zu schließen ist.

Psittakose — Ornithose

Wenn hier die eigentliche Psittakose (Papageienkrankheit) und die Ornithose (durch einheimisches Geflügel übertragen) als Einheit aufgefaßt und abgehandelt werden kann [MEYER (*80*)]; HAUSSMANN et al. (*45*), so wird damit dem Fehlen serologischer und klinischer Unterscheidungsmerkmale entsprochen. Nachdem MEYER und EDDIE (*79*) den Nachweis der komplementbindenden Antikörper bei Psittakoseinfektionen führen konnten, setzte sich die KBR als diagnostische Methode durch. Zur Antigenherstellung wird das Virus im Dottersack des Bruteies gezüchtet [BEVERIDGE und BURNET (*11*)], wobei von den gebildeten Komponenten die hitzestabile, gruppenspezifische Komponente verwertet wird. Wie eingangs erwähnt, bleiben die Antikörper-Titer bei der Psittakose häufig lange Zeit (Monate bis Jahre) hindurch hoch — ein Nachteil für ihre diagnostische Verwertung —, sie steigen jedoch zu Beginn der Erkrankung steil an [HAUSSMANN et al. (*45*)], so daß der Reaktionsausfall bei einer 2. Probe eine Aussage über die Phase der Infektion ermöglicht. Den Wert der serodiagnostischen Reaktion bei der Psittakose erhellt die Gegenüberstellung der Ergebnisse von Komplementbindungsreaktionen und Virusisolierungen. Während MEYER und EDDIE (*81*) bei klinisch und serologisch gesicherten Psittakose-Erkrankungen in weniger als 20% (62 von 364) der Fälle tatsächlich Virus isolieren konnten, erhielten HAUSSMANN et al. (*44*) in 100% der klinisch und anamnestisch sicheren Psittakose-Infektionen eine positive Seroreaktion. Die Mitreaktionen bei Infektionen mit Viren der gleichen Gruppe sind bei der Psittakose-Ornithose praktisch von geringer Bedeutung, da die unterschiedlichen klinischen Krankheitsbilder eine Verwechslung von vorneherein unwahrscheinlich machen. Es sei dennoch nur der Vollständigkeit halber darauf hingewiesen, daß eine positive Psittakose-Ornithose-Reaktion in Fällen von Lymphogranuloma inguinale, Trachom, Katzenkratzkrankheit (Virus-Lymphadenitis) [DEBRE et al. (*29*)] auftritt, während bei der Einschlußblennorrhoe keine gruppenspezifischen komplementbindenden Antikörper auftreten, wofür möglicherweise das Fehlen einer Virämie bei der letztgenannten Erkrankung verantwortlich zu machen ist.

Andere Viruserkrankungen des Respirationstraktes

Q-Fieber

Obwohl das Q-Fieber als Rickettsiose nicht zu den Viruserkrankungen gehört, hat es sich eingebürgert, das Q-Fieber im Zusammenhang mit den übrigen nichtbakteriellen Krankheitsbildern zu besprechen. Dabei ist es an dieser Stelle nicht möglich, alle diejenigen Konsequenzen aufzuzeigen, die sich aus der Tatsache einer Infektion mit Rickettsien im Gegensatz zu einer solchen durch Viren ergeben.

Für die Diagnose des Q-Fiebers ist die serologische Untersuchung unerläßlich, da die außerordentlich hohe Infektiosität der Rickettsia burneti jeden Versuch einer Isolierung des Erregers zu einer Gefahr für das Personal macht und daher besonderen Instituten vorbehalten bleiben sollte.

Methodisch ergeben sich für die Serodiagnose 2 Möglichkeiten und zwar einmal die KBR und zum anderen die Agglutination. Als Methode der Wahl ist auch hier die KBR anzusehen, die infolge der guten Qualität von Dottersack-antigenen an Spezifität nichts zu wünschen übrigläßt [Bengtson (*9*)]. Auch beim Q-Fieber sind unbedingt 2 Untersuchungen zu fordern, da einerseits die komplementbindenden Antikörper bei Rickettsiosen grundsätzlich persistieren, zum andern aber nicht selten erst in der 3. Woche ein positiver Reaktionsausfall erfolgt. Eine für die Serodiagnose angegebene Modifikation der üblichen KBR besteht im Nachweis der Konglutination von Hammelblutkörperchen statt deren Hämolyse [Wolfe et al. (*120*)].

Die Agglutination von Rickettsiensuspensionen bei Q-Fieber [Burnet und Freeman (*21*)] ergibt keine frühere Nachweismöglichkeit, so daß sie hinter der KBR zurücktritt. Bei beiden Methoden ergibt sich als Vorteil, daß keine Mitreaktion verwandter Rickettsienarten erfolgt, und daß die Weil-Felix-Reaktion negativ ausfällt.

Adenoviren (APC)

Bei den außerordentlich häufigen Erkrankungen der Atmungsorgane durch virusartige Erreger wird es verständlich, daß ätiologisch neben dem Influenza-Virus und den schon erwähnten Erregern auch noch mit dem Vorkommen anderer Virusarten zu rechnen ist. Für die spezifische Serodiagnostik bestehen hier jedoch nur spärliche und unbefriedigende Möglichkeiten, da es bisher nicht gelang, die Ätiologie der meisten dieser Erkrankungen restlos zu klären. Als ein Fortschritt darf die Entdeckung derjenigen Viren angesehen werden, die sich nach der ersten Isolierung bei einer primär atypischen Pneumonie [Hillemann und Werner (*55*)] als sog. APC-Viren zusammenfassen ließen [Huebner et al. (*61*)]. Hier scheint eine Erweiterung der Diagnostik möglich, jedoch dürfen die Erwartungen wohl nicht zu hoch gespannt werden, denn Agentien, welche von Affektionen der Lungen, der Mandeln, des Pharynx, der Conjunctiven, der Hornhaut usw. stammen, werden diagnostisch wegen ihres nahezu ubiquitären Vorkommens kaum die Spezifität erlangen können, welche beispielsweise die Influenza-Diagnostik erreicht hat. In einer orientierenden Studie über die Beteiligung der APC-Viren an Erkrankungen des Respirationstraktes fanden Berge et al. (*10*) nur in rund 40% der Fälle Antikörper gegen diese Agentien, obwohl bei rund doppelt soviel Patienten wegen der gleichartigen klinischen Diagnose solche zu erwarten gewesen wären. Über die tatsächliche Brauchbarkeit dieser Virusarten in der Diagnostik kann daher noch nichts Abschließendes gesagt werden. Einer serologischen Verwendung würde auch bei den APC-Viren die Tatsache entgegenkommen, daß die verschiedenen Stämme, welche bisher aufgefunden wurden, ein gemeinsames komplementbindendes Antigen haben, während gegen die infektiösen Partikel stammspezifische AK gebildet werden, so daß die antigenen Eigenschaften dieser neuaufgefundenen Virusart denen der bekannten Viren in ihrem Verhalten ähneln. Das komplementbindende Antigen liegt bei den Adenoviren, wie sie seit kurzem benannt werden [Enders et al. (*33*)], offenbar nicht in der gleichen, leicht zugänglichen Art vor, wie dies von anderen Virusarten bekannt ist. Die Gewinnung des Antigens gelingt erst nach Verreiben der Zelltrümmer von Gewebekulturen im Mörser und häufigem Frieren und Tauen [Balducci et al. (*5*)]. Wie aus den Untersuchungen von Glander et al. (*40*) hervorgeht, scheint das Vorkommen von Adenovirusinfektionen vor allem bei epidemischem Auftreten von leichteren Affektionen der Atemwege eine Rolle zu spielen.

Epidemische Keratoconjunctivitis

Obwohl der epidemische Charakter dieser Erkrankung schon seit nahezu 70 Jahren bekannt ist, besteht bis heute noch keine einheitliche Auffassung über den Erreger, dessen Virusnatur allerdings unbestritten ist. Die Skepsis gegenüber den verschiedenen Berichten von erfolgreichen Virusisolierungen beruht auf den unterschiedlichen Ergebnissen der Rückübertragung der gewonnenen Agentien auf menschliche oder tierische Hornhaut, vor allem aber wird diese Skepsis genährt durch die unbefriedigenden serologischen Befunde, die mit den isolierten Virusstämmen erhoben wurden. Bei der Bedeutung der ep. K.C. [TRÜB (*118*)] erscheint es daher begrüßenswert, daß von verschiedenen Autoren eine neuerliche Untersuchung des Erregers mit modernen virologischen Methoden in Angriff genommen wurde, wobei hier nur auf die Arbeiten von JAWETZ et al. (*62*) und die von RHODE und WOLFFERSDORFF (*89*) hingewiesen werden soll. Die erstgenannten Autoren berichten über einen neuen Adenovirustyp, den sie für den Erreger der ep. KC halten, während die letztgenannten auf Kükencorneagewebekulturen ein andersartiges Virus als Ursache dieser Erkrankung fanden und den Beweis dafür durch Rückübertragung antreten konnten. Mit diesen Befunden zeichnen sich demnach Möglichkeiten ab, zu einer serologischen Diagnostik zu kommen, da die Frage der Viruszüchtung und Antigenherstellung damit grundsätzlich gelöst erscheint.

Virusarten der Pockengruppe

Schon bevor es gelungen war, das Brutei zur Züchtung von Variola- [LAZARUS et al. (*67*)] und Vacciniavirus [GOODPASTURE et al. (*41*)] heranzuziehen, konnte gezeigt werden, daß Pustelinhalt als Antigen mit Patientenserum unter Bindung von Komplement reagiert [BEINTKER (*8*)]. Dabei ist hervorzuheben, daß serologisch keine Unterscheidung zwischen echten Pocken und Kuhpocken möglich ist. Die serologische Diagnostik gestattet daher keine Abgrenzung der Krankheitsbilder, die aus einer Infektion durch in der Natur vorkommende Ansteckungsherde oder nach Inoculation durch Impfung resultieren. Der Wert einer Serodiagnose liegt daher in der Unterscheidungsmöglichkeit gegenüber Virusarten anderer Gruppen. Über die Brauchbarkeit serologischer Methoden bei Impfencephalitiden liegen Erfahrungen nicht vor, die Klärung der serologischen Verhältnisse bei Impfschäden dieser Art scheint jedoch einer Prüfung wert.

Da die menschenpathogenen Virusarten der Pockengruppe (Variola, Alastrim, Vaccinia, Kuhpocken, Moll. contagiosum) die Fähigkeit zur Agglutination von Erythrocyten besitzen, außerdem die Größe der Partikel sowie deren relativ einfache Züchtung hochkonzentrierte Suspensionen ermöglichen, lassen sich bei den Pocken-Viren eine Reihe von serodiagnostischen Methoden durchführen, von denen hier nur Hämagglutinations-Hemmung, Komplementbindungsreaktion und Flockungstest besprochen werden sollen.

Beim Hämagglutinations-Hemmungs-Test wird das Hämagglutinin benutzt, das sich bei dieser Virusgruppe im Gegensatz zur Influenzagruppe als unabhängig von den Viruspartikeln und vom löslichen (komplementbindenden) Antigen erwiesen hat [CHU (*26*)]. Die Reaktion gestattet eine frühe Diagnose durch den Nachweis der die Hämagglutination hemmenden Antikörper [COLLIER et al. (*27*), SALCHOW (*98*)], die schon in den ersten Tagen der Erkrankung auftreten.

Trotzdem wird meist die Komplementbindungsreaktion bevorzugt, und zwar in einem Verfahren, das umgekehrt zur üblichen KBR durchgeführt wird. Bei der Pockengruppe läßt sich aus Pustelinhalt und Krusten ein Antigen herstellen, das mit Kaninchenimmunserum spezifisch reagiert [GORDON (*42*)], so daß die serologische Diagnose unmittelbar nach Auftreten der ersten Eruptionen gestellt werden kann. Der Nachweis komplementbindender Antikörper im Patienten-Serum mit Standardantigen tritt demgegenüber zurück.

Der Flockungstest beruht auf dem Vorkommen präzipitierender Antikörper im Patientenserum [TANAKA (*113*)]. Die Reaktion besitzt zwar nur einen Bruchteil der Spezifität der KBR, jedoch hat sie sich bei kleineren Epidemien als wertvoll erwiesen [BURGESS et al. (*20*)].

Die serologische Diagnose von Molluscum contagiosum kann grundsätzlich in der soeben beschriebenen Weise erfolgen. Nach der Natur des Krankheitsbildes kommt ihr jedoch lediglich akademisches Interesse zu.

Varicellen, Herpes zoster, Herpes simplex

Wenn im folgenden Abschnitt so verschiedenartige Krankheitsbilder wie Varicellen, Herpes zoster und Herpes simplex besprochen werden sollen, so geschieht dies auf Grund einer Reihe von Übereinstimmungen, welche diese 3 Virusarten aufweisen.

Den erwähnten Viren ist beispielsweise gemein, daß sie in der Mehrzahl der Fälle die Haut befallen, und daß sie in den befallenen Zellen intranucleäre Einschlüsse hervorrufen [BUDDINGH (*19*)], während die Viren der Pockengruppe solche im Cytoplasma verursachen.

Während die Serodiagnose von Varicellen und Herpes zoster nicht möglich ist, da sich bisher kein empfänglicher Wirtsorganismus auffinden ließ [STOKES (*110*)], treten der serologischen Methodik bei Herpes simplex andere Schwierigkeiten entgegen, die indes nicht für dieses Virus typisch sind, sondern auch als epidemiologisches Phänomen Erwähnung verdienen.

Nach der heutigen Vorstellung über die Epidemiologie des Herpes simplex-Virus [SCOTT (*106*)] muß damit gerechnet werden, daß jeder Infizierte lange, wenn nicht immer, latent infiziert bleibt, also ein Virusreservoir darstellt. Nach Antikörperbestimmungen wird wahrscheinlich, daß ein großer Teil der Bevölkerung latent mit Herpes simplex-Virus behaftet ist. Als Erklärung für die häufigen Wiedererkrankungen der gleichen Patienten ist an eine Re-Manifestation, hervorgerufen durch physische oder psychische Traumata, zu denken.

Das ubiquitäre Vorkommen von Herpes simplex-Virus einerseits und der nicht selten beträchtliche Antikörperspiegel bei klinisch gesunden Personen hat einen außerordentlich beschränkten Aussagewert der Serodiagnose zur Folge, so daß nur 2 Untersuchungen mit ausreichendem zeitlichem Abstand Erfolg versprechen, wenn in der üblichen Weise ein bekanntes Antigen gegen Patientenserum geprüft wird. Da eine Reihe von Wirtsorganismen für Herpes simplex-Virus empfindlich ist, läßt sich sowohl der Virusneutralisationstest im Tierversuch oder der Gewebekultur als auch die KBR mit hochkonzentrierten Antigenen durchführen. Für die Serodiagnose empfiehlt sich wegen der grundsätzlichen diagnostischen Vorteile und aus praktischen Gründen nur die KBR

[BEDSON und BLAND (*7*)]. Zufriedenstellende Antigene können aus infizierten Bruteiern [BURNET (*22*)] gewonnen werden [MARTINEZ et al. (*76*), MODI et al. (*83*)], jedoch geben BARSKI et al. (*6*) die Antigenherstellung in Kaninchen-Nieren-Gewebekultur als besonders brauchbar an. Bei Bewertung der Ergebnisse mit der gebotenen Vorsicht, die sich aus den oben angegebenen epidemiologischen Fakten ergibt, sind brauchbare Resultate vor allem bei den atypisch verlaufenden Herpes simplex-Infektionen zu erwarten.

Maul- und Klauenseuche

Die Diagnose der Maul- und Klauenseuche bedarf in den meisten Fällen keiner serologischen Stütze, da sie auf Grund des klinischen Befundes und der anamnestischen Angaben gewöhnlich ohne Laboratoriumshilfe gestellt werden kann. Wo eine serologische Diagnose trotzdem notwendig ist, läßt sie sich insofern relativ einfach als KBR durchführen, als sich Antigen aus Bläscheninhalt verwenden läßt [KANYO und OLAH (*64*)] oder aber Gewebekulturvirus [FRENKEL (*36*)] wie auch solches aus Bruteiern [TRAUB und SCHNEIDER (*116*)]. Auch beim MKS-Virus konnte ein komplementbindendes Antigen nachgewiesen werden, das unabhängig ist von den infektiösen Viruspartikeln [HOBOHM et al (*58*), HIRTZ (*57*)].

Poliomyelitis

Die Serodiagnose der Poliomyelitis ist trotz der intensiven Bearbeitung dieses Gebietes in den letzten Jahrzehnten vornehmlich durch amerikanische Autoren noch relativ jungen Datums. Die Schwierigkeiten der Laboratoriumsdiagnostik lagen bei der Poliomyelitis in der Eigenart des Virus selbst, das erst 1949 durch ENDERS (*32*) in Gewebekulturen extraneuralen Ursprungs gezüchtet werden konnte, während bis dahin als Ausgangsmaterial für die Antigenherstellung nur ZNS-Suspensionen infizierter Versuchstiere (Affen und Nager) zur Verfügung standen. Darüber hinaus brach sich erst um die gleiche Zeit die Erkenntnis von der Existenz dreier serologisch unterschiedlicher Typen des Poliomyelitis-Virus Bahn [Com. of Typing (28)], so daß es verständlich scheint, wenn bis zu diesem Zeitpunkt jede serologische Diagnose bei der Poliomyelitis mißlingen mußte. Der Nachweis von komplementbindenden Antikörpern gegenüber ZNS-Antigenen wurde zunächst im Tierversuch geführt [CASALS et al. (*24*)], während sich die bereits bekannten, virusneutralisierenden Antikörper als persistierend erwiesen [WINSSER und SABIN (*119*)] und daher für die Diagnostik als ungünstig angesehen werden mußten. Mit Gewebekultur-Antigenen gelangten SVEDMYR et al. (*111*) zu diagnostisch zufriedenstellenden Resultaten, jedoch waren diese Autoren gezwungen, das Rohantigen — infizierte Gewebekulturflüssigkeit — durch Ultrafiltration zu konzentrieren. BLACK und MELNICK (*14*) erhielten mit nicht aufgearbeiteten Antigenen von höherem Ausgangstiter verwertbare Ergebnisse, wenn sie bei Durchführung der KBR die Komplement-Avidität des Serums berücksichtigten; schließlich berichtete HENNESSEN (*49*) über serodiagnostische Untersuchungen bei Poliomyelitis-Infektionen, ohne daß besondere Einschränkungen gegenüber dem Vorgehen bei anderen Virusinfektionen erforderlich gewesen wären. LE BOUVIER et al. (*68*) zeigten, daß sich auch bei der Poliomyelitis die umgekehrte KBR — unbekanntes Virus gegen bekanntes (Standard-)

Serum — zur Identifizierung frisch isolierter Virusstämme verwenden läßt, wie dies in grundsätzlich ähnlicher Weise bei den Pockenviren ebenfalls möglich ist (s. oben). Mit der umgekehrten KBR läßt sich eine beträchtliche Beschleunigung der Typisierung von Virusstämmen im Vergleich zum Neutralisationstest erreichen. In allen Fällen bedienten sich die genannten Autoren der eingangs beschriebenen Mikro-Komplementbindungsreaktion nach FULTON und DUMBELL (*37*), die jeweils für die Erfordernisse der Praxis modifiziert wurde. Nach eigenen Erfahrungen bei der Untersuchung von einigen tausend Polio-Infektionen erscheint es nicht notwendig, die Reaktionen mit aktiven Virussuspensionen durchzuführen. Veränderungen durch Hitzeinaktivierung [LE BOUVIER (*68*)] können demnach für die praktische Serodiagnostik vernachlässigt werden, obschon ihre Bedeutung für das Studium von Antigen-Strukturen nicht verkannt werden soll. Das Auftreten von heterotypen Reaktionsausfällen konnte in weniger als 1% der zahlreichen eigenen Untersuchungen beobachtet werden. Obwohl beim Polio-Virus eine Trennung von Viruspartikeln und etwaigen anderen Antigenen infolge der Kleinheit der erstgenannten — Partikelgröße 27 mμ [SABIN et al. (*97*)] — nicht wie etwa bei Influenza ohne größte Schwierigkeiten gelingt, konnte der Nachweis löslicher Polio-Antigene trotzdem geführt werden [BLACK et al (*15*); SELZER et al. (*107*)], so daß auch bei diesem Virus die Partikel als vermutliche Träger der Infektiosität doch nicht die einzigen und kleinsten Elemente von Virussuspensionen darstellen.

Wenn im folgenden die Serumbefunde bei Poliomyelitis eine eingehendere Darstellung erfahren sollen als die bei anderen Virusinfektionen, so geschieht dies wegen der Bedeutung der exakten Poliomyelitis-Diagnose einerseits und der Neuartigkeit dieser Art der Diagnostik andererseits.

Die mit Hilfe der KBR erzielten Ergebnisse der Polio-Diagnostik besagen, daß in rund 90% der gesicherten Polio-Fälle ein positiver Reaktionsausfall mit Typendifferenzierung gelingt [HENNESSEN et al. (*50*)], wobei in $^9/_{10}$ der Fälle eine einmalige Untersuchung ausreicht. Damit erweist sich auch hier die Serodiagnostik als weitaus leistungsfähiger als die Virusisolierung.

Ein optimaler Zeitpunkt für den serologischen Nachweis der Polio-Infektion läßt sich schwer angeben, weil dieser die Kenntnis von Infektionstermin, Infektionsdosis und Reaktionslage des Patienten bei und nach der Infektion voraussetzen würde, also drei individuelle Größen, die sämtlich unbekannt sind, von denen aber naturgemäß sowohl die Dauer der Inkubationszeit als auch die Antikörper-Höhe bei Beginn der manifesten Erkrankung abhängt. Es ließ sich aber nachweisen, daß in $^9/_{10}$ derjenigen Fälle, die überhaupt serologisch erfaßbar sind, die Reaktionen dann schon positiv werden, wenn klinisch die Diagnose Poliomyelitis in Erwägung gezogen wird, d. h. zu Beginn des paralytischen Stadiums. Abweichungen mit späterem, aber auch mit früherem Antikörper-Anstieg kommen vor.

Lymphocytäre Choriomeningitis

Die Besprechung der Serodiagnose der lymphocytären Choriomeningitis (LCM) verursachte insofern Unbehagen, als mit dem klinischen Begriff der LCM etwas gänzlich anderes ausgedrückt wird als mit dem virologischen. Während klinisch die meisten abakteriellen Meningitiden, die mit einer Lymphocytenvermehrung im Liquor einhergehen, als LCM bezeichnet werden, ist die virologische Definition sehr viel einschränkender und bezieht sich nur auf diejenigen Zustände, welche durch Infektion mit dem LCM-Virus [Armstrong (*3*)] hervorgerufen werden. Da eine solche einen engen anamnestischen Kontakt mit Mäusen voraussetzt [Scheid et al. (*103*)], wird es verständlich, daß diese Erkrankung in Ländern mit ausreichender Wohnhygiene eine Rarität darstellt.

Als diagnostische Methode hat sich auch hier die KBR bewährt. Die Gewinnung brauchbarer Antigene kann auf verschiedene Art erfolgen [Smadel (*108*)], wobei für die Viruszüchtung Meerschweinchen bevorzugt werden, obwohl die Kultur im Brutei ebenfalls gelingt. Das Virus der LCM besitzt wie die meisten bekannten Virusarten ein lösliches (S-)Antigen, das nach Ultrazentrifugation im virusfreien Überstand von Virussuspensionen enthalten ist. Dieses S-Antigen eignet sich besonders für die Serodiagnostik. Das Auftreten komplementbindender Antikörper ist etwa eine Woche nach Beginn der Erkrankung zu erwarten, während die Angaben über die Persistenz der genannten Antikörper differieren, denn Olitsky (*86*) gibt ihr Absinken nach 6 Wochen an, doch Scheid (*103*) spricht von weit längeren Zeiträumen. Die Autoren sind sich jedoch einig über das späte Auftreten und die Persistenz über Jahre, welche die neutralisierenden Antikörper aufweisen.

Coxsackiegruppe

Die relativ einfach mögliche Isolierung von Coxsackieviren durch Inoculation von Babymäusen hat zur Folge gehabt, daß es nur vereinzelt zur rein serologischen Diagnosestellung bei klinisch verdächtigen Krankheitsfällen kam. Das Vorgehen bei Coxsackieinfektion war vielmehr meist von der Überlegung bestimmt, daß die Diagnose unbedingt eine Virusisolierung erfordere, und daß die isolierten Virusstämme hernach serologisch zu identifizieren seien. Obwohl die grundsätzlichen Bedenken gegen diese Form der Diagnostik eingangs bei Erwähnung der inapparenten Infektion schon für alle Virusinfekte angemeldet wurden, muß gerade für die Coxsackieviren der Einwand erneut erhoben werden, daß die Virusisolierung nicht auch der Nachweis der Virusinfektion sein kann. Als sichere Infektion kann auch bei Coxsackieviren nur derjenige Fall angesehen werden, dessen Serum die Auseinandersetzung mit dem infizierenden Virus in Form der charakteristischen Antikörper-Spiegelbewegungen anzeigt. Es gibt — außer der Virusisolierung aus dem Blut und ihren bekannten Schwierigkeiten — keine andere Möglichkeit, das Stadium des Virusausscheiders oder Virusträgers von dem des Virusinfizierten zu unterscheiden.

Die serologische Situation bei Coxsackieviren, soweit sie diagnostisch von Bedeutung ist, ähnelt derjenigen anderer Virusarten. Die Vielzahl der Typen dieser wenig einheitlichen Virusart kompliziert das Bild zwar, doch scheinen diese Typen vornehmlich für den Neutralisationstest entscheidend zu sein, während Melnick (*77*), Lepine (*71*), Vivell (*118*) u. a. darauf hinweisen, daß in der diagnostisch bedeutsameren KBR entweder die gruppenspezifische Mitreaktion bei Prototypantigenen oder aber Mischantigene brauchbare Resultate zu liefern imstande sind, die selbstverständlich nur die Aussage über das Vorliegen von Coxsackie A- oder B-Infektionen zulassen. Die epidemiologisch interessierende

Aufschlüsselung nach Typen kann dann durch Neutralisationstest und/oder KBR gegenüber dem gesamten Typenspektrum erfolgen.

Die Herstellung von Antigenen erfolgt bei Coxsackievirus gewöhnlich nach der von CASALS (*25*) angegebenen Methode, die von der Skeletmuskulatur infizierter Babymäuse ausgeht und durch wiederholte Aceton-Ätherextraktion zu wirksamem Antigen gelangt. Die damit gewonnenen Resultate sind befriedigend und halten sich im Rahmen der von anderen Viruskrankheiten bekannten Ergebnisse.

Gelbfieber

Die serologische Diagnose von Gelbfieber stellt insofern eine Seltenheit dar, als in den meisten Fällen die klinischen Symptome genügen, die Erkrankung zu erkennen. Die gut ausgebaute Gelbfieberserologie hat deshalb weniger zu diagnostischen als vielmehr zu epidemiologischen Studien Verwendung gefunden.

Die gebräuchlichste serologische Methode ist beim Gelbfieber der Maus-Schutz-Test [mouse protection test SAWYER und LLOYD (*100*)] geworden, obwohl sein diagnostischer Aussagewert wegen der Persistenz der damit nachweisbaren neutralisierenden Antikörper begrenzt ist und seine Verwendung mehr für Durchseuchungsstudien geeignet wäre. Das Prinzip dieser Methode liegt im Nachweis von Antikörpern im Patientenserum, das in fallenden Dosen mit einem Standardvirus gemischt und dann intraperitoneal in Mäuse gegeben wird. Die ebenfalls mögliche KBR bedient sich der Organextrakte infizierter Versuchstiere (Leber oder ZNS) als Antigen. Das Auftreten von klompementbindenden Antikörpern ist nach THEILER (*115*) erst nach demjenigen von neutralisierenden zu erwarten, jedoch fallen die erstgenannten auch beim Gelbfieber im Anschluß an die Infektion wieder ab, während die neutralisierenden Antikörper jahrelang, wenn nicht lebenslang nachweisbar bleiben.

Virusencephalitiden

Obwohl keineswegs Klarheit und Einmütigkeit darüber besteht, welche encephalitischen Krankheitsbilder hier zu besprechen sind, da eine ganze Reihe von Virusarten potentiell zur Infektion des Encephalon befähigt sind, jedoch nur selten eine Encephalitis hervorrufen, und da die Virusgenese anderer Encephalitiden nur vermutet, nicht aber bewiesen werden kann, scheint doch im ganzen genügend Material vorzuliegen, um einen besonderen Abschnitt zu rechtfertigen. Die Themastellung verbietet jedoch über diejenigen Encephalitiden hinauszugehen, deren Virusätiologie klar ist, da nur diese einer serologischen Diagnostik zugänglich sind. Die Einteilung der verschiedenen Encephalitisviren nach geographischen Gesichtspunkten, wie sie LEPINE (*70*) vornimmt, geht davon aus, daß in den heißen Klimazonen aller Kontinente Infektionen dieser Art vorkommen, die sich im Übertragungsmodus durch Insektenstiche gleichen. Für die serologische Diagnose interessiert, daß die Verwandtschaft der in Frage kommenden Virusarten sich nicht auf gleichartige Krankheitsbilder beschränkt, sondern eine stark übergreifende ist [OLITSKY und CASALS (*86*)]. Hierdurch und durch die Ausweitung, welche dieses Gebiet durch Forschungen der letzten Jahre sowohl in der Anzahl der isolierten Virusarten als auch in den zugehörigen Erkrankungen erfahren hat, steigerte sich die Unübersichtlichkeit dieser Gruppe von Virusinfektionen, die weiter unten aufgeführt werden.

Einen besonderen Hinweis verdienen in diesem Zusammenhang diejenigen Encephalitiden, für die eine klimatische Gebundenheit anscheinend nicht gegeben ist und die auch in gemäßigten bis kalten Zonen anzutreffen sind. Erwähnt seien hier die erst kürzlich beschriebenen Meningoencephalitiden [Holzer und Steinbäcker (*59*)], deren Übertragung durch Zecken erfolgt, und eine Encephalitis völlig unklarer Genese, die als Infektionskrankheit in der Pfalz beobachtet wurde [Bader und Hengel (*4*)].

Während in allen Fällen, deren Virusätiologie klar ist, die KBR die serologische Methode der Wahl darstellt [Sabin (*94*)], wobei ZNS-Material das vorteilhafteste Antigen darstellt, konnte bei einer Reihe von Virusencephalitiden ein Hämagglutinin isoliert werden, das mit den Viruspartikeln nicht identisch ist. Derartige Hämagglutinine erlauben eine zusätzliche serologische Untersuchungstechnik, da nach Infektionen Antikörper gegen dieses Hämagglutinin gebildet werden [Sabin (*95*)]. Die serologische Methodik für die verschiedenen Erkrankungen, insbesondere die Technik der Antigengewinnung wurde von Lepine und Sohier (*71*) so erschöpfend dargestellt, daß hier auch aus räumlichen Gründen auf eine Einzeldarstellung verzichtet werden kann.

Im Rahmen der Virusencephalitiden verdienen schließlich die postvaccinalen Affektionen Erwähnung, die nach Pocken- oder Tollwut-Impfungen in seltenen Fällen auftreten können. Sowenig die verschiedenen Theorien ihrer Entstehung als restlos bewiesen oder widerlegt gelten können, und sowenig diese hier besprochen werden sollen, sosehr liegt aber in dieser Unklarheit die Ursache dafür, daß mit einer serologischen Diagnose in diesen Fällen nicht zu rechnen ist. Es versteht sich von selbst, daß serologische Reaktionen zwar bei diesen Encephalitiden positiv ausfallen können und werden, daß aber in diesem Reaktionsausfall keinerlei Fingerzeig für die Ursache der jeweiligen Encephalitis liegen kann, da identische Befunde auch bei erfolgreich Geimpften ohne Encephalitis zu erwarten sind.

Systematik der insektenübertragenen Virusenzephalitiden [nach Lepine (70)]

Amerikanische Gruppe

- Westliche Pferde-Encephalomyelitis
  (Western equine encephalomyelitis)
- Östliche Pferde-Encephalomyelitis
  (Eastern equine encephalomyelitis)
- Venezuela-Pferde-Encephalomyelitis
  (Venezuelan equine encephalomyelitis)
- St. Louis-Encephalitis

Fernöstliche Gruppe

- Japanische B-Encephalitis
  (Japanese B encephalitis)
  mit Australischer X-Krankheit
- Mandschurische Encephalitis

Eurasiatische Gruppe

- Russische Frühjahrs-Sommer-Encephalitis
  (Russian spring-summer encephalitis)
- Mährische Zeckenencephalitis
  (Moravian tick encephalitis)
- Pfälzer Encephalitis
- Drehkrankheit der Schafe (louping ill)

Afrikanische Gruppe

- Mengo Virus-Encephalitis
- Bwamba Virus-Encephalitis
- Semliki Forest-Encephalitis
- Bunyamwera Virus-Encephalitis
- West Nil Virus-Encephalitis und andere
  noch nicht identifizierte Virusarten

Dengue- und Pappataci-Fieber

Obwohl serologische Methoden, insbesondere die KBR mit ZNS-Antigenen von Sabin (*96*) als die erfolgversprechendste Technik für diagnostisches Vorgehen bei den genannten Viruskrankheiten bezeichnet wurden, sind diese doch bis heute nur selten angewendet worden. Es erscheint müßig, den Gründen hierfür nachzugehen, da hier nur von Bedeutung ist, daß eine serologische Diagnose möglich ist, und daß sie eine größere Sicherheit bietet als die Virusisolierung. Die Serologie des Dengue-Fiebers bietet insofern Besonderheiten gegenüber anderen Erkrankungen, als einerseits die Durchführung von Neutralisationstests in Mäusen dadurch behindert wird, daß im Patientenserum das Virus in großen Mengen vorkommen kann. Dieses — nicht an Mäuse adaptierte — Virus ruft mit dem Mäusevirus-Teststamm Interferenzerscheinungen hervor, die eine Bewertung bis zur Unmöglichkeit erschweren können. Ein anderer Punkt von Bedeutung liegt in der Isolierung eines (nicht Partikel) Hämagglutinins bei Dengue-Virus [Sweet et al. (*112*)], womit die Dengue-Serologie um die Methode der Hämagglutinations-Hemmung bereichert wurde.

Die serologische Situation des Pappataci-Fiebers ist von der des Dengue völlig verschieden, da es bisher nicht gelang, brauchbare Versuchstiere für das erstgenannte Virus zu finden. Die Diagnose beschränkt sich aus diesem Grunde auf die klinischen Symptome.

Tollwut

Der Erreger der Tollwut, das Rabies- oder Lyssavirus, gehört zu den ersten Virusarten, die im Laboratorium bearbeitet wurden, nachdem Pasteur an diesem Virus wichtige Prinzipien der Impfung demonstrieren konnte.

Trotzdem existiert keine serologische Diagnostik für Tollwutinfektionen, da die Entwicklung der Antikörper — es ist nur der Nachweis der neutralisierenden bekannt — so langsam vor sich geht, daß sie diagnostisch nicht verwertet werden können. Aus diesem Grunde beruht die Diagnose nach wie vor auf dem anamnestisch zu erhebenden Kontakt mit tollwütigen Tieren. Der Nachweis Negrischer Körperchen oder die Infektion von Kaninchen oder Meerschweinchen mit verdächtigem Tiermaterial ist die einzige sichere diagnostische Methode geblieben [Johnson (*63*)].

Serodiagnostisch nicht faßbare Viruserkrankungen

Schnupfen, Masern, Röteln, primäre atypische Pneumonie, Hepatitis

Die Besprechung dieser so unterschiedlichen Virusarten und der zugehörigen Krankheiten geschieht, dem gemeinsamen Fehlen serodiagnostischer Möglichkeiten entsprechend, am Schluß der serologischen Betrachtung. Bei allen diesen Viruskrankheiten ist es bisher nicht gelungen, den Erreger in einer Weise zu züchten, die eine, wie immer auch geartete Antigen-Antikörper-Reaktion in vitro erlauben könnte. Während serologische Reaktion und dadurch erhöhte diagnostische Sicherheit keineswegs bei allen der genannten Krankheiten einem echten Bedürfnis entgegenkommen würden, wäre dies für die primäre atypische Pneumonie und die Hepatitis in so großem Maße der Fall, daß sich ohne Übertreibung sagen läßt, daß mit serologischen Reaktionen hierfür wesentliche Probleme der Virologie überhaupt gelöst würden. Bisher zeichnet sich für beide Krankheiten keine auch noch so vage Möglichkeit ab, da das Wirtsspektrum der Erreger in beiden Fällen scharf auf den Menschen begrenzt ist. Günstigere Aussichten für eine serologische Diagnostik scheinen dagegen bei den Masern zu bestehen, da die Adaptation des Virus an menschliche Gewebekulturen erreicht werden [RUCKLE (*92*)] und auch ARAKAWA (*2*) über die Maus-Adaptation berichten konnte. Die Untersuchungen sind jedoch in einem zu frühen Stadium, als daß sich etwas über diagnostische Verfahren aussagen ließe, es ist bisher nur die grundsätzliche Möglichkeit einer Serodiagnostik erkennbar. Das Fehlen spezifischer Nachweismethoden bei diesen Virusinfektionen macht es notwendig, sich einiger Reaktionen zu bedienen, die sich — obwohl unspezifisch — als Stütze der Diagnostik erwiesen haben. Diese Verfahren sind jedoch nur für die primäre atypische Pneumonie bekannt, bei der die Kälteagglutination und die MG-Streptokokken-Agglutination angewandt werden können [SMADEL (*109*)]. Bei beiden Reaktionen ist es unklar geblieben, was das zugrunde liegende Phänomen auslöst, es hat sich jedoch empirisch gezeigt, daß sich bei der primären atypischen Pneumonie bei etwa 50—60% der Patienten ein Anstieg der mit den genannten Reaktionen faßbaren Titer nachweisen läßt.

C. Schlußbetrachtung

Im Laufe der vorliegenden Ausführungen wurde der Versuch unternommen, für die Virusinfektionen zu zeigen, daß in der serologischen Diagnostik der entscheidende Infektionsnachweis zu sehen ist. Die Ursache der Beweiskraft der serologischen Reaktion im Vergleich zur direkten Virusisolierung liegt im häufigen Vorkommen von Virusinfektionen, welches in einer hohen Durchseuchung der Bevölkerung durch inapparente Infektionen resultiert. Die besonderen Eigenschaften der Virusinfektionen formen eine diagnostische Methodik, die von der übrigen Serologie insofern abweichen muß, als sie aus dem Antigenmuster eines Virus jeweils dasjenige herauszugreifen gezwungen ist, welches die diagnostisch verwendbaren Antikörper des Patienten-Serums erfassen läßt. Die Strukturanalyse der Virusantigene konnte derartige Körper aufzeigen und machte damit eine Serodiagnostik im engeren Sinne erst möglich, da der Nachweis der Antikörper, welche die oft lebenslange Virusimmunität anzeigen, diagnostisch nicht brauchbar sein konnte. In den komplementbindenden Antikörpern ließ sich der Antikörper-Anteil erfassen, der im parallelen Verlauf zur Infektion diese frühzeitig erkennen läßt.

Die beträchtliche Anzahl von Virusinfektionen, deren serodiagnostische Bearbeitung auf diese Weise ermöglicht wurde, sollte aber nicht über alle die zahllosen Infektionen hinwegtäuschen, die sich nach wie vor dem Laboratorium und damit auch einer Serodiagnose entziehen. Daß unter diesen nicht zugänglichen Virusinfektionen eine Anzahl der bedeutsamsten Infektionskrankheiten zu finden ist, zeigt mit aller Deutlichkeit, wie sehr dieses Gebiet noch am Anfang steht.

Literatur

1. ANDREWES, C. H.: The Rio congress decisions with regard to study of selected groups of viruses. Ann. New York Acad. Sci. **56**, 428 (1953).

2. ARAKAWA, S.: Weitere Untersuchungen über das Virus der Masern. Z. Hyg. **139**, 227 (1954).

3. ARMSTRONG, C., and R. D. LILLIE: Experimental lymphocytic choriomeningitis of monkeys and mice produced by a virus encountered in studies of the 1933 St. Louis encephalitis epidemic. Publ. Health Rep. **1934**, 1019.

4. Bader, R. E., et R. Hengel: Recherches epidemiologic sur l'epidemie d'encephalite survenue dans le Palatinat de 1947 à 1949. Ann. Inst. Pasteur **78**, 481 (1950).
5. Balducci, D., E. Zaiman and D. A. J. Tyrrell: Laboratory studies of APC and influenza C viruses. Brit. J. Exper. Path. **37**, 205 (1956).
6. Barski, G., M. Lamy et P. Lepine: Culture de cellule trypsinees de rein de lapin et leur application à l'etude de virus du group herpetique. Ann. Inst. Pasteur **89**, 415 (1955).
7. Bedson, S. P., and J. O. W. Bland: Complement fixation with filterable viruses and their antisera. Brit. J. Exper. Path. **10**, 393 (1929).
8. Beintker, H.: Über das Verhalten der Bordetschen Reaktion bei Variola. Zbl. Bakter. **48**, 500 (1908).
9. Bengtson, I. A.: CF in Q-fever. Proc. Soc. Exper. Biol. a. Med. **46**, 665 (1941).
10. Berge, T. O., B. England, C. Mauris, H. E. Shuey and E. H. Leunette: Etiology of acute respiratory disease among service personnel at Ft. Ord, Ca. Amer. J. Hyg. **62**, 283 (1955).
11. Beveridge, W. I. B., and F. M. Burnet: The cultivation of viruses and rickettsiae in the chick embryo. Med. Res. Counc., Spec. Rep. **1946**, 256.
12. — P. E. Lind and S. G. Anderson: Mumps, 1. isolation and cultivation of the virus in the chick embryo. Austral. J. Exper. Biol. a. Med. Sci. **24**, 15 (1946).
13. Bieling, R.: Zit. bei H. Schmidt, Fortschritte der Serologie, S. 487.
14. Black, F. L., and J. P. Melnick: Specifity of the CFT in polio. Yale J. Biol. a. Med. **26**, 385 (1954).
15. — — Appearance of soluble and cross reactive CF antigens on treatment of poliovirus with formalin. Proc. Soc. Exper. Biol. a. Med. **89**, 353 (1955).
16. Bodian, D.: Pathogenesis of poliomyelitis. Amer. J. Publ. Health **42**, 1388 (1952).
17. Bozzo, A.: Studies of the antigenic composition of influenza B viruses. Bull. World Health Org. **5**, 149 (1952).
18. Brans, L. M., and J. Mulder: Studies on the antigenic composition of influenza virus B strains. Leiden (Holland): St. Kruse 1952.
19. Buddingh, G. J.: Nomenclature and classification of the pox group. Ann. New York Acad. Sci. **56**, 561 (1953).
20. Burgess, W. L., J. Craigie and W. J. Tulloch: Med. Res. Counc. No **143**, London 1929.
21. Burnet, F. M., and M. Freeman: Rickettsia of Q-fever. Med. J. Austral. **1938**, 296.
22. —, and S. W. Williams: Herpes simplex, a new point of view. Med. J. Austral. **1**, 637 (1939).
23. — Principles of animal virology. New York: Acad. Press Inc. Publishers 1955.
24. Casals, J., and P. K. Olitsky: CFT for poliomyelitisvirus. Proc. Soc. Exper. Biol. a. Med. **75**, 315 (1950).
25. — — and L. C. Murphy: Hemagglutination and complement fixation with types I and II Albany strains of coxsackie virus. Proc. Soc. Exper. Biol. a. Med. **72**, 636 (1949).
26. Chu, C. M.: Studies on vaccinia hemagglutinin. J. of Hyg. **46**, 42 (1948).
27. Collier, W. A., A. M. Smit u. A. F. v. Heerde: Der Nachweis von Antihämagglutininen bei Variolapatienten als diagnostisches Hilfsmittel. Z. Hyg. **131**, 555 (1950).
28. Committee on Typing of the N.F.I.P.: Immunnologic classification of polioviruses. Amer. J. Hyg.. **54**, 191 (1951).
29. Debré, R., M. Lamy, M. L. Jammet, L. Costil et P. Mozziconacci: La maladie des griffes de chat. Bull. Soc. méd. Hôp. Paris **66**, 76 (1950).
30. Doerr, R., u. C. Hallauer: Handbuch der Virusforschung. Wien: Springer 1939.
31. Enders, J. F., and S. Cohen: Detection of antibody by complement fixation in sera of man and monkey convalescent from mumps. Proc. Soc. Exper. Biol. a. Med. **50**, 180 (1942).
32. Enders, J. F.: 1949 Zit. bei J. F. Enders, T. M. Weller and F. C. Robbins, General preface to studies on the cultivation of polioviruses in tissue culture. J. of Immun. **69**, 639 (1952).
33. — et al.: Adenoviruses. Science (Lancaster, Pa.) **124**, 119 (1956).
34. Fenner, F.: Clinical features and pathogenesis of mouse pox. J. of Path. **60**, 529 (1948).
35. Fox, J. P., H. L. Gelfand, D. R. Leblanc and D. P. Conwell: A continuing study of the acquisition of natural immunity to polio in representative Louisiana households. Amer. J. Publ. Health **46**, 283 (1956).

36. FRENKEL, H. S.: Research on foot and mouth disease. Amer. J. Vet. Res. **11**, 371 (1950).
37. FULTON, F., and K. R. DUMBELL: Serological comparison of influenza virus. J. Gen. Microbiol. **3**, 97 (1948).
38. GÄDECKE, R.: Die inapparente Virusinfektion und ihre Bedeutung für die Klinik. Heidelberg: Springer 1956.
39. GILDEMEISTER, E., E. HAAGEN u. O. WALDMANN: Handbuch der Viruskrankheiten. Jena: Gustav Fischer 1939.
40. GLANDER, R., G. A. v. HARNACK u. H. LIPPELT: Eine durch das APC-Virus hervorgerufene Epidemie. Dtsch. med. Wschr. **1956**, 1147.
41. GOODPASTURE, E. W., and G. J. BUDDINGH: Human immunization with dermal vaccine cultivated on membranes of chick embryo. Science (Lancaster, Pa.) **78**, 484 (1933).
42. GORDON, M. H.: Studies of the viruses of vaccinia and variola. Med. Res. Counc. Spec. Rep. Ser. **98**, (1925).
43. GRUMBACH, A., u. W. KIKUTH: Menschliche Infektionskrankheiten und ihre Erreger. Stuttgart: Georg Thieme 1957.
44. HAUSSMANN, H. G., R. SIEGERT u. H. SCHWEINSBERG: Komplementbindungsreaktion zur Influenzadiagnostik. Z. Hyg. **135**, 235 (1952).
45. — — W. UNGAR u. H. RUOFF: Beiträge zur Psittakose-Ornithose-Infektion des Menschen. Arch. f. Hyg. **140**, 52 (1956).
46. HENGEL, R., u. E. MESSMER: Klinische Untersuchungen über Krankheitsverläufe bei Zusammentreffen mehrerer Infektionen. Z. Hyg. **141**, 439 (1955).
47. HENNESSEN, W.: Antigenic analysis of influenza B strains isolated in 1952. Bull. World Health Org. **6**, 481 (1952).
48. — Influenzakomplementbindungsreaktion für die Praxis. Z. Hyg. **141**, 557 (1955).
49. — Serologische Diagnose der Poliomyelitis. Dtsch. med. Wschr. **1955**, 1044.
50. — W. JACOB u. H. SEICHTER: Typenbestimmung bei Polio.-Erkrankungen mit der KBR. Dtsch. med. Wschr. **1955**, 1565.
51. — Zur serologischen Diagnostik der Viruserkrankungen. Dtsch. med. Wschr. **1956**, 933.
52. HENLE, G., S. HENLE and S. HARRIS: Serological differentiation of mumps CF-antigen. Proc. Soc. Exper. Biol. a. Med. **64**, 290 (1947).
53. — J. STOKES, J. S. BURGOON, W. J. BASHE, C. F. BURGOON, W. HENLE and G. R. HUNT: Studies on the prevention of mumps. J. of Immun. **66**, 535 (1951).
54. HENOCH, E.: Lehrbuch der Kinderkrankheiten, 2. Auflage. Berlin 1887.
55. HILLEMANN, M. R., and J. H. WERNER: Recovery of new agents from patients with acute respiratory illness. Proc. Soc. Exper. Biol. a. Med. **85**, 183 (1954).
56. HIRST, E. K.: Agglutination of red cells by allantoic fluid of chick embryos infected with influenza virus. Science (Lancaster, Pa.) **94**, 22 (1941).
57. HIRTZ, J.: Electrophoretic properties of foot and mouth disease virus. Arch. Virusforsch. (Wien) **6**, 124 (1955).
58. HOBOHM, K. O., H. G. PETERMANN u. G. E. FOREST: Elution und Nachweis des komplementbindenden Antigens in Adsorbatvaccinen gegen Maul- und Klauenseuche. Z. Bakter. **157**, 556 (1952).
59. HOLZER, W., u. A. STEINBÄCKER: Epidemie von Meningitis und Meningo-Enzephalitis. Wien. med. Wschr. **1950**, 233.
60. HOYLE, L., and R. W. FAIRBROTHER: Antigenic structure of influenza viruses. J. of Hyg. **37**, 512 (1937).
61. HUEBNER, R. J., W. P. ROWE, T. G. WARD, R. H. PAROTT and I. A. BELL: Adenoidal-pharyngeal-conjunctival agents. New. England J. Med. **251**, 1077 (1954).
62. JAWETZ, E., S. KIMURA, A. N. NICHOLAS, P. THYGESON and L. HANNA: New type of APC virus from epidemic keratoconjunctivitis. Science (Lancaster, Pa.) **122**, 1190 (1955).
63. JOHNSON, H. N.: Rabies. Viral and rickettsial infect. of man, S. 267. Philadelphia: J. B. Lippincott Company 1952.
64. KANYO, B., u. G. OLAH: Komplementbindungsversuche in einem Fall von Maul- und Klauenseucheinfektion bei Menschen. Z. Immunforsch. **92**, 20 (1938).
65. KLÖNE, W.: Laboratoriums-Diagnose menschlicher Virusinfektion. Berlin: Springer 1953.
66. KOLMER, I. A., and F. BOERNERS: Approv. labor. technic. New York: Appelton, Century & Co. 1941.

67. LAZARUS, A. S., E. EDDIE and K. F. MEYER: Propagation of variola virus in the developing egg. Proc. Soc. Exper. Biol. a. Med. **36**, 7 (1937).
68. LE BOUVIER, G. L., G. D. LAURENCE, E. M. PARFITT, M. G. JENNENS and A. P. GOFFE: Typing of polio. Viruses by CF. Lancet **1954, 11**, 531.
69. — Modification of polio. Virus antigen by heat and UV light. Lancet **1955**, 12, 1013.
70. LEPINE, P.: Nomenclature and classification of arthropodoborne encephalitides. Ann. New York Acad. Sci. **56**, 574 (1953).
71. —, et R. SOHIER: Diagnostic des maladies a virus. Paris: Masson & Cie. 1954.
72. LEVENS, J. H., and J. F. ENDERS: Hemagglutinative properties of amniotic fluid from embryonated eggs infected with mumps virus. Science (Lancaster, Pa.) **102**, 117 (1945).
73. LIPPELT, H., u. G. BRAND: KBR in der Diagnostik der Ornithose-Psittakose. Dtsch. med. Wschr. **1955**, 110.
74. LURIA, S. E.: General virology, S. 118. New York: J. Wiley & Sons, Inc. 1953.
75. MARIS, E. P., J. F. ENDERS, J. STOKES and L. W. KANE: Immunity in mumps. J. of Exper. Med. **84**, 323 (1946).
76. MARTINEZ, J. S., and E. H. LENETTE: Studies on a CFT for herpes simplex. J. Bacter. **70**, 205 (1955).
77. MELNICK, J. L.: Coxsackie group of viruses. Ann. New York Acad. Sci. **56**, 594 (1953).
78. — M. RAMOS-ALVAREZ, L. F. BLACK, J. A. GIRARDI and D. NAGAKI: Poliomyelitis viruses in tissue culture, VII. Yale J. Biol. a. Med. **26**, 465 (1954).
79. MEYER, K. F., and B. EDDIE: Value of CFT in diagnosis of psittakosis. J. Inf. Dis. **65**, 225 (1939).
80. — Psittakosis. In: Viral and rickettsial diseases of man, S. 440. Philadelphia: J. B. Lippincott Company 1952.
81. —, and B. EDDIE: Human carrier of the psittakosis virus. J. Inf. Dis. 88, 109 (1951).
82. MINER, R. W.: Virus and rickettsial classification and nomenclature. Ann. New York Acad. Sci. **56**, 381 (1953).
83. MODI, N. L., and J. O'H. TOBIN: CFT in the diagnosis of herpes simplex infections. J. Clin. Path. **8**, 292 (1955).
84. MÜLLER, F., u. G. BRAND: Beobachtungen bei der Herstellung, Auswertung und Anwendung komplementbindender Mumpsantigene. Arch. Virusforsch. (Wien) **5**, 289 (1954).
85. NICOLLE, CH.: Destin des maladies infectieuses. Paris: F. Alcan 1933.
86. OLITSKY, P. K., and J. CASALS: Viral encephalitides. In: Viral and rickettsial diseases of man, S. 214. Philadelphia: J. B. Lippincott 1952.
87. POLLARD, E. C.: The physics of viruses, S. **124**. New York: Acad. Press Inc. Publ. 1953.
88. RHODE, W., u. H. v. WOLFFERSDORFF: Studien zur Isolierung und Identifizierung infektiöser Virusagentien bei einer Epidemie von Keratoconjunctivitis epidemica. Z. Hyg. **143**, 93 (1956).
89. RIVERS, T. H.: Viruses and Koch's postulates. J. Bacter. **33**, 1 (1938).
90. — Viral and ricketts. Diseases of man. Philadelphia: J. B. Lippincott Company 1952.
91. ROOYEN, C. E. VAN, and A. J. RHODES: Virus diseases of man. New York: Th. Nelson & Sons 1948.
92. RUCKLE u. J. F. ENDERS: Nach W. SCHÄFER, persönliche Mitteilung 1956.
93. RUSKA, H.: In Handbuch der Virusforschung, Bd. II. Wien: Springer 1950.
94. SABIN, A. B.: Epid. encephalitis in military personnel. J. Amer. Med. Assoc. **133**, 281 (1947).
95. — Hemagglutination by viruses affecting the human nervous system. Federat. Proc. **10**, 573 (1951).
96. — Dengue, Phlebotomus. In: Viral and rickettsial infect. of man, S. 556 u. 569. Philadelphia: J. B. Lippincott Company 1952.
97. — W. HENNESSEN and J. WARREN: Ultrafiltration and electron microscopy of the three types of poliomyelitis virus propagated in tissue culture. Proc. Soc. Exper. Biol. a. Med. **85**, 359 (1954).
98. SALCHOW, U. A.: Über diagnostische Möglichkeiten bei Infektionen mit einem Virus der Variola-Vaccinia-Gruppe. Arch. Hyg. **139**, 608 (1955).
99. SALK, J. E.: A simplified procedure of titrating the hemagglutinating capacity of influenza virus and corresponding antibody. J. of Immun. **49**, 87 (1944).

100. SAWYER, W. A., and W. LLOYD: Use of mice in tests of immunity against yellow fever. J. of Exper. Med. **54**, 533 (1931).

101. SCHÄFER, W., u. W. ZILLIG: Über den Aufbau des Viruselementarteilchens der klassischen Gepflügelpest. Z. Naturforsch. **9**, 779 (1954).

102. — Vergleichende sero-immunologische Untersuchungen über die Viren der Influenza und klassischen Gepflügelpest. Z. Naturforsch. **10**, 81 (1955).

103. SCHEID, W., K. A. JOCHHEIM: Infektionen mit dem Virus der lymphozytären Choriomeningitis in Deutschland. Dtsch. med. Wschr. **1956**, 700.

104. SCHMIDT, H.: Fortschritte der Serologie, S. 121, 486, 496 u. 506. Frankfurt a. M.: D. Steinkopf 1950.

105. SCHMIDT, N., and H. B. HARDING: Investigation of the specifity of substances in some human sera which are inhibitory in certain viral and rickettsial complement fixing antigen-antibody systems. J. Bacter. **71**, 223 (1956).

106. SCOTT, T. F. MCN.: Diaseases caused by the virus of herpes simplex. In: Viral and rickettsial infect. of man, S. 499. Philadelphia: J. B. Lippincott Company 1952.

107. SELZER, G., and M. V. D. ENDE: Experiments with soluble antigen of poliomyelitis virus. J. of Hyg. **54**, 1 (1956).

108. SMADEL, J. E., and M. J. WALL: Identification of the virus of lymphocytic choriomeningitis. J. Bacter. **41**, 421 (1941).

109. — Serologic reactions in viral and rickettsial infections. In: Viral and rickettsial infect. of man, S. 88. Philadelphia: J. B. Lippincott Company 1952.

110. STOKES, J.: Varicella-herpes zoster group. In: Viral and rickettsial infect. of man, S. 506. Philadelphia: J. B. Lippincott Company 1952.

111. SVEDMYR, A., J. F. ENDERS and A. HOLLOWAY: Complement fixation with Brunhilde and Lansing polio viruses propagated in tissue culture. Proc. Soc. Exper. Biol. a. Med. **79**, 296 (1952).

112. SWEET, B. H., R. M. CHANOCK and A. B. SABIN: Recovery and characterization of hemagglutinins from two immunologically distinct types of Dengue virus. Federat. Proc. **12**, No 1520 (1953).

113. TANAKA, K.: Über die Untersuchung des Pockenerregers. Z. Bakter. **32**, 726 (1902).

114. TERZIN, A. L., M. N. BORDJOSKI, M. V. MILOVANOVIC, L. V. STOJKOVIC and M. M. DIMIC: Some viral, rickettsial and leptospiral infections diagnosed in Serbia. J. of Hyg. **52**, 129 (1954).

115. THEILER, M.: Yellow fever. In: Viral and rickettsial infect. of man, S. 531. Philadelphia: J. B. Lippincott Company 1952.

116. TRAUB, E., u. B. SCHNEIDER: Z. Naturforsch. **3**, 178 (1948). Zit. bei G. SCHRAMM, Biochemie der Viren, S. 221. Heidelberg: Springer 1954.

117. TRÜB, C. P. L.: Verbreitung der epidemischen Keratoconjunctivitis etc. Klin. Mbl. Augenheilk. 126, 180 (1955).

118. VIVELL, O.: Typenbestimmungen bei in Deutschland in den Jahren 1951—1954 isolierten Coxsackievirusstämmen der A-Gruppe. Arch. Virusforsch. (Wien) **4**, 329 (1955).

119. WINSSER, J., and A. B. SABIN: Levels of homotypic neutralizing antibody in human poliomyelitis three years after infection. J. of Exper. Med. **96**, 477 (1952).

120. WOLFE, D. M., and L. KORNFELD: Conglutination complement absorbtion test compared with hemolytic complement fixation reactions using Q-fever immune bovine serum. Proc. Soc. Exper. Biol. a. Med. **69**, 251 (1948).

Namenverzeichnis

Die *kursiv* gedruckten Seitenzahlen beziehen sich auf die Literatur

Sachverzeichnis

ERGEBNISSE DER MIKROBIOLOGIE IMMUNITÄTSFORSCHUNG UND EXPERIMENTELLEN THERAPIE

FORTSETZUNG DER
ERGEBNISSE DER HYGIENE, BAKTERIOLOGIE, IMMUNITÄTSFORSCHUNG
UND EXPERIMENTELLEN THERAPIE

HERAUSGEGEBEN VON

W. KIKUTH
DÜSSELDORF

K. F. MEYER
SAN FRANCISCO

E. G. NAUCK
HAMBURG

A. M. PAPPENHEIMER JR.
NEW YORK

J. TOMCSIK
BASEL

DREISSIGSTER BAND

SONDERABDRUCK

K. A. BISSET

MULTICELLULARITY IN BACTERIA

WITH 11 FIGURES AND 2 PLATES

NICHT IM HANDEL

SPRINGER-VERLAG
BERLIN · GÖTTINGEN · HEIDELBERG
1957

ERGEBNISSE DER MIKROBIOLOGIE IMMUNITÄTSFORSCHUNG UND EXPERIMENTELLEN THERAPIE

FORTSETZUNG DER
ERGEBNISSE DER HYGIENE, BAKTERIOLOGIE, IMMUNITÄTSFORSCHUNG
UND EXPERIMENTELLEN THERAPIE

HERAUSGEGEBEN VON

W. KIKUTH
DÜSSELDORF

K. F. MEYER
SAN FRANCISCO

E. G. NAUCK
HAMBURG

A. M. PAPPENHEIMER JR.
NEW YORK

J. TOMCSIK
BASEL

DREISSIGSTER BAND

SONDERABDRUCK

ADRIANUS PIJPER

BACTERIAL FLAGELLA AND MOTILITY

WITH 38 FIGURES

NICHT IM HANDEL

SPRINGER-VERLAG
BERLIN · GÖTTINGEN · HEIDELBERG
1957

ERGEBNISSE DER MIKROBIOLOGIE IMMUNITÄTSFORSCHUNG UND EXPERIMENTELLEN THERAPIE

FORTSETZUNG DER
ERGEBNISSE DER HYGIENE, BAKTERIOLOGIE, IMMUNITÄTSFORSCHUNG
UND EXPERIMENTELLEN THERAPIE

DREISSIGSTER BAND

SONDERABDRUCK

HENNING BRANDIS

DIE ANWENDUNG VON PHAGEN IN DER BAKTERIOLOGISCHEN DIAGNOSTIK MIT BESONDERER BERÜCKSICHTIGUNG DER TYPISIERUNG VON TYPHUS- UND PARATYPHUS B-BAKTERIEN SOWIE STAPHYLOKOKKEN

MIT 10 ABBILDUNGEN

NICHT IM HANDEL

SPRINGER-VERLAG
BERLIN · GÖTTINGEN · HEIDELBERG
1957

ERGEBNISSE DER MIKROBIOLOGIE IMMUNITÄTSFORSCHUNG UND EXPERIMENTELLEN THERAPIE

FORTSETZUNG DER
ERGEBNISSE DER HYGIENE, BAKTERIOLOGIE, IMMUNITÄTSFORSCHUNG UND EXPERIMENTELLEN THERAPIE

HERAUSGEGEBEN VON

W. KIKUTH
DÜSSELDORF

K. F. MEYER
SAN FRANCISCO

E. G. NAUCK
HAMBURG

A. M. PAPPENHEIMER JR.
NEW YORK

J. TOMCSIK
BASEL

DREISSIGSTER BAND

SONDERABDRUCK

F. KAUFFMANN

DAS KAUFFMANN-WHITE-SCHEMA

NICHT IM HANDEL

SPRINGER-VERLAG
BERLIN · GÖTTINGEN · HEIDELBERG
1957

ERGEBNISSE DER MIKROBIOLOGIE IMMUNITÄTSFORSCHUNG UND EXPERIMENTELLEN THERAPIE

FORTSETZUNG DER
ERGEBNISSE DER HYGIENE, BAKTERIOLOGIE, IMMUNITÄTSFORSCHUNG
UND EXPERIMENTELLEN THERAPIE

HERAUSGEGEBEN VON

W. KIKUTH
DÜSSELDORF

K. F. MEYER
SAN FRANCISCO

E. G. NAUCK
HAMBURG

A. M. PAPPENHEIMER JR.
NEW YORK

J. TOMCSIK
BASEL

DREISSIGSTER BAND

SONDERABDRUCK

MAURICE WELSCH

ACTIVITÉS BACTÉRIOLYTIQUES DES MICROORGANISMES

NICHT IM HANDEL

SPRINGER-VERLAG
BERLIN · GÖTTINGEN · HEIDELBERG
1957

ERGEBNISSE DER MIKROBIOLOGIE IMMUNITÄTSFORSCHUNG UND EXPERIMENTELLEN THERAPIE

FORTSETZUNG DER
ERGEBNISSE DER HYGIENE, BAKTERIOLOGIE, IMMUNITÄTSFORSCHUNG
UND EXPERIMENTELLEN THERAPIE

HERAUSGEGEBEN VON

W. KIKUTH
DÜSSELDORF

K. F. MEYER
SAN FRANCISCO

E. G. NAUCK
HAMBURG

A. M. PAPPENHEIMER JR.
NEW YORK

J. TOMCSIK
BASEL

DREISSIGSTER BAND

SONDERABDRUCK

HELMUT RUSKA

DIE BAKTERIENCYTOLOGIE IM VERGLEICH ZUM FEINBAU HÖHER DIFFERENZIERTER ZELLEN

NICHT IM HANDEL

SPRINGER-VERLAG
BERLIN · GÖTTINGEN · HEIDELBERG
1957

ERGEBNISSE DER MIKROBIOLOGIE IMMUNITÄTSFORSCHUNG UND EXPERIMENTELLEN THERAPIE

FORTSETZUNG DER
ERGEBNISSE DER HYGIENE, BAKTERIOLOGIE, IMMUNITÄTSFORSCHUNG
UND EXPERIMENTELLEN THERAPIE

DREISSIGSTER BAND

SONDERABDRUCK

WALTER HENNESSEN

DIE SEROLOGISCHE DIAGNOSTIK DER VIRUSERKRANKUNGEN DES MENSCHEN

MIT 2 ABBILDUNGEN

NICHT IM HANDEL

SPRINGER-VERLAG
BERLIN · GÖTTINGEN · HEIDELBERG
1957

Infektionskrankheiten

Bearbeitet von R. Aschenbrenner, H. Baur, K. Bingold, H. Eyer, G. Fanconi, L. Fischer, E. Glanzmann, O. Gsell, F. O. Höring, H. Hormann †, A. Hottinger, H. Kleinschmidt, F. Linder, H. Lippelt, W. Löffler, F. Lüthy, R. Massini, W. Minning, W. Mohr, D. Moroni, E. G. Nauck, E. Reichenow, H. Schlossberger, H. Schulten, H. Vogel, G. Walther, F. Weyer.

(Band 1 von Handbuch der inneren Medizin, vierte Auflage. Herausgegeben von Professor Dr. **G. von Bergmann** †, Professor Dr. **W. Frey**-Bern und Professor Dr. **H. Schwiegk**-Marburg a. d. Lahn. In 9 Bänden. Jeder Band ist einzeln käuflich.)

1. Teil: Mit 417 zum Teil farbigen Abbildungen. XVII, 1536 Seiten Gr.-8⁰. 1952.

2. Teil: Mit 293 zum Teil farbigen Abbildungen. XV, 1225 Seiten Gr.-8⁰. 1952.

Gesamtpreis Ganzleinen DM 374.—

Bei Verpflichtung zur Abnahme des gesamten Handbuches Subskriptionspreis.

Ganzleinen DM 299.—

Die inapparente Virusinfektion und ihre Bedeutung für die Klinik

Von Dr. **Roland Gädeke**, Dozent der Kinderheilkunde, Freiburg i. Br. Mit 6 Abbildungen in 36 Einzeldarstellungen. VIII, 158 Seiten Gr.-8⁰. 1957. Steif geheftet DM 24.80

Bakteriologische Grundlagen der chemotherapeutischen Laboratoriumspraxis

Von Dr. med. **Paul Klein**, Privatdozent für Hygiene und Mikrobiologie an der Medizinischen Akademie, Düsseldorf. Mit etwa 52 Abbildungen. Etwa 232 Seiten Gr.-8⁰. 1957.

Ganzleinen etwa DM 39.60

Lehrbuch der Parasitologie

unter besonderer Berücksichtigung der Parasiten des Menschen.

Von Dr. **Gerhard Piekarski**, apl. Professor der med. Parasitologie und Mikrobiologie, Abteilungsleiter am Hygiene-Institut der Universität Bonn. Mit 411 zum Teil farbigen Abbildungen. XII, 760 Seiten Gr.-8⁰. 1954. Ganzleinen DM 108.—

Die Brucellose als Anthropo-Zoonose (Febris undulans)

Eine zusammenfassende Darstellung für Ärzte und Tierärzte. Von Professor Dr. **W. Löffler**, Direktor der Med. Univ.-Klinik Zürich, Dr. **D. L. Moroni**, Med. Univ.-Klinik Zürich, und Professor Dr. **W. Frei**, ehem. Direktor des Vet.-Path. Instituts der Universität Zürich. Mit 67 Abbildungen. XII, 193 Seiten Gr.-8⁰. 1955.

Steif geheftet DM 29.60

Laboratoriumsdiagnose menschlicher Virus- und Rickettsieninfektionen

Ein Leitfaden. Von Dr. med. **Wilhelm Klöne**, Laboratorium der Stiftung zur Erforschung der spinalen Kinderlähmung, Universitätskrankenhaus Hamburg-Eppendorf. Mit 10 Abbildungen. VII, 161 Seiten Gr.-8⁰. 1953. Steif geheftet DM 16.80